Elementary Linear Algebra

NINTH EDITION ABRIDGED

HOWARD ANTON
CHRIS RORRES

Drexel University

John Wiley & Sons Canada, Ltd.

Library and Archives Canada Cataloguing in Publication

Anton, Howard
    Elementary linear algebra / Howard Anton, Chris Rorres. – 9th ed. abridged

Includes index.
ISBN-13: 978-0470-83724-5
ISBN-10: 0-470-83724-1

    1. Algebras, Linear–Textbooks. I. Rorres, Chris II. Title.

A184.A57 2005   512'.5   C2005-903931-0

Production Credits

Publisher: John Horne
Editorial Manager: Karen Staudinger
Publishing Services Director: Karen Bryan
Marketing Manager: Sean O'Reilly
Design and Layout: Techsetters, Inc.
Printing and Binding: Tri-Graphic Printing Limited

Printed and bound in Canada

10 9 8 7 6 5 4 3 2 1

John Wiley & Sons Canada, Ltd.
6045 Freemont Blvd.
Mississauga, Ontario L5R 4J3

Visit our website at www.wiley.ca

This textbook is an abridged version of *Elementary Linear Algebra, Applications Version*, Ninth Edition, by Howard Anton and Chris Rorres. Because we wanted to ensure that the text could be used for both science and social science students, we have introduced a new sixth chapter that presents nine applications of linear algebra drawn from business, economics, and computer science. It was created in consultation with CEGEP professors to meet the needs of their course. The applications are independent of one another and each comes with a list of mathematical prerequisites. Thus, each instructor has the flexibility to choose those applications that are suitable for his or her students and to incorporate each application anywhere in the course after the mathematical prerequisites have been satisfied.

This edition of *Elementary Linear Algebra*, like those that have preceded it, gives an elementary treatment of linear algebra that is suitable for students in the CEGEP system. The aim is to present the fundamentals of linear algebra in the clearest possible way; pedagogy is the main consideration. Calculus is not a prerequisite, but there are clearly labeled exercises and examples for students who have studied calculus. Those exercises can be omitted without loss of continuity. Technology is also not required, but for those who would like to use MATLAB, Maple, *Mathematica*, or calculators with linear algebra capabilities, exercises have been included at the ends of the chapters that allow for further exploration of that chapter's contents.

SUMMARY OF CHANGES IN THIS EDITION

This edition contains organizational changes and additional material suggested by users of the text. The entire text has also been reviewed for accuracy, typographical errors, and areas where the exposition could be improved or additional examples are needed. The following changes have been made:

- A new Section 4.4 Spaces of Polynomials has been added to further smooth the transition to general linear transformations.

- Chapter 2 has been reorganized by switching Section 2.1 with Section 2.4. The cofactor expansion approach to determinants is now covered first and the combinatorial approach is now at the end of the chapter.

- Additional exercises, including Discussion and Discovery, Supplementary, and Technology exercises, have been added throughout the text.

- In response to instructors' requests, the number of exercises that have answers in the back of the book has been reduced considerably.

- The page design has been modified to enhance the readability of the text.

Hallmark Features

- **Relationships Between Concepts:** One of the important goals of a course in linear algebra is to establish the intricate thread of relationships between systems of linear equations, matrices, determinants, vectors, linear transformations, and eigenvalues. That thread of relationships is developed through the following crescendo of theorems that link each new idea with ideas that preceded it: 1.5.3, 1.6.4, 2.3.6, 4.3.4, 5.6.9. These theorems bring a coherence to the linear algebra landscape and also serve as a constant source of review.

- **Smooth Transition to Abstraction:** The transition from R^n to general vector spaces is often difficult for students. To smooth out that transition, the underlying geometry of R^n is emphasized and key ideas are developed in R^n before proceeding to general vector spaces.

- **Early Exposure to Linear Transformations and Eigenvalues:** To ensure that the material on linear transformations and eigenvalues does not get lost at the end of the course, some of the basic concepts relating to those topics are developed early in the text and then reviewed and expanded on when the topic is treated in more depth later in the text. For example, characteristic equations are discussed briefly in the chapter on determinants, and linear transformations from R^n to R^m are discussed immediately after R^n is introduced, then reviewed later in the context of general linear transformations.

About the Exercises

Each section exercise set begins with routine drill problems, progresses to problems with more substance, and concludes with theoretical problems. In most sections, the main part of the exercise set is followed by the *Discussion and Discovery* problems described above. Most chapters end with a set of supplementary exercises that tend to be more challenging and force the student to draw on ideas from the entire chapter rather than a specific section. The technology exercises follow the supplementary exercises and are classified according to the section in which we suggest that they be assigned. Data for these exercises in MATLAB, Maple, and *Mathematica* formats can be downloaded from **www.wiley.com/college/anton**.

About Chapter 6

This chapter consists of nine applications of linear algebra. The selection was determined in consultation with faculty teaching the course in the social science stream. Special thanks go to Jean-François Deslandes of Marianopolis College for the creation of the section on the Simplex Method. Each application is in its own independent section, so that sections can be deleted or permuted freely to fit individual needs and interests. Each topic begins with a list of linear algebra prerequisites so that a reader can tell in advance if he or she has sufficient background to read the section.

Because the topics vary considerably in difficulty, we have included a subjective rating of each topic—easy or more difficult as follows:

Easy: The average student who has met the stated prerequisites should be able to read the material with no help from the instructor.

More Difficult: The average student who has met the stated prerequisites may require a little help from the instructor.

	1	2	3	4	5	6	7	8	9
Easy	x	x	x	x					
More Difficult					x	x	x	x	x

Our evaluation is based more on the intrinsic difficulty of the material rather than the number of prerequisites; thus, a topic requiring fewer mathematical prerequisites may be rated harder than one requiring more prerequisites.

Because our primary objective is to present applications of linear algebra, proofs are often omitted. We assume that the reader has met the linear algebra prerequisites and whenever results from other fields are needed, they are stated precisely (with motivation where possible), but usually without proof.

Supplementary Materials for Students

Data for Technology Exercises is provided in MATLAB, Maple, and *Mathematica* formats. This data can be downloaded from **www.wiley.com/canada/anton**.

Supplementary Materials for Instructors

Instructor's Solutions Manual—This new supplement provides solutions to all exercises in the text.

Test Bank—This includes approximately 50 free-form questions, five essay questions for each chapter, and a sample cumulative final examination.

Web Resources—More information about this text and its resources can be obtained from your Wiley representative or from **www.wiley.com/canada/anton**.

CONTENTS

CHAPTER 6   Applications of Linear Algebra   295

Systems of Linear Equations and Matrices

CHAPTER CONTENTS

INTRODUCTION: Information in science and mathematics is often organized into rows and columns to form rectangular arrays, called "matrices" (plural of "matrix"). Matrices are often tables of numerical data that arise from physical observations, but they also occur in various mathematical contexts. For example, we shall see in this chapter that to solve a system of equations such as

$$5x + y = 3$$
$$2x - y = 4$$

all of the information required for the solution is embodied in the matrix

$$\begin{bmatrix} 5 & 1 & 3 \\ 2 & -1 & 4 \end{bmatrix}$$

and that the solution can be obtained by performing appropriate operations on this matrix. This is particularly important in developing computer programs to solve systems of linear equations because computers are well suited for manipulating arrays of numerical information. However, matrices are not simply a notational tool for solving systems of equations; they can be viewed as mathematical objects in their own right, and there is a rich and important theory associated with them that has a wide variety of applications. In this chapter we will begin the study of matrices.

1.1
INTRODUCTION TO SYSTEMS OF LINEAR EQUATIONS

Systems of linear algebraic equations and their solutions constitute one of the major topics studied in the course known as "linear algebra." In this first section we shall introduce some basic terminology and discuss a method for solving such systems.

Linear Equations

Any straight line in the xy-plane can be represented algebraically by an equation of the form

$$a_1 x + a_2 y = b$$

where a_1, a_2, and b are real constants and a_1 and a_2 are not both zero. An equation of this form is called a linear equation in the variables x and y. More generally, we define a *linear equation* in the n variables $x_1, x_2, \ldots, x_n$ to be one that can be expressed in the form

$$a_1 x_1 + a_2 x_2 + \cdots + a_n x_n = b$$

where $a_1, a_2, \ldots, a_n$, and b are real constants. The variables in a linear equation are sometimes called *unknowns*.

EXAMPLE 1 Linear Equations

The equations

$$x + 3y = 7, \quad y = \tfrac{1}{2}x + 3z + 1, \quad \text{and} \quad x_1 - 2x_2 - 3x_3 + x_4 = 7$$

are linear. Observe that a linear equation does not involve any products or roots of variables. All variables occur only to the first power and do not appear as arguments for trigonometric, logarithmic, or exponential functions. The equations

$$x + 3\sqrt{y} = 5, \quad 3x + 2y - z + xz = 4, \quad \text{and} \quad y = \sin x$$

are *not* linear. ◆

A *solution* of a linear equation $a_1 x_1 + a_2 x_2 + \cdots + a_n x_n = b$ is a sequence of n numbers $s_1, s_2, \ldots, s_n$ such that the equation is satisfied when we substitute $x_1 = s_1$, $x_2 = s_2, \ldots, x_n = s_n$. The set of all solutions of the equation is called its *solution set* or sometimes the *general solution* of the equation.

EXAMPLE 2 Finding a Solution Set

Find the solution set of (a) $4x - 2y = 1$, and (b) $x_1 - 4x_2 + 7x_3 = 5$.

Solution (a)

To find solutions of (a), we can assign an arbitrary value to x and solve for y, or choose an arbitrary value for y and solve for x. If we follow the first approach and assign x an arbitrary value t, we obtain

$$x = t, \quad y = 2t - \tfrac{1}{2}$$

These formulas describe the solution set in terms of an arbitrary number t, called a *parameter*. Particular numerical solutions can be obtained by substituting specific values for t. For example, $t = 3$ yields the solution $x = 3$, $y = \frac{11}{2}$; and $t = -\frac{1}{2}$ yields the solution $x = -\frac{1}{2}$, $y = -\frac{3}{2}$.

If we follow the second approach and assign y the arbitrary value t, we obtain

$$x = \tfrac{1}{2}t + \tfrac{1}{4}, \qquad y = t$$

Although these formulas are different from those obtained above, they yield the same solution set as t varies over all possible real numbers. For example, the previous formulas gave the solution $x = 3$, $y = \frac{11}{2}$ when $t = 3$, whereas the formulas immediately above yield that solution when $t = \frac{11}{2}$.

Solution (b)

To find the solution set of (b), we can assign arbitrary values to any two variables and solve for the third variable. In particular, if we assign arbitrary values s and t to x_2 and x_3, respectively, and solve for x_1, we obtain

$$x_1 = 5 + 4s - 7t, \qquad x_2 = s, \qquad x_3 = t \quad \blacklozenge$$

Linear Systems

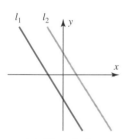

(*a*) No solution

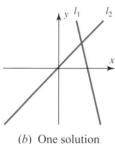

(*b*) One solution

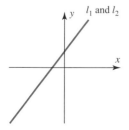

(*c*) Infinitely many solutions

Figure 1.1.1

A finite set of linear equations in the variables $x_1, x_2, \ldots, x_n$ is called a **system of linear equations** or a **linear system**. A sequence of numbers $s_1, s_2, \ldots, s_n$ is called a **solution** of the system if $x_1 = s_1, x_2 = s_2, \ldots, x_n = s_n$ is a solution of every equation in the system. For example, the system

$$4x_1 - x_2 + 3x_3 = -1$$
$$3x_1 + x_2 + 9x_3 = -4$$

has the solution $x_1 = 1$, $x_2 = 2$, $x_3 = -1$ since these values satisfy both equations. However, $x_1 = 1, x_2 = 8, x_3 = 1$ is not a solution since these values satisfy only the first equation in the system.

Not all systems of linear equations have solutions. For example, if we multiply the second equation of the system

$$x + y = 4$$
$$2x + 2y = 6$$

by $\frac{1}{2}$, it becomes evident that there are no solutions since the resulting equivalent system

$$x + y = 4$$
$$x + y = 3$$

has contradictory equations.

A system of equations that has no solutions is said to be **inconsistent**; if there is at least one solution of the system, it is called **consistent**. To illustrate the possibilities that can occur in solving systems of linear equations, consider a general system of two linear equations in the unknowns x and y:

$$a_1 x + b_1 y = c_1 \quad (a_1, b_1 \text{ not both zero})$$
$$a_2 x + b_2 y = c_2 \quad (a_2, b_2 \text{ not both zero})$$

The graphs of these equations are lines; call them l_1 and l_2. Since a point (x, y) lies on a line if and only if the numbers x and y satisfy the equation of the line, the solutions of the system of equations correspond to points of intersection of l_1 and l_2. There are three possibilities, illustrated in Figure 1.1.1:

• The lines l_1 and l_2 may be parallel, in which case there is no intersection and consequently no solution to the system.

- The lines l_1 and l_2 may intersect at only one point, in which case the system has exactly one solution.

- The lines l_1 and l_2 may coincide, in which case there are infinitely many points of intersection and consequently infinitely many solutions to the system.

Although we have considered only two equations with two unknowns here, we will show later that the same three possibilities hold for arbitrary linear systems:

Every system of linear equations has no solutions, or has exactly one solution, or has infinitely many solutions.

An arbitrary system of m linear equations in n unknowns can be written as

$$
\begin{aligned}
a_{11}x_1 + a_{12}x_2 + \cdots + a_{1n}x_n &= b_1 \\
a_{21}x_1 + a_{22}x_2 + \cdots + a_{2n}x_n &= b_2 \\
\vdots \qquad \vdots \qquad\quad \vdots \qquad\ \vdots& \\
a_{m1}x_1 + a_{m2}x_2 + \cdots + a_{mn}x_n &= b_m
\end{aligned}
$$

where $x_1, x_2, \ldots, x_n$ are the unknowns and the subscripted a's and b's denote constants. For example, a general system of three linear equations in four unknowns can be written as

$$
\begin{aligned}
a_{11}x_1 + a_{12}x_2 + a_{13}x_3 + a_{14}x_4 &= b_1 \\
a_{21}x_1 + a_{22}x_2 + a_{23}x_3 + a_{24}x_4 &= b_2 \\
a_{31}x_1 + a_{32}x_2 + a_{33}x_3 + a_{34}x_4 &= b_3
\end{aligned}
$$

The double subscripting on the coefficients of the unknowns is a useful device that is used to specify the location of the coefficient in the system. The first subscript on the coefficient a_{ij} indicates the equation in which the coefficient occurs, and the second subscript indicates which unknown it multiplies. Thus, a_{12} is in the first equation and multiplies unknown x_2.

Augmented Matrices

If we mentally keep track of the location of the $+$'s, the x's, and the $=$'s, a system of m linear equations in n unknowns can be abbreviated by writing only the rectangular array of numbers:

$$
\begin{bmatrix}
a_{11} & a_{12} & \cdots & a_{1n} & b_1 \\
a_{21} & a_{22} & \cdots & a_{2n} & b_2 \\
\vdots & \vdots & & \vdots & \vdots \\
a_{m1} & a_{m2} & \cdots & a_{mn} & b_m
\end{bmatrix}
$$

This is called the ***augmented matrix*** for the system. (The term *matrix* is used in mathematics to denote a rectangular array of numbers. Matrices arise in many contexts, which we will consider in more detail in later sections.) For example, the augmented matrix for the system of equations

$$
\begin{aligned}
x_1 + \ x_2 + 2x_3 &= 9 \\
2x_1 + 4x_2 - 3x_3 &= 1 \\
3x_1 + 6x_2 - 5x_3 &= 0
\end{aligned}
$$

is

$$
\begin{bmatrix}
1 & 1 & 2 & 9 \\
2 & 4 & -3 & 1 \\
3 & 6 & -5 & 0
\end{bmatrix}
$$

REMARK When constructing an augmented matrix, we must write the unknowns in the same order in each equation, and the constants must be on the right.

The basic method for solving a system of linear equations is to replace the given system by a new system that has the same solution set but is easier to solve. This new system is generally obtained in a series of steps by applying the following three types of operations to eliminate unknowns systematically:

1. Multiply an equation through by a nonzero constant.
2. Interchange two equations.
3. Add a multiple of one equation to another.

Since the rows (horizontal lines) of an augmented matrix correspond to the equations in the associated system, these three operations correspond to the following operations on the rows of the augmented matrix:

1. Multiply a row through by a nonzero constant.
2. Interchange two rows.
3. Add a multiple of one row to another row.

Elementary Row Operations

These are called ***elementary row operations***. The following example illustrates how these operations can be used to solve systems of linear equations. Since a systematic procedure for finding solutions will be derived in the next section, it is not necessary to worry about how the steps in this example were selected. The main effort at this time should be devoted to understanding the computations and the discussion.

EXAMPLE 3   Using Elementary Row Operations

In the left column below we solve a system of linear equations by operating on the equations in the system, and in the right column we solve the same system by operating on the rows of the augmented matrix.

$$\begin{aligned} x + y + 2z &= 9 \\ 2x + 4y - 3z &= 1 \\ 3x + 6y - 5z &= 0 \end{aligned} \qquad \begin{bmatrix} 1 & 1 & 2 & 9 \\ 2 & 4 & -3 & 1 \\ 3 & 6 & -5 & 0 \end{bmatrix}$$

Add -2 times the first equation to the second to obtain

Add -2 times the first row to the second to obtain

$$\begin{aligned} x + y + 2z &= 9 \\ 2y - 7z &= -17 \\ 3x + 6y - 5z &= 0 \end{aligned} \qquad \begin{bmatrix} 1 & 1 & 2 & 9 \\ 0 & 2 & -7 & -17 \\ 3 & 6 & -5 & 0 \end{bmatrix}$$

Add -3 times the first equation to the third to obtain

Add -3 times the first row to the third to obtain

$$\begin{aligned} x + y + 2z &= 9 \\ 2y - 7z &= -17 \\ 3y - 11z &= -27 \end{aligned} \qquad \begin{bmatrix} 1 & 1 & 2 & 9 \\ 0 & 2 & -7 & -17 \\ 0 & 3 & -11 & -27 \end{bmatrix}$$

Multiply the second equation by $\frac{1}{2}$ to obtain

Multiply the second row by $\frac{1}{2}$ to obtain

$$\begin{aligned} x + y + 2z &= 9 \\ y - \tfrac{7}{2}z &= -\tfrac{17}{2} \\ 3y - 11z &= -27 \end{aligned} \qquad \begin{bmatrix} 1 & 1 & 2 & 9 \\ 0 & 1 & -\tfrac{7}{2} & -\tfrac{17}{2} \\ 0 & 3 & -11 & -27 \end{bmatrix}$$

Add −3 times the second equation to the third to obtain

$$\begin{aligned} x + y + 2z &= 9 \\ y - \tfrac{7}{2}z &= -\tfrac{17}{2} \\ -\tfrac{1}{2}z &= -\tfrac{3}{2} \end{aligned}$$

Add −3 times the second row to the third to obtain

$$\begin{bmatrix} 1 & 1 & 2 & 9 \\ 0 & 1 & -\tfrac{7}{2} & -\tfrac{17}{2} \\ 0 & 0 & -\tfrac{1}{2} & -\tfrac{3}{2} \end{bmatrix}$$

Multiply the third equation by −2 to obtain

$$\begin{aligned} x + y + 2z &= 9 \\ y - \tfrac{7}{2}z &= -\tfrac{17}{2} \\ z &= 3 \end{aligned}$$

Multiply the third row by −2 to obtain

$$\begin{bmatrix} 1 & 1 & 2 & 9 \\ 0 & 1 & -\tfrac{7}{2} & -\tfrac{17}{2} \\ 0 & 0 & 1 & 3 \end{bmatrix}$$

Add −1 times the second equation to the first to obtain

$$\begin{aligned} x \quad\;\; + \tfrac{11}{2}z &= \tfrac{35}{2} \\ y - \tfrac{7}{2}z &= -\tfrac{17}{2} \\ z &= 3 \end{aligned}$$

Add −1 times the second row to the first to obtain

$$\begin{bmatrix} 1 & 0 & \tfrac{11}{2} & \tfrac{35}{2} \\ 0 & 1 & -\tfrac{7}{2} & -\tfrac{17}{2} \\ 0 & 0 & 1 & 3 \end{bmatrix}$$

Add $-\tfrac{11}{2}$ times the third equation to the first and $\tfrac{7}{2}$ times the third equation to the second to obtain

$$\begin{aligned} x \quad\;\;\;\;\;\; &= 1 \\ y \quad\;\; &= 2 \\ z &= 3 \end{aligned}$$

Add $-\tfrac{11}{2}$ times the third row to the first and $\tfrac{7}{2}$ times the third row to the second to obtain

$$\begin{bmatrix} 1 & 0 & 0 & 1 \\ 0 & 1 & 0 & 2 \\ 0 & 0 & 1 & 3 \end{bmatrix}$$

The solution $x = 1$, $y = 2$, $z = 3$ is now evident. ◆

EXERCISE SET
1.1

1. Which of the following are linear equations in x_1, x_2, and x_3?

   (a) $x_1 + 5x_2 - \sqrt{2}x_3 = 1$ (b) $x_1 + 3x_2 + x_1x_3 = 2$

   (c) $x_1 = -7x_2 + 3x_3$ (d) $x_1^{-2} + x_2 + 8x_3 = 5$

   (e) $x_1^{3/5} - 2x_2 + x_3 = 4$ (f) $\pi x_1 - \sqrt{2}x_2 + \tfrac{1}{3}x_3 = 7^{1/3}$

2. Given that k is a constant, which of the following are linear equations?

   (a) $x_1 - x_2 + x_3 = \sin k$ (b) $kx_1 - \dfrac{1}{k}x_2 = 9$ (c) $2^k x_1 + 7x_2 - x_3 = 0$

3. Find the solution set of each of the following linear equations.

   (a) $7x - 5y = 3$ (b) $3x_1 - 5x_2 + 4x_3 = 7$

   (c) $-8x_1 + 2x_2 - 5x_3 + 6x_4 = 1$ (d) $3v - 8w + 2x - y + 4z = 0$

4. Find the augmented matrix for each of the following systems of linear equations.

   (a) $\begin{aligned} 3x_1 - 2x_2 &= -1 \\ 4x_1 + 5x_2 &= 3 \\ 7x_1 + 3x_2 &= 2 \end{aligned}$ (b) $\begin{aligned} 2x_1 \quad\;\; + 2x_3 &= 1 \\ 3x_1 - x_2 + 4x_3 &= 7 \\ 6x_1 + x_2 - x_3 &= 0 \end{aligned}$

   (c) $\begin{aligned} x_1 + 2x_2 \quad\;\; - x_4 + x_5 &= 1 \\ 3x_2 + x_3 \quad\;\; - x_5 &= 2 \\ x_3 + 7x_4 \quad\;\; &= 1 \end{aligned}$ (d) $\begin{aligned} x_1 \quad\;\;\;\;\;\; &= 1 \\ x_2 \quad\;\; &= 2 \\ x_3 &= 3 \end{aligned}$

5. Find a system of linear equations corresponding to the augmented matrix.

(a) $\begin{bmatrix} 2 & 0 & 0 \\ 3 & -4 & 0 \\ 0 & 1 & 1 \end{bmatrix}$
(b) $\begin{bmatrix} 3 & 0 & -2 & 5 \\ 7 & 1 & 4 & -3 \\ 0 & -2 & 1 & 7 \end{bmatrix}$

(c) $\begin{bmatrix} 7 & 2 & 1 & -3 & 5 \\ 1 & 2 & 4 & 0 & 1 \end{bmatrix}$
(d) $\begin{bmatrix} 1 & 0 & 0 & 0 & 7 \\ 0 & 1 & 0 & 0 & -2 \\ 0 & 0 & 1 & 0 & 3 \\ 0 & 0 & 0 & 1 & 4 \end{bmatrix}$

6. (a) Find a linear equation in the variables x and y that has the general solution $x = 5 + 2t$, $y = t$.

(b) Show that $x = t$, $y = \frac{1}{2}t - \frac{5}{2}$ is also the general solution of the equation in part (a).

7. The curve $y = ax^2 + bx + c$ shown in the accompanying figure passes through the points (x_1, y_1), (x_2, y_2), and (x_3, y_3). Show that the coefficients a, b, and c are a solution of the system of linear equations whose augmented matrix is

$$\begin{bmatrix} x_1^2 & x_1 & 1 & y_1 \\ x_2^2 & x_2 & 1 & y_2 \\ x_3^2 & x_3 & 1 & y_3 \end{bmatrix}$$

Figure Ex-7

8. Consider the system of equations

$$\begin{aligned} x + y + 2z &= a \\ x + z &= b \\ 2x + y + 3z &= c \end{aligned}$$

Show that for this system to be consistent, the constants a, b, and c must satisfy $c = a + b$.

9. Show that if the linear equations $x_1 + kx_2 = c$ and $x_1 + lx_2 = d$ have the same solution set, then the equations are identical.

10. Show that the elementary row operations do not affect the solution set of a linear system.

Discussion & Discovery

11. For which value(s) of the constant k does the system

$$\begin{aligned} x - y &= 3 \\ 2x - 2y &= k \end{aligned}$$

have no solutions? Exactly one solution? Infinitely many solutions? Explain your reasoning.

12. Consider the system of equations

$$\begin{aligned} ax + by &= k \\ cx + dy &= l \\ ex + fy &= m \end{aligned}$$

Indicate what we can say about the relative positions of the lines $ax + by = k$, $cx + dy = l$, and $ex + fy = m$ when

(a) the system has no solutions.

(b) the system has exactly one solution.

(c) the system has infinitely many solutions.

13. If the system of equations in Exercise 12 is consistent, explain why at least one equation can be discarded from the system without altering the solution set.

14. If $k = l = m = 0$ in Exercise 12, explain why the system must be consistent. What can be said about the point of intersection of the three lines if the system has exactly one solution?

15. We could also define elementary column operations in analogy with the elementary row operations. What can you say about the effect of elementary column operations on the solution set of a linear system? How would you interpret the effects of elementary column operations?

1.2
GAUSSIAN ELIMINATION

In this section we shall develop a systematic procedure for solving systems of linear equations. The procedure is based on the idea of reducing the augmented matrix of a system to another augmented matrix that is simple enough that the solution of the system can be found by inspection.

Echelon Forms

In Example 3 of the last section, we solved a linear system in the unknowns x, y, and z by reducing the augmented matrix to the form

$$\begin{bmatrix} 1 & 0 & 0 & 1 \\ 0 & 1 & 0 & 2 \\ 0 & 0 & 1 & 3 \end{bmatrix}$$

from which the solution $x = 1$, $y = 2$, $z = 3$ became evident. This is an example of a matrix that is in ***reduced row-echelon form***. To be of this form, a matrix must have the following properties:

1. If a row does not consist entirely of zeros, then the first nonzero number in the row is a 1. We call this a ***leading 1***.
2. If there are any rows that consist entirely of zeros, then they are grouped together at the bottom of the matrix.
3. In any two successive rows that do not consist entirely of zeros, the leading 1 in the lower row occurs farther to the right than the leading 1 in the higher row.
4. Each column that contains a leading 1 has zeros everywhere else in that column.

A matrix that has the first three properties is said to be in ***row-echelon form***. (Thus, a matrix in reduced row-echelon form is of necessity in row-echelon form, but not conversely.)

EXAMPLE 1 Row-Echelon and Reduced Row-Echelon Form

The following matrices are in reduced row-echelon form.

$$\begin{bmatrix} 1 & 0 & 0 & 4 \\ 0 & 1 & 0 & 7 \\ 0 & 0 & 1 & -1 \end{bmatrix}, \quad \begin{bmatrix} 1 & 0 & 0 \\ 0 & 1 & 0 \\ 0 & 0 & 1 \end{bmatrix}, \quad \begin{bmatrix} 0 & 1 & -2 & 0 & 1 \\ 0 & 0 & 0 & 1 & 3 \\ 0 & 0 & 0 & 0 & 0 \\ 0 & 0 & 0 & 0 & 0 \end{bmatrix}, \quad \begin{bmatrix} 0 & 0 \\ 0 & 0 \end{bmatrix}$$

The following matrices are in row-echelon form.

$$\begin{bmatrix} 1 & 4 & -3 & 7 \\ 0 & 1 & 6 & 2 \\ 0 & 0 & 1 & 5 \end{bmatrix}, \quad \begin{bmatrix} 1 & 1 & 0 \\ 0 & 1 & 0 \\ 0 & 0 & 0 \end{bmatrix}, \quad \begin{bmatrix} 0 & 1 & 2 & 6 & 0 \\ 0 & 0 & 1 & -1 & 0 \\ 0 & 0 & 0 & 0 & 1 \end{bmatrix}$$

We leave it for you to confirm that each of the matrices in this example satisfies all of the requirements for its stated form. ◆

EXAMPLE 2 More on Row-Echelon and Reduced Row-Echelon Form

As the last example illustrates, a matrix in row-echelon form has zeros below each leading 1, whereas a matrix in reduced row-echelon form has zeros below *and above* each leading 1. Thus, with any real numbers substituted for the *'s, all matrices of the following types are in row-echelon form:

$$\begin{bmatrix} 1 & * & * & * \\ 0 & 1 & * & * \\ 0 & 0 & 1 & * \\ 0 & 0 & 0 & 1 \end{bmatrix}, \quad \begin{bmatrix} 1 & * & * & * \\ 0 & 1 & * & * \\ 0 & 0 & 1 & * \\ 0 & 0 & 0 & 0 \end{bmatrix},$$

$$\begin{bmatrix} 1 & * & * & * \\ 0 & 1 & * & * \\ 0 & 0 & 0 & 0 \\ 0 & 0 & 0 & 0 \end{bmatrix}, \quad \begin{bmatrix} 0 & 1 & * & * & * & * & * & * & * & * \\ 0 & 0 & 0 & 1 & * & * & * & * & * & * \\ 0 & 0 & 0 & 0 & 1 & * & * & * & * & * \\ 0 & 0 & 0 & 0 & 0 & 1 & * & * & * & * \\ 0 & 0 & 0 & 0 & 0 & 0 & 0 & 0 & 1 & * \end{bmatrix}$$

Moreover, all matrices of the following types are in reduced row-echelon form:

$$\begin{bmatrix} 1 & 0 & 0 & 0 \\ 0 & 1 & 0 & 0 \\ 0 & 0 & 1 & 0 \\ 0 & 0 & 0 & 1 \end{bmatrix}, \quad \begin{bmatrix} 1 & 0 & 0 & * \\ 0 & 1 & 0 & * \\ 0 & 0 & 1 & * \\ 0 & 0 & 0 & 0 \end{bmatrix},$$

$$\begin{bmatrix} 1 & 0 & * & * \\ 0 & 1 & * & * \\ 0 & 0 & 0 & 0 \\ 0 & 0 & 0 & 0 \end{bmatrix}, \quad \begin{bmatrix} 0 & 1 & * & 0 & 0 & 0 & * & * & 0 & * \\ 0 & 0 & 0 & 1 & 0 & 0 & * & * & 0 & * \\ 0 & 0 & 0 & 0 & 1 & 0 & * & * & 0 & * \\ 0 & 0 & 0 & 0 & 0 & 1 & * & * & 0 & * \\ 0 & 0 & 0 & 0 & 0 & 0 & 0 & 0 & 1 & * \end{bmatrix} \quad ◆$$

If, by a sequence of elementary row operations, the augmented matrix for a system of linear equations is put in reduced row-echelon form, then the solution set of the system will be evident by inspection or after a few simple steps. The next example illustrates this situation.

EXAMPLE 3 Solutions of Four Linear Systems

Suppose that the augmented matrix for a system of linear equations has been reduced by row operations to the given reduced row-echelon form. Solve the system.

(a) $\begin{bmatrix} 1 & 0 & 0 & 5 \\ 0 & 1 & 0 & -2 \\ 0 & 0 & 1 & 4 \end{bmatrix}$ (b) $\begin{bmatrix} 1 & 0 & 0 & 4 & -1 \\ 0 & 1 & 0 & 2 & 6 \\ 0 & 0 & 1 & 3 & 2 \end{bmatrix}$

(c) $\begin{bmatrix} 1 & 6 & 0 & 0 & 4 & -2 \\ 0 & 0 & 1 & 0 & 3 & 1 \\ 0 & 0 & 0 & 1 & 5 & 2 \\ 0 & 0 & 0 & 0 & 0 & 0 \end{bmatrix}$ (d) $\begin{bmatrix} 1 & 0 & 0 & 0 \\ 0 & 1 & 2 & 0 \\ 0 & 0 & 0 & 1 \end{bmatrix}$

Solution (a)

The corresponding system of equations is

$$
\begin{aligned}
x_1 \quad\quad\quad &= 5 \\
x_2 \quad\quad &= -2 \\
x_3 &= 4
\end{aligned}
$$

By inspection, $x_1 = 5$, $x_2 = -2$, $x_3 = 4$.

Solution (b)

The corresponding system of equations is

$$
\begin{aligned}
x_1 \quad\quad\quad\quad + 4x_4 &= -1 \\
x_2 \quad\quad + 2x_4 &= 6 \\
x_3 + 3x_4 &= 2
\end{aligned}
$$

Since x_1, x_2, and x_3 correspond to leading 1's in the augmented matrix, we call them *leading variables* or *pivots*. The nonleading variables (in this case x_4) are called *free variables*. Solving for the leading variables in terms of the free variable gives

$$
\begin{aligned}
x_1 &= -1 - 4x_4 \\
x_2 &= 6 - 2x_4 \\
x_3 &= 2 - 3x_4
\end{aligned}
$$

From this form of the equations we see that the free variable x_4 can be assigned an arbitrary value, say t, which then determines the values of the leading variables x_1, x_2, and x_3. Thus there are infinitely many solutions, and the general solution is given by the formulas

$$
x_1 = -1 - 4t, \qquad x_2 = 6 - 2t, \qquad x_3 = 2 - 3t, \qquad x_4 = t
$$

Solution (c)

The row of zeros leads to the equation $0x_1 + 0x_2 + 0x_3 + 0x_4 + 0x_5 = 0$, which places no restrictions on the solutions (why?). Thus, we can omit this equation and write the corresponding system as

$$
\begin{aligned}
x_1 + 6x_2 \quad\quad\quad + 4x_5 &= -2 \\
x_3 \quad\quad + 3x_5 &= 1 \\
x_4 + 5x_5 &= 2
\end{aligned}
$$

Here the leading variables are x_1, x_3, and x_4, and the free variables are x_2 and x_5. Solving for the leading variables in terms of the free variables gives

$$
\begin{aligned}
x_1 &= -2 - 6x_2 - 4x_5 \\
x_3 &= 1 - 3x_5 \\
x_4 &= 2 - 5x_5
\end{aligned}
$$

Since x_5 can be assigned an arbitrary value, t, and x_2 can be assigned an arbitrary value, s, there are infinitely many solutions. The general solution is given by the formulas

$$
x_1 = -2 - 6s - 4t, \qquad x_2 = s, \qquad x_3 = 1 - 3t, \qquad x_4 = 2 - 5t, \qquad x_5 = t
$$

Solution (d)

The last equation in the corresponding system of equations is

$$
0x_1 + 0x_2 + 0x_3 = 1
$$

Since this equation cannot be satisfied, there is no solution to the system. ◆

Elimination Methods

We have just seen how easy it is to solve a system of linear equations once its augmented matrix is in reduced row-echelon form. Now we shall give a step-by-step *elimination* procedure that can be used to reduce any matrix to reduced row-echelon form. As we state each step in the procedure, we shall illustrate the idea by reducing the following matrix to reduced row-echelon form.

$$\begin{bmatrix} 0 & 0 & -2 & 0 & 7 & 12 \\ 2 & 4 & -10 & 6 & 12 & 28 \\ 2 & 4 & -5 & 6 & -5 & -1 \end{bmatrix}$$

Step 1. Locate the leftmost column that does not consist entirely of zeros.

$$\begin{bmatrix} 0 & 0 & -2 & 0 & 7 & 12 \\ 2 & 4 & -10 & 6 & 12 & 28 \\ 2 & 4 & -5 & 6 & -5 & -1 \end{bmatrix}$$

$\uparrow$
Leftmost nonzero column

Step 2. Interchange the top row with another row, if necessary, to bring a nonzero entry to the top of the column found in Step 1.

$$\begin{bmatrix} 2 & 4 & -10 & 6 & 12 & 28 \\ 0 & 0 & -2 & 0 & 7 & 12 \\ 2 & 4 & -5 & 6 & -5 & -1 \end{bmatrix}$$

$\longleftarrow$ The first and second rows in the preceding matrix were interchanged.

Step 3. If the entry that is now at the top of the column found in Step 1 is a, multiply the first row by $1/a$ in order to introduce a leading 1.

$$\begin{bmatrix} 1 & 2 & -5 & 3 & 6 & 14 \\ 0 & 0 & -2 & 0 & 7 & 12 \\ 2 & 4 & -5 & 6 & -5 & -1 \end{bmatrix}$$

$\longleftarrow$ The first row of the preceding matrix was multiplied by $\frac{1}{2}$.

Step 4. Add suitable multiples of the top row to the rows below so that all entries below the leading 1 become zeros.

$$\begin{bmatrix} 1 & 2 & -5 & 3 & 6 & 14 \\ 0 & 0 & -2 & 0 & 7 & 12 \\ 0 & 0 & 5 & 0 & -17 & -29 \end{bmatrix}$$

$\longleftarrow$ -2 times the first row of the preceding matrix was added to the third row.

Step 5. Now cover the top row in the matrix and begin again with Step 1 applied to the submatrix that remains. Continue in this way until the *entire* matrix is in row-echelon form.

$$\begin{bmatrix} 1 & 2 & -5 & 3 & 6 & 14 \\ 0 & 0 & -2 & 0 & 7 & 12 \\ 0 & 0 & 5 & 0 & -17 & -29 \end{bmatrix}$$

$\uparrow$
**Leftmost nonzero column
in the submatrix**

$$\begin{bmatrix} 1 & 2 & -5 & 3 & 6 & 14 \\ 0 & 0 & 1 & 0 & -\frac{7}{2} & -6 \\ 0 & 0 & 5 & 0 & -17 & -29 \end{bmatrix}$$

$\longleftarrow$ The first row in the submatrix was multiplied by $-\frac{1}{2}$ to introduce a leading 1.

$$\begin{bmatrix} 1 & 2 & -5 & 3 & 6 & 14 \\ 0 & 0 & 1 & 0 & -\frac{7}{2} & -6 \\ 0 & 0 & 0 & 0 & \frac{1}{2} & 1 \end{bmatrix}$$

◄——— −5 times the first row of the submatrix was added to the second row of the submatrix to introduce a zero below the leading 1.

$$\begin{bmatrix} 1 & 2 & -5 & 3 & 6 & 14 \\ 0 & 0 & 1 & 0 & -\frac{7}{2} & -6 \\ 0 & 0 & 0 & 0 & \frac{1}{2} & 1 \end{bmatrix}$$

◄——— The top row in the submatrix was covered, and we returned again to Step 1.

└—— **Leftmost nonzero column in the new submatrix**

$$\begin{bmatrix} 1 & 2 & -5 & 3 & 6 & 14 \\ 0 & 0 & 1 & 0 & -\frac{7}{2} & -6 \\ 0 & 0 & 0 & 0 & 1 & 2 \end{bmatrix}$$

◄——— The first (and only) row in the new submatrix was multiplied by 2 to introduce a leading 1.

The *entire* matrix is now in row-echelon form. To find the reduced row-echelon form we need the following additional step.

Step 6. Beginning with the last nonzero row and working upward, add suitable multiples of each row to the rows above to introduce zeros above the leading 1's.

$$\begin{bmatrix} 1 & 2 & -5 & 3 & 6 & 14 \\ 0 & 0 & 1 & 0 & 0 & 1 \\ 0 & 0 & 0 & 0 & 1 & 2 \end{bmatrix}$$

◄——— $\frac{7}{2}$ times the third row of the preceding matrix was added to the second row.

$$\begin{bmatrix} 1 & 2 & -5 & 3 & 0 & 2 \\ 0 & 0 & 1 & 0 & 0 & 1 \\ 0 & 0 & 0 & 0 & 1 & 2 \end{bmatrix}$$

◄——— −6 times the third row was added to the first row.

$$\begin{bmatrix} 1 & 2 & 0 & 3 & 0 & 7 \\ 0 & 0 & 1 & 0 & 0 & 1 \\ 0 & 0 & 0 & 0 & 1 & 2 \end{bmatrix}$$

◄——— 5 times the second row was added to the first row.

The last matrix is in reduced row-echelon form.

If we use only the first five steps, the above procedure produces a row-echelon form and is called ***Gaussian elimination***. Carrying the procedure through to the sixth step and producing a matrix in reduced row-echelon form is called ***Gauss–Jordan elimination***.

REMARK It can be shown that *every matrix has a unique reduced row-echelon form*; that is, one will arrive at the same reduced row-echelon form for a given matrix no matter how the row operations are varied. (A proof of this result can be found in the article "The Reduced Row Echelon Form of a Matrix Is Unique: A Simple Proof," by Thomas Yuster, *Mathematics Magazine*, Vol. 57, No. 2, 1984, pp. 93–94.) In contrast, *a row-echelon form of a given matrix is not unique*: different sequences of row operations can produce different row-echelon forms.

EXAMPLE 4 Gauss–Jordan Elimination

Solve by Gauss–Jordan elimination.

$$\begin{aligned} x_1 + 3x_2 - 2x_3 \qquad\quad + 2x_5 \qquad\qquad &= 0 \\ 2x_1 + 6x_2 - 5x_3 - 2x_4 + 4x_5 - 3x_6 &= -1 \\ 5x_3 + 10x_4 \qquad + 15x_6 &= 5 \\ 2x_1 + 6x_2 \qquad\quad + 8x_4 + 4x_5 + 18x_6 &= 6 \end{aligned}$$

Karl Friedrich Gauss

Wilhelm Jordan

Karl Friedrich Gauss *(1777–1855)* was a German mathematician and scientist. Sometimes called the "prince of mathematicians," Gauss ranks with Isaac Newton and Archimedes as one of the three greatest mathematicians who ever lived. In the entire history of mathematics there may never have been a child so precocious as Gauss—by his own account he worked out the rudiments of arithmetic before he could talk. One day, before he was even three years old, his genius became apparent to his parents in a very dramatic way. His father was preparing the weekly payroll for the laborers under his charge while the boy watched quietly from a corner. At the end of the long and tedious calculation, Gauss informed his father that there was an error in the result and stated the answer, which he had worked out in his head. To the astonishment of his parents, a check of the computations showed Gauss to be correct!

In his doctoral dissertation Gauss gave the first complete proof of the fundamental theorem of algebra, which states that every polynomial equation has as many solutions as its degree. At age 19 he solved a problem that baffled Euclid, inscribing a regular polygon of seventeen sides in a circle using straightedge and compass; and in 1801, at age 24, he published his first masterpiece, *Disquisitiones Arithmeticae*, considered by many to be one of the most brilliant achievements in mathematics. In that paper Gauss systematized the study of number theory (properties of the integers) and formulated the basic concepts that form the foundation of the subject.

Among his myriad achievements, Gauss discovered the Gaussian or "bell-shaped" curve that is fundamental in probability, gave the first geometric interpretation of complex numbers and established their fundamental role in mathematics, developed methods of characterizing surfaces intrinsically by means of the curves that they contain, developed the theory of conformal (angle-preserving) maps, and discovered non-Euclidean geometry 30 years before the ideas were published by others. In physics he made major contributions to the theory of lenses and capillary action, and with Wilhelm Weber he did fundamental work in electromagnetism. Gauss invented the heliotrope, bifilar magnetometer, and an electrotelegraph.

Gauss, who was deeply religious and aristocratic in demeanor, mastered foreign languages with ease, read extensively, and enjoyed mineralogy and botany as hobbies. He disliked teaching and was usually cool and discouraging to other mathematicians, possibly because he had already anticipated their work. It has been said that if Gauss had published all of his discoveries, the current state of mathematics would be advanced by 50 years. He was without a doubt the greatest mathematician of the modern era.

Wilhelm Jordan *(1842–1899)* was a German engineer who specialized in geodesy. His contribution to solving linear systems appeared in his popular book, *Handbuch der Vermessungskunde* (*Handbook of Geodesy*), in 1888.

Solution

The augmented matrix for the system is

$$\begin{bmatrix} 1 & 3 & -2 & 0 & 2 & 0 & 0 \\ 2 & 6 & -5 & -2 & 4 & -3 & -1 \\ 0 & 0 & 5 & 10 & 0 & 15 & 5 \\ 2 & 6 & 0 & 8 & 4 & 18 & 6 \end{bmatrix}$$

Adding -2 times the first row to the second and fourth rows gives

$$\begin{bmatrix} 1 & 3 & -2 & 0 & 2 & 0 & 0 \\ 0 & 0 & -1 & -2 & 0 & -3 & -1 \\ 0 & 0 & 5 & 10 & 0 & 15 & 5 \\ 0 & 0 & 4 & 8 & 0 & 18 & 6 \end{bmatrix}$$

Multiplying the second row by -1 and then adding -5 times the new second row to the third row and -4 times the new second row to the fourth row gives

$$\begin{bmatrix} 1 & 3 & -2 & 0 & 2 & 0 & 0 \\ 0 & 0 & 1 & 2 & 0 & 3 & 1 \\ 0 & 0 & 0 & 0 & 0 & 0 & 0 \\ 0 & 0 & 0 & 0 & 0 & 6 & 2 \end{bmatrix}$$

Interchanging the third and fourth rows and then multiplying the third row of the resulting matrix by $\frac{1}{6}$ gives the row-echelon form

$$\begin{bmatrix} 1 & 3 & -2 & 0 & 2 & 0 & 0 \\ 0 & 0 & 1 & 2 & 0 & 3 & 1 \\ 0 & 0 & 0 & 0 & 0 & 1 & \frac{1}{3} \\ 0 & 0 & 0 & 0 & 0 & 0 & 0 \end{bmatrix}$$

Adding -3 times the third row to the second row and then adding 2 times the second row of the resulting matrix to the first row yields the reduced row-echelon form

$$\begin{bmatrix} 1 & 3 & 0 & 4 & 2 & 0 & 0 \\ 0 & 0 & 1 & 2 & 0 & 0 & 0 \\ 0 & 0 & 0 & 0 & 0 & 1 & \frac{1}{3} \\ 0 & 0 & 0 & 0 & 0 & 0 & 0 \end{bmatrix}$$

The corresponding system of equations is

$$\begin{aligned} x_1 + 3x_2 \quad + 4x_4 + 2x_5 \quad &= 0 \\ x_3 + 2x_4 \quad &= 0 \\ x_6 &= \tfrac{1}{3} \end{aligned}$$

(We have discarded the last equation, $0x_1 + 0x_2 + 0x_3 + 0x_4 + 0x_5 + 0x_6 = 0$, since it will be satisfied automatically by the solutions of the remaining equations.) Solving for the leading variables, we obtain

$$\begin{aligned} x_1 &= -3x_2 - 4x_4 - 2x_5 \\ x_3 &= -2x_4 \\ x_6 &= \tfrac{1}{3} \end{aligned}$$

If we assign the free variables x_2, x_4, and x_5 arbitrary values r, s, and t, respectively, the general solution is given by the formulas

$$x_1 = -3r - 4s - 2t, \qquad x_2 = r, \qquad x_3 = -2s, \qquad x_4 = s, \qquad x_5 = t, \qquad x_6 = \tfrac{1}{3} \; \blacklozenge$$

Back-Substitution

It is sometimes preferable to solve a system of linear equations by using Gaussian elimination to bring the augmented matrix into row-echelon form without continuing all the way to the reduced row-echelon form. When this is done, the corresponding system of equations can be solved by a technique called ***back-substitution***. The next example illustrates the idea.

EXAMPLE 5 Example 4 Solved by Back-Substitution

From the computations in Example 4, a row-echelon form of the augmented matrix is

$$\begin{bmatrix} 1 & 3 & -2 & 0 & 2 & 0 & 0 \\ 0 & 0 & 1 & 2 & 0 & 3 & 1 \\ 0 & 0 & 0 & 0 & 0 & 1 & \frac{1}{3} \\ 0 & 0 & 0 & 0 & 0 & 0 & 0 \end{bmatrix}$$

To solve the corresponding system of equations

$$\begin{aligned} x_1 + 3x_2 - 2x_3 \qquad + 2x_5 \qquad &= 0 \\ x_3 + 2x_4 \qquad + 3x_6 &= 1 \\ x_6 &= \tfrac{1}{3} \end{aligned}$$

we proceed as follows:

Step 1.   Solve the equations for the leading variables.

$$\begin{aligned} x_1 &= -3x_2 + 2x_3 - 2x_5 \\ x_3 &= 1 - 2x_4 - 3x_6 \\ x_6 &= \tfrac{1}{3} \end{aligned}$$

Step 2.   Beginning with the bottom equation and working upward, successively substitute each equation into all the equations above it.
Substituting $x_6 = \tfrac{1}{3}$ into the second equation yields

$$\begin{aligned} x_1 &= -3x_2 + 2x_3 - 2x_5 \\ x_3 &= -2x_4 \\ x_6 &- \tfrac{1}{3} \end{aligned}$$

Substituting $x_3 = -2x_4$ into the first equation yields

$$\begin{aligned} x_1 &= -3x_2 - 4x_4 - 2x_5 \\ x_3 &= -2x_4 \\ x_6 &= \tfrac{1}{3} \end{aligned}$$

Step 3.   Assign arbitrary values to the free variables, if any.

If we assign x_2, x_4, and x_5 the arbitrary values r, s, and t, respectively, the general solution is given by the formulas

$$x_1 = -3r - 4s - 2t, \qquad x_2 = r, \qquad x_3 = -2s, \qquad x_4 = s, \qquad x_5 = t, \qquad x_6 = \tfrac{1}{3}$$

This agrees with the solution obtained in Example 4.   ◆

REMARK   The arbitrary values that are assigned to the free variables are often called **parameters**. Although we shall generally use the letters $r, s, t, \ldots$ for the parameters, any letters that do not conflict with the variable names may be used.

EXAMPLE 6   Gaussian Elimination

Solve

$$\begin{aligned} x + y + 2z &= 9 \\ 2x + 4y - 3z &= 1 \\ 3x + 6y - 5z &= 0 \end{aligned}$$

by Gaussian elimination and back-substitution.

Solution

This is the system in Example 3 of Section 1.1. In that example we converted the augmented matrix

$$\begin{bmatrix} 1 & 1 & 2 & 9 \\ 2 & 4 & -3 & 1 \\ 3 & 6 & -5 & 0 \end{bmatrix}$$

to the row-echelon form

$$\begin{bmatrix} 1 & 1 & 2 & 9 \\ 0 & 1 & -\frac{7}{2} & -\frac{17}{2} \\ 0 & 0 & 1 & 3 \end{bmatrix}$$

The system corresponding to this matrix is

$$\begin{aligned} x + y + 2z &= 9 \\ y - \tfrac{7}{2}z &= -\tfrac{17}{2} \\ z &= 3 \end{aligned}$$

Solving for the leading variables yields

$$\begin{aligned} x &= 9 - y - 2z \\ y &= -\tfrac{17}{2} + \tfrac{7}{2}z \\ z &= 3 \end{aligned}$$

Substituting the bottom equation into those above yields

$$\begin{aligned} x &= 3 - y \\ y &= 2 \\ z &= 3 \end{aligned}$$

and substituting the second equation into the top yields $x = 1$, $y = 2$, $z = 3$. This agrees with the result found by Gauss–Jordan elimination in Example 3 of Section 1.1. ◆

Homogeneous Linear Systems

A system of linear equations is said to be *homogeneous* if the constant terms are all zero; that is, the system has the form

$$\begin{aligned} a_{11}x_1 + a_{12}x_2 + \cdots + a_{1n}x_n &= 0 \\ a_{21}x_1 + a_{22}x_2 + \cdots + a_{2n}x_n &= 0 \\ \vdots \qquad \vdots \qquad\quad \vdots \quad\; \vdots \\ a_{m1}x_1 + a_{m2}x_2 + \cdots + a_{mn}x_n &= 0 \end{aligned}$$

Every homogeneous system of linear equations is consistent, since all such systems have $x_1 = 0, x_2 = 0, \ldots, x_n = 0$ as a solution. This solution is called the *trivial solution*; if there are other solutions, they are called *nontrivial solutions*.

Because a homogeneous linear system always has the trivial solution, there are only two possibilities for its solutions:

• The system has only the trivial solution.

• The system has infinitely many solutions in addition to the trivial solution.

In the special case of a homogeneous linear system of two equations in two unknowns, say

$$a_1 x + b_1 y = 0 \quad \text{(} a_1, b_1 \text{ not both zero)}$$
$$a_2 x + b_2 y = 0 \quad \text{(} a_2, b_2 \text{ not both zero)}$$

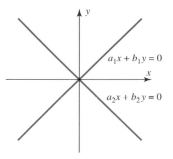

(*a*) Only the trivial solution

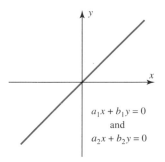

(*b*) Infinitely many solutions

Figure 1.2.1

the graphs of the equations are lines through the origin, and the trivial solution corresponds to the point of intersection at the origin (Figure 1.2.1).

There is one case in which a homogeneous system is assured of having nontrivial solutions—namely, whenever the system involves more unknowns than equations. To see why, consider the following example of four equations in five unknowns.

EXAMPLE 7 Gauss–Jordan Elimination

Solve the following homogeneous system of linear equations by using Gauss–Jordan elimination.

$$
\begin{aligned}
2x_1 + 2x_2 - x_3 \quad\quad + x_5 &= 0 \\
-x_1 - x_2 + 2x_3 - 3x_4 + x_5 &= 0 \\
x_1 + x_2 - 2x_3 \quad\quad - x_5 &= 0 \\
x_3 + x_4 + x_5 &= 0
\end{aligned}
\tag{1}
$$

Solution

The augmented matrix for the system is

$$
\begin{bmatrix}
2 & 2 & -1 & 0 & 1 & 0 \\
-1 & -1 & 2 & -3 & 1 & 0 \\
1 & 1 & -2 & 0 & -1 & 0 \\
0 & 0 & 1 & 1 & 1 & 0
\end{bmatrix}
$$

Reducing this matrix to reduced row-echelon form, we obtain

$$
\begin{bmatrix}
1 & 1 & 0 & 0 & 1 & 0 \\
0 & 0 & 1 & 0 & 1 & 0 \\
0 & 0 & 0 & 1 & 0 & 0 \\
0 & 0 & 0 & 0 & 0 & 0
\end{bmatrix}
$$

The corresponding system of equations is

$$
\begin{aligned}
x_1 + x_2 \quad\quad + x_5 &= 0 \\
x_3 \quad + x_5 &= 0 \\
x_4 \quad &= 0
\end{aligned}
\tag{2}
$$

Solving for the leading variables yields

$$
\begin{aligned}
x_1 &= -x_2 - x_5 \\
x_3 &= -x_5 \\
x_4 &= 0
\end{aligned}
$$

Thus, the general solution is

$$
x_1 = -s - t, \quad x_2 = s, \quad x_3 = -t, \quad x_4 = 0, \quad x_5 = t
$$

Note that the trivial solution is obtained when $s = t = 0$. ◆

Example 7 illustrates two important points about solving homogeneous systems of linear equations. First, none of the three elementary row operations alters the final column of zeros in the augmented matrix, so the system of equations corresponding to the reduced row-echelon form of the augmented matrix must also be a homogeneous system [see system (2)]. Second, depending on whether the reduced row-echelon form of the augmented matrix has any zero rows, the number of equations in the reduced system is the same as or less than the number of equations in the original system [compare systems (1) and (2)]. Thus, if the given homogeneous system has m equations in n unknowns with $m < n$, and if there are r nonzero rows in the reduced row-echelon form of the augmented matrix, we will have $r < n$. It follows that the system of equations corresponding to the reduced row-echelon form of the augmented matrix will have the form

$$
\begin{aligned}
\cdots x_{k_1} \hspace{4em} + \Sigma(\ ) &= 0 \\
\cdots x_{k_2} \hspace{3em} + \Sigma(\ ) &= 0 \\
\cdots \ \cdot \ \cdot \hspace{5em} \vdots \hspace{2em} & \\
x_{k_r} + \Sigma(\ ) &= 0
\end{aligned}
\tag{3}
$$

where $x_{k_1}, x_{k_2}, \ldots, x_{k_r}$ are the leading variables and $\Sigma(\ )$ denotes sums (possibly all different) that involve the $n - r$ free variables [compare system (3) with system (2) above]. Solving for the leading variables gives

$$
\begin{aligned}
x_{k_1} &= -\Sigma(\ ) \\
x_{k_2} &= -\Sigma(\ ) \\
&\vdots \\
x_{k_r} &= -\Sigma(\ )
\end{aligned}
$$

As in Example 7, we can assign arbitrary values to the free variables on the right-hand side and thus obtain infinitely many solutions to the system.

In summary, we have the following important theorem.

THEOREM 1.2.1

> *A homogeneous system of linear equations with more unknowns than equations has infinitely many solutions.*

REMARK Note that Theorem 1.2.1 applies only to homogeneous systems. A nonhomogeneous system with more unknowns than equations need not be consistent (Exercise 28); however, if the system is consistent, it will have infinitely many solutions. This will be proved later.

Computer Solution of Linear Systems

In applications it is not uncommon to encounter large linear systems that must be solved by computer. Most computer algorithms for solving such systems are based on Gaussian elimination or Gauss–Jordan elimination, but the basic procedures are often modified to deal with such issues as

- Reducing roundoff errors
- Minimizing the use of computer memory space
- Solving the system with maximum speed

Some of these matters will be considered in Chapter 9. For hand computations, fractions are an annoyance that often cannot be avoided. However, in some cases it is possible to avoid them by varying the elementary row operations in the right way. Thus, once the methods of Gaussian elimination and Gauss–Jordan elimination have been mastered, the reader may wish to vary the steps in specific problems to avoid fractions (see Exercise 18).

REMARK Since Gauss–Jordan elimination avoids the use of back-substitution, it would seem that this method would be the more efficient of the two methods we have considered.

It can be argued that this statement is true for solving small systems by hand since Gauss–Jordan elimination actually involves less writing. However, for large systems of equations, it has been shown that the Gauss–Jordan elimination method requires about 50% more operations than Gaussian elimination. This is an important consideration when one is working on computers.

**EXERCISE SET
1.2**

1. Which of the following 3×3 matrices are in reduced row-echelon form?

(a) $\begin{bmatrix} 1 & 0 & 0 \\ 0 & 1 & 0 \\ 0 & 0 & 1 \end{bmatrix}$ (b) $\begin{bmatrix} 1 & 0 & 0 \\ 0 & 1 & 0 \\ 0 & 0 & 0 \end{bmatrix}$ (c) $\begin{bmatrix} 0 & 1 & 0 \\ 0 & 0 & 1 \\ 0 & 0 & 0 \end{bmatrix}$ (d) $\begin{bmatrix} 1 & 0 & 0 \\ 0 & 0 & 1 \\ 0 & 0 & 0 \end{bmatrix}$

(e) $\begin{bmatrix} 1 & 0 & 0 \\ 0 & 0 & 0 \\ 0 & 0 & 1 \end{bmatrix}$ (f) $\begin{bmatrix} 0 & 1 & 0 \\ 1 & 0 & 0 \\ 0 & 0 & 0 \end{bmatrix}$ (g) $\begin{bmatrix} 1 & 1 & 0 \\ 0 & 1 & 0 \\ 0 & 0 & 0 \end{bmatrix}$ (h) $\begin{bmatrix} 1 & 0 & 2 \\ 0 & 1 & 3 \\ 0 & 0 & 0 \end{bmatrix}$

(i) $\begin{bmatrix} 0 & 0 & 1 \\ 0 & 0 & 0 \\ 0 & 0 & 0 \end{bmatrix}$ (j) $\begin{bmatrix} 0 & 0 & 0 \\ 0 & 0 & 0 \\ 0 & 0 & 0 \end{bmatrix}$

2. Which of the following 3×3 matrices are in row-echelon form?

(a) $\begin{bmatrix} 1 & 0 & 0 \\ 0 & 1 & 0 \\ 0 & 0 & 1 \end{bmatrix}$ (b) $\begin{bmatrix} 1 & 2 & 0 \\ 0 & 1 & 0 \\ 0 & 0 & 0 \end{bmatrix}$ (c) $\begin{bmatrix} 1 & 0 & 0 \\ 0 & 1 & 0 \\ 0 & 2 & 0 \end{bmatrix}$

(d) $\begin{bmatrix} 1 & 3 & 4 \\ 0 & 0 & 1 \\ 0 & 0 & 0 \end{bmatrix}$ (e) $\begin{bmatrix} 1 & 5 & -3 \\ 0 & 1 & 1 \\ 0 & 0 & 0 \end{bmatrix}$ (f) $\begin{bmatrix} 1 & 2 & 3 \\ 0 & 0 & 0 \\ 0 & 0 & 1 \end{bmatrix}$

3. In each part determine whether the matrix is in row-echelon form, reduced row-echelon form, both, or neither.

(a) $\begin{bmatrix} 1 & 2 & 0 & 3 & 0 \\ 0 & 0 & 1 & 1 & 0 \\ 0 & 0 & 0 & 0 & 1 \\ 0 & 0 & 0 & 0 & 0 \end{bmatrix}$ (b) $\begin{bmatrix} 1 & 0 & 0 & 5 \\ 0 & 0 & 1 & 3 \\ 0 & 1 & 0 & 4 \end{bmatrix}$ (c) $\begin{bmatrix} 1 & 0 & 3 & 1 \\ 0 & 1 & 2 & 4 \end{bmatrix}$

(d) $\begin{bmatrix} 1 & -7 & 5 & 5 \\ 0 & 1 & 3 & 2 \end{bmatrix}$ (e) $\begin{bmatrix} 1 & 3 & 0 & 2 & 0 \\ 1 & 0 & 2 & 2 & 0 \\ 0 & 0 & 0 & 0 & 1 \\ 0 & 0 & 0 & 0 & 0 \end{bmatrix}$ (f) $\begin{bmatrix} 0 & 0 \\ 0 & 0 \\ 0 & 0 \end{bmatrix}$

4. In each part suppose that the augmented matrix for a system of linear equations has been reduced by row operations to the given reduced row-echelon form. Solve the system.

(a) $\begin{bmatrix} 1 & 0 & 0 & -3 \\ 0 & 1 & 0 & 0 \\ 0 & 0 & 1 & 7 \end{bmatrix}$ (b) $\begin{bmatrix} 1 & 0 & 0 & -7 & 8 \\ 0 & 1 & 0 & 3 & 2 \\ 0 & 0 & 1 & 1 & -5 \end{bmatrix}$

(c) $\begin{bmatrix} 1 & -6 & 0 & 0 & 3 & -2 \\ 0 & 0 & 1 & 0 & 4 & 7 \\ 0 & 0 & 0 & 1 & 5 & 8 \\ 0 & 0 & 0 & 0 & 0 & 0 \end{bmatrix}$ (d) $\begin{bmatrix} 1 & -3 & 0 & 0 \\ 0 & 0 & 1 & 0 \\ 0 & 0 & 0 & 1 \end{bmatrix}$

5. In each part suppose that the augmented matrix for a system of linear equations has been reduced by row operations to the given row-echelon form. Solve the system.

(a) $\begin{bmatrix} 1 & -3 & 4 & 7 \\ 0 & 1 & 2 & 2 \\ 0 & 0 & 1 & 5 \end{bmatrix}$

(b) $\begin{bmatrix} 1 & 0 & 8 & -5 & 6 \\ 0 & 1 & 4 & -9 & 3 \\ 0 & 0 & 1 & 1 & 2 \end{bmatrix}$

(c) $\begin{bmatrix} 1 & 7 & -2 & 0 & -8 & -3 \\ 0 & 0 & 1 & 1 & 6 & 5 \\ 0 & 0 & 0 & 1 & 3 & 9 \\ 0 & 0 & 0 & 0 & 0 & 0 \end{bmatrix}$

(d) $\begin{bmatrix} 1 & -3 & 7 & 1 \\ 0 & 1 & 4 & 0 \\ 0 & 0 & 0 & 1 \end{bmatrix}$

6. Solve each of the following systems by Gauss–Jordan elimination.

(a)
$$\begin{aligned} x_1 + x_2 + 2x_3 &= 8 \\ -x_1 - 2x_2 + 3x_3 &= 1 \\ 3x_1 - 7x_2 + 4x_3 &= 10 \end{aligned}$$

(b)
$$\begin{aligned} 2x_1 + 2x_2 + 2x_3 &= 0 \\ -2x_1 + 5x_2 + 2x_3 &= 1 \\ 8x_1 + x_2 + 4x_3 &= -1 \end{aligned}$$

(c)
$$\begin{aligned} x - y + 2z - w &= -1 \\ 2x + y - 2z - 2w &= -2 \\ -x + 2y - 4z + w &= 1 \\ 3x - 3w &= -3 \end{aligned}$$

(d)
$$\begin{aligned} -2b + 3c &= 1 \\ 3a + 6b - 3c &= -2 \\ 6a + 6b + 3c &= 5 \end{aligned}$$

7. Solve each of the systems in Exercise 6 by Gaussian elimination.

8. Solve each of the following systems by Gauss–Jordan elimination.

(a)
$$\begin{aligned} 2x_1 - 3x_2 &= -2 \\ 2x_1 + x_2 &= 1 \\ 3x_1 + 2x_2 &= 1 \end{aligned}$$

(b)
$$\begin{aligned} 3x_1 + 2x_2 - x_3 &= -15 \\ 5x_1 + 3x_2 + 2x_3 &= 0 \\ 3x_1 + x_2 + 3x_3 &= 11 \\ -6x_1 - 4x_2 + 2x_3 &= 30 \end{aligned}$$

(c)
$$\begin{aligned} 4x_1 - 8x_2 &= 12 \\ 3x_1 - 6x_2 &= 9 \\ -2x_1 + 4x_2 &= -6 \end{aligned}$$

(d)
$$\begin{aligned} 10y - 4z + w &= 1 \\ x + 4y - z + w &= 2 \\ 3x + 2y + z + 2w &= 5 \\ -2x - 8y + 2z - 2w &= -4 \\ x - 6y + 3z &= 1 \end{aligned}$$

9. Solve each of the systems in Exercise 8 by Gaussian elimination.

10. Solve each of the following systems by Gauss–Jordan elimination.

(a)
$$\begin{aligned} 5x_1 - 2x_2 + 6x_3 &= 0 \\ -2x_1 + x_2 + 3x_3 &= 1 \end{aligned}$$

(b)
$$\begin{aligned} x_1 - 2x_2 + x_3 - 4x_4 &= 1 \\ x_1 + 3x_2 + 7x_3 + 2x_4 &= 2 \\ x_1 - 12x_2 - 11x_3 - 16x_4 &= 5 \end{aligned}$$

(c)
$$\begin{aligned} w + 2x - y &= 4 \\ x - y &= 3 \\ w + 3x - 2y &= 7 \\ 2u + 4v + w + 7x &= 7 \end{aligned}$$

11. Solve each of the systems in Exercise 10 by Gaussian elimination.

12. Without using pencil and paper, determine which of the following homogeneous systems have nontrivial solutions.

(a)
$$\begin{aligned} 2x_1 - 3x_2 + 4x_3 - x_4 &= 0 \\ 7x_1 + x_2 - 8x_3 + 9x_4 &= 0 \\ 2x_1 + 8x_2 + x_3 - x_4 &= 0 \end{aligned}$$

(b)
$$\begin{aligned} x_1 + 3x_2 - x_3 &= 0 \\ x_2 - 8x_3 &= 0 \\ 4x_3 &= 0 \end{aligned}$$

(c)
$$\begin{aligned} a_{11}x_1 + a_{12}x_2 + a_{13}x_3 &= 0 \\ a_{21}x_1 + a_{22}x_2 + a_{23}x_3 &= 0 \end{aligned}$$

(d)
$$\begin{aligned} 3x_1 - 2x_2 &= 0 \\ 6x_1 - 4x_2 &= 0 \end{aligned}$$

13. Solve the following homogeneous systems of linear equations by any method.

(a)
$$\begin{aligned} 2x_1 + x_2 + 3x_3 &= 0 \\ x_1 + 2x_2 &= 0 \\ x_2 + x_3 &= 0 \end{aligned}$$

(b)
$$\begin{aligned} 3x_1 + x_2 + x_3 + x_4 &= 0 \\ 5x_1 - x_2 + x_3 - x_4 &= 0 \end{aligned}$$

(c)
$$
\begin{aligned}
2x + 2y + 4z &= 0 \\
w \quad - y - 3z &= 0 \\
2w + 3x + y + z &= 0 \\
-2w + x + 3y - 2z &= 0
\end{aligned}
$$

14. Solve the following homogeneous systems of linear equations by any method.

(a)
$$
\begin{aligned}
2x - y - 3z &= 0 \\
-x + 2y - 3z &= 0 \\
x + y + 4z &= 0
\end{aligned}
$$

(b)
$$
\begin{aligned}
v + 3w - 2x &= 0 \\
2u + v - 4w + 3x &= 0 \\
2u + 3v + 2w - x &= 0 \\
-4u - 3v + 5w - 4x &= 0
\end{aligned}
$$

(c)
$$
\begin{aligned}
x_1 + 3x_2 \quad + x_4 &= 0 \\
x_1 + 4x_2 + 2x_3 \quad &= 0 \\
- 2x_2 - 2x_3 - x_4 &= 0 \\
2x_1 - 4x_2 + x_3 + x_4 &= 0 \\
x_1 - 2x_2 - x_3 + x_4 &= 0
\end{aligned}
$$

15. Solve the following systems by any method.

(a)
$$
\begin{aligned}
2I_1 - I_2 + 3I_3 + 4I_4 &= 9 \\
I_1 \quad - 2I_3 + 7I_4 &= 11 \\
3I_1 - 3I_2 + I_3 + 5I_4 &= 8 \\
2I_1 + I_2 + 4I_3 + 4I_4 &= 10
\end{aligned}
$$

(b)
$$
\begin{aligned}
Z_3 + Z_4 + Z_5 &= 0 \\
-Z_1 - Z_2 + 2Z_3 - 3Z_4 + Z_5 &= 0 \\
Z_1 + Z_2 - 2Z_3 \quad - Z_5 &= 0 \\
2Z_1 + 2Z_2 - Z_3 \quad + Z_5 &= 0
\end{aligned}
$$

16. Solve the following systems, where a, b, and c are constants.

(a)
$$
\begin{aligned}
2x + y &= a \\
3x + 6y &= b
\end{aligned}
$$

(b)
$$
\begin{aligned}
x_1 + x_2 + x_3 &= a \\
2x_1 \quad + 2x_3 &= b \\
3x_2 + 3x_3 &= c
\end{aligned}
$$

17. For which values of a will the following system have no solutions? Exactly one solution? Infinitely many solutions?
$$
\begin{aligned}
x + 2y - \quad 3z &= 4 \\
3x - y + \quad 5z &= 2 \\
4x + y + (a^2 - 14)z &= a + 2
\end{aligned}
$$

18. Reduce
$$
\begin{bmatrix}
2 & 1 & 3 \\
0 & -2 & -29 \\
3 & 4 & 5
\end{bmatrix}
$$
to reduced row-echelon form.

19. Find two different row-echelon forms of
$$
\begin{bmatrix}
1 & 3 \\
2 & 7
\end{bmatrix}
$$

20. Solve the following system of nonlinear equations for the unknown angles α, β, and γ, where $0 \le \alpha \le 2\pi, 0 \le \beta \le 2\pi$, and $0 \le \gamma < \pi$.
$$
\begin{aligned}
2 \sin \alpha - \cos \beta + 3 \tan \gamma &= 3 \\
4 \sin \alpha + 2 \cos \beta - 2 \tan \gamma &= 2 \\
6 \sin \alpha - 3 \cos \beta + \tan \gamma &= 9
\end{aligned}
$$

21. Show that the following nonlinear system has 18 solutions if $0 \le \alpha \le 2\pi, 0 \le \beta \le 2\pi$, and $0 \le \gamma < 2\pi$.
$$
\begin{aligned}
\sin \alpha + 2 \cos \beta + 3 \tan \gamma &= 0 \\
2 \sin \alpha + 5 \cos \beta + 3 \tan \gamma &= 0 \\
- \sin \alpha - 5 \cos \beta + 5 \tan \gamma &= 0
\end{aligned}
$$

22. For which value(s) of λ does the system of equations
$$
\begin{aligned}
(\lambda - 3)x + y &= 0 \\
x + (\lambda - 3)y &= 0
\end{aligned}
$$
have nontrivial solutions?

23. Solve the system

$$2x_1 - x_2 \qquad\quad = \lambda x_1$$
$$2x_1 - x_2 + x_3 = \lambda x_2$$
$$-2x_1 + 2x_2 + x_3 = \lambda x_3$$

for x_1, x_2, and x_3 in the two cases $\lambda = 1$, $\lambda = 2$.

24. Solve the following system for x, y, and z.

$$\frac{1}{x} + \frac{2}{y} - \frac{4}{z} = 1$$
$$\frac{2}{x} + \frac{3}{y} + \frac{8}{z} = 0$$
$$-\frac{1}{x} + \frac{9}{y} + \frac{10}{z} = 5$$

25. Find the coefficients a, b, c, and d so that the curve shown in the accompanying figure is the graph of the equation $y = ax^3 + bx^2 + cx + d$.

26. Find coefficients a, b, c, and d so that the curve shown in the accompanying figure is given by the equation $ax^2 + ay^2 + bx + cy + d = 0$.

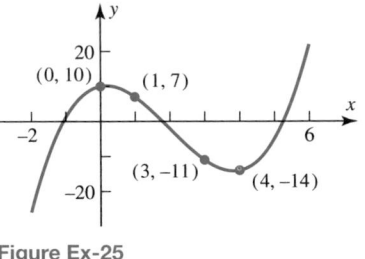

Figure Ex-25

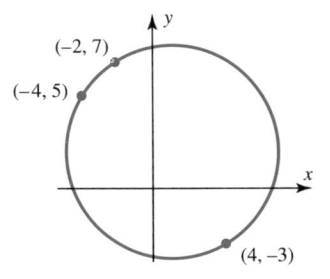

Figure Ex-26

27. (a) Show that if $ad - bc \neq 0$, then the reduced row-echelon form of

$$\begin{bmatrix} a & b \\ c & d \end{bmatrix} \quad \text{is} \quad \begin{bmatrix} 1 & 0 \\ 0 & 1 \end{bmatrix}$$

(b) Use part (a) to show that the system

$$ax + by = k$$
$$cx + dy = l$$

has exactly one solution when $ad - bc \neq 0$.

28. Find an inconsistent linear system that has more unknowns than equations.

Discussion & Discovery

29. Indicate all possible reduced row-echelon forms of

(a) $\begin{bmatrix} a & b & c \\ d & e & f \\ g & h & i \end{bmatrix}$ (b) $\begin{bmatrix} a & b & c & d \\ e & f & g & h \\ i & j & k & l \\ m & n & p & q \end{bmatrix}$

30. Consider the system of equations

$$ax + by = 0$$
$$cx + dy = 0$$
$$ex + fy = 0$$

Discuss the relative positions of the lines $ax + by = 0$, $cx + dy = 0$, and $ex + fy = 0$ when (a) the system has only the trivial solution, and (b) the system has nontrivial solutions.

31. Indicate whether the statement is always true or sometimes false. Justify your answer by giving a logical argument or a counterexample.

    (a) If a matrix is reduced to reduced row-echelon form by two different sequences of elementary row operations, the resulting matrices will be different.

    (b) If a matrix is reduced to row-echelon form by two different sequences of elementary row operations, the resulting matrices might be different.

    (c) If the reduced row-echelon form of the augmented matrix for a linear system has a row of zeros, then the system must have infinitely many solutions.

    (d) If three lines in the xy-plane are sides of a triangle, then the system of equations formed from their equations has three solutions, one corresponding to each vertex.

32. Indicate whether the statement is always true or sometimes false. Justify your answer by giving a logical argument or a counterexample.

    (a) A linear system of three equations in five unknowns must be consistent.

    (b) A linear system of five equations in three unknowns cannot be consistent.

    (c) If a linear system of n equations in n unknowns has n leading 1's in the reduced row-echelon form of its augmented matrix, then the system has exactly one solution.

    (d) If a linear system of n equations in n unknowns has two equations that are multiples of one another, then the system is inconsistent.

1.3
MATRICES AND MATRIX OPERATIONS

Rectangular arrays of real numbers arise in many contexts other than as augmented matrices for systems of linear equations. In this section we begin our study of matrix theory by giving some of the fundamental definitions of the subject. We shall see how matrices can be combined through the arithmetic operations of addition, subtraction, and multiplication.

Matrix Notation and Terminology

In Section 1.2 we used rectangular arrays of numbers, called *augmented matrices*, to abbreviate systems of linear equations. However, rectangular arrays of numbers occur in other contexts as well. For example, the following rectangular array with three rows and seven columns might describe the number of hours that a student spent studying three subjects during a certain week:

	Mon.	Tues.	Wed.	Thurs.	Fri.	Sat.	Sun.
Math	2	3	2	4	1	4	2
History	0	3	1	4	3	2	2
Language	4	1	3	1	0	0	2

If we suppress the headings, then we are left with the following rectangular array of numbers with three rows and seven columns, called a "matrix":

$$\begin{bmatrix} 2 & 3 & 2 & 4 & 1 & 4 & 2 \\ 0 & 3 & 1 & 4 & 3 & 2 & 2 \\ 4 & 1 & 3 & 1 & 0 & 0 & 2 \end{bmatrix}$$

More generally, we make the following definition.

DEFINITION

A *matrix* is a rectangular array of numbers. The numbers in the array are called the *entries* in the matrix.

EXAMPLE 1 Examples of Matrices

Some examples of matrices are

$$\begin{bmatrix} 1 & 2 \\ 3 & 0 \\ -1 & 4 \end{bmatrix}, \qquad [2 \quad 1 \quad 0 \quad -3], \qquad \begin{bmatrix} e & \pi & -\sqrt{2} \\ 0 & \frac{1}{2} & 1 \\ 0 & 0 & 0 \end{bmatrix}, \qquad \begin{bmatrix} 1 \\ 3 \end{bmatrix}, \qquad [4] \quad \blacklozenge$$

The *size* of a matrix is described in terms of the number of rows (horizontal lines) and columns (vertical lines) it contains. For example, the first matrix in Example 1 has three rows and two columns, so its size is 3 by 2 (written 3×2). In a size description, the first number always denotes the number of rows, and the second denotes the number of columns. The remaining matrices in Example 1 have sizes 1×4, 3×3, 2×1, and 1×1, respectively. A matrix with only one column is called a *column matrix* (or a *column vector*), and a matrix with only one row is called a *row matrix* (or a *row vector*). Thus, in Example 1 the 2×1 matrix is a column matrix, the 1×4 matrix is a row matrix, and the 1×1 matrix is both a row matrix and a column matrix. (The term *vector* has another meaning that we will discuss in subsequent chapters.)

REMARK It is common practice to omit the brackets from a 1×1 matrix. Thus we might write 4 rather than [4]. Although this makes it impossible to tell whether 4 denotes the number "four" or the 1×1 matrix whose entry is "four," this rarely causes problems, since it is usually possible to tell which is meant from the context in which the symbol appears.

We shall use capital letters to denote matrices and lowercase letters to denote numerical quantities; thus we might write

$$A = \begin{bmatrix} 2 & 1 & 7 \\ 3 & 4 & 2 \end{bmatrix} \quad \text{or} \quad C = \begin{bmatrix} a & b & c \\ d & e & f \end{bmatrix}$$

When discussing matrices, it is common to refer to numerical quantities as *scalars*. Unless stated otherwise, *scalars will be real numbers*; complex scalars will be considered in Chapter 10.

The entry that occurs in row i and column j of a matrix A will be denoted by a_{ij}. Thus a general 3×4 matrix might be written as

$$A = \begin{bmatrix} a_{11} & a_{12} & a_{13} & a_{14} \\ a_{21} & a_{22} & a_{23} & a_{24} \\ a_{31} & a_{32} & a_{33} & a_{34} \end{bmatrix}$$

and a general $m \times n$ matrix as

$$A = \begin{bmatrix} a_{11} & a_{12} & \cdots & a_{1n} \\ a_{21} & a_{22} & \cdots & a_{2n} \\ \vdots & \vdots & & \vdots \\ a_{m1} & a_{m2} & \cdots & a_{mn} \end{bmatrix} \tag{1}$$

When compactness of notation is desired, the preceding matrix can be written as

$$[a_{ij}]_{m \times n} \quad \text{or} \quad [a_{ij}]$$

the first notation being used when it is important in the discussion to know the size, and the second being used when the size need not be emphasized. Usually, we shall match the letter denoting a matrix with the letter denoting its entries; thus, for a matrix B we would generally use b_{ij} for the entry in row i and column j, and for a matrix C we would use the notation c_{ij}.

The entry in row i and column j of a matrix A is also commonly denoted by the symbol $(A)_{ij}$. Thus, for matrix (1) above, we have

$$(A)_{ij} = a_{ij}$$

and for the matrix

$$A = \begin{bmatrix} 2 & -3 \\ 7 & 0 \end{bmatrix}$$

we have $(A)_{11} = 2$, $(A)_{12} = -3$, $(A)_{21} = 7$, and $(A)_{22} = 0$.

Row and column matrices are of special importance, and it is common practice to denote them by boldface lowercase letters rather than capital letters. For such matrices, double subscripting of the entries is unnecessary. Thus a general $1 \times n$ row matrix $\mathbf{a}$ and a general $m \times 1$ column matrix $\mathbf{b}$ would be written as

$$\mathbf{a} = [a_1 \quad a_2 \quad \cdots \quad a_n] \quad \text{and} \quad \mathbf{b} = \begin{bmatrix} b_1 \\ b_2 \\ \vdots \\ b_m \end{bmatrix}$$

A matrix A with n rows and n columns is called a **square matrix of order n**, and the shaded entries $a_{11}, a_{22}, \ldots, a_{nn}$ in (2) are said to be on the **main diagonal** of A.

$$\begin{bmatrix} a_{11} & a_{12} & \cdots & a_{1n} \\ a_{21} & a_{22} & \cdots & a_{2n} \\ \vdots & \vdots & & \vdots \\ a_{n1} & a_{n2} & \cdots & a_{nn} \end{bmatrix} \qquad (2)$$

Operations on Matrices

So far, we have used matrices to abbreviate the work in solving systems of linear equations. For other applications, however, it is desirable to develop an "arithmetic of matrices" in which matrices can be added, subtracted, and multiplied in a useful way. The remainder of this section will be devoted to developing this arithmetic.

DEFINITION

Two matrices are defined to be **equal** if they have the same size and their corresponding entries are equal.

In matrix notation, if $A = [a_{ij}]$ and $B = [b_{ij}]$ have the same size, then $A = B$ if and only if $(A)_{ij} = (B)_{ij}$, or, equivalently, $a_{ij} = b_{ij}$ for all i and j.

EXAMPLE 2 Equality of Matrices

Consider the matrices

$$A = \begin{bmatrix} 2 & 1 \\ 3 & x \end{bmatrix}, \qquad B = \begin{bmatrix} 2 & 1 \\ 3 & 5 \end{bmatrix}, \qquad C = \begin{bmatrix} 2 & 1 & 0 \\ 3 & 4 & 0 \end{bmatrix}$$

If $x = 5$, then $A = B$, but for all other values of x the matrices A and B are not equal, since not all of their corresponding entries are equal. There is no value of x for which $A = C$ since A and C have different sizes. ◆

> **DEFINITION**
>
> If A and B are matrices of the same size, then the *sum* $A + B$ is the matrix obtained by adding the entries of B to the corresponding entries of A, and the *difference* $A - B$ is the matrix obtained by subtracting the entries of B from the corresponding entries of A. Matrices of different sizes cannot be added or subtracted.

In matrix notation, if $A = [a_{ij}]$ and $B = [b_{ij}]$ have the same size, then

$$(A + B)_{ij} = (A)_{ij} + (B)_{ij} = a_{ij} + b_{ij} \quad \text{and} \quad (A - B)_{ij} = (A)_{ij} - (B)_{ij} = a_{ij} - b_{ij}$$

EXAMPLE 3 Addition and Subtraction

Consider the matrices

$$A = \begin{bmatrix} 2 & 1 & 0 & 3 \\ -1 & 0 & 2 & 4 \\ 4 & -2 & 7 & 0 \end{bmatrix}, \qquad B = \begin{bmatrix} -4 & 3 & 5 & 1 \\ 2 & 2 & 0 & -1 \\ 3 & 2 & -4 & 5 \end{bmatrix}, \qquad C = \begin{bmatrix} 1 & 1 \\ 2 & 2 \end{bmatrix}$$

Then

$$A + B = \begin{bmatrix} -2 & 4 & 5 & 4 \\ 1 & 2 & 2 & 3 \\ 7 & 0 & 3 & 5 \end{bmatrix} \quad \text{and} \quad A - B = \begin{bmatrix} 6 & -2 & -5 & 2 \\ -3 & -2 & 2 & 5 \\ 1 & -4 & 11 & -5 \end{bmatrix}$$

The expressions $A + C$, $B + C$, $A - C$, and $B - C$ are undefined. ◆

> **DEFINITION**
>
> If A is any matrix and c is any scalar, then the *product* cA is the matrix obtained by multiplying each entry of the matrix A by c. The matrix cA is said to be a *scalar multiple* of A.

In matrix notation, if $A = [a_{ij}]$, then

$$(cA)_{ij} = c(A)_{ij} = ca_{ij}$$

EXAMPLE 4 Scalar Multiples

For the matrices

$$A = \begin{bmatrix} 2 & 3 & 4 \\ 1 & 3 & 1 \end{bmatrix}, \qquad B = \begin{bmatrix} 0 & 2 & 7 \\ -1 & 3 & -5 \end{bmatrix}, \qquad C = \begin{bmatrix} 9 & -6 & 3 \\ 3 & 0 & 12 \end{bmatrix}$$

we have

$$2A = \begin{bmatrix} 4 & 6 & 8 \\ 2 & 6 & 2 \end{bmatrix}, \qquad (-1)B = \begin{bmatrix} 0 & -2 & -7 \\ 1 & -3 & 5 \end{bmatrix}, \qquad \tfrac{1}{3}C = \begin{bmatrix} 3 & -2 & 1 \\ 1 & 0 & 4 \end{bmatrix}$$

It is common practice to denote $(-1)B$ by $-B$. ◆

   If $A_1, A_2, \ldots, A_n$ are matrices of the same size and $c_1, c_2, \ldots, c_n$ are scalars, then an expression of the form

$$c_1 A_1 + c_2 A_2 + \cdots + c_n A_n$$

is called a ***linear combination*** of $A_1, A_2, \ldots, A_n$ with ***coefficients*** $c_1, c_2, \ldots, c_n$. For example, if A, B, and C are the matrices in Example 4, then

$$2A - B + \tfrac{1}{3}C = 2A + (-1)B + \tfrac{1}{3}C$$
$$= \begin{bmatrix} 4 & 6 & 8 \\ 2 & 6 & 2 \end{bmatrix} + \begin{bmatrix} 0 & -2 & -7 \\ 1 & -3 & 5 \end{bmatrix} + \begin{bmatrix} 3 & -2 & 1 \\ 1 & 0 & 4 \end{bmatrix} = \begin{bmatrix} 7 & 2 & 2 \\ 4 & 3 & 11 \end{bmatrix}$$

is the linear combination of A, B, and C with scalar coefficients 2, -1, and $\tfrac{1}{3}$.

   Thus far we have defined multiplication of a matrix by a scalar but not the multiplication of two matrices. Since matrices are added by adding corresponding entries and subtracted by subtracting corresponding entries, it would seem natural to define multiplication of matrices by multiplying corresponding entries. However, it turns out that such a definition would not be very useful for most problems. Experience has led mathematicians to the following more useful definition of matrix multiplication.

DEFINITION

If A is an $m \times r$ matrix and B is an $r \times n$ matrix, then the ***product*** AB is the $m \times n$ matrix whose entries are determined as follows. To find the entry in row i and column j of AB, single out row i from the matrix A and column j from the matrix B. Multiply the corresponding entries from the row and column together, and then add up the resulting products.

EXAMPLE 5 Multiplying Matrices

Consider the matrices

$$A = \begin{bmatrix} 1 & 2 & 4 \\ 2 & 6 & 0 \end{bmatrix}, \qquad B = \begin{bmatrix} 4 & 1 & 4 & 3 \\ 0 & -1 & 3 & 1 \\ 2 & 7 & 5 & 2 \end{bmatrix}$$

   Since A is a 2×3 matrix and B is a 3×4 matrix, the product AB is a 2×4 matrix. To determine, for example, the entry in row 2 and column 3 of AB, we single out row 2 from A and column 3 from B. Then, as illustrated below, we multiply corresponding entries together and add up these products.

$$\begin{bmatrix} 1 & 2 & 4 \\ 2 & 6 & 0 \end{bmatrix} \begin{bmatrix} 4 & 1 & 4 & 3 \\ 0 & -1 & 3 & 1 \\ 2 & 7 & 5 & 2 \end{bmatrix} = \begin{bmatrix} \square & \square & \square & \square \\ \square & \square & 26 & \square \end{bmatrix}$$

$$(2 \cdot 4) + (6 \cdot 3) + (0 \cdot 5) = 26$$

The entry in row 1 and column 4 of AB is computed as follows:

$$\begin{bmatrix} 1 & 2 & 4 \\ 2 & 6 & 0 \end{bmatrix} \begin{bmatrix} 4 & 1 & 4 & 3 \\ 0 & -1 & 3 & 1 \\ 2 & 7 & 5 & 2 \end{bmatrix} = \begin{bmatrix} \square & \square & \square & 13 \\ \square & \square & \square & \square \end{bmatrix}$$

$$(1 \cdot 3) + (2 \cdot 1) + (4 \cdot 2) = 13$$

The computations for the remaining entries are

$$(1 \cdot 4) + (2 \cdot 0) + (4 \cdot 2) = 12$$
$$(1 \cdot 1) - (2 \cdot 1) + (4 \cdot 7) = 27$$
$$(1 \cdot 4) + (2 \cdot 3) + (4 \cdot 5) = 30$$
$$(2 \cdot 4) + (6 \cdot 0) + (0 \cdot 2) = 8 \qquad AB = \begin{bmatrix} 12 & 27 & 30 & 13 \\ 8 & -4 & 26 & 12 \end{bmatrix} \blacklozenge$$
$$(2 \cdot 1) - (6 \cdot 1) + (0 \cdot 7) = -4$$
$$(2 \cdot 3) + (6 \cdot 1) + (0 \cdot 2) = 12$$

The definition of matrix multiplication requires that the number of columns of the first factor A be the same as the number of rows of the second factor B in order to form the product AB. If this condition is not satisfied, the product is undefined. A convenient way to determine whether a product of two matrices is defined is to write down the size of the first factor and, to the right of it, write down the size of the second factor. If, as in (3), the inside numbers are the same, then the product is defined. The outside numbers then give the size of the product.

$$\begin{array}{ccccccc} A & & & B & & AB \\ m \times r & & r \times n & = & m \times n \end{array}$$

Inside

Outside

(3)

EXAMPLE 6 Determining Whether a Product Is Defined

Suppose that A, B, and C are matrices with the following sizes:

$$\begin{array}{ccc} A & B & C \\ 3 \times 4 & 4 \times 7 & 7 \times 3 \end{array}$$

Then by (3), AB is defined and is a 3×7 matrix; BC is defined and is a 4×3 matrix; and CA is defined and is a 7×4 matrix. The products AC, CB, and BA are all undefined. $\blacklozenge$

In general, if $A = [a_{ij}]$ is an $m \times r$ matrix and $B = [b_{ij}]$ is an $r \times n$ matrix, then, as illustrated by the shading in (4),

$$AB = \begin{bmatrix} a_{11} & a_{12} & \cdots & a_{1r} \\ a_{21} & a_{22} & \cdots & a_{2r} \\ \vdots & \vdots & & \vdots \\ a_{i1} & a_{i2} & \cdots & a_{ir} \\ \vdots & \vdots & & \vdots \\ a_{m1} & a_{m2} & \cdots & a_{mr} \end{bmatrix} \begin{bmatrix} b_{11} & b_{12} & \cdots & b_{1j} & \cdots & b_{1n} \\ b_{21} & b_{22} & \cdots & b_{2j} & \cdots & b_{2n} \\ \vdots & \vdots & & \vdots & & \vdots \\ b_{r1} & b_{r2} & \cdots & b_{rj} & \cdots & b_{rn} \end{bmatrix} \tag{4}$$

the entry $(AB)_{ij}$ in row i and column j of AB is given by

$$(AB)_{ij} = a_{i1}b_{1j} + a_{i2}b_{2j} + a_{i3}b_{3j} + \cdots + a_{ir}b_{rj} \tag{5}$$

Partitioned Matrices

A matrix can be subdivided or *partitioned* into smaller matrices by inserting horizontal and vertical rules between selected rows and columns. For example, the following are three possible partitions of a general 3×4 matrix A—the first is a partition of A into four *submatrices* A_{11}, A_{12}, A_{21}, and A_{22}; the second is a partition of A into its row matrices $\mathbf{r}_1$, $\mathbf{r}_2$, and $\mathbf{r}_3$; and the third is a partition of A into its column matrices $\mathbf{c}_1$, $\mathbf{c}_2$, $\mathbf{c}_3$, and $\mathbf{c}_4$:

$$A = \left[\begin{array}{ccc|c} a_{11} & a_{12} & a_{13} & a_{14} \\ a_{21} & a_{22} & a_{23} & a_{24} \\ \hline a_{31} & a_{32} & a_{33} & a_{34} \end{array} \right] = \begin{bmatrix} A_{11} & A_{12} \\ A_{21} & A_{22} \end{bmatrix}$$

$$A = \left[\begin{array}{cccc} a_{11} & a_{12} & a_{13} & a_{14} \\ \hline a_{21} & a_{22} & a_{23} & a_{24} \\ \hline a_{31} & a_{32} & a_{33} & a_{34} \end{array} \right] = \begin{bmatrix} \mathbf{r}_1 \\ \mathbf{r}_2 \\ \mathbf{r}_3 \end{bmatrix}$$

$$A = \left[\begin{array}{c|c|c|c} a_{11} & a_{12} & a_{13} & a_{14} \\ a_{21} & a_{22} & a_{23} & a_{24} \\ a_{31} & a_{32} & a_{33} & a_{34} \end{array} \right] = \begin{bmatrix} \mathbf{c}_1 & \mathbf{c}_2 & \mathbf{c}_3 & \mathbf{c}_4 \end{bmatrix}$$

Matrix Multiplication by Columns and by Rows

Sometimes it may be desirable to find a particular row or column of a matrix product AB without computing the entire product. The following results, whose proofs are left as exercises, are useful for that purpose:

$$j\text{th column matrix of } AB = A[\,j\text{th column matrix of } B] \tag{6}$$

$$i\text{th row matrix of } AB = [i\text{th row matrix of } A]B \tag{7}$$

EXAMPLE 7   Example 5 Revisited

If A and B are the matrices in Example 5, then from (6) the second column matrix of AB can be obtained by the computation

$$\begin{bmatrix} 1 & 2 & 4 \\ 2 & 6 & 0 \end{bmatrix} \begin{bmatrix} 1 \\ -1 \\ 7 \end{bmatrix} = \begin{bmatrix} 27 \\ -4 \end{bmatrix}$$

↑   ↑

Second column   Second column
of B of AB

and from (7) the first row matrix of AB can be obtained by the computation

$$\rightarrow \begin{bmatrix} 1 & 2 & 4 \end{bmatrix} \begin{bmatrix} 4 & 1 & 4 & 3 \\ 0 & -1 & 3 & 1 \\ 2 & 7 & 5 & 2 \end{bmatrix} = \begin{bmatrix} 12 & 27 & 30 & 13 \end{bmatrix} \leftarrow \blacklozenge$$

First row of A First row of AB

If $\mathbf{a}_1, \mathbf{a}_2, \ldots, \mathbf{a}_m$ denote the row matrices of A and $\mathbf{b}_1, \mathbf{b}_2, \ldots, \mathbf{b}_n$ denote the column matrices of B, then it follows from Formulas (6) and (7) that

$$AB = A[\mathbf{b}_1 \quad \mathbf{b}_2 \quad \cdots \quad \mathbf{b}_n] = [A\mathbf{b}_1 \quad A\mathbf{b}_2 \quad \cdots \quad A\mathbf{b}_n] \tag{8}$$

(*AB* computed column by column)

$$AB = \begin{bmatrix} \mathbf{a}_1 \\ \mathbf{a}_2 \\ \vdots \\ \mathbf{a}_m \end{bmatrix} B = \begin{bmatrix} \mathbf{a}_1 B \\ \mathbf{a}_2 B \\ \vdots \\ \mathbf{a}_m B \end{bmatrix} \tag{9}$$

(*AB* computed row by row)

REMARK Formulas (8) and (9) are special cases of a more general procedure for multiplying partitioned matrices (see Exercises 15–17).

Matrix Products as Linear Combinations

Row and column matrices provide an alternative way of thinking about matrix multiplication. For example, suppose that

$$A = \begin{bmatrix} a_{11} & a_{12} & \cdots & a_{1n} \\ a_{21} & a_{22} & \cdots & a_{2n} \\ \vdots & \vdots & & \vdots \\ a_{m1} & a_{m2} & \cdots & a_{mn} \end{bmatrix} \quad \text{and} \quad \mathbf{x} = \begin{bmatrix} x_1 \\ x_2 \\ \vdots \\ x_n \end{bmatrix}$$

Then

$$A\mathbf{x} = \begin{bmatrix} a_{11}x_1 + a_{12}x_2 + \cdots + a_{1n}x_n \\ a_{21}x_1 + a_{22}x_2 + \cdots + a_{2n}x_n \\ \vdots & \vdots & & \vdots \\ a_{m1}x_1 + a_{m2}x_2 + \cdots + a_{mn}x_n \end{bmatrix} = x_1 \begin{bmatrix} a_{11} \\ a_{21} \\ \vdots \\ a_{m1} \end{bmatrix} + x_2 \begin{bmatrix} a_{12} \\ a_{22} \\ \vdots \\ a_{m2} \end{bmatrix} + \cdots + x_n \begin{bmatrix} a_{1n} \\ a_{2n} \\ \vdots \\ a_{mn} \end{bmatrix} \tag{10}$$

In words, (10) tells us that *the product $A\mathbf{x}$ of a matrix A with a column matrix $\mathbf{x}$ is a linear combination of the column matrices of A with the coefficients coming from the matrix $\mathbf{x}$.* In the exercises we ask the reader to show that *the product $\mathbf{y}A$ of a $1 \times m$*

matrix **y** *with an m* × *n matrix A is a linear combination of the row matrices of A with scalar coefficients coming from* **y.**

EXAMPLE 8   Linear Combinations

The matrix product

$$\begin{bmatrix} -1 & 3 & 2 \\ 1 & 2 & -3 \\ 2 & 1 & -2 \end{bmatrix} \begin{bmatrix} 2 \\ -1 \\ 3 \end{bmatrix} = \begin{bmatrix} 1 \\ -9 \\ -3 \end{bmatrix}$$

can be written as the linear combination of column matrices

$$2 \begin{bmatrix} -1 \\ 1 \\ 2 \end{bmatrix} - 1 \begin{bmatrix} 3 \\ 2 \\ 1 \end{bmatrix} + 3 \begin{bmatrix} 2 \\ -3 \\ -2 \end{bmatrix} = \begin{bmatrix} 1 \\ -9 \\ -3 \end{bmatrix}$$

The matrix product

$$\begin{bmatrix} 1 & -9 & -3 \end{bmatrix} \begin{bmatrix} -1 & 3 & 2 \\ 1 & 2 & -3 \\ 2 & 1 & -2 \end{bmatrix} = \begin{bmatrix} -16 & -18 & 35 \end{bmatrix}$$

can be written as the linear combination of row matrices

$$1 \begin{bmatrix} -1 & 3 & 2 \end{bmatrix} - 9 \begin{bmatrix} 1 & 2 & -3 \end{bmatrix} - 3 \begin{bmatrix} 2 & 1 & -2 \end{bmatrix} = \begin{bmatrix} -16 & -18 & 35 \end{bmatrix} \quad \blacklozenge$$

It follows from (8) and (10) that *the jth column matrix of a product AB is a linear combination of the column matrices of A with the coefficients coming from the jth column of B.*

EXAMPLE 9   Columns of a Product *AB* as Linear Combinations

We showed in Example 5 that

$$AB = \begin{bmatrix} 1 & 2 & 4 \\ 2 & 6 & 0 \end{bmatrix} \begin{bmatrix} 4 & 1 & 4 & 3 \\ 0 & -1 & 3 & 1 \\ 2 & 7 & 5 & 2 \end{bmatrix} = \begin{bmatrix} 12 & 27 & 30 & 13 \\ 8 & -4 & 26 & 12 \end{bmatrix}$$

The column matrices of AB can be expressed as linear combinations of the column matrices of *A* as follows:

$$\begin{bmatrix} 12 \\ 8 \end{bmatrix} = 4 \begin{bmatrix} 1 \\ 2 \end{bmatrix} + 0 \begin{bmatrix} 2 \\ 6 \end{bmatrix} + 2 \begin{bmatrix} 4 \\ 0 \end{bmatrix}$$

$$\begin{bmatrix} 27 \\ -4 \end{bmatrix} = \begin{bmatrix} 1 \\ 2 \end{bmatrix} - \begin{bmatrix} 2 \\ 6 \end{bmatrix} + 7 \begin{bmatrix} 4 \\ 0 \end{bmatrix}$$

$$\begin{bmatrix} 30 \\ 26 \end{bmatrix} = 4 \begin{bmatrix} 1 \\ 2 \end{bmatrix} + 3 \begin{bmatrix} 2 \\ 6 \end{bmatrix} + 5 \begin{bmatrix} 4 \\ 0 \end{bmatrix}$$

$$\begin{bmatrix} 13 \\ 12 \end{bmatrix} = 3 \begin{bmatrix} 1 \\ 2 \end{bmatrix} + \begin{bmatrix} 2 \\ 6 \end{bmatrix} + 2 \begin{bmatrix} 4 \\ 0 \end{bmatrix} \quad \blacklozenge$$

Matrix Form of a Linear System

Matrix multiplication has an important application to systems of linear equations. Consider any system of *m* linear equations in *n* unknowns.

$$a_{11}x_1 + a_{12}x_2 + \cdots + a_{1n}x_n = b_1$$
$$a_{21}x_1 + a_{22}x_2 + \cdots + a_{2n}x_n = b_2$$
$$\vdots \qquad \vdots \qquad \qquad \vdots \qquad \vdots$$
$$a_{m1}x_1 + a_{m2}x_2 + \cdots + a_{mn}x_n = b_m$$

Since two matrices are equal if and only if their corresponding entries are equal, we can replace the m equations in this system by the single matrix equation

$$\begin{bmatrix} a_{11}x_1 + a_{12}x_2 + \cdots + a_{1n}x_n \\ a_{21}x_1 + a_{22}x_2 + \cdots + a_{2n}x_n \\ \vdots \qquad \vdots \qquad \qquad \vdots \\ a_{m1}x_1 + a_{m2}x_2 + \cdots + a_{mn}x_n \end{bmatrix} = \begin{bmatrix} b_1 \\ b_2 \\ \vdots \\ b_m \end{bmatrix}$$

The $m \times 1$ matrix on the left side of this equation can be written as a product to give

$$\begin{bmatrix} a_{11} & a_{12} & \cdots & a_{1n} \\ a_{21} & a_{22} & \cdots & a_{2n} \\ \vdots & \vdots & & \vdots \\ a_{m1} & a_{m2} & \cdots & a_{mn} \end{bmatrix} \begin{bmatrix} x_1 \\ x_2 \\ \vdots \\ x_n \end{bmatrix} = \begin{bmatrix} b_1 \\ b_2 \\ \vdots \\ b_m \end{bmatrix}$$

If we designate these matrices by A, $\mathbf{x}$, and $\mathbf{b}$, respectively, then the original system of m equations in n unknowns has been replaced by the single matrix equation

$$\mathbf{Ax = b}$$

The matrix A in this equation is called the ***coefficient matrix*** of the system. The augmented matrix for the system is obtained by adjoining $\mathbf{b}$ to A as the last column; thus the augmented matrix is

$$[A \mid \mathbf{b}] = \begin{bmatrix} a_{11} & a_{12} & \cdots & a_{1n} & b_1 \\ a_{21} & a_{22} & \cdots & a_{2n} & b_2 \\ \vdots & \vdots & & \vdots & \vdots \\ a_{m1} & a_{m2} & \cdots & a_{mn} & b_m \end{bmatrix}$$

Matrices Defining Functions

The equation $Ax = b$ with A and b given defines a linear system to be solved for x. But we could also write this equation as $y = Ax$, where A and x are given. In this case, we want to compute y. If A is $m \times n$, then this is a function that associates with every $n \times 1$ column vector x an $m \times 1$ column vector y, and we may view A as defining a rule that shows how a given x is mapped into a corresponding y. This idea is discussed in more detail starting in Section 4.2.

EXAMPLE 10   A Function Using Matrices

Consider the following matrices.

$$A = \begin{bmatrix} 1 & 0 \\ 0 & -1 \end{bmatrix}, \qquad x = \begin{bmatrix} a \\ b \end{bmatrix}$$

The product $y = Ax$ is

$$y = \begin{bmatrix} 1 & 0 \\ 0 & -1 \end{bmatrix} \begin{bmatrix} a \\ b \end{bmatrix} = \begin{bmatrix} a \\ -b \end{bmatrix}$$

so the effect of multiplying A by a column vector is to change the sign of the second entry of the column vector. For the matrix

$$B = \begin{bmatrix} 0 & 1 \\ -1 & 0 \end{bmatrix}$$

the product $y = Bx$ is

$$y = \begin{bmatrix} 0 & 1 \\ -1 & 0 \end{bmatrix} \begin{bmatrix} a \\ b \end{bmatrix} = \begin{bmatrix} b \\ -a \end{bmatrix}$$

so the effect of multiplying B by a column vector is to interchange the first and second entries of the column vector, also changing the sign of the first entry.

If we view the column vector x as locating a point (a, b) in the plane, then the effect of A is to reflect the point about the x-axis (Figure 1.3.1a) whereas the effect of B is to rotate the line segment from the origin to the point through a right angle (Figure 1.3.1b).

◆

Transpose of a Matrix

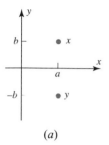

(a)

(b)

Figure 1.3.1

We conclude this section by defining two matrix operations that have no analogs in the real numbers.

> **DEFINITION**
>
> If A is any $m \times n$ matrix, then the **transpose of A**, denoted by A^T, is defined to be the $n \times m$ matrix that results from interchanging the rows and columns of A; that is, the first column of A^T is the first row of A, the second column of A^T is the second row of A, and so forth.

EXAMPLE 11 Some Transposes

The following are some examples of matrices and their transposes.

$$A = \begin{bmatrix} a_{11} & a_{12} & a_{13} & a_{14} \\ a_{21} & a_{22} & a_{23} & a_{24} \\ a_{31} & a_{32} & a_{33} & a_{34} \end{bmatrix}, \qquad B = \begin{bmatrix} 2 & 3 \\ 1 & 4 \\ 5 & 6 \end{bmatrix}, \qquad C = [1 \quad 3 \quad 5], \qquad D = [4]$$

$$A^T = \begin{bmatrix} a_{11} & a_{21} & a_{31} \\ a_{12} & a_{22} & a_{32} \\ a_{13} & a_{23} & a_{33} \\ a_{14} & a_{24} & a_{34} \end{bmatrix}, \qquad B^T = \begin{bmatrix} 2 & 1 & 5 \\ 3 & 4 & 6 \end{bmatrix}, \qquad C^T = \begin{bmatrix} 1 \\ 3 \\ 5 \end{bmatrix}, \qquad D^T = [4] \quad ◆$$

Observe that not only are the columns of A^T the rows of A, but the rows of A^T are the columns of A. Thus the entry in row i and column j of A^T is the entry in row j and column i of A; that is,

$$(A^T)_{ij} = (A)_{ji} \tag{11}$$

Note the reversal of the subscripts.

In the special case where A is a square matrix, the transpose of A can be obtained by interchanging entries that are symmetrically positioned about the main diagonal. In

(12) it is shown that A^T can also be obtained by "reflecting" A about its main diagonal.

$$A = \begin{bmatrix} 1 & -2 & 4 \\ 3 & 7 & 0 \\ -5 & 8 & 6 \end{bmatrix} \rightarrow \begin{bmatrix} 1 & -2 & 4 \\ 3 & 7 & 0 \\ -5 & 8 & 6 \end{bmatrix} \rightarrow A^T = \begin{bmatrix} 1 & 3 & -5 \\ -2 & 7 & 8 \\ 4 & 0 & 6 \end{bmatrix} \quad (12)$$

Interchange entries that are
symmetrically positioned
about the main diagonal.

DEFINITION

If A is a square matrix, then the ***trace of A***, denoted by $\operatorname{tr}(A)$, is defined to be the sum of the entries on the main diagonal of A. The trace of A is undefined if A is not a square matrix.

EXAMPLE 12   Trace of a Matrix

The following are examples of matrices and their traces.

$$A = \begin{bmatrix} a_{11} & a_{12} & a_{13} \\ a_{21} & a_{22} & a_{23} \\ a_{31} & a_{32} & a_{33} \end{bmatrix}, \qquad B = \begin{bmatrix} -1 & 2 & 7 & 0 \\ 3 & 5 & -8 & 4 \\ 1 & 2 & 7 & -3 \\ 4 & -2 & 1 & 0 \end{bmatrix}$$

$$\operatorname{tr}(A) = a_{11} + a_{22} + a_{33} \qquad \operatorname{tr}(B) = -1 + 5 + 7 + 0 = 11 \quad \blacklozenge$$

EXERCISE SET
1.3

1. Suppose that A, B, C, D, and E are matrices with the following sizes:

$$
\begin{array}{ccccc}
A & B & C & D & E \\
(4 \times 5) & (4 \times 5) & (5 \times 2) & (4 \times 2) & (5 \times 4)
\end{array}
$$

Determine which of the following matrix expressions are defined. For those that are defined, give the size of the resulting matrix.

(a) BA (b) $AC + D$ (c) $AE + B$ (d) $AB + B$

(e) $E(A + B)$ (f) $E(AC)$ (g) $E^T A$ (h) $(A^T + E)D$

2. Solve the following matrix equation for a, b, c, and d.

$$\begin{bmatrix} a - b & b + c \\ 3d + c & 2a - 4d \end{bmatrix} = \begin{bmatrix} 8 & 1 \\ 7 & 6 \end{bmatrix}$$

3. Consider the matrices

$$A = \begin{bmatrix} 3 & 0 \\ -1 & 2 \\ 1 & 1 \end{bmatrix}, \qquad B = \begin{bmatrix} 4 & -1 \\ 0 & 2 \end{bmatrix}, \qquad C = \begin{bmatrix} 1 & 4 & 2 \\ 3 & 1 & 5 \end{bmatrix},$$

$$D = \begin{bmatrix} 1 & 5 & 2 \\ -1 & 0 & 1 \\ 3 & 2 & 4 \end{bmatrix}, \qquad E = \begin{bmatrix} 6 & 1 & 3 \\ -1 & 1 & 2 \\ 4 & 1 & 3 \end{bmatrix}$$

Compute the following (where possible).

(a) $D + E$ (b) $D - E$ (c) $5A$ (d) $-7C$

(e) $2B - C$ (f) $4E - 2D$ (g) $-3(D + 2E)$ (h) $A - A$

(i) $\text{tr}(D)$ (j) $\text{tr}(D - 3E)$ (k) $4\,\text{tr}(7B)$ (l) $\text{tr}(A)$

4. Using the matrices in Exercise 3, compute the following (where possible).

(a) $2A^T + C$ (b) $D^T - E^T$ (c) $(D - E)^T$ (d) $B^T + 5C^T$

(e) $\tfrac{1}{2}C^T - \tfrac{1}{4}A$ (f) $B - B^T$ (g) $2E^T - 3D^T$ (h) $(2E^T - 3D^T)^T$

5. Using the matrices in Exercise 3, compute the following (where possible).

(a) AB (b) BA (c) $(3E)D$ (d) $(AB)C$

(e) $A(BC)$ (f) CC^T (g) $(DA)^T$ (h) $(C^TB)A^T$

(i) $\text{tr}(DD^T)$ (j) $\text{tr}(4E^T - D)$ (k) $\text{tr}(C^TA^T + 2E^T)$

6. Using the matrices in Exercise 3, compute the following (where possible).

(a) $(2D^T - E)A$ (b) $(4B)C + 2B$ (c) $(-AC)^T + 5D^T$

(d) $(BA^T - 2C)^T$ (e) $B^T(CC^T - A^TA)$ (f) $D^TE^T - (ED)^T$

7. Let

$$A = \begin{bmatrix} 3 & -2 & 7 \\ 6 & 5 & 4 \\ 0 & 4 & 9 \end{bmatrix} \quad \text{and} \quad B = \begin{bmatrix} 6 & -2 & 4 \\ 0 & 1 & 3 \\ 7 & 7 & 5 \end{bmatrix}$$

Use the method of Example 7 to find

(a) the first row of AB (b) the third row of AB

(c) the second column of AB (d) the first column of BA

(e) the third row of AA (f) the third column of AA

8. Let A and B be the matrices in Exercise 7. Use the method of Example 9 to

(a) express each column matrix of AB as a linear combination of the column matrices of A

(b) express each column matrix of BA as a linear combination of the column matrices of B

9. Let

$$\mathbf{y} = [y_1 \quad y_2 \quad \cdots \quad y_m] \quad \text{and} \quad A = \begin{bmatrix} a_{11} & a_{12} & \cdots & a_{1n} \\ a_{21} & a_{22} & \cdots & a_{2n} \\ \vdots & \vdots & & \vdots \\ a_{m1} & a_{m2} & \cdots & a_{mn} \end{bmatrix}$$

(a) Show that the product $\mathbf{y}A$ can be expressed as a linear combination of the row matrices of A with the scalar coefficients coming from $\mathbf{y}$.

(b) Relate this to the method of Example 8.

Hint Use the transpose operation.

10. Let A and B be the matrices in Exercise 7.

(a) Use the result in Exercise 9 to express each row matrix of AB as a linear combination of the row matrices of B.

(b) Use the result in Exercise 9 to express each row matrix of BA as a linear combination of the row matrices of A.

11. Let C, D, and E be the matrices in Exercise 3. Using as few computations as possible, determine the entry in row 2 and column 3 of $C(DE)$.

12. (a) Show that if AB and BA are both defined, then AB and BA are square matrices.

(b) Show that if A is an $m \times n$ matrix and $A(BA)$ is defined, then B is an $n \times m$ matrix.

13. In each part, find matrices A, $\mathbf{x}$, and $\mathbf{b}$ that express the given system of linear equations as a single matrix equation $A\mathbf{x} = \mathbf{b}$.

(a) $\begin{aligned} 2x_1 - 3x_2 + 5x_3 &= 7 \\ 9x_1 - x_2 + x_3 &= -1 \\ x_1 + 5x_2 + 4x_3 &= 0 \end{aligned}$

(b) $\begin{aligned} 4x_1 \quad\quad - 3x_3 + x_4 &= 1 \\ 5x_1 + x_2 \quad\quad - 8x_4 &= 3 \\ 2x_1 - 5x_2 + 9x_3 - x_4 &= 0 \\ 3x_2 - x_3 + 7x_4 &= 2 \end{aligned}$

14. In each part, express the matrix equation as a system of linear equations.

(a) $\begin{bmatrix} 3 & -1 & 2 \\ 4 & 3 & 7 \\ -2 & 1 & 5 \end{bmatrix} \begin{bmatrix} x_1 \\ x_2 \\ x_3 \end{bmatrix} = \begin{bmatrix} 2 \\ -1 \\ 4 \end{bmatrix}$

(b) $\begin{bmatrix} 3 & -2 & 0 & 1 \\ 5 & 0 & 2 & -2 \\ 3 & 1 & 4 & 7 \\ -2 & 5 & 1 & 6 \end{bmatrix} \begin{bmatrix} w \\ x \\ y \\ z \end{bmatrix} = \begin{bmatrix} 0 \\ 0 \\ 0 \\ 0 \end{bmatrix}$

15. If A and B are partitioned into submatrices, for example,

$$A = \left[\begin{array}{c|c} A_{11} & A_{12} \\ \hline A_{21} & A_{22} \end{array} \right] \quad \text{and} \quad B = \left[\begin{array}{c|c} B_{11} & B_{12} \\ \hline B_{21} & B_{22} \end{array} \right]$$

then AB can be expressed as

$$AB = \left[\begin{array}{c|c} A_{11}B_{11} + A_{12}B_{21} & A_{11}B_{12} + A_{12}B_{22} \\ \hline A_{21}B_{11} + A_{22}B_{21} & A_{21}B_{12} + A_{22}B_{22} \end{array} \right]$$

provided the sizes of the submatrices of A and B are such that the indicated operations can be performed. This method of multiplying partitioned matrices is called **_block multiplication_**. In each part, compute the product by block multiplication. Check your results by multiplying directly.

(a) $A = \left[\begin{array}{cc|cc} -1 & 2 & 1 & 5 \\ 0 & -3 & 4 & 2 \\ \hline 1 & 5 & 6 & 1 \end{array} \right]$, $B = \left[\begin{array}{cc|c} 2 & 1 & 4 \\ -3 & 5 & 2 \\ \hline 7 & -1 & 5 \\ 0 & 3 & -3 \end{array} \right]$

(b) $A = \left[\begin{array}{ccc|c} -1 & 2 & 1 & 5 \\ 0 & -3 & 4 & 2 \\ 1 & 5 & 6 & 1 \end{array} \right]$, $B = \left[\begin{array}{cc|c} 2 & 1 & 4 \\ -3 & 5 & 2 \\ 7 & -1 & 5 \\ \hline 0 & 3 & -3 \end{array} \right]$

16. Adapt the method of Exercise 15 to compute the following products by block multiplication.

(a) $\left[\begin{array}{cc|cc} 3 & -1 & 0 & -3 \\ 2 & 1 & 4 & 5 \end{array} \right] \left[\begin{array}{ccc} 2 & -4 & 1 \\ 3 & 0 & 2 \\ \hline 1 & -3 & 5 \\ 2 & 1 & 4 \end{array} \right]$

(b) $\left[\begin{array}{c|c} 2 & -5 \\ 1 & 3 \\ 0 & 5 \\ \hline 1 & 4 \end{array} \right] \left[\begin{array}{cc|cc} 2 & -1 & 3 & -4 \\ 0 & 1 & 5 & 7 \end{array} \right]$

(c) $\left[\begin{array}{ccc|cc} 1 & 0 & 0 & 0 & 0 \\ 0 & 1 & 0 & 0 & 0 \\ 0 & 0 & 1 & 0 & 0 \\ \hline 0 & 0 & 0 & 2 & 0 \\ 0 & 0 & 0 & -1 & 2 \end{array} \right] \left[\begin{array}{cc} 3 & 3 \\ -1 & 4 \\ 1 & 5 \\ \hline 2 & -2 \\ 1 & 6 \end{array} \right]$

17. In each part, determine whether block multiplication can be used to compute AB from the given partitions. If so, compute the product by block multiplication.

Note See Exercise 15.

(a) $A = \begin{bmatrix} -1 & 2 & 1 & 5 \\ 0 & -3 & 4 & 2 \\ \hline 1 & 5 & 6 & 1 \end{bmatrix}$, $B = \begin{bmatrix} 2 & 1 & 4 \\ -3 & 5 & 2 \\ \hline 7 & -1 & 5 \\ 0 & 3 & -3 \end{bmatrix}$

(b) $A = \begin{bmatrix} -1 & 2 & 1 & 5 \\ 0 & -3 & 4 & 2 \\ \hline 1 & 5 & 6 & 1 \end{bmatrix}$, $B = \begin{bmatrix} 2 & 1 & 4 \\ -3 & 5 & 2 \\ \hline 7 & -1 & 5 \\ 0 & 3 & -3 \end{bmatrix}$

18. (a) Show that if A has a row of zeros and B is any matrix for which AB is defined, then AB also has a row of zeros.

   (b) Find a similar result involving a column of zeros.

19. Let A be any $m \times n$ matrix and let 0 be the $m \times n$ matrix each of whose entries is zero. Show that if $kA = 0$, then $k = 0$ or $A = 0$.

20. Let I be the $n \times n$ matrix whose entry in row i and column j is

$$\begin{cases} 1 & \text{if} \quad i = j \\ 0 & \text{if} \quad i \neq j \end{cases}$$

Show that $AI = IA = A$ for every $n \times n$ matrix A.

21. In each part, find a 6×6 matrix $[a_{ij}]$ that satisfies the stated condition. Make your answers as general as possible by using letters rather than specific numbers for the nonzero entries.

   (a) $a_{ij} = 0$ if $i \neq j$ (b) $a_{ij} = 0$ if $i > j$
   (c) $a_{ij} = 0$ if $i < j$ (d) $a_{ij} = 0$ if $|i - j| > 1$

22. Find the 4×4 matrix $A = [a_{ij}]$ whose entries satisfy the stated condition.

   (a) $a_{ij} = i + j$ (b) $a_{ij} = i^{j-1}$ (c) $a_{ij} = \begin{cases} 1 & \text{if} \quad |i - j| > 1 \\ -1 & \text{if} \quad |i - j| \leq 1 \end{cases}$

23. Consider the function $y = f(x)$ defined for 2×1 matrices x by $y = Ax$, where

$$A = \begin{bmatrix} 1 & 1 \\ 0 & 1 \end{bmatrix}$$

Plot $f(x)$ together with x in each case below. How would you describe the action of f?

   (a) $x = \begin{pmatrix} 1 \\ 1 \end{pmatrix}$ (b) $x = \begin{pmatrix} 2 \\ 0 \end{pmatrix}$ (c) $x = \begin{pmatrix} 4 \\ 3 \end{pmatrix}$ (d) $x = \begin{pmatrix} 2 \\ -2 \end{pmatrix}$

24. Let A be a $n \times m$ matrix. Show that if the function $y = f(x)$ defined for $m \times 1$ matrices x by $y = Ax$ satisfies the linearity property, then $f(\alpha w + \beta z) = \alpha f(w) + \beta f(z)$ for any real numbers α and β and any $m \times 1$ matrices w and z.

25. Prove: If A and B are $n \times n$ matrices, then $\text{tr}(A + B) = \text{tr}(A) + \text{tr}(B)$.

Discussion & Discovery

26. Describe three different methods for computing a matrix product, and illustrate the methods by computing some product AB three different ways.

27. How many 3×3 matrices A can you find such that

$$A \begin{bmatrix} x \\ y \\ z \end{bmatrix} = \begin{bmatrix} x + y \\ x - y \\ 0 \end{bmatrix}$$

for all choices of x, y, and z?

28. How many 3×3 matrices A can you find such that

$$A \begin{bmatrix} x \\ y \\ z \end{bmatrix} = \begin{bmatrix} xy \\ 0 \\ 0 \end{bmatrix}$$

for all choices of x, y, and z?

29. A matrix B is said to be a ***square root*** of a matrix A if $BB = A$.

    (a)   Find two square roots of $A = \begin{bmatrix} 2 & 2 \\ 2 & 2 \end{bmatrix}$.

    (b)   How many different square roots can you find of $A = \begin{bmatrix} 5 & 0 \\ 0 & 9 \end{bmatrix}$?

    (c)   Do you think that every 2×2 matrix has at least one square root? Explain your reasoning.

30. Let 0 denote a 2×2 matrix, each of whose entries is zero.

    (a)   Is there a 2×2 matrix A such that $A \neq 0$ and $AA = 0$? Justify your answer.

    (b)   Is there a 2×2 matrix A such that $A \neq 0$ and $AA = A$? Justify your answer.

31. Indicate whether the statement is always true or sometimes false. Justify your answer with a logical argument or a counterexample.

    (a)   The expressions $\operatorname{tr}(AA^T)$ and $\operatorname{tr}(A^T A)$ are always defined, regardless of the size of A.

    (b) $\operatorname{tr}(AA^T) = \operatorname{tr}(A^T A)$ for every matrix A.

    (c)   If the first column of A has all zeros, then so does the first column of every product AB.

    (d)   If the first row of A has all zeros, then so does the first row of every product AB.

32. Indicate whether the statement is always true or sometimes false. Justify your answer with a logical argument or a counterexample.

    (a)   If A is a square matrix with two identical rows, then AA has two identical rows.

    (b)   If A is a square matrix and AA has a column of zeros, then A must have a column of zeros.

    (c)   If B is an $n \times n$ matrix whose entries are positive even integers, and if A is an $n \times n$ matrix whose entries are positive integers, then the entries of AB and BA are positive even integers.

    (d)   If the matrix sum $AB + BA$ is defined, then A and B must be square.

33. Suppose the array

$$\begin{bmatrix} 4 & 3 & 3 \\ 2 & 1 & 0 \\ 4 & 4 & 2 \end{bmatrix}$$

represents the orders placed by three individuals at a fast-food restaurant. The first person orders 4 burgers, 3 sodas, and 3 fries; the second orders 2 burgers and 1 soda, and the third orders 4 burgers, 4 sodas, and 2 fries. Burgers cost \$2 each, sodas \$1 each, and fries \$1.50 each.

    (a)   Argue that the amounts owed by these persons may be represented as a function $y = f(x)$, where $f(x)$ is equal to the array given above times a certain vector.

    (b)   Compute the amounts owed in this case by performing the appropriate multiplication.

    (c)   Change the matrix for the case in which the second person orders an additional soda and 2 fries, and recompute the costs.

1.4

INVERSES; RULES OF MATRIX ARITHMETIC

In this section we shall discuss some properties of the arithmetic operations on matrices. We shall see that many of the basic rules of arithmetic for real numbers also hold for matrices, but a few do not.

Properties of Matrix Operations

For real numbers a and b, we always have $ab = ba$, which is called the *commutative law for multiplication*. For matrices, however, AB and BA need not be equal. Equality can fail to hold for three reasons: It can happen that the product AB is defined but BA is undefined. For example, this is the case if A is a 2×3 matrix and B is a 3×4 matrix. Also, it can happen that AB and BA are both defined but have different sizes. This is the situation if A is a 2×3 matrix and B is a 3×2 matrix. Finally, as Example 1 shows, it is possible to have $AB \neq BA$ even if both AB and BA are defined and have the same size.

EXAMPLE 1 *AB* and *BA* Need Not Be Equal

Consider the matrices

$$A = \begin{bmatrix} -1 & 0 \\ 2 & 3 \end{bmatrix}, \qquad B = \begin{bmatrix} 1 & 2 \\ 3 & 0 \end{bmatrix}$$

Multiplying gives

$$AB = \begin{bmatrix} -1 & -2 \\ 11 & 4 \end{bmatrix}, \qquad BA = \begin{bmatrix} 3 & 6 \\ -3 & 0 \end{bmatrix}$$

Thus, $AB \neq BA$. ◆

Although the commutative law for multiplication is not valid in matrix arithmetic, many familiar laws of arithmetic are valid for matrices. Some of the most important ones and their names are summarized in the following theorem.

THEOREM 1.4.1

Properties of Matrix Arithmetic

Assuming that the sizes of the matrices are such that the indicated operations can be performed, the following rules of matrix arithmetic are valid.

(a) $A + B = B + A$ **(Commutative law for addition)**

(b) $A + (B + C) = (A + B) + C$ **(Associative law for addition)**

(c) $A(BC) = (AB)C$ **(Associative law for multiplication)**

(d) $A(B + C) = AB + AC$ **(Left distributive law)**

(e) $(B + C)A = BA + CA$ **(Right distributive law)**

(f) $A(B - C) = AB - AC$ (j) $(a + b)C = aC + bC$

(g) $(B - C)A = BA - CA$ (k) $(a - b)C = aC - bC$

(h) $a(B + C) = aB + aC$ (l) $a(bC) = (ab)C$

(i) $a(B - C) = aB - aC$ (m) $a(BC) = (aB)C = B(aC)$

To prove the equalities in this theorem, we must show that the matrix on the left side has the same size as the matrix on the right side and that corresponding entries on the two sides are equal. With the exception of the associative law in part (c), the proofs all follow the same general pattern. We shall prove part (d) as an illustration. The proof of the associative law, which is more complicated, is outlined in the exercises.

Proof (d) We must show that $A(B + C)$ and $AB + AC$ have the same size and that corresponding entries are equal. To form $A(B + C)$, the matrices B and C must have the same size, say $m \times n$, and the matrix A must then have m columns, so its size must be of the form $r \times m$. This makes $A(B + C)$ an $r \times n$ matrix. It follows that $AB + AC$ is also an $r \times n$ matrix and, consequently, $A(B + C)$ and $AB + AC$ have the same size.

Suppose that $A = [a_{ij}]$, $B = [b_{ij}]$, and $C = [c_{ij}]$. We want to show that corresponding entries of $A(B + C)$ and $AB + AC$ are equal; that is,

$$[A(B + C)]_{ij} = [AB + AC]_{ij}$$

for all values of i and j. But from the definitions of matrix addition and matrix multiplication, we have

$$[A(B + C)]_{ij} = a_{i1}(b_{1j} + c_{1j}) + a_{i2}(b_{2j} + c_{2j}) + \cdots + a_{im}(b_{mj} + c_{mj})$$
$$= (a_{i1}b_{1j} + a_{i2}b_{2j} + \cdots + a_{im}b_{mj}) + (a_{i1}c_{1j} + a_{i2}c_{2j} + \cdots + a_{im}c_{mj})$$
$$= [AB]_{ij} + [AC]_{ij} = [AB + AC]_{ij} \qquad \blacksquare$$

REMARK Although the operations of matrix addition and matrix multiplication were defined for pairs of matrices, associative laws (*b*) and (*c*) enable us to denote sums and products of three matrices as $A + B + C$ and ABC without inserting any parentheses. This is justified by the fact that no matter how parentheses are inserted, the associative laws guarantee that the same end result will be obtained. In general, *given any sum or any product of matrices, pairs of parentheses can be inserted or deleted anywhere within the expression without affecting the end result.*

EXAMPLE 2 Associativity of Matrix Multiplication

As an illustration of the associative law for matrix multiplication, consider

$$A = \begin{bmatrix} 1 & 2 \\ 3 & 4 \\ 0 & 1 \end{bmatrix}, \qquad B = \begin{bmatrix} 4 & 3 \\ 2 & 1 \end{bmatrix}, \qquad C = \begin{bmatrix} 1 & 0 \\ 2 & 3 \end{bmatrix}$$

Then

$$AB = \begin{bmatrix} 1 & 2 \\ 3 & 4 \\ 0 & 1 \end{bmatrix}\begin{bmatrix} 4 & 3 \\ 2 & 1 \end{bmatrix} = \begin{bmatrix} 8 & 5 \\ 20 & 13 \\ 2 & 1 \end{bmatrix} \quad \text{and} \quad BC = \begin{bmatrix} 4 & 3 \\ 2 & 1 \end{bmatrix}\begin{bmatrix} 1 & 0 \\ 2 & 3 \end{bmatrix} = \begin{bmatrix} 10 & 9 \\ 4 & 3 \end{bmatrix}$$

Thus

$$(AB)C = \begin{bmatrix} 8 & 5 \\ 20 & 13 \\ 2 & 1 \end{bmatrix}\begin{bmatrix} 1 & 0 \\ 2 & 3 \end{bmatrix} = \begin{bmatrix} 18 & 15 \\ 46 & 39 \\ 4 & 3 \end{bmatrix}$$

and

$$A(BC) = \begin{bmatrix} 1 & 2 \\ 3 & 4 \\ 0 & 1 \end{bmatrix}\begin{bmatrix} 10 & 9 \\ 4 & 3 \end{bmatrix} = \begin{bmatrix} 18 & 15 \\ 46 & 39 \\ 4 & 3 \end{bmatrix}$$

so $(AB)C = A(BC)$, as guaranteed by Theorem 1.4.1c. ◆

Zero Matrices

A matrix, all of whose entries are zero, such as

$$\begin{bmatrix} 0 & 0 \\ 0 & 0 \end{bmatrix}, \qquad \begin{bmatrix} 0 & 0 & 0 \\ 0 & 0 & 0 \\ 0 & 0 & 0 \end{bmatrix}, \qquad \begin{bmatrix} 0 & 0 & 0 & 0 \\ 0 & 0 & 0 & 0 \end{bmatrix}, \qquad \begin{bmatrix} 0 \\ 0 \\ 0 \\ 0 \end{bmatrix}, \qquad [0]$$

is called a **zero matrix**. A zero matrix will be denoted by *0*; if it is important to emphasize the size, we shall write $0_{m \times n}$ for the $m \times n$ zero matrix. Moreover, in keeping with our convention of using boldface symbols for matrices with one column, we will denote a zero matrix with one column by **0**.

If A is any matrix and 0 is the zero matrix with the same size, it is obvious that $A + 0 = 0 + A = A$. The matrix 0 plays much the same role in these matrix equations as the number 0 plays in the numerical equations $a + 0 = 0 + a = a$.

Since we already know that some of the rules of arithmetic for real numbers do not carry over to matrix arithmetic, it would be foolhardy to assume that all the properties of the real number zero carry over to zero matrices. For example, consider the following two standard results in the arithmetic of real numbers.

- If $ab = ac$ and $a \neq 0$, then $b = c$. (This is called the *cancellation law*.)
- If $ad = 0$, then at least one of the factors on the left is 0.

As the next example shows, the corresponding results are not generally true in matrix arithmetic.

EXAMPLE 3 The Cancellation Law Does Not Hold

Consider the matrices

$$A = \begin{bmatrix} 0 & 1 \\ 0 & 2 \end{bmatrix}, \qquad B = \begin{bmatrix} 1 & 1 \\ 3 & 4 \end{bmatrix}, \qquad C = \begin{bmatrix} 2 & 5 \\ 3 & 4 \end{bmatrix}, \qquad D = \begin{bmatrix} 3 & 7 \\ 0 & 0 \end{bmatrix}$$

You should verify that

$$AB = AC = \begin{bmatrix} 3 & 4 \\ 6 & 8 \end{bmatrix} \quad \text{and} \quad AD = \begin{bmatrix} 0 & 0 \\ 0 & 0 \end{bmatrix}$$

Thus, although $A \neq 0$, it is *incorrect* to cancel the A from both sides of the equation $AB = AC$ and write $B = C$. Also, $AD = 0$, yet $A \neq 0$ and $D \neq 0$. Thus, the cancellation law is not valid for matrix multiplication, and it is possible for a product of matrices to be zero without either factor being zero. ◆

In spite of the above example, there are a number of familiar properties of the real number 0 that *do* carry over to zero matrices. Some of the more important ones are summarized in the next theorem. The proofs are left as exercises.

THEOREM 1.4.2

Properties of Zero Matrices

Assuming that the sizes of the matrices are such that the indicated operations can be performed, the following rules of matrix arithmetic are valid.

(a) $A + 0 = 0 + A = A$ (c) $0 - A = -A$

(b) $A - A = 0$ (d) $A0 = 0; \quad 0A = 0$

Identity Matrices

Of special interest are square matrices with 1's on the main diagonal and 0's off the main diagonal, such as

$$\begin{bmatrix} 1 & 0 \\ 0 & 1 \end{bmatrix}, \qquad \begin{bmatrix} 1 & 0 & 0 \\ 0 & 1 & 0 \\ 0 & 0 & 1 \end{bmatrix}, \qquad \begin{bmatrix} 1 & 0 & 0 & 0 \\ 0 & 1 & 0 & 0 \\ 0 & 0 & 1 & 0 \\ 0 & 0 & 0 & 1 \end{bmatrix}, \qquad \text{and so on.}$$

A matrix of this form is called an ***identity matrix*** and is denoted by I. If it is important to emphasize the size, we shall write I_n for the $n \times n$ identity matrix.

If A is an $m \times n$ matrix, then, as illustrated in the next example,

$$AI_n = A \quad \text{and} \quad I_m A = A$$

Thus, an identity matrix plays much the same role in matrix arithmetic that the number 1 plays in the numerical relationships $a \cdot 1 = 1 \cdot a = a$.

EXAMPLE 4 Multiplication by an Identity Matrix

Consider the matrix

$$A = \begin{bmatrix} a_{11} & a_{12} & a_{13} \\ a_{21} & a_{22} & a_{23} \end{bmatrix}$$

Then

$$I_2 A = \begin{bmatrix} 1 & 0 \\ 0 & 1 \end{bmatrix} \begin{bmatrix} a_{11} & a_{12} & a_{13} \\ a_{21} & a_{22} & a_{23} \end{bmatrix} = \begin{bmatrix} a_{11} & a_{12} & a_{13} \\ a_{21} & a_{22} & a_{23} \end{bmatrix} = A$$

and

$$AI_3 = \begin{bmatrix} a_{11} & a_{12} & a_{13} \\ a_{21} & a_{22} & a_{23} \end{bmatrix} \begin{bmatrix} 1 & 0 & 0 \\ 0 & 1 & 0 \\ 0 & 0 & 1 \end{bmatrix} = \begin{bmatrix} a_{11} & a_{12} & a_{13} \\ a_{21} & a_{22} & a_{23} \end{bmatrix} = A \quad \blacklozenge$$

As the next theorem shows, identity matrices arise naturally in studying reduced row-echelon forms of *square* matrices.

THEOREM 1.4.3

If R is the reduced row-echelon form of an $n \times n$ matrix A, then either R has a row of zeros or R is the identity matrix I_n.

Proof Suppose that the reduced row-echelon form of A is

$$R = \begin{bmatrix} r_{11} & r_{12} & \cdots & r_{1n} \\ r_{21} & r_{22} & \cdots & r_{2n} \\ \vdots & \vdots & & \vdots \\ r_{n1} & r_{n2} & \cdots & r_{nn} \end{bmatrix}$$

Either the last row in this matrix consists entirely of zeros or it does not. If not, the matrix contains no zero rows, and consequently each of the n rows has a leading entry of 1. Since these leading 1's occur progressively farther to the right as we move down the matrix, each of these 1's must occur on the main diagonal. Since the other entries in the same column as one of these 1's are zero, R must be I_n. Thus, either R has a row of zeros or $R = I_n$. ∎

> **DEFINITION**
>
> If A is a square matrix, and if a matrix B of the same size can be found such that $AB = BA = I$, then A is said to be **invertible** and B is called an **inverse** of A. If no such matrix B can be found, then A is said to be **singular**.

EXAMPLE 5   Verifying the Inverse Requirements

The matrix

$$B = \begin{bmatrix} 3 & 5 \\ 1 & 2 \end{bmatrix} \quad \text{is an inverse of} \quad A = \begin{bmatrix} 2 & -5 \\ -1 & 3 \end{bmatrix}$$

since

$$AB = \begin{bmatrix} 2 & -5 \\ -1 & 3 \end{bmatrix} \begin{bmatrix} 3 & 5 \\ 1 & 2 \end{bmatrix} = \begin{bmatrix} 1 & 0 \\ 0 & 1 \end{bmatrix} = I$$

and

$$BA = \begin{bmatrix} 3 & 5 \\ 1 & 2 \end{bmatrix} \begin{bmatrix} 2 & -5 \\ -1 & 3 \end{bmatrix} = \begin{bmatrix} 1 & 0 \\ 0 & 1 \end{bmatrix} = I \quad \blacklozenge$$

EXAMPLE 6   A Matrix with No Inverse

The matrix

$$A = \begin{bmatrix} 1 & 4 & 0 \\ 2 & 5 & 0 \\ 3 & 6 & 0 \end{bmatrix}$$

is singular. To see why, let

$$B = \begin{bmatrix} b_{11} & b_{12} & b_{13} \\ b_{21} & b_{22} & b_{23} \\ b_{31} & b_{32} & b_{33} \end{bmatrix}$$

be any 3×3 matrix. The third column of BA is

$$\begin{bmatrix} b_{11} & b_{12} & b_{13} \\ b_{21} & b_{22} & b_{23} \\ b_{31} & b_{32} & b_{33} \end{bmatrix} \begin{bmatrix} 0 \\ 0 \\ 0 \end{bmatrix} = \begin{bmatrix} 0 \\ 0 \\ 0 \end{bmatrix}$$

Thus

$$BA \neq I = \begin{bmatrix} 1 & 0 & 0 \\ 0 & 1 & 0 \\ 0 & 0 & 1 \end{bmatrix} \quad \blacklozenge$$

Properties of Inverses

It is reasonable to ask whether an invertible matrix can have more than one inverse. The next theorem shows that the answer is no—*an invertible matrix has exactly one inverse.*

THEOREM 1.4.4

> If B and C are both inverses of the matrix A, then $B = C$.

Proof   Since B is an inverse of A, we have $BA = I$. Multiplying both sides on the right by C gives $(BA)C = IC = C$. But $(BA)C = B(AC) = BI = B$, so $C = B$. ■

As a consequence of this important result, we can now speak of "the" inverse of an invertible matrix. If A is invertible, then its inverse will be denoted by the symbol A^{-1}. Thus,

$$AA^{-1} = I \quad \text{and} \quad A^{-1}A = I$$

The inverse of A plays much the same role in matrix arithmetic that the reciprocal a^{-1} plays in the numerical relationships $aa^{-1} = 1$ and $a^{-1}a = 1$.

In the next section we shall develop a method for finding inverses of invertible matrices of any size; however, the following theorem gives conditions under which a 2×2 matrix is invertible and provides a simple formula for the inverse.

THEOREM 1.4.5

The matrix
$$A = \begin{bmatrix} a & b \\ c & d \end{bmatrix}$$

is invertible if $ad - bc \neq 0$, in which case the inverse is given by the formula

$$A^{-1} = \frac{1}{ad - bc} \begin{bmatrix} d & -b \\ -c & a \end{bmatrix} = \begin{bmatrix} \dfrac{d}{ad - bc} & -\dfrac{b}{ad - bc} \\ -\dfrac{c}{ad - bc} & \dfrac{a}{ad - bc} \end{bmatrix}$$

Proof   We leave it for the reader to verify that $AA^{-1} = I_2$ and $A^{-1}A = I_2$. ■

THEOREM 1.4.6

If A and B are invertible matrices of the same size, then AB is invertible and
$$(AB)^{-1} = B^{-1}A^{-1}$$

Proof   If we can show that $(AB)(B^{-1}A^{-1}) = (B^{-1}A^{-1})(AB) = I$, then we will have simultaneously shown that the matrix AB is invertible and that $(AB)^{-1} = B^{-1}A^{-1}$. But $(AB)(B^{-1}A^{-1}) = A(BB^{-1})A^{-1} = AIA^{-1} = AA^{-1} = I$. A similar argument shows that $(B^{-1}A^{-1})(AB) = I$. ■

Although we will not prove it, this result can be extended to include three or more factors; that is,

A product of any number of invertible matrices is invertible, and the inverse of the product is the product of the inverses in the reverse order.

EXAMPLE 7   Inverse of a Product

Consider the matrices

$$A = \begin{bmatrix} 1 & 2 \\ 1 & 3 \end{bmatrix}, \qquad B = \begin{bmatrix} 3 & 2 \\ 2 & 2 \end{bmatrix}, \qquad AB = \begin{bmatrix} 7 & 6 \\ 9 & 8 \end{bmatrix}$$

Applying the formula in Theorem 1.4.5, we obtain

$$A^{-1} = \begin{bmatrix} 3 & -2 \\ -1 & 1 \end{bmatrix}, \qquad B^{-1} = \begin{bmatrix} 1 & -1 \\ -1 & \frac{3}{2} \end{bmatrix}, \qquad (AB)^{-1} = \begin{bmatrix} 4 & -3 \\ -\frac{9}{2} & \frac{7}{2} \end{bmatrix}$$

Also,

$$B^{-1}A^{-1} = \begin{bmatrix} 1 & -1 \\ -1 & \frac{3}{2} \end{bmatrix} \begin{bmatrix} 3 & -2 \\ -1 & 1 \end{bmatrix} = \begin{bmatrix} 4 & -3 \\ -\frac{9}{2} & \frac{7}{2} \end{bmatrix}$$

Therefore, $(AB)^{-1} = B^{-1}A^{-1}$, as guaranteed by Theorem 1.4.6. ◆

Powers of a Matrix

Next, we shall define powers of a square matrix and discuss their properties.

DEFINITION

If A is a square matrix, then we define the nonnegative integer powers of A to be

$$A^0 = I \qquad A^n = \underbrace{AA \cdots A}_{n \text{ factors}} \qquad (n > 0)$$

Moreover, if A is invertible, then we define the negative integer powers to be

$$A^{-n} = (A^{-1})^n = \underbrace{A^{-1}A^{-1} \cdots A^{-1}}_{n \text{ factors}}$$

Because this definition parallels that for real numbers, the usual laws of exponents hold. (We omit the details.)

THEOREM 1.4.7

Laws of Exponents

If A is a square matrix and r and s are integers, then

$$A^r A^s = A^{r+s}, \qquad (A^r)^s = A^{rs}$$

The next theorem provides some useful properties of negative exponents.

THEOREM 1.4.8

Laws of Exponents

If A is an invertible matrix, then:

(a) A^{-1} is invertible and $(A^{-1})^{-1} = A$.

(b) A^n is invertible and $(A^n)^{-1} = (A^{-1})^n$ for $n = 0, 1, 2, \ldots$.

(c) For any nonzero scalar k, the matrix kA is invertible and $(kA)^{-1} = \dfrac{1}{k}A^{-1}$.

Proof

(a) Since $AA^{-1} = A^{-1}A = I$, the matrix A^{-1} is invertible and $(A^{-1})^{-1} = A$.

(b) This part is left as an exercise.

(c) If k is any nonzero scalar, results (l) and (m) of Theorem 1.4.1 enable us to write

$$(kA)\left(\frac{1}{k}A^{-1}\right) = \frac{1}{k}(kA)A^{-1} = \left(\frac{1}{k}k\right)AA^{-1} = (1)I = I$$

Similarly, $\left(\frac{1}{k}A^{-1}\right)(kA) = I$ so that kA is invertible and $(kA)^{-1} = \frac{1}{k}A^{-1}$. ∎

EXAMPLE 8 Powers of a Matrix

Let A and A^{-1} be as in Example 7; that is,

$$A = \begin{bmatrix} 1 & 2 \\ 1 & 3 \end{bmatrix} \quad \text{and} \quad A^{-1} = \begin{bmatrix} 3 & -2 \\ -1 & 1 \end{bmatrix}$$

Then

$$A^3 = \begin{bmatrix} 1 & 2 \\ 1 & 3 \end{bmatrix}\begin{bmatrix} 1 & 2 \\ 1 & 3 \end{bmatrix}\begin{bmatrix} 1 & 2 \\ 1 & 3 \end{bmatrix} = \begin{bmatrix} 11 & 30 \\ 15 & 41 \end{bmatrix}$$

$$A^{-3} = (A^{-1})^3 = \begin{bmatrix} 3 & -2 \\ -1 & 1 \end{bmatrix}\begin{bmatrix} 3 & -2 \\ -1 & 1 \end{bmatrix}\begin{bmatrix} 3 & -2 \\ -1 & 1 \end{bmatrix} = \begin{bmatrix} 41 & -30 \\ -15 & 11 \end{bmatrix} \quad \blacklozenge$$

Polynomial Expressions Involving Matrices

If A is a square matrix, say $m \times m$, and if

$$p(x) = a_0 + a_1 x + \cdots + a_n x^n \tag{1}$$

is any polynomial, then we define

$$p(A) = a_0 I + a_1 A + \cdots + a_n A^n$$

where I is the $m \times m$ identity matrix. In words, $p(A)$ is the $m \times m$ matrix that results when A is substituted for x in (1) and a_0 is replaced by $a_0 I$.

EXAMPLE 9 Matrix Polynomial

If

$$p(x) = 2x^2 - 3x + 4 \quad \text{and} \quad A = \begin{bmatrix} -1 & 2 \\ 0 & 3 \end{bmatrix}$$

then

$$p(A) = 2A^2 - 3A + 4I = 2\begin{bmatrix} -1 & 2 \\ 0 & 3 \end{bmatrix}^2 - 3\begin{bmatrix} -1 & 2 \\ 0 & 3 \end{bmatrix} + 4\begin{bmatrix} 1 & 0 \\ 0 & 1 \end{bmatrix}$$

$$= \begin{bmatrix} 2 & 8 \\ 0 & 18 \end{bmatrix} - \begin{bmatrix} -3 & 6 \\ 0 & 9 \end{bmatrix} + \begin{bmatrix} 4 & 0 \\ 0 & 4 \end{bmatrix} = \begin{bmatrix} 9 & 2 \\ 0 & 13 \end{bmatrix} \quad \blacklozenge$$

Properties of the Transpose

The next theorem lists the main properties of the transpose operation.

THEOREM 1.4.9

> ## Properties of the Transpose
>
> *If the sizes of the matrices are such that the stated operations can be performed, then*
>
> (*a*) $((A)^T)^T = A$
> (*b*) $(A + B)^T = A^T + B^T$ *and* $(A - B)^T = A^T - B^T$
> (*c*) $(kA)^T = kA^T$, *where k is any scalar*
> (*d*) $(AB)^T = B^T A^T$

If we keep in mind that transposing a matrix interchanges its rows and columns, parts (*a*), (*b*), and (*c*) should be self-evident. For example, part (*a*) states that interchanging rows and columns twice leaves a matrix unchanged; part (*b*) asserts that adding and then interchanging rows and columns yields the same result as first interchanging rows and columns and then adding; and part (*c*) asserts that multiplying by a scalar and then interchanging rows and columns yields the same result as first interchanging rows and columns and then multiplying by the scalar. Part (*d*) is not so obvious, so we give its proof.

Proof (*d*) Let $A = [a_{ij}]_{m \times r}$ and $B = [b_{ij}]_{r \times n}$ so that the products AB and $B^T A^T$ can both be formed. We leave it for the reader to check that $(AB)^T$ and $B^T A^T$ have the same size, namely $n \times m$. Thus it only remains to show that corresponding entries of $(AB)^T$ and $B^T A^T$ are the same; that is,

$$\left((AB)^T\right)_{ij} = (B^T A^T)_{ij} \tag{2}$$

Applying Formula (11) of Section 1.3 to the left side of this equation and using the definition of matrix multiplication, we obtain

$$\left((AB)^T\right)_{ij} = (AB)_{ji} = a_{j1}b_{1i} + a_{j2}b_{2i} + \cdots + a_{jr}b_{ri} \tag{3}$$

To evaluate the right side of (2), it will be convenient to let a'_{ij} and b'_{ij} denote the ijth entries of A^T and B^T, respectively, so

$$a'_{ij} = a_{ji} \quad \text{and} \quad b'_{ij} = b_{ji}$$

From these relationships and the definition of matrix multiplication, we obtain

$$(B^T A^T)_{ij} = b'_{i1}a'_{1j} + b'_{i2}a'_{2j} + \cdots + b'_{ir}a'_{rj}$$
$$= b_{1i}a_{j1} + b_{2i}a_{j2} + \cdots + b_{ri}a_{jr}$$
$$= a_{j1}b_{1i} + a_{j2}b_{2i} + \cdots + a_{jr}b_{ri}$$

This, together with (3), proves (2). ∎

Although we shall not prove it, part (*d*) of this theorem can be extended to include three or more factors; that is,

The transpose of a product of any number of matrices is equal to the product of their transposes in the reverse order.

REMARK Note the similarity between this result and the result following Theorem 1.4.6 about the inverse of a product of matrices.

Invertibility of a Transpose

The following theorem establishes a relationship between the inverse of an invertible matrix and the inverse of its transpose.

THEOREM 1.4.10

If A is an invertible matrix, then A^T is also invertible and

$$(A^T)^{-1} = (A^{-1})^T \tag{4}$$

Proof We can prove the invertibility of A^T and obtain (4) by showing that

$$A^T(A^{-1})^T = (A^{-1})^T A^T = I$$

But from part (d) of Theorem 1.4.9 and the fact that $I^T = I$, we have

$$A^T(A^{-1})^T = (A^{-1}A)^T = I^T = I$$
$$(A^{-1})^T A^T = (AA^{-1})^T = I^T = I$$

which completes the proof. ∎

EXAMPLE 10 Verifying Theorem 1.4.10

Consider the matrices

$$A = \begin{bmatrix} -5 & -3 \\ 2 & 1 \end{bmatrix}, \qquad A^T = \begin{bmatrix} -5 & 2 \\ -3 & 1 \end{bmatrix}$$

Applying Theorem 1.4.5 yields

$$A^{-1} = \begin{bmatrix} 1 & 3 \\ -2 & -5 \end{bmatrix}, \qquad (A^{-1})^T = \begin{bmatrix} 1 & -2 \\ 3 & -5 \end{bmatrix}, \qquad (A^T)^{-1} = \begin{bmatrix} 1 & -2 \\ 3 & -5 \end{bmatrix}$$

As guaranteed by Theorem 1.4.10, these matrices satisfy (4). ◆

EXERCISE SET 1.4

1. Let

$$A = \begin{bmatrix} 2 & -1 & 3 \\ 0 & 4 & 5 \\ -2 & 1 & 4 \end{bmatrix}, \qquad B = \begin{bmatrix} 8 & -3 & -5 \\ 0 & 1 & 2 \\ 4 & -7 & 6 \end{bmatrix},$$

$$C = \begin{bmatrix} 0 & -2 & 3 \\ 1 & 7 & 4 \\ 3 & 5 & 9 \end{bmatrix}, \qquad a = 4, \qquad b = -7$$

Show that
(a) $A + (B + C) = (A + B) + C$ (b) $(AB)C = A(BC)$
(c) $(a + b)C = aC + bC$ (d) $a(B - C) = aB - aC$

2. Using the matrices and scalars in Exercise 1, verify that
(a) $a(BC) = (aB)C = B(aC)$ (b) $A(B - C) = AB - AC$
(c) $(B + C)A = BA + CA$ (d) $a(bC) = (ab)C$

3. Using the matrices and scalars in Exercise 1, verify that
(a) $(A^T)^T = A$ (b) $(A + B)^T = A^T + B^T$
(c) $(aC)^T = aC^T$ (d) $(AB)^T = B^TA^T$

4. Use Theorem 1.4.5 to compute the inverses of the following matrices.

(a) $A = \begin{bmatrix} 3 & 1 \\ 5 & 2 \end{bmatrix}$ (b) $B = \begin{bmatrix} 2 & -3 \\ 4 & 4 \end{bmatrix}$

(c) $C = \begin{bmatrix} 6 & 4 \\ -2 & -1 \end{bmatrix}$ (d) $D = \begin{bmatrix} 2 & 0 \\ 0 & 3 \end{bmatrix}$

5. Use the matrices A and B in Exercise 4 to verify that
   (a) $(A^{-1})^{-1} = A$ (b) $(B^T)^{-1} = (B^{-1})^T$

6. Use the matrices A, B, and C in Exercise 4 to verify that
   (a) $(AB)^{-1} = B^{-1}A^{-1}$ (b) $(ABC)^{-1} = C^{-1}B^{-1}A^{-1}$

7. In each part, use the given information to find A.

   (a) $A^{-1} = \begin{bmatrix} 2 & -1 \\ 3 & 5 \end{bmatrix}$ (b) $(7A)^{-1} = \begin{bmatrix} -3 & 7 \\ 1 & -2 \end{bmatrix}$

   (c) $(5A^T)^{-1} = \begin{bmatrix} -3 & -1 \\ 5 & 2 \end{bmatrix}$ (d) $(I + 2A)^{-1} = \begin{bmatrix} -1 & 2 \\ 4 & 5 \end{bmatrix}$

8. Let A be the matrix
   $$\begin{bmatrix} 2 & 0 \\ 4 & 1 \end{bmatrix}$$
   Compute A^3, A^{-3}, and $A^2 - 2A + I$.

9. Let A be the matrix
   $$\begin{bmatrix} 3 & 1 \\ 2 & 1 \end{bmatrix}$$
   In each part, find $p(A)$.
   (a) $p(x) = x - 2$ (b) $p(x) = 2x^2 - x + 1$ (c) $p(x) = x^3 - 2x + 4$

10. Let $p_1(x) = x^2 - 9$, $p_2(x) = x + 3$, and $p_3(x) = x - 3$.
    (a) Show that $p_1(A) = p_2(A)p_3(A)$ for the matrix A in Exercise 9.
    (b) Show that $p_1(A) = p_2(A)p_3(A)$ for any square matrix A.

11. Find the inverse of
    $$\begin{bmatrix} \cos\theta & \sin\theta \\ -\sin\theta & \cos\theta \end{bmatrix}$$

12. Find the inverse of
    $$\begin{bmatrix} \frac{1}{2}(e^x + e^{-x}) & \frac{1}{2}(e^x - e^{-x}) \\ \frac{1}{2}(e^x - e^{-x}) & \frac{1}{2}(e^x + e^{-x}) \end{bmatrix}$$

13. Consider the matrix
    $$A = \begin{bmatrix} a_{11} & 0 & \cdots & 0 \\ 0 & a_{22} & \cdots & 0 \\ \vdots & \vdots & & \vdots \\ 0 & 0 & \cdots & a_{nn} \end{bmatrix}$$
    where $a_{11}a_{22}\cdots a_{nn} \neq 0$. Show that A is invertible and find its inverse.

14. Show that if a square matrix A satisfies $A^2 - 3A + I = 0$, then $A^{-1} = 3I - A$.

15. (a) Show that a matrix with a row of zeros cannot have an inverse.
    (b) Show that a matrix with a column of zeros cannot have an inverse.

16. Is the sum of two invertible matrices necessarily invertible?

17. Let A and B be square matrices such that $AB = 0$. Show that if A is invertible, then $B = 0$.

18. Let A, B, and 0 be 2×2 matrices. Assuming that A is invertible, find a matrix C such that
    $$\left[\begin{array}{c|c} A^{-1} & 0 \\ \hline C & A^{-1} \end{array}\right]$$
    is the inverse of the partitioned matrix
    $$\left[\begin{array}{c|c} A & 0 \\ \hline B & A \end{array}\right]$$
    (See Exercise 15 of the preceding section.)

19. Use the result in Exercise 18 to find the inverses of the following matrices.

(a) $\begin{bmatrix} 1 & 1 & 0 & 0 \\ -1 & 1 & 0 & 0 \\ 1 & 1 & 1 & 1 \\ 1 & 1 & -1 & 1 \end{bmatrix}$ (b) $\begin{bmatrix} 1 & 1 & 0 & 0 \\ 0 & 1 & 0 & 0 \\ 0 & 0 & 1 & 1 \\ 0 & 0 & 0 & 1 \end{bmatrix}$

20. (a) Find a nonzero 3×3 matrix A such that $A^T = A$.

(b) Find a nonzero 3×3 matrix A such that $A^T = -A$.

21. A square matrix A is called **symmetric** if $A^T = A$ and **skew-symmetric** if $A^T = -A$. Show that if B is a square matrix, then

(a) BB^T and $B + B^T$ are symmetric   (b) $B - B^T$ is skew-symmetric

22. If A is a square matrix and n is a positive integer, is it true that $(A^n)^T = (A^T)^n$? Justify your answer.

23. Let A be the matrix

$$\begin{bmatrix} 1 & 0 & 1 \\ 1 & 1 & 0 \\ 0 & 1 & 1 \end{bmatrix}$$

Determine whether A is invertible, and if so, find its inverse.

Hint Solve $AX = I$ by equating corresponding entries on the two sides.

24. Prove:
(a) part (*b*) of Theorem 1.4.1   (b) part (*i*) of Theorem 1.4.1
(c) part (*m*) of Theorem 1.4.1

25. Apply parts (*d*) and (*m*) of Theorem 1.4.1 to the matrices A, B, and $(-1)C$ to derive the result in part (*f*).

26. Prove Theorem 1.4.2.

27. Consider the laws of exponents $A^r A^s = A^{r+s}$ and $(A^r)^s = A^{rs}$.

(a) Show that if A is any square matrix, then these laws are valid for all nonnegative integer values of r and s.

(b) Show that if A is invertible, then these laws hold for all negative integer values of r and s.

28. Show that if A is invertible and k is any nonzero scalar, then $(kA)^n = k^n A^n$ for all integer values of n.

29. (a) Show that if A is invertible and $AB = AC$, then $B = C$.

(b) Explain why part (a) and Example 3 do not contradict one another.

30. Prove part (*c*) of Theorem 1.4.1.

Hint Assume that A is $m \times n$, B is $n \times p$, and C is $p \times q$. The ijth entry on the left side is $l_{ij} = a_{i1}[BC]_{1j} + a_{i2}[BC]_{2j} + \cdots + a_{in}[BC]_{nj}$ and the ijth entry on the right side is $r_{ij} = [AB]_{i1}c_{1j} + [AB]_{i2}c_{2j} + \cdots + [AB]_{ip}c_{pj}$. Verify that $l_{ij} = r_{ij}$.

Discussion & Discovery

31. Let A and B be square matrices with the same size.

(a) Give an example in which $(A + B)^2 \neq A^2 + 2AB + B^2$.

(b) Fill in the blank to create a matrix identity that is valid for all choices of A and B.
$(A + B)^2 = A^2 + B^2 + $ _____.

32. Let A and B be square matrices with the same size.

(a) Give an example in which $(A + B)(A - B) \neq A^2 - B^2$.

(b) Let A and B be square matrices with the same size. Fill in the blank to create a matrix identity that is valid for all choices of A and B. $(A + B)(A - B) =$ _____.

33. In the real number system the equation $a^2 = 1$ has exactly two solutions. Find at least eight different 3×3 matrices that satisfy the equation $A^2 = I_3$.

Hint Look for solutions in which all entries off the main diagonal are zero.

34. A statement of the form "If p, then q" is logically equivalent to the statement "If not q, then not p." (The second statement is called the ***logical contrapositive*** of the first.) For example, the logical contrapositive of the statement "If it is raining, then the ground is wet" is "If the ground is not wet, then it is not raining."

(a) Find the logical contrapositive of the following statement: If A^T is singular, then A is singular.

(b) Is the statement true or false? Explain.

35. Let A and B be $n \times n$ matrices. Indicate whether the statement is always true or sometimes false. Justify each answer.

(a) $(AB)^2 = A^2B^2$ (b) $(A - B)^2 = (B - A)^2$

(c) $(AB^{-1})(BA^{-1}) = I_n$ (d) $AB \neq BA$.

36. Assuming that all matrices are $n \times n$ and invertible, solve for D.

$$ABC^TDBA^TC = AB^T$$

1.5

ELEMENTARY MATRICES AND A METHOD FOR FINDING A^{-1}

In this section we shall develop an algorithm for finding the inverse of an invertible matrix. We shall also discuss some of the basic properties of invertible matrices.

We begin with the definition of a special type of matrix that can be used to carry out an elementary row operation by matrix multiplication.

> **DEFINITION**
>
> An $n \times n$ matrix is called an ***elementary matrix*** if it can be obtained from the $n \times n$ identity matrix I_n by performing a single elementary row operation.

EXAMPLE 1 Elementary Matrices and Row Operations

Listed below are four elementary matrices and the operations that produce them.

$$\begin{bmatrix} 1 & 0 \\ 0 & -3 \end{bmatrix} \qquad \begin{bmatrix} 1 & 0 & 0 & 0 \\ 0 & 0 & 0 & 1 \\ 0 & 0 & 1 & 0 \\ 0 & 1 & 0 & 0 \end{bmatrix} \qquad \begin{bmatrix} 1 & 0 & 3 \\ 0 & 1 & 0 \\ 0 & 0 & 1 \end{bmatrix} \qquad \begin{bmatrix} 1 & 0 & 0 \\ 0 & 1 & 0 \\ 0 & 0 & 1 \end{bmatrix} \blacklozenge$$

↑	↑	↑	↑
Multiply the second row of I_2 by -3.	Interchange the second and fourth rows of I_4.	Add 3 times the third row of I_3 to the first row.	Multiply the first row of I_3 by 1.

When a matrix A is multiplied on the *left* by an elementary matrix E, the effect is to perform an elementary row operation on A. This is the content of the following theorem, the proof of which is left for the exercises.

THEOREM 1.5.1

Row Operations by Matrix Multiplication

If the elementary matrix E results from performing a certain row operation on I_m and if A is an $m \times n$ matrix, then the product EA is the matrix that results when this same row operation is performed on A.

EXAMPLE 2   Using Elementary Matrices

Consider the matrix

$$A = \begin{bmatrix} 1 & 0 & 2 & 3 \\ 2 & -1 & 3 & 6 \\ 1 & 4 & 4 & 0 \end{bmatrix}$$

and consider the elementary matrix

$$E = \begin{bmatrix} 1 & 0 & 0 \\ 0 & 1 & 0 \\ 3 & 0 & 1 \end{bmatrix}$$

which results from adding 3 times the first row of I_3 to the third row. The product EA is

$$EA = \begin{bmatrix} 1 & 0 & 2 & 3 \\ 2 & -1 & 3 & 6 \\ 4 & 4 & 10 & 9 \end{bmatrix}$$

which is precisely the same matrix that results when we add 3 times the first row of A to the third row. ◆

REMARK   Theorem 1.5.1 is primarily of theoretical interest and will be used for developing some results about matrices and systems of linear equations. Computationally, it is preferable to perform row operations directly rather than multiplying on the left by an elementary matrix.

If an elementary row operation is applied to an identity matrix I to produce an elementary matrix E, then there is a second row operation that, when applied to E, produces I back again. For example, if E is obtained by multiplying the ith row of I by a nonzero constant c, then I can be recovered if the ith row of E is multiplied by $1/c$. The various possibilities are listed in Table 1. The operations on the right side of this table are called the ***inverse operations*** of the corresponding operations on the left.

EXAMPLE 3   Row Operations and Inverse Row Operations

In each of the following, an elementary row operation is applied to the 2×2 identity matrix to obtain an elementary matrix E, then E is restored to the identity matrix by

Table 1

Row Operation on I That Produces E	Row Operation on E That Reproduces I
Multiply row i by $c \neq 0$	Multiply row i by $1/c$
Interchange rows i and j	Interchange rows i and j
Add c times row i to row j	Add $-c$ times row i to row j

applying the inverse row operation.

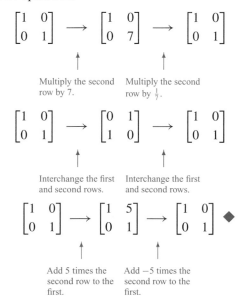

The next theorem gives an important property of elementary matrices.

THEOREM 1.5.2

Every elementary matrix is invertible, and the inverse is also an elementary matrix.

Proof If E is an elementary matrix, then E results from performing some row operation on I. Let E_0 be the matrix that results when the inverse of this operation is performed on I. Applying Theorem 1.5.1 and using the fact that inverse row operations cancel the effect of each other, it follows that

$$E_0 E = I \quad \text{and} \quad E E_0 = I$$

Thus, the elementary matrix E_0 is the inverse of E. ∎

The next theorem establishes some fundamental relationships among invertibility, homogeneous linear systems, reduced row-echelon forms, and elementary matrices. These results are extremely important and will be used many times in later sections.

THEOREM 1.5.3

Equivalent Statements

If A is an $n \times n$ matrix, then the following statements are equivalent, that is, all true or all false.

> (a) A is invertible.
> (b) $A\mathbf{x} = \mathbf{0}$ has only the trivial solution.
> (c) The reduced row-echelon form of A is I_n.
> (d) A is expressible as a product of elementary matrices.

Proof We shall prove the equivalence by establishing the chain of implications: $(a) \Rightarrow (b) \Rightarrow (c) \Rightarrow (d) \Rightarrow (a)$.

(a) $\Rightarrow$ (b) Assume A is invertible and let $\mathbf{x}_0$ be any solution of $A\mathbf{x} = \mathbf{0}$; thus $A\mathbf{x}_0 = \mathbf{0}$. Multiplying both sides of this equation by the matrix A^{-1} gives $A^{-1}(A\mathbf{x}_0) = A^{-1}\mathbf{0}$, or $(A^{-1}A)\mathbf{x}_0 = \mathbf{0}$, or $I\mathbf{x}_0 = \mathbf{0}$, or $\mathbf{x}_0 = \mathbf{0}$. Thus, $A\mathbf{x} = \mathbf{0}$ has only the trivial solution.

(b) $\Rightarrow$ (c) Let $A\mathbf{x} = \mathbf{0}$ be the matrix form of the system

$$\begin{aligned} a_{11}x_1 + a_{12}x_2 + \cdots + a_{1n}x_n &= 0 \\ a_{21}x_1 + a_{22}x_2 + \cdots + a_{2n}x_n &= 0 \\ \vdots \qquad \vdots \qquad\qquad \vdots \quad\ \ \vdots \\ a_{n1}x_1 + a_{n2}x_2 + \cdots + a_{nn}x_n &= 0 \end{aligned} \tag{1}$$

and assume that the system has only the trivial solution. If we solve by Gauss–Jordan elimination, then the system of equations corresponding to the reduced row-echelon form of the augmented matrix will be

$$\begin{aligned} x_1 \qquad\qquad\qquad &= 0 \\ x_2 \qquad\qquad &= 0 \\ \ddots \qquad\quad \\ x_n &= 0 \end{aligned} \tag{2}$$

Thus the augmented matrix

$$\begin{bmatrix} a_{11} & a_{12} & \cdots & a_{1n} & 0 \\ a_{21} & a_{22} & \cdots & a_{2n} & 0 \\ \vdots & \vdots & & \vdots & \vdots \\ a_{n1} & a_{n2} & \cdots & a_{nn} & 0 \end{bmatrix}$$

for (1) can be reduced to the augmented matrix

$$\begin{bmatrix} 1 & 0 & 0 & \cdots & 0 & 0 \\ 0 & 1 & 0 & \cdots & 0 & 0 \\ 0 & 0 & 1 & \cdots & 0 & 0 \\ \vdots & \vdots & \vdots & & \vdots & \vdots \\ 0 & 0 & 0 & \cdots & 1 & 0 \end{bmatrix}$$

for (2) by a sequence of elementary row operations. If we disregard the last column (of zeros) in each of these matrices, we can conclude that the reduced row-echelon form of A is I_n.

(c) $\Rightarrow$ (d) Assume that the reduced row-echelon form of A is I_n, so that A can be reduced to I_n by a finite sequence of elementary row operations. By Theorem 1.5.1, each of these operations can be accomplished by multiplying on the left by an appropriate elementary matrix. Thus we can find elementary matrices $E_1, E_2, \ldots, E_k$ such that

$$E_k \cdots E_2 E_1 A = I_n \tag{3}$$

By Theorem 1.5.2, $E_1, E_2, \ldots, E_k$ are invertible. Multiplying both sides of Equation (3) on the left successively by $E_k^{-1}, \ldots, E_2^{-1}, E_1^{-1}$ we obtain

$$A = E_1^{-1} E_2^{-1} \cdots E_k^{-1} I_n = E_1^{-1} E_2^{-1} \cdots E_k^{-1} \qquad (4)$$

By Theorem 1.5.2, this equation expresses A as a product of elementary matrices.

(d) ⇒ **(a)** If A is a product of elementary matrices, then from Theorems 1.4.6 and 1.5.2, the matrix A is a product of invertible matrices and hence is invertible. ∎

Row Equivalence

If a matrix B can be obtained from a matrix A by performing a finite sequence of elementary row operations, then obviously we can get from B back to A by performing the inverses of these elementary row operations in reverse order. Matrices that can be obtained from one another by a finite sequence of elementary row operations are said to be ***row equivalent***. With this terminology, it follows from parts (*a*) and (*c*) of Theorem 3 that an $n \times n$ matrix A is invertible if and only if it is row equivalent to the $n \times n$ identity matrix.

A Method for Inverting Matrices

As our first application of Theorem 3, we shall establish a method for determining the inverse of an invertible matrix. Multiplying (3) on the right by A^{-1} yields

$$A^{-1} = E_k \cdots E_2 E_1 I_n \qquad (5)$$

which tells us that A^{-1} can be obtained by multiplying I_n successively on the left by the elementary matrices $E_1, E_2, \ldots, E_k$. Since each multiplication on the left by one of these elementary matrices performs a row operation, it follows, by comparing Equations (3) and (5), that *the sequence of row operations that reduces A to I_n will reduce I_n to A^{-1}.* Thus we have the following result:

> *To find the inverse of an invertible matrix A, we must find a sequence of elementary row operations that reduces A to the identity and then perform this same sequence of operations on I_n to obtain A^{-1}.*

A simple method for carrying out this procedure is given in the following example.

EXAMPLE 4 Using Row Operations to Find A^{-1}

Find the inverse of

$$A = \begin{bmatrix} 1 & 2 & 3 \\ 2 & 5 & 3 \\ 1 & 0 & 8 \end{bmatrix}$$

Solution

We want to reduce A to the identity matrix by row operations and simultaneously apply these operations to I to produce A^{-1}. To accomplish this we shall adjoin the identity matrix to the right side of A, thereby producing a matrix of the form

$$[A \mid I]$$

Then we shall apply row operations to this matrix until the left side is reduced to I; these operations will convert the right side to A^{-1}, so the final matrix will have the form

$$[I \mid A^{-1}]$$

The computations are as follows:

$$\left[\begin{array}{ccc|ccc} 1 & 2 & 3 & 1 & 0 & 0 \\ 2 & 5 & 3 & 0 & 1 & 0 \\ 1 & 0 & 8 & 0 & 0 & 1 \end{array}\right]$$

$$\left[\begin{array}{ccc|ccc} 1 & 2 & 3 & 1 & 0 & 0 \\ 0 & 1 & -3 & -2 & 1 & 0 \\ 0 & -2 & 5 & -1 & 0 & 1 \end{array}\right] \longleftarrow \text{We added } -2 \text{ times the first row to the second and } -1 \text{ times the first row to the third.}$$

$$\left[\begin{array}{ccc|ccc} 1 & 2 & 3 & 1 & 0 & 0 \\ 0 & 1 & -3 & -2 & 1 & 0 \\ 0 & 0 & -1 & -5 & 2 & 1 \end{array}\right] \longleftarrow \text{We added 2 times the second row to the third.}$$

$$\left[\begin{array}{ccc|ccc} 1 & 2 & 3 & 1 & 0 & 0 \\ 0 & 1 & -3 & -2 & 1 & 0 \\ 0 & 0 & 1 & 5 & -2 & -1 \end{array}\right] \longleftarrow \text{We multiplied the third row by } -1.$$

$$\left[\begin{array}{ccc|ccc} 1 & 2 & 0 & -14 & 6 & 3 \\ 0 & 1 & 0 & 13 & -5 & -3 \\ 0 & 0 & 1 & 5 & -2 & -1 \end{array}\right] \longleftarrow \text{We added 3 times the third row to the second and } -3 \text{ times the third row to the first.}$$

$$\left[\begin{array}{ccc|ccc} 1 & 0 & 0 & -40 & 16 & 9 \\ 0 & 1 & 0 & 13 & -5 & -3 \\ 0 & 0 & 1 & 5 & -2 & -1 \end{array}\right] \longleftarrow \text{We added } -2 \text{ times the second row to the first.}$$

Thus,

$$A^{-1} = \left[\begin{array}{ccc} -40 & 16 & 9 \\ 13 & -5 & -3 \\ 5 & -2 & -1 \end{array}\right] \; \blacklozenge$$

Often it will not be known in advance whether a given matrix is invertible. If an $n \times n$ matrix A is not invertible, then it cannot be reduced to I_n by elementary row operations [part (c) of Theorem 3]. Stated another way, the reduced row-echelon form of A has at least one row of zeros. Thus, if the procedure in the last example is attempted on a matrix that is not invertible, then at some point in the computations a row of zeros will occur on the *left side*. It can then be concluded that the given matrix is not invertible, and the computations can be stopped.

EXAMPLE 5   Showing That a Matrix Is Not Invertible

Consider the matrix

$$A = \left[\begin{array}{ccc} 1 & 6 & 4 \\ 2 & 4 & -1 \\ -1 & 2 & 5 \end{array}\right]$$

Applying the procedure of Example 4 yields

$$\begin{bmatrix} 1 & 6 & 4 & | & 1 & 0 & 0 \\ 2 & 4 & -1 & | & 0 & 1 & 0 \\ -1 & 2 & 5 & | & 0 & 0 & 1 \end{bmatrix}$$

$$\begin{bmatrix} 1 & 6 & 4 & | & 1 & 0 & 0 \\ 0 & -8 & -9 & | & -2 & 1 & 0 \\ 0 & 8 & 9 & | & 1 & 0 & 1 \end{bmatrix} \quad \longleftarrow \quad \text{We added } -2 \text{ times the first row to the second and added the first row to the third.}$$

$$\begin{bmatrix} 1 & 6 & 4 & | & 1 & 0 & 0 \\ 0 & -8 & -9 & | & -2 & 1 & 0 \\ 0 & 0 & 0 & | & -1 & 1 & 1 \end{bmatrix} \quad \longleftarrow \quad \text{We added the second row to the third.}$$

Since we have obtained a row of zeros on the left side, A is not invertible. ◆

EXAMPLE 6   A Consequence of Invertibility

In Example 4 we showed that

$$A = \begin{bmatrix} 1 & 2 & 3 \\ 2 & 5 & 3 \\ 1 & 0 & 8 \end{bmatrix}$$

is an invertible matrix. From Theorem 3, it follows that the homogeneous system

$$x_1 + 2x_2 + 3x_3 = 0$$
$$2x_1 + 5x_2 + 3x_3 = 0$$
$$x_1 \qquad + 8x_3 = 0$$

has only the trivial solution. ◆

EXERCISE SET
1.5

1. Which of the following are elementary matrices?

(a) $\begin{bmatrix} 1 & 0 \\ -5 & 1 \end{bmatrix}$ 　　(b) $\begin{bmatrix} -5 & 1 \\ 1 & 0 \end{bmatrix}$ 　　(c) $\begin{bmatrix} 1 & 0 \\ 0 & \sqrt{3} \end{bmatrix}$ 　　(d) $\begin{bmatrix} 0 & 0 & 1 \\ 0 & 1 & 0 \\ 1 & 0 & 0 \end{bmatrix}$

(e) $\begin{bmatrix} 1 & 1 & 0 \\ 0 & 0 & 1 \\ 0 & 0 & 0 \end{bmatrix}$ 　　(f) $\begin{bmatrix} 1 & 0 & 0 \\ 0 & 1 & 9 \\ 0 & 0 & 1 \end{bmatrix}$ 　　(g) $\begin{bmatrix} 2 & 0 & 0 & 2 \\ 0 & 1 & 0 & 0 \\ 0 & 0 & 1 & 0 \\ 0 & 0 & 0 & 1 \end{bmatrix}$

2. Find a row operation that will restore the given elementary matrix to an identity matrix.

(a) $\begin{bmatrix} 1 & 0 \\ -3 & 1 \end{bmatrix}$ 　　(b) $\begin{bmatrix} 1 & 0 & 0 \\ 0 & 1 & 0 \\ 0 & 0 & 3 \end{bmatrix}$ 　　(c) $\begin{bmatrix} 0 & 0 & 0 & 1 \\ 0 & 1 & 0 & 0 \\ 0 & 0 & 1 & 0 \\ 1 & 0 & 0 & 0 \end{bmatrix}$ 　　(d) $\begin{bmatrix} 1 & 0 & -\frac{1}{7} & 0 \\ 0 & 1 & 0 & 0 \\ 0 & 0 & 1 & 0 \\ 0 & 0 & 0 & 1 \end{bmatrix}$

3. Consider the matrices

$$A = \begin{bmatrix} 3 & 4 & 1 \\ 2 & -7 & -1 \\ 8 & 1 & 5 \end{bmatrix}, \qquad B = \begin{bmatrix} 8 & 1 & 5 \\ 2 & -7 & -1 \\ 3 & 4 & 1 \end{bmatrix}, \qquad C = \begin{bmatrix} 3 & 4 & 1 \\ 2 & -7 & -1 \\ 2 & -7 & 3 \end{bmatrix}$$

Find elementary matrices E_1, E_2, E_3, and E_4 such that

(a) $E_1 A = B$ (b) $E_2 B = A$ (c) $E_3 A = C$ (d) $E_4 C = A$

4. In Exercise 3 is it possible to find an elementary matrix E such that $EB = C$? Justify your answer.

5. If a 2×2 matrix is multiplied on the left by the given matrices, what elementary row operation is performed on that matrix?

(a) $\begin{bmatrix} 0 & 1 \\ 1 & 0 \end{bmatrix}$ (b) $\begin{bmatrix} 2 & 0 \\ 0 & -3 \end{bmatrix}$ (c) $\begin{bmatrix} 1 & 0 \\ -2 & 1 \end{bmatrix}$

In Exercises 6–8 use the method shown in Examples 4 and 5 to find the inverse of the given matrix if the matrix is invertible, and check your answer by multiplication.

6. (a) $\begin{bmatrix} 1 & 4 \\ 2 & 7 \end{bmatrix}$ (b) $\begin{bmatrix} -3 & 6 \\ 4 & 5 \end{bmatrix}$ (c) $\begin{bmatrix} 6 & -4 \\ -3 & 2 \end{bmatrix}$

7. (a) $\begin{bmatrix} 3 & 4 & -1 \\ 1 & 0 & 3 \\ 2 & 5 & -4 \end{bmatrix}$ (b) $\begin{bmatrix} -1 & 3 & -4 \\ 2 & 4 & 1 \\ -4 & 2 & -9 \end{bmatrix}$ (c) $\begin{bmatrix} 1 & 0 & 1 \\ 0 & 1 & 1 \\ 1 & 1 & 0 \end{bmatrix}$

(d) $\begin{bmatrix} 2 & 6 & 6 \\ 2 & 7 & 6 \\ 2 & 7 & 7 \end{bmatrix}$ (e) $\begin{bmatrix} 1 & 0 & 1 \\ -1 & 1 & 1 \\ 0 & 1 & 0 \end{bmatrix}$

8. (a) $\begin{bmatrix} \frac{1}{5} & \frac{1}{5} & -\frac{2}{5} \\ \frac{1}{5} & \frac{1}{5} & \frac{1}{10} \\ \frac{1}{5} & -\frac{4}{5} & \frac{1}{10} \end{bmatrix}$ (b) $\begin{bmatrix} \sqrt{2} & 3\sqrt{2} & 0 \\ -4\sqrt{2} & \sqrt{2} & 0 \\ 0 & 0 & 1 \end{bmatrix}$ (c) $\begin{bmatrix} 1 & 0 & 0 & 0 \\ 1 & 3 & 0 & 0 \\ 1 & 3 & 5 & 0 \\ 1 & 3 & 5 & 7 \end{bmatrix}$

(d) $\begin{bmatrix} -8 & 17 & 2 & \frac{1}{3} \\ 4 & 0 & \frac{2}{5} & -9 \\ 0 & 0 & 0 & 0 \\ -1 & 13 & 4 & 2 \end{bmatrix}$ (e) $\begin{bmatrix} 0 & 0 & 2 & 0 \\ 1 & 0 & 0 & 1 \\ 0 & -1 & 3 & 0 \\ 2 & 1 & 5 & -3 \end{bmatrix}$

9. Find the inverse of each of the following 4×4 matrices, where k_1, k_2, k_3, k_4, and k are all nonzero.

(a) $\begin{bmatrix} k_1 & 0 & 0 & 0 \\ 0 & k_2 & 0 & 0 \\ 0 & 0 & k_3 & 0 \\ 0 & 0 & 0 & k_4 \end{bmatrix}$ (b) $\begin{bmatrix} 0 & 0 & 0 & k_1 \\ 0 & 0 & k_2 & 0 \\ 0 & k_3 & 0 & 0 \\ k_4 & 0 & 0 & 0 \end{bmatrix}$ (c) $\begin{bmatrix} k & 0 & 0 & 0 \\ 1 & k & 0 & 0 \\ 0 & 1 & k & 0 \\ 0 & 0 & 1 & k \end{bmatrix}$

10. Consider the matrix

$$A = \begin{bmatrix} 1 & 0 \\ -5 & 2 \end{bmatrix}$$

(a) Find elementary matrices E_1 and E_2 such that $E_2 E_1 A = I$.

(b) Write A^{-1} as a product of two elementary matrices.

(c) Write A as a product of two elementary matrices.

11. In each part, perform the stated row operation on

$$\begin{bmatrix} 2 & -1 & 0 \\ 4 & 5 & -3 \\ 1 & -4 & 7 \end{bmatrix}$$

by multiplying A on the left by a suitable elementary matrix. Check your answer in each case by performing the row operation directly on A.

(a) Interchange the first and third rows.

(b) Multiply the second row by $\frac{1}{3}$.

(c) Add twice the second row to the first row.

12. Write the matrix

$$\begin{bmatrix} 3 & -2 \\ 3 & -1 \end{bmatrix}$$

as a product of elementary matrices.

Note There is more than one correct solution.

13. Let

$$\begin{bmatrix} 1 & 0 & -2 \\ 0 & 4 & 3 \\ 0 & 0 & 1 \end{bmatrix}$$

(a) Find elementary matrices E_1, E_2, and E_3 such that $E_3 E_2 E_1 A = I_3$.

(b) Write A as a product of elementary matrices.

14. Express the matrix

$$A = \begin{bmatrix} 0 & 1 & 7 & 8 \\ 1 & 3 & 3 & 8 \\ -2 & -5 & 1 & -8 \end{bmatrix}$$

in the form $A = EFGR$, where E, F, and G are elementary matrices and R is in row-echelon form.

15. Show that if

$$A = \begin{bmatrix} 1 & 0 & 0 \\ 0 & 1 & 0 \\ a & b & c \end{bmatrix}$$

is an elementary matrix, then at least one entry in the third row must be a zero.

16. Show that

$$A = \begin{bmatrix} 0 & a & 0 & 0 & 0 \\ b & 0 & c & 0 & 0 \\ 0 & d & 0 & e & 0 \\ 0 & 0 & f & 0 & g \\ 0 & 0 & 0 & h & 0 \end{bmatrix}$$

is not invertible for any values of the entries.

17. Prove that if A is an $m \times n$ matrix, there is an invertible matrix C such that CA is in reduced row-echelon form.

18. Prove that if A is an invertible matrix and B is row equivalent to A, then B is also invertible.

19. (a) Prove: If A and B are $m \times n$ matrices, then A and B are row equivalent if and only if A and B have the same reduced row-echelon form.

(b) Show that A and B are row equivalent, and find a sequence of elementary row operations that produces B from A.

$$A = \begin{bmatrix} 1 & 2 & 3 \\ 1 & 4 & 1 \\ 2 & 1 & 9 \end{bmatrix}, \qquad B = \begin{bmatrix} 1 & 0 & 5 \\ 0 & 2 & -2 \\ 1 & 1 & 4 \end{bmatrix}$$

20. Prove Theorem 1.5.1.

21. Suppose that A is some unknown invertible matrix, but you know of a sequence of elementary row operations that produces the identity matrix when applied in succession to A. Explain how you can use the known information to find A.

22. Indicate whether the statement is always true or sometimes false. Justify your answer with a logical argument or a counterexample.

    (a) Every square matrix can be expressed as a product of elementary matrices.

    (b) The product of two elementary matrices is an elementary matrix.

    (c) If A is invertible and a multiple of the first row of A is added to the second row, then the resulting matrix is invertible.

    (d) If A is invertible and $AB = 0$, then it must be true that $B = 0$.

23. Indicate whether the statement is always true or sometimes false. Justify your answer with a logical argument or a counterexample.

    (a) If A is a singular $n \times n$ matrix, then $A\mathbf{x} = \mathbf{0}$ has infinitely many solutions.

    (b) If A is a singular $n \times n$ matrix, then the reduced row-echelon form of A has at least one row of zeros.

    (c) If A^{-1} is expressible as a product of elementary matrices, then the homogeneous linear system $A\mathbf{x} = \mathbf{0}$ has only the trivial solution.

    (d) If A is a singular $n \times n$ matrix, and B results by interchanging two rows of A, then B may or may not be singular.

24. Do you think that there is a 2×2 matrix A such that

$$A \begin{bmatrix} a & b \\ c & d \end{bmatrix} = \begin{bmatrix} b & d \\ a & c \end{bmatrix}$$

for all values of a, b, c, and d? Explain your reasoning.

1.6
FURTHER RESULTS ON SYSTEMS OF EQUATIONS AND INVERTIBILITY

In this section we shall establish more results about systems of linear equations and invertibility of matrices. Our work will lead to a new method for solving n equations in n unknowns.

A Basic Theorem

In Section 1.1 we made the statement (based on Figure 1.1.1) that every linear system has no solutions, or has one solution, or has infinitely many solutions. We are now in a position to prove this fundamental result.

THEOREM 1.6.1

Every system of linear equations has no solutions, or has exactly one solution, or has infinitely many solutions.

Proof If $A\mathbf{x} = \mathbf{b}$ is a system of linear equations, exactly one of the following is true: (a) the system has no solutions, (b) the system has exactly one solution, or (c) the system has more than one solution. The proof will be complete if we can show that the system has infinitely many solutions in case (c).

Assume that $A\mathbf{x} = \mathbf{b}$ has more than one solution, and let $\mathbf{x}_0 = \mathbf{x}_1 - \mathbf{x}_2$, where $\mathbf{x}_1$ and $\mathbf{x}_2$ are any two distinct solutions. Because $\mathbf{x}_1$ and $\mathbf{x}_2$ are distinct, the matrix $\mathbf{x}_0$ is nonzero; moreover,

$$A\mathbf{x}_0 = A(\mathbf{x}_1 - \mathbf{x}_2) = A\mathbf{x}_1 - A\mathbf{x}_2 = \mathbf{b} - \mathbf{b} = \mathbf{0}$$

If we now let k be any scalar, then

$$A(\mathbf{x}_1 + k\mathbf{x}_0) = A\mathbf{x}_1 + A(k\mathbf{x}_0) = A\mathbf{x}_1 + k(A\mathbf{x}_0)$$
$$= \mathbf{b} + k\mathbf{0} = \mathbf{b} + \mathbf{0} = \mathbf{b}$$

But this says that $\mathbf{x}_1 + k\mathbf{x}_0$ is a solution of $A\mathbf{x} = \mathbf{b}$. Since $\mathbf{x}_0$ is nonzero and there are infinitely many choices for k, the system $A\mathbf{x} = \mathbf{b}$ has infinitely many solutions. ∎

Solving Linear Systems by Matrix Inversion

Thus far, we have studied two methods for solving linear systems: Gaussian elimination and Gauss–Jordan elimination. The following theorem provides a new method for solving certain linear systems.

THEOREM 1.6.2

> *If A is an invertible $n \times n$ matrix, then for each $n \times 1$ matrix $\mathbf{b}$, the system of equations $A\mathbf{x} = \mathbf{b}$ has exactly one solution, namely, $\mathbf{x} = A^{-1}\mathbf{b}$.*

Proof   Since $A(A^{-1}\mathbf{b}) = \mathbf{b}$, it follows that $\mathbf{x} = A^{-1}\mathbf{b}$ is a solution of $A\mathbf{x} = \mathbf{b}$. To show that this is the only solution, we will assume that $\mathbf{x}_0$ is an arbitrary solution and then show that $\mathbf{x}_0$ must be the solution $A^{-1}\mathbf{b}$.

If $\mathbf{x}_0$ is any solution, then $A\mathbf{x}_0 = \mathbf{b}$. Multiplying both sides by A^{-1}, we obtain $\mathbf{x}_0 = A^{-1}\mathbf{b}$. ∎

EXAMPLE 1   Solution of a Linear System Using A^{-1}

Consider the system of linear equations

$$
\begin{aligned}
x_1 + 2x_2 + 3x_3 &= 5 \\
2x_1 + 5x_2 + 3x_3 &= 3 \\
x_1 \qquad\quad + 8x_3 &= 17
\end{aligned}
$$

In matrix form this system can be written as $A\mathbf{x} = \mathbf{b}$, where

$$
A = \begin{bmatrix} 1 & 2 & 3 \\ 2 & 5 & 3 \\ 1 & 0 & 8 \end{bmatrix}, \qquad
\mathbf{x} = \begin{bmatrix} x_1 \\ x_2 \\ x_3 \end{bmatrix}, \qquad
\mathbf{b} = \begin{bmatrix} 5 \\ 3 \\ 17 \end{bmatrix}
$$

In Example 4 of the preceding section, we showed that A is invertible and

$$
A^{-1} = \begin{bmatrix} -40 & 16 & 9 \\ 13 & -5 & -3 \\ 5 & -2 & -1 \end{bmatrix}
$$

By Theorem 1.6.2, the solution of the system is

$$
\mathbf{x} = A^{-1}\mathbf{b} = \begin{bmatrix} -40 & 16 & 9 \\ 13 & -5 & -3 \\ 5 & -2 & -1 \end{bmatrix} \begin{bmatrix} 5 \\ 3 \\ 17 \end{bmatrix} = \begin{bmatrix} 1 \\ -1 \\ 2 \end{bmatrix}
$$

or $x_1 = 1, x_2 = -1, x_3 = 2$. ◆

REMARK Note that the method of Example 1 applies only when the system has as many equations as unknowns and the coefficient matrix is invertible. This method is less efficient, computationally, than Gaussian elimination, but it is important in the analysis of equations involving matrices.

Linear Systems with a Common Coefficient Matrix

Frequently, one is concerned with solving a sequence of systems

$$A\mathbf{x} = \mathbf{b}_1, \quad A\mathbf{x} = \mathbf{b}_2, \quad A\mathbf{x} = \mathbf{b}_3, \ldots, \quad A\mathbf{x} = \mathbf{b}_k$$

each of which has the same square coefficient matrix A. If A is invertible, then the solutions

$$\mathbf{x}_1 = A^{-1}\mathbf{b}_1, \quad \mathbf{x}_2 = A^{-1}\mathbf{b}_2, \quad \mathbf{x}_3 = A^{-1}\mathbf{b}_3, \ldots, \quad \mathbf{x}_k = A^{-1}\mathbf{b}_k$$

can be obtained with one matrix inversion and k matrix multiplications. Once again, however, a more efficient method is to form the matrix

$$[A \mid \mathbf{b}_1 \mid \mathbf{b}_2 \mid \cdots \mid \mathbf{b}_k] \qquad (1)$$

in which the coefficient matrix A is "augmented" by all k of the matrices $\mathbf{b}_1, \mathbf{b}_2, \ldots, \mathbf{b}_k$, and then reduce (1) to reduced row-echelon form by Gauss–Jordan elimination. In this way we can solve all k systems at once. This method has the added advantage that it applies even when A is not invertible.

EXAMPLE 2 Solving Two Linear Systems at Once

Solve the systems

(a) $\quad x_1 + 2x_2 + 3x_3 = 4$ (b) $\quad x_1 + 2x_2 + 3x_3 = \quad 1$

$\qquad\quad 2x_1 + 5x_2 + 3x_3 = 5$ $\quad 2x_1 + 5x_2 + 3x_3 = \quad 6$

$\qquad\quad x_1 \qquad\quad + 8x_3 = 9$ $\quad x_1 \qquad\quad + 8x_3 = -6$

Solution

The two systems have the same coefficient matrix. If we augment this coefficient matrix with the columns of constants on the right sides of these systems, we obtain

$$\begin{bmatrix} 1 & 2 & 3 & 4 & 1 \\ 2 & 5 & 3 & 5 & 6 \\ 1 & 0 & 8 & 9 & -6 \end{bmatrix}$$

Reducing this matrix to reduced row-echelon form yields (verify)

$$\begin{bmatrix} 1 & 0 & 0 & 1 & 2 \\ 0 & 1 & 0 & 0 & 1 \\ 0 & 0 & 1 & 1 & -1 \end{bmatrix}$$

It follows from the last two columns that the solution of system (a) is $x_1 = 1$, $x_2 = 0$, $x_3 = 1$ and the solution of system (b) is $x_1 = 2$, $x_2 = 1$, $x_3 = -1$. ◆

Properties of Invertible Matrices

Up to now, to show that an $n \times n$ matrix A is invertible, it has been necessary to find an $n \times n$ matrix B such that

$$AB = I \quad \text{and} \quad BA = I$$

The next theorem shows that if we produce an $n \times n$ matrix B satisfying *either* condition, then the other condition holds automatically.

THEOREM 1.6.3

> *Let A be a square matrix.*
>
> (*a*) *If B is a square matrix satisfying BA = I, then B = A^{-1}.*
> (*b*) *If B is a square matrix satisfying AB = I, then B = A^{-1}.*

We shall prove part (*a*) and leave part (*b*) as an exercise.

***Proof** (a)*   Assume that $BA = I$. If we can show that A is invertible, the proof can be completed by multiplying $BA = I$ on both sides by A^{-1} to obtain

$$BAA^{-1} = IA^{-1} \quad \text{or} \quad BI = IA^{-1} \quad \text{or} \quad B = A^{-1}$$

To show that A is invertible, it suffices to show that the system $A\mathbf{x} = \mathbf{0}$ has only the trivial solution (see Theorem 3). Let $\mathbf{x}_0$ be any solution of this system. If we multiply both sides of $A\mathbf{x}_0 = \mathbf{0}$ on the left by B, we obtain $BA\mathbf{x}_0 = B\mathbf{0}$ or $I\mathbf{x}_0 = \mathbf{0}$ or $\mathbf{x}_0 = \mathbf{0}$. Thus, the system of equations $A\mathbf{x} = \mathbf{0}$ has only the trivial solution. ∎

We are now in a position to add two more statements that are equivalent to the four given in Theorem 3.

THEOREM 1.6.4

> **Equivalent Statements**
>
> *If A is an n × n matrix, then the following are equivalent.*
>
> (*a*)   *A is invertible.*
> (*b*)   *A$\mathbf{x}$ = $\mathbf{0}$ has only the trivial solution.*
> (*c*)   *The reduced row-echelon form of A is I$_n$.*
> (*d*)   *A is expressible as a product of elementary matrices.*
> (*e*)   *A$\mathbf{x}$ = $\mathbf{b}$ is consistent for every n × 1 matrix $\mathbf{b}$.*
> (*f*)   *A$\mathbf{x}$ = $\mathbf{b}$ has exactly one solution for every n × 1 matrix $\mathbf{b}$.*

Proof   Since we proved in Theorem 3 that (*a*), (*b*), (*c*), and (*d*) are equivalent, it will be sufficient to prove that $(a) \Rightarrow (f) \Rightarrow (e) \Rightarrow (a)$.

(*a*) ⇒ (*f*)   This was already proved in Theorem 1.6.2.

(*f*) ⇒ (*e*)   This is self-evident: If $A\mathbf{x} = \mathbf{b}$ has exactly one solution for every $n \times 1$ matrix $\mathbf{b}$, then $A\mathbf{x} = \mathbf{b}$ is consistent for every $n \times 1$ matrix $\mathbf{b}$.

(*e*) ⇒ (*a*)   If the system $A\mathbf{x} = \mathbf{b}$ is consistent for every $n \times 1$ matrix $\mathbf{b}$, then in particular, the systems

$$A\mathbf{x} = \begin{bmatrix} 1 \\ 0 \\ 0 \\ \vdots \\ 0 \end{bmatrix}, \quad A\mathbf{x} = \begin{bmatrix} 0 \\ 1 \\ 0 \\ \vdots \\ 0 \end{bmatrix}, \dots, \quad A\mathbf{x} = \begin{bmatrix} 0 \\ 0 \\ 0 \\ \vdots \\ 1 \end{bmatrix}$$

are consistent. Let $\mathbf{x}_1, \mathbf{x}_2, \dots, \mathbf{x}_n$ be solutions of the respective systems, and let us form an $n \times n$ matrix C having these solutions as columns. Thus C has the form

$$C = [\mathbf{x}_1 \mid \mathbf{x}_2 \mid \cdots \mid \mathbf{x}_n]$$

As discussed in Section 1.3, the successive columns of the product AC will be

$$A\mathbf{x}_1, A\mathbf{x}_2, \dots, A\mathbf{x}_n$$

Thus

$$AC = [A\mathbf{x}_1 \mid A\mathbf{x}_2 \mid \cdots \mid A\mathbf{x}_n] = \begin{bmatrix} 1 & 0 & \cdots & 0 \\ 0 & 1 & \cdots & 0 \\ 0 & 0 & \cdots & 0 \\ \vdots & \vdots & & \vdots \\ 0 & 0 & \cdots & 1 \end{bmatrix} = I$$

By part (b) of Theorem 1.6.3, it follows that $C = A^{-1}$. Thus, A is invertible. ■

We know from earlier work that invertible matrix factors produce an invertible product. The following theorem, which will be proved later, looks at the converse: It shows that if the product of square matrices is invertible, then the factors themselves must be invertible.

THEOREM 1.6.5

Let A and B be square matrices of the same size. If AB is invertible, then A and B must also be invertible.

In our later work the following fundamental problem will occur frequently in various contexts.

A Fundamental Problem: Let A be a fixed $m \times n$ matrix. Find all $m \times 1$ matrices **b** such that the system of equations $A\mathbf{x} = \mathbf{b}$ is consistent.

If A is an invertible matrix, Theorem 1.6.2 completely solves this problem by asserting that for *every* $m \times 1$ matrix **b**, the linear system $A\mathbf{x} = \mathbf{b}$ has the unique solution $\mathbf{x} = A^{-1}\mathbf{b}$. If A is not square, or if A is square but not invertible, then Theorem 1.6.2 does not apply. In these cases the matrix **b** must usually satisfy certain conditions in order for $A\mathbf{x} = \mathbf{b}$ to be consistent. The following example illustrates how the elimination methods of Section 1.2 can be used to determine such conditions.

EXAMPLE 3 Determining Consistency by Elimination

What conditions must b_1, b_2, and b_3 satisfy in order for the system of equations

$$\begin{aligned} x_1 + x_2 + 2x_3 &= b_1 \\ x_1 \quad\quad + x_3 &= b_2 \\ 2x_1 + x_2 + 3x_3 &= b_3 \end{aligned}$$

to be consistent?

Solution

The augmented matrix is

$$\begin{bmatrix} 1 & 1 & 2 & b_1 \\ 1 & 0 & 1 & b_2 \\ 2 & 1 & 3 & b_3 \end{bmatrix}$$

which can be reduced to row-echelon form as follows:

$$\begin{bmatrix} 1 & 1 & 2 & b_1 \\ 0 & -1 & -1 & b_2 - b_1 \\ 0 & -1 & -1 & b_3 - 2b_1 \end{bmatrix}$$

◄——— −1 times the first row was added to the second and −2 times the first row was added to the third.

$$\begin{bmatrix} 1 & 1 & 2 & b_1 \\ 0 & 1 & 1 & b_1 - b_2 \\ 0 & -1 & -1 & b_3 - 2b_1 \end{bmatrix}$$

◄——— The second row was multiplied by −1.

$$\begin{bmatrix} 1 & 1 & 2 & b_1 \\ 0 & 1 & 1 & b_1 - b_2 \\ 0 & 0 & 0 & b_3 - b_2 - b_1 \end{bmatrix}$$

◄——— The second row was added to the third.

It is now evident from the third row in the matrix that the system has a solution if and only if b_1, b_2, and b_3 satisfy the condition

$$b_3 - b_2 - b_1 = 0 \quad \text{or} \quad b_3 = b_1 + b_2$$

To express this condition another way, $A\mathbf{x} = \mathbf{b}$ is consistent if and only if $\mathbf{b}$ is a matrix of the form

$$\mathbf{b} = \begin{bmatrix} b_1 \\ b_2 \\ b_1 + b_2 \end{bmatrix}$$

where b_1 and b_2 are arbitrary. ◆

EXAMPLE 4 Determining Consistency by Elimination

What conditions must b_1, b_2, and b_3 satisfy in order for the system of equations

$$x_1 + 2x_2 + 3x_3 = b_1$$
$$2x_1 + 5x_2 + 3x_3 = b_2$$
$$x_1 \qquad\quad + 8x_3 = b_3$$

to be consistent?

Solution

The augmented matrix is

$$\begin{bmatrix} 1 & 2 & 3 & b_1 \\ 2 & 5 & 3 & b_2 \\ 1 & 0 & 8 & b_3 \end{bmatrix}$$

Reducing this to reduced row-echelon form yields (verify)

$$\begin{bmatrix} 1 & 0 & 0 & -40b_1 + 16b_2 + 9b_3 \\ 0 & 1 & 0 & 13b_1 - 5b_2 - 3b_3 \\ 0 & 0 & 1 & 5b_1 - 2b_2 - b_3 \end{bmatrix} \tag{2}$$

In this case there are no restrictions on b_1, b_2, and b_3; that is, the given system $A\mathbf{x} = \mathbf{b}$ has the unique solution

$$x_1 = -40b_1 + 16b_2 + 9b_3, \quad x_2 = 13b_1 - 5b_2 - 3b_3, \quad x_3 = 5b_1 - 2b_2 - b_3 \tag{3}$$

for all $\mathbf{b}$. ◆

REMARK Because the system $A\mathbf{x} = \mathbf{b}$ in the preceding example is consistent for all $\mathbf{b}$, it follows from Theorem 1.6.4 that A is invertible. We leave it for the reader to verify that the formulas in (3) can also be obtained by calculating $\mathbf{x} = A^{-1}\mathbf{b}$.

EXERCISE SET 1.6

In Exercises 1–8 solve the system by inverting the coefficient matrix and using Theorem 1.6.2.

1. $\begin{aligned} x_1 + x_2 &= 2 \\ 5x_1 + 6x_2 &= 9 \end{aligned}$
2. $\begin{aligned} 4x_1 - 3x_2 &= -3 \\ 2x_1 - 5x_2 &= 9 \end{aligned}$
3. $\begin{aligned} x_1 + 3x_2 + x_3 &= 4 \\ 2x_1 + 2x_2 + x_3 &= -1 \\ 2x_1 + 3x_2 + x_3 &= 3 \end{aligned}$

4. $\begin{aligned} 5x_1 + 3x_2 + 2x_3 &= 4 \\ 3x_1 + 3x_2 + 2x_3 &= 2 \\ x_2 + x_3 &= 5 \end{aligned}$
5. $\begin{aligned} x + y + z &= 5 \\ x + y - 4z &= 10 \\ -4x + y + z &= 0 \end{aligned}$
6. $\begin{aligned} -x - 2y - 3z &= 0 \\ w + x + 4y + 4z &= 7 \\ w + 3x + 7y + 9z &= 4 \\ -w - 2x - 4y - 6z &= 6 \end{aligned}$

7. $\begin{aligned} 3x_1 + 5x_2 &= b_1 \\ x_1 + 2x_2 &= b_2 \end{aligned}$
8. $\begin{aligned} x_1 + 2x_2 + 3x_3 &= b_1 \\ 2x_1 + 5x_2 + 5x_3 &= b_2 \\ 3x_1 + 5x_2 + 8x_3 &= b_3 \end{aligned}$

9. Solve the following general system by inverting the coefficient matrix and using Theorem 1.6.2.

$$x_1 + 2x_2 + x_3 = b_1$$
$$x_1 - x_2 + x_3 = b_2$$
$$x_1 + x_2 \quad\;\; = b_3$$

Use the resulting formulas to find the solution if

(a) $b_1 = -1, \quad b_2 = 3, \quad b_3 = 4$ (b) $b_1 = 5, \quad b_2 = 0, \quad b_3 = 0$

(c) $b_1 = -1, \quad b_2 = -1, \quad b_3 = 3$

10. Solve the three systems in Exercise 9 using the method of Example 2.

In Exercises 11–14 use the method of Example 2 to solve the systems in all parts simultaneously.

11. $\begin{aligned} x_1 - 5x_2 &= b_1 \\ 3x_1 + 2x_2 &= b_2 \end{aligned}$

  (a) $b_1 = 1, \quad b_2 = 4$

  (b) $b_1 = -2, \quad b_2 = 5$

12. $\begin{aligned} -x_1 + 4x_2 + x_3 &= b_1 \\ x_1 + 9x_2 - 2x_3 &= b_2 \\ 6x_1 + 4x_2 - 8x_3 &= b_3 \end{aligned}$

  (a) $b_1 = 0, \quad b_2 = 1, \quad b_3 = 0$

  (b) $b_1 = -3, \quad b_2 = 4, \quad b_3 = -5$

13. $\begin{aligned} 4x_1 - 7x_2 &= b_1 \\ x_1 + 2x_2 &= b_2 \end{aligned}$

  (a) $b_1 = 0, \quad b_2 = 1$

  (b) $b_1 = -4, \quad b_2 = 6$

  (c) $b_1 = -1, \quad b_2 = 3$

  (d) $b_1 = -5, \quad b_2 = 1$

14. $\begin{aligned} x_1 + 3x_2 + 5x_3 &= b_1 \\ -x_1 - 2x_2 \quad\;\; &= b_2 \\ 2x_1 + 5x_2 + 4x_3 &= b_3 \end{aligned}$

  (a) $b_1 = 1, \quad b_2 = 0, \quad b_3 = -1$

  (b) $b_1 = 0, \quad b_2 = 1, \quad b_3 = 1$

  (c) $b_1 = -1, \quad b_2 = -1, \quad b_3 = 0$

15. The method of Example 2 can be used for linear systems with infinitely many solutions. Use that method to solve the systems in both parts at the same time.

(a) $\begin{aligned} x_1 - 2x_2 + x_3 &= -2 \\ 2x_1 - 5x_2 + x_3 &= 1 \\ 3x_1 - 7x_2 + 2x_3 &= -1 \end{aligned}$ (b) $\begin{aligned} x_1 - 2x_2 + x_3 &= 1 \\ 2x_1 - 5x_2 + x_3 &= -1 \\ 3x_1 - 7x_2 + 2x_3 &= 0 \end{aligned}$

In Exercises 16–19 find conditions that the b's must satisfy for the system to be consistent.

16. $\begin{aligned} 6x_1 - 4x_2 &= b_1 \\ 3x_1 - 2x_2 &= b_2 \end{aligned}$
17. $\begin{aligned} x_1 - 2x_2 + 5x_3 &= b_1 \\ 4x_1 - 5x_2 + 8x_3 &= b_2 \\ -3x_1 + 3x_2 - 3x_3 &= b_3 \end{aligned}$

18. $\begin{aligned} x_1 - 2x_2 - x_3 &= b_1 \\ -4x_1 + 5x_2 + 2x_3 &= b_2 \\ -4x_1 + 7x_2 + 4x_3 &= b_3 \end{aligned}$ **19.** $\begin{aligned} x_1 - x_2 + 3x_3 + 2x_4 &= b_1 \\ -2x_1 + x_2 + 5x_3 + x_4 &= b_2 \\ -3x_1 + 2x_2 + 2x_3 - x_4 &= b_3 \\ 4x_1 - 3x_2 + x_3 + 3x_4 &= b_4 \end{aligned}$

20. Consider the matrices

$$A = \begin{bmatrix} 2 & 1 & 2 \\ 2 & 2 & -2 \\ 3 & 1 & 1 \end{bmatrix} \quad \text{and} \quad \mathbf{x} = \begin{bmatrix} x_1 \\ x_2 \\ x_3 \end{bmatrix}$$

(a) Show that the equation $A\mathbf{x} = \mathbf{x}$ can be rewritten as $(A - I)\mathbf{x} = \mathbf{0}$ and use this result to solve $A\mathbf{x} = \mathbf{x}$ for $\mathbf{x}$.

(b) Solve $A\mathbf{x} = 4\mathbf{x}$.

21. Solve the following matrix equation for X.

$$\begin{bmatrix} 1 & -1 & 1 \\ 2 & 3 & 0 \\ 0 & 2 & -1 \end{bmatrix} X = \begin{bmatrix} 2 & -1 & 5 & 7 & 8 \\ 4 & 0 & -3 & 0 & 1 \\ 3 & 5 & -7 & 2 & 1 \end{bmatrix}$$

22. In each part, determine whether the homogeneous system has a nontrivial solution (without using pencil and paper); then state whether the given matrix is invertible.

(a) $\begin{aligned} 2x_1 + x_2 - 3x_3 + x_4 &= 0 \\ 5x_2 + 4x_3 + 3x_4 &= 0 \\ x_3 + 2x_4 &= 0 \\ 3x_4 &= 0 \end{aligned}$ $\begin{bmatrix} 2 & 1 & -3 & 1 \\ 0 & 5 & 4 & 3 \\ 0 & 0 & 1 & 2 \\ 0 & 0 & 0 & 3 \end{bmatrix}$

(b) $\begin{aligned} 5x_1 + x_2 + 4x_3 + x_4 &= 0 \\ 2x_3 - x_4 &= 0 \\ x_3 + x_4 &= 0 \\ 7x_4 &= 0 \end{aligned}$ $\begin{bmatrix} 5 & 1 & 4 & 1 \\ 0 & 0 & 2 & -1 \\ 0 & 0 & 1 & 1 \\ 0 & 0 & 0 & 7 \end{bmatrix}$

23. Let $A\mathbf{x} = \mathbf{0}$ be a homogeneous system of n linear equations in n unknowns that has only the trivial solution. Show that if k is any positive integer, then the system $A^k\mathbf{x} = \mathbf{0}$ also has only the trivial solution.

24. Let $A\mathbf{x} = \mathbf{0}$ be a homogeneous system of n linear equations in n unknowns, and let Q be an invertible $n \times n$ matrix. Show that $A\mathbf{x} = \mathbf{0}$ has just the trivial solution if and only if $(QA)\mathbf{x} = \mathbf{0}$ has just the trivial solution.

25. Let $A\mathbf{x} = \mathbf{b}$ be any consistent system of linear equations, and let $\mathbf{x}_1$ be a fixed solution. Show that every solution to the system can be written in the form $\mathbf{x} = \mathbf{x}_1 + \mathbf{x}_0$, where $\mathbf{x}_0$ is a solution to $A\mathbf{x} = \mathbf{0}$. Show also that every matrix of this form is a solution.

26. Use part (*a*) of Theorem 1.6.3 to prove part (*b*).

27. What restrictions must be placed on x and y for the following matrices to be invertible?

(a) $\begin{bmatrix} x & y \\ x & x \end{bmatrix}$ (b) $\begin{bmatrix} x & 0 \\ y & y \end{bmatrix}$ (c) $\begin{bmatrix} x & y \\ y & x \end{bmatrix}$

Discussion
Discovery
&

28. (a) If A is an $n \times n$ matrix and if $\mathbf{b}$ is an $n \times 1$ matrix, what conditions would you impose to ensure that the equation $\mathbf{x} = A\mathbf{x} + \mathbf{b}$ has a unique solution for $\mathbf{x}$?

(b) Assuming that your conditions are satisfied, find a formula for the solution in terms of an appropriate inverse.

29. Suppose that A is an invertible $n \times n$ matrix. Must the system of equations $A\mathbf{x} = \mathbf{x}$ have a unique solution? Explain your reasoning.

30. Is it possible to have $AB = I$ without B being the inverse of A? Explain your reasoning.

31. Create a theorem by rewriting Theorem 1.6.5 in contrapositive form (see Exercise 34 of Section 1.4).

1.7
DIAGONAL, TRIANGULAR, AND SYMMETRIC MATRICES

In this section we shall consider certain classes of matrices that have special forms. The matrices that we study in this section are among the most important kinds of matrices encountered in linear algebra and will arise in many different settings throughout the text.

Diagonal Matrices

A square matrix in which all the entries off the main diagonal are zero is called a ***diagonal matrix***. Here are some examples:

$$\begin{bmatrix} 2 & 0 \\ 0 & -5 \end{bmatrix}, \qquad \begin{bmatrix} 1 & 0 & 0 \\ 0 & 1 & 0 \\ 0 & 0 & 1 \end{bmatrix}, \qquad \begin{bmatrix} 6 & 0 & 0 & 0 \\ 0 & -4 & 0 & 0 \\ 0 & 0 & 0 & 0 \\ 0 & 0 & 0 & 8 \end{bmatrix}$$

A general $n \times n$ diagonal matrix D can be written as

$$D = \begin{bmatrix} d_1 & 0 & \cdots & 0 \\ 0 & d_2 & \cdots & 0 \\ \vdots & \vdots & & \vdots \\ 0 & 0 & \cdots & d_n \end{bmatrix} \tag{1}$$

A diagonal matrix is invertible if and only if all of its diagonal entries are nonzero; in this case the inverse of (1) is

$$D^{-1} = \begin{bmatrix} 1/d_1 & 0 & \cdots & 0 \\ 0 & 1/d_2 & \cdots & 0 \\ \vdots & \vdots & & \vdots \\ 0 & 0 & \cdots & 1/d_n \end{bmatrix}$$

The reader should verify that $DD^{-1} = D^{-1}D = I$.

Powers of diagonal matrices are easy to compute; we leave it for the reader to verify that if D is the diagonal matrix (1) and k is a positive integer, then

$$D^k = \begin{bmatrix} d_1^k & 0 & \cdots & 0 \\ 0 & d_2^k & \cdots & 0 \\ \vdots & \vdots & & \vdots \\ 0 & 0 & \cdots & d_n^k \end{bmatrix}$$

EXAMPLE 1 Inverses and Powers of Diagonal Matrices

If

$$A = \begin{bmatrix} 1 & 0 & 0 \\ 0 & -3 & 0 \\ 0 & 0 & 2 \end{bmatrix}$$

then

$$A^{-1} = \begin{bmatrix} 1 & 0 & 0 \\ 0 & -\frac{1}{3} & 0 \\ 0 & 0 & \frac{1}{2} \end{bmatrix}, \quad A^5 = \begin{bmatrix} 1 & 0 & 0 \\ 0 & -243 & 0 \\ 0 & 0 & 32 \end{bmatrix}, \quad A^{-5} = \begin{bmatrix} 1 & 0 & 0 \\ 0 & -\frac{1}{243} & 0 \\ 0 & 0 & \frac{1}{32} \end{bmatrix}$$

◆

Matrix products that involve diagonal factors are especially easy to compute. For example,

$$\begin{bmatrix} d_1 & 0 & 0 \\ 0 & d_2 & 0 \\ 0 & 0 & d_3 \end{bmatrix} \begin{bmatrix} a_{11} & a_{12} & a_{13} & a_{14} \\ a_{21} & a_{22} & a_{23} & a_{24} \\ a_{31} & a_{32} & a_{33} & a_{34} \end{bmatrix} = \begin{bmatrix} d_1a_{11} & d_1a_{12} & d_1a_{13} & d_1a_{14} \\ d_2a_{21} & d_2a_{22} & d_2a_{23} & d_2a_{24} \\ d_3a_{31} & d_3a_{32} & d_3a_{33} & d_3a_{34} \end{bmatrix}$$

$$\begin{bmatrix} a_{11} & a_{12} & a_{13} \\ a_{21} & a_{22} & a_{23} \\ a_{31} & a_{32} & a_{33} \\ a_{41} & a_{42} & a_{43} \end{bmatrix} \begin{bmatrix} d_1 & 0 & 0 \\ 0 & d_2 & 0 \\ 0 & 0 & d_3 \end{bmatrix} = \begin{bmatrix} d_1a_{11} & d_2a_{12} & d_3a_{13} \\ d_1a_{21} & d_2a_{22} & d_3a_{23} \\ d_1a_{31} & d_2a_{32} & d_3a_{33} \\ d_1a_{41} & d_2a_{42} & d_3a_{43} \end{bmatrix}$$

In words, *to multiply a matrix A on the left by a diagonal matrix D, one can multiply successive rows of A by the successive diagonal entries of D, and to multiply A on the right by D, one can multiply successive columns of A by the successive diagonal entries of D.*

Triangular Matrices

A square matrix in which all the entries above the main diagonal are zero is called ***lower triangular***, and a square matrix in which all the entries below the main diagonal are zero is called ***upper triangular***. A matrix that is either upper triangular or lower triangular is called ***triangular***.

EXAMPLE 2 Upper and Lower Triangular Matrices

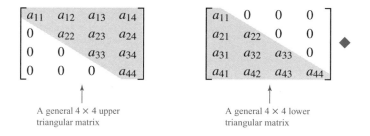

A general 4 × 4 upper
triangular matrix

A general 4 × 4 lower
triangular matrix

◆

REMARK Observe that diagonal matrices are both upper triangular and lower triangular since they have zeros below and above the main diagonal. Observe also that a *square* matrix in row-echelon form is upper triangular since it has zeros below the main diagonal.

The following are four useful characterizations of triangular matrices. The reader will find it instructive to verify that the matrices in Example 2 have the stated properties.

- A square matrix $A = [a_{ij}]$ is upper triangular if and only if the ith row starts with at least $i - 1$ zeros.

- A square matrix $A = [a_{ij}]$ is lower triangular if and only if the jth column starts with at least $j - 1$ zeros.

- A square matrix $A = [a_{ij}]$ is upper triangular if and only if $a_{ij} = 0$ for $i > j$.

- A square matrix $A = [a_{ij}]$ is lower triangular if and only if $a_{ij} = 0$ for $i < j$.

The following theorem lists some of the basic properties of triangular matrices.

THEOREM 1.7.1

(a) *The transpose of a lower triangular matrix is upper triangular, and the transpose of an upper triangular matrix is lower triangular.*

(b) *The product of lower triangular matrices is lower triangular, and the product of upper triangular matrices is upper triangular.*

(c) *A triangular matrix is invertible if and only if its diagonal entries are all nonzero.*

(d) *The inverse of an invertible lower triangular matrix is lower triangular, and the inverse of an invertible upper triangular matrix is upper triangular.*

Part (a) is evident from the fact that transposing a square matrix can be accomplished by reflecting the entries about the main diagonal; we omit the formal proof. We will prove (b), but we will defer the proofs of (c) and (d) to the next chapter, where we will have the tools to prove those results more efficiently.

Proof (b)   We will prove the result for lower triangular matrices; the proof for upper triangular matrices is similar. Let $A = [a_{ij}]$ and $B = [b_{ij}]$ be lower triangular $n \times n$ matrices, and let $C = [c_{ij}]$ be the product $C = AB$. From the remark preceding this theorem, we can prove that C is lower triangular by showing that $c_{ij} = 0$ for $i < j$. But from the definition of matrix multiplication,

$$c_{ij} = a_{i1}b_{1j} + a_{i2}b_{2j} + \cdots + a_{in}b_{nj}$$

If we assume that $i < j$, then the terms in this expression can be grouped as follows:

$$c_{ij} = \underbrace{a_{i1}b_{1j} + a_{i2}b_{2j} + \cdots + a_{i(j-1)}b_{(j-1)j}}_{\substack{\text{Terms in which the row} \\ \text{number of } b \text{ is less than the} \\ \text{column number of } b}} + \underbrace{a_{ij}b_{jj} + \cdots + a_{in}b_{nj}}_{\substack{\text{Terms in which the row} \\ \text{number of } a \text{ is less than} \\ \text{the column number of } a}}$$

In the first grouping all of the b factors are zero since B is lower triangular, and in the second grouping all of the a factors are zero since A is lower triangular. Thus, $c_{ij} = 0$, which is what we wanted to prove. ∎

EXAMPLE 3   Upper Triangular Matrices

Consider the upper triangular matrices

$$A = \begin{bmatrix} 1 & 3 & -1 \\ 0 & 2 & 4 \\ 0 & 0 & 5 \end{bmatrix}, \qquad B = \begin{bmatrix} 3 & -2 & 2 \\ 0 & 0 & -1 \\ 0 & 0 & 1 \end{bmatrix}$$

The matrix A is invertible, since its diagonal entries are nonzero, but the matrix B is not. We leave it for the reader to calculate the inverse of A by the method of Section 1.5 and show that

$$A^{-1} = \begin{bmatrix} 1 & -\frac{3}{2} & \frac{7}{5} \\ 0 & \frac{1}{2} & -\frac{2}{5} \\ 0 & 0 & \frac{1}{5} \end{bmatrix}$$

This inverse is upper triangular, as guaranteed by part (d) of Theorem 1.7.1. We also leave it for the reader to check that the product AB is

$$AB = \begin{bmatrix} 3 & -2 & -2 \\ 0 & 0 & 2 \\ 0 & 0 & 5 \end{bmatrix}$$

This product is upper triangular, as guaranteed by part (b) of Theorem 1.7.1. ◆

Symmetric Matrices

A square matrix A is called **symmetric** if $A = A^T$.

EXAMPLE 4 Symmetric Matrices

The following matrices are symmetric, since each is equal to its own transpose (verify).

$$\begin{bmatrix} 7 & -3 \\ -3 & 5 \end{bmatrix}, \quad \begin{bmatrix} 1 & 4 & 5 \\ 4 & -3 & 0 \\ 5 & 0 & 7 \end{bmatrix}, \quad \begin{bmatrix} d_1 & 0 & 0 & 0 \\ 0 & d_2 & 0 & 0 \\ 0 & 0 & d_3 & 0 \\ 0 & 0 & 0 & d_4 \end{bmatrix} ◆$$

It is easy to recognize symmetric matrices by inspection: The entries on the main diagonal may be arbitrary, but as shown in (2), "mirror images" of entries across the main diagonal must be equal.

$$\begin{bmatrix} 1 & 4 & 5 \\ 4 & -3 & 0 \\ 5 & 0 & 7 \end{bmatrix} \tag{2}$$

This follows from the fact that transposing a square matrix can be accomplished by interchanging entries that are symmetrically positioned about the main diagonal. Expressed in terms of the individual entries, a matrix $A = [a_{ij}]$ is symmetric if and only if $a_{ij} = a_{ji}$ for all values of i and j. As illustrated in Example 4, all diagonal matrices are symmetric.

The following theorem lists the main algebraic properties of symmetric matrices. The proofs are direct consequences of Theorem 1.4.9 and are left for the reader.

THEOREM 1.7.2

If A and B are symmetric matrices with the same size, and if k is any scalar, then:

(a) A^T *is symmetric.*

(b) $A + B$ *and* $A - B$ *are symmetric.*

(c) kA *is symmetric.*

REMARK   It is not true, in general, that the product of symmetric matrices is symmetric. To see why this is so, let A and B be symmetric matrices with the same size. Then from part (d) of Theorem 1.4.9 and the symmetry, we have

$$(AB)^T = B^T A^T = BA$$

Since AB and BA are not usually equal, it follows that AB will not usually be symmetric. However, in the special case where $AB = BA$, the product AB will be symmetric. If A and B are matrices such that $AB = BA$, then we say that A and B **commute**. In summary: *The product of two symmetric matrices is symmetric if and only if the matrices commute.*

EXAMPLE 5   Products of Symmetric Matrices

The first of the following equations shows a product of symmetric matrices that *is not* symmetric, and the second shows a product of symmetric matrices that *is* symmetric. We conclude that the factors in the first equation do not commute, but those in the second equation do. We leave it for the reader to verify that this is so.

$$\begin{bmatrix} 1 & 2 \\ 2 & 3 \end{bmatrix} \begin{bmatrix} -4 & 1 \\ 1 & 0 \end{bmatrix} = \begin{bmatrix} -2 & 1 \\ -5 & 2 \end{bmatrix}$$

$$\begin{bmatrix} 1 & 2 \\ 2 & 3 \end{bmatrix} \begin{bmatrix} -4 & 3 \\ 3 & -1 \end{bmatrix} = \begin{bmatrix} 2 & 1 \\ 1 & 3 \end{bmatrix} \quad \blacklozenge$$

In general, a symmetric matrix need not be invertible; for example, a square zero matrix is symmetric, but not invertible. However, if a symmetric matrix is invertible, then that inverse is also symmetric.

THEOREM 1.7.3

If A is an invertible symmetric matrix, then A^{-1} is symmetric.

Proof   Assume that A is symmetric and invertible. From Theorem 1.4.10 and the fact that $A = A^T$, we have

$$(A^{-1})^T = (A^T)^{-1} = A^{-1}$$

which proves that A^{-1} is symmetric. $\blacklozenge$

Products AA^T and $A^T A$

Matrix products of the form AA^T and $A^T A$ arise in a variety of applications. If A is an $m \times n$ matrix, then A^T is an $n \times m$ matrix, so the products AA^T and $A^T A$ are both square matrices—the matrix AA^T has size $m \times m$, and the matrix $A^T A$ has size $n \times n$. Such products are always symmetric since

$$(AA^T)^T = (A^T)^T A^T = AA^T \quad \text{and} \quad (A^T A)^T = A^T (A^T)^T = A^T A$$

EXAMPLE 6   The Product of a Matrix and Its Transpose Is Symmetric

Let A be the 2×3 matrix

$$A = \begin{bmatrix} 1 & -2 & 4 \\ 3 & 0 & -5 \end{bmatrix}$$

Then

$$A^T A = \begin{bmatrix} 1 & 3 \\ -2 & 0 \\ 4 & -5 \end{bmatrix} \begin{bmatrix} 1 & -2 & 4 \\ 3 & 0 & -5 \end{bmatrix} = \begin{bmatrix} 10 & -2 & -11 \\ -2 & 4 & -8 \\ -11 & -8 & 41 \end{bmatrix}$$

$$A A^T = \begin{bmatrix} 1 & -2 & 4 \\ 3 & 0 & -5 \end{bmatrix} \begin{bmatrix} 1 & 3 \\ -2 & 0 \\ 4 & -5 \end{bmatrix} = \begin{bmatrix} 21 & -17 \\ -17 & 34 \end{bmatrix}$$

Observe that $A^T A$ and $A A^T$ are symmetric as expected. ◆

Later in this text, we will obtain general conditions on A under which $A A^T$ and $A^T A$ are invertible. However, in the special case where A is *square*, we have the following result.

THEOREM 1.7.4

> *If A is an invertible matrix, then $A A^T$ and $A^T A$ are also invertible.*

Proof Since A is invertible, so is A^T by Theorem 1.4.10. Thus $A A^T$ and $A^T A$ are invertible, since they are the products of invertible matrices. ■

EXERCISE SET 1.7

1. Determine whether the matrix is invertible; if so, find the inverse by inspection.

(a) $\begin{bmatrix} 2 & 0 \\ 0 & -5 \end{bmatrix}$ (b) $\begin{bmatrix} 4 & 0 & 0 \\ 0 & 0 & 0 \\ 0 & 0 & 5 \end{bmatrix}$ (c) $\begin{bmatrix} -1 & 0 & 0 \\ 0 & 2 & 0 \\ 0 & 0 & \frac{1}{3} \end{bmatrix}$

2. Compute the product by inspection.

(a) $\begin{bmatrix} 3 & 0 & 0 \\ 0 & -1 & 0 \\ 0 & 0 & 2 \end{bmatrix} \begin{bmatrix} 2 & 1 \\ -4 & 1 \\ 2 & 5 \end{bmatrix}$ (b) $\begin{bmatrix} 2 & 0 & 0 \\ 0 & -1 & 0 \\ 0 & 0 & 4 \end{bmatrix} \begin{bmatrix} 4 & -1 & 3 \\ 1 & 2 & 0 \\ -5 & 1 & -2 \end{bmatrix} \begin{bmatrix} -3 & 0 & 0 \\ 0 & 5 & 0 \\ 0 & 0 & 2 \end{bmatrix}$

3. Find A^2, A^{-2}, and A^{-k} by inspection.

(a) $A = \begin{bmatrix} 1 & 0 \\ 0 & -2 \end{bmatrix}$ (b) $A = \begin{bmatrix} \frac{1}{2} & 0 & 0 \\ 0 & \frac{1}{3} & 0 \\ 0 & 0 & \frac{1}{4} \end{bmatrix}$

4. Which of the following matrices are symmetric?

(a) $\begin{bmatrix} 2 & -1 \\ 1 & 2 \end{bmatrix}$ (b) $\begin{bmatrix} 3 & 4 \\ 4 & 0 \end{bmatrix}$ (c) $\begin{bmatrix} 2 & -1 & 3 \\ -1 & 5 & 1 \\ 3 & 1 & 7 \end{bmatrix}$ (d) $\begin{bmatrix} 0 & 0 & 1 \\ 0 & 2 & 0 \\ 3 & 0 & 0 \end{bmatrix}$

5. By inspection, determine whether the given triangular matrix is invertible.

(a) $\begin{bmatrix} -1 & 2 & 4 \\ 0 & 3 & 0 \\ 0 & 0 & 5 \end{bmatrix}$ (b) $\begin{bmatrix} 0 & 1 & -2 & 5 \\ 0 & 1 & 5 & 6 \\ 0 & 0 & -3 & 1 \\ 0 & 0 & 0 & 5 \end{bmatrix}$

6. Find all values of a, b, and c for which A is symmetric.

$$A = \begin{bmatrix} 2 & a - 2b + 2c & 2a + b + c \\ 3 & 5 & a + c \\ 0 & -2 & 7 \end{bmatrix}$$

7. Find all values of a and b for which A and B are both not invertible.

$$A = \begin{bmatrix} a + b - 1 & 0 \\ 0 & 3 \end{bmatrix}, \qquad B = \begin{bmatrix} 5 & 0 \\ 0 & 2a - 3b - 7 \end{bmatrix}$$

8. Use the given equation to determine by inspection whether the matrices on the left commute.

(a) $\begin{bmatrix} 1 & -3 \\ -3 & 2 \end{bmatrix} \begin{bmatrix} 4 & 1 \\ 1 & 2 \end{bmatrix} = \begin{bmatrix} 1 & -5 \\ -10 & 1 \end{bmatrix}$ (b) $\begin{bmatrix} 2 & -1 \\ -1 & 3 \end{bmatrix} \begin{bmatrix} 3 & 2 \\ 2 & 1 \end{bmatrix} = \begin{bmatrix} 4 & 3 \\ 3 & 1 \end{bmatrix}$

9. Show that A and B commute if $a - d = 7b$.

$$A = \begin{bmatrix} 2 & 1 \\ 1 & -5 \end{bmatrix}, \qquad B = \begin{bmatrix} a & b \\ b & d \end{bmatrix}$$

10. Find a diagonal matrix A that satisfies

(a) $A^5 = \begin{bmatrix} 1 & 0 & 0 \\ 0 & -1 & 0 \\ 0 & 0 & -1 \end{bmatrix}$ (b) $A^{-2} = \begin{bmatrix} 9 & 0 & 0 \\ 0 & 4 & 0 \\ 0 & 0 & 1 \end{bmatrix}$

11. (a) Factor A into the form $A = BD$, where D is a diagonal matrix.

$$A = \begin{bmatrix} 3a_{11} & 5a_{12} & 7a_{13} \\ 3a_{21} & 5a_{22} & 7a_{23} \\ 3a_{31} & 5a_{32} & 7a_{33} \end{bmatrix}$$

(b) Is your factorization the only one possible? Explain.

12. Verify Theorem 1.7.1b for the product AB, where

$$A = \begin{bmatrix} -1 & 2 & 5 \\ 0 & 1 & 3 \\ 0 & 0 & -4 \end{bmatrix}, \qquad B = \begin{bmatrix} 2 & -8 & 0 \\ 0 & 2 & 1 \\ 0 & 0 & 3 \end{bmatrix}$$

13. Verify Theorem 1.7.1d for the matrices A and B in Exercise 12.

14. Verify Theorem 1.7.3 for the given matrix A.

(a) $A = \begin{bmatrix} 2 & -1 \\ -1 & 3 \end{bmatrix}$ (b) $A = \begin{bmatrix} 1 & -2 & 3 \\ -2 & 1 & -7 \\ 3 & -7 & 4 \end{bmatrix}$

15. Let A be an $n \times n$ symmetric matrix.

(a) Show that A^2 is symmetric.

(b) Show that $2A^2 - 3A + I$ is symmetric.

16. Let A be an $n \times n$ symmetric matrix.

(a) Show that A^k is symmetric if k is any nonnegative integer.

(b) If $p(x)$ is a polynomial, is $p(A)$ necessarily symmetric? Explain.

17. Let A be an $n \times n$ upper triangular matrix, and let $p(x)$ be a polynomial. Is $p(A)$ necessarily upper triangular? Explain.

18. Prove: If $A^T A = A$, then A is symmetric and $A = A^2$.

19. Find all 3×3 diagonal matrices A that satisfy $A^2 - 3A - 4I = 0$.

20. Let $A = [a_{ij}]$ be an $n \times n$ matrix. Determine whether A is symmetric.

    (a) $a_{ij} = i^2 + j^2$ (b) $a_{ij} = i^2 - j^2$

    (c) $a_{ij} = 2i + 2j$ (d) $a_{ij} = 2i^2 + 2j^3$

21. On the basis of your experience with Exercise 20, devise a general test that can be applied to a formula for a_{ij} to determine whether $A = [a_{ij}]$ is symmetric.

22. A square matrix A is called **skew-symmetric** if $A^T = -A$. Prove:

    (a) If A is an invertible skew-symmetric matrix, then A^{-1} is skew-symmetric.

    (b) If A and B are skew-symmetric, then so are A^T, $A + B$, $A - B$, and kA for any scalar k.

    (c) Every square matrix A can be expressed as the sum of a symmetric matrix and a skew-symmetric matrix.

    Hint Note the identity $A = \frac{1}{2}(A + A^T) + \frac{1}{2}(A - A^T)$.

23. We showed in the text that the product of symmetric matrices is symmetric if and only if the matrices commute. Is the product of commuting skew-symmetric matrices skew-symmetric? Explain.

    Note See Exercise 22 for terminology.

24. If the $n \times n$ matrix A can be expressed as $A = LU$, where L is a lower triangular matrix and U is an upper triangular matrix, then the linear system $A\mathbf{x} = \mathbf{b}$ can be expressed as $LU\mathbf{x} = \mathbf{b}$ and can be solved in two steps:

    Step 1. Let $U\mathbf{x} = \mathbf{y}$, so that $LU\mathbf{x} = \mathbf{b}$ can be expressed as $L\mathbf{y} = \mathbf{b}$. Solve this system.

    Step 2. Solve the system $U\mathbf{x} = \mathbf{y}$ for $\mathbf{x}$.

    In each part, use this two-step method to solve the given system.

    (a) $\begin{bmatrix} 1 & 0 & 0 \\ -2 & 3 & 0 \\ 2 & 4 & 1 \end{bmatrix} \begin{bmatrix} 2 & -1 & 3 \\ 0 & 1 & 2 \\ 0 & 0 & 4 \end{bmatrix} \begin{bmatrix} x_1 \\ x_2 \\ x_3 \end{bmatrix} = \begin{bmatrix} 1 \\ -2 \\ 0 \end{bmatrix}$

    (b) $\begin{bmatrix} 2 & 0 & 0 \\ 4 & 1 & 0 \\ -3 & -2 & 3 \end{bmatrix} \begin{bmatrix} 3 & -5 & 2 \\ 0 & 4 & 1 \\ 0 & 0 & 2 \end{bmatrix} \begin{bmatrix} x_1 \\ x_2 \\ x_3 \end{bmatrix} = \begin{bmatrix} 4 \\ -5 \\ 2 \end{bmatrix}$

25. Find an upper triangular matrix that satisfies

    $$A^3 = \begin{bmatrix} 1 & 30 \\ 0 & -8 \end{bmatrix}$$

Discussion & Discovery

26. What is the maximum number of distinct entries that an $n \times n$ symmetric matrix can have? Explain your reasoning.

27. Invent and prove a theorem that describes how to multiply two diagonal matrices.

28. Suppose that A is a square matrix and D is a diagonal matrix such that $AD = I$. What can you say about the matrix A? Explain your reasoning.

29. (a) Make up a consistent linear system of five equations in five unknowns that has a lower triangular coefficient matrix with no zeros on or below the main diagonal.

    (b) Devise an efficient procedure for solving your system by hand.

    (c) Invent an appropriate name for your procedure.

30. Indicate whether the statement is always true or sometimes false. Justify each answer.

    (a) If AA^T is singular, then so is A.

    (b) If $A + B$ is symmetric, then so are A and B.

    (c) If A is an $n \times n$ matrix and $A\mathbf{x} = \mathbf{0}$ has only the trivial solution, then so does $A^T\mathbf{x} = \mathbf{0}$.

    (d) If A^2 is symmetric, then so is A.

CHAPTER 1
Supplementary Exercises

1. Use Gauss–Jordan elimination to solve for x' and y' in terms of x and y.

$$x = \tfrac{3}{5}x' - \tfrac{4}{5}y'$$
$$y = \tfrac{4}{5}x' + \tfrac{3}{5}y'$$

2. Use Gauss–Jordan elimination to solve for x' and y' in terms of x and y.

$$x = x'\cos\theta - y'\sin\theta$$
$$y = x'\sin\theta + y'\cos\theta$$

3. Find a homogeneous linear system with two equations that are not multiples of one another and such that

$$x_1 = 1, \quad x_2 = -1, \quad x_3 = 1, \quad x_4 = 2$$

and

$$x_1 = 2, \quad x_2 = 0, \quad x_3 = 3, \quad x_4 = -1$$

are solutions of the system.

4. A box containing pennies, nickels, and dimes has 13 coins with a total value of 83 cents. How many coins of each type are in the box?

5. Find positive integers that satisfy

$$x + y + z = 9$$
$$x + 5y + 10z = 44$$

6. For which value(s) of a does the following system have zero solutions? One solution? Infinitely many solutions?

$$x_1 + x_2 + x_3 = 4$$
$$x_3 = 2$$
$$(a^2 - 4)x_3 = a - 2$$

7. Let

$$\begin{bmatrix} a & 0 & b & 2 \\ a & a & 4 & 4 \\ 0 & a & 2 & b \end{bmatrix}$$

be the augmented matrix for a linear system. Find for what values of a and b the system has

    (a) a unique solution.   (b) a one-parameter solution.

    (c) a two-parameter solution.   (d) no solution.

8. Solve for x, y, and z.

$$xy - 2\sqrt{y} + 3zy = 8$$
$$2xy - 3\sqrt{y} + 2zy = 7$$
$$-xy + \sqrt{y} + 2zy = 4$$

9. Find a matrix K such that $AKB = C$ given that

$$A = \begin{bmatrix} 1 & 4 \\ -2 & 3 \\ 1 & -2 \end{bmatrix}, \quad B = \begin{bmatrix} 2 & 0 & 0 \\ 0 & 1 & -1 \end{bmatrix}, \quad C = \begin{bmatrix} 8 & 6 & -6 \\ 6 & -1 & 1 \\ -4 & 0 & 0 \end{bmatrix}$$

10. How should the coefficients a, b, and c be chosen so that the system

$$ax + by - 3z = -3$$
$$-2x - by + cz = -1$$
$$ax + 3y - cz = -3$$

has the solution $x = 1$, $y = -1$, and $z = 2$?

11. In each part, solve the matrix equation for X.

(a) $X \begin{bmatrix} -1 & 0 & 1 \\ 1 & 1 & 0 \\ 3 & 1 & -1 \end{bmatrix} = \begin{bmatrix} 1 & 2 & 0 \\ -3 & 1 & 5 \end{bmatrix}$

(b) $X \begin{bmatrix} 1 & -1 & 2 \\ 3 & 0 & 1 \end{bmatrix} = \begin{bmatrix} -5 & -1 & 0 \\ 6 & -3 & 7 \end{bmatrix}$

(c) $\begin{bmatrix} 3 & 1 \\ -1 & 2 \end{bmatrix} X - X \begin{bmatrix} 1 & 4 \\ 2 & 0 \end{bmatrix} = \begin{bmatrix} 2 & -2 \\ 5 & 4 \end{bmatrix}$

12. (a) Express the equations

$$y_1 = x_1 - x_2 + x_3$$
$$y_2 = 3x_1 + x_2 - 4x_3 \qquad \text{and} \qquad \begin{aligned} z_1 &= 4y_1 - y_2 + y_3 \\ z_2 &= -3y_1 + 5y_2 - y_3 \end{aligned}$$
$$y_3 = -2x_1 - 2x_2 + 3x_3$$

in the matrix forms $Y = AX$ and $Z = BY$. Then use these to obtain a direct relationship $Z = CX$ between Z and X.

(b) Use the equation $Z = CX$ obtained in (a) to express z_1 and z_2 in terms of x_1, x_2, and x_3.

(c) Check the result in (b) by directly substituting the equations for y_1, y_2, and y_3 into the equations for z_1 and z_2 and then simplifying.

13. If A is $m \times n$ and B is $n \times p$, how many multiplication operations and how many addition operations are needed to calculate the matrix product AB?

14. Let A be a square matrix.

(a) Show that $(I - A)^{-1} = I + A + A^2 + A^3$ if $A^4 = 0$.

(b) Show that $(I - A)^{-1} = I + A + A^2 + \cdots + A^n$ if $A^{n+1} = 0$.

15. Find values of a, b, and c such that the graph of the polynomial $p(x) = ax^2 + bx + c$ passes through the points $(1, 2)$, $(-1, 6)$, and $(2, 3)$.

16. **(For Readers Who Have Studied Calculus)** Find values of a, b, and c such that the graph of the polynomial $p(x) = ax^2 + bx + c$ passes through the point $(-1, 0)$ and has a horizontal tangent at $(2, -9)$.

17. Let J_n be the $n \times n$ matrix each of whose entries is 1. Show that if $n > 1$, then

$$(I - J_n)^{-1} = I - \frac{1}{n-1} J_n$$

18. Show that if a square matrix A satisfies $A^3 + 4A^2 - 2A + 7I = 0$, then so does A^T.

19. Prove: If B is invertible, then $AB^{-1} = B^{-1}A$ if and only if $AB = BA$.

20. Prove: If A is invertible, then $A + B$ and $I + BA^{-1}$ are both invertible or both not invertible.

21. Prove that if A and B are $n \times n$ matrices, then

(a) $\text{tr}(A + B) = \text{tr}(A) + \text{tr}(B)$ (b) $\text{tr}(kA) = k \, \text{tr}(A)$

(c) $\text{tr}(A^T) = \text{tr}(A)$ (d) $\text{tr}(AB) = \text{tr}(BA)$

22. Use Exercise 21 to show that there are no square matrices A and B such that

$$AB - BA = I$$

23. Prove: If A is an $m \times n$ matrix and B is the $n \times 1$ matrix each of whose entries is $1/n$, then

$$AB = \begin{bmatrix} \bar{r}_1 \\ \bar{r}_2 \\ \vdots \\ \bar{r}_m \end{bmatrix}$$

where $\bar{r}_i$ is the average of the entries in the ith row of A.

24. **(For Readers Who Have Studied Calculus)** If the entries of the matrix

$$C = \begin{bmatrix} c_{11}(x) & c_{12}(x) & \cdots & c_{1n}(x) \\ c_{21}(x) & c_{22}(x) & \cdots & c_{2n}(x) \\ \vdots & \vdots & & \vdots \\ c_{m1}(x) & c_{m2}(x) & \cdots & c_{mn}(x) \end{bmatrix}$$

are differentiable functions of x, then we define

$$\frac{dC}{dx} = \begin{bmatrix} c'_{11}(x) & c'_{12}(x) & \cdots & c'_{1n}(x) \\ c'_{21}(x) & c'_{22}(x) & \cdots & c'_{2n}(x) \\ \vdots & \vdots & & \vdots \\ c'_{m1}(x) & c'_{m2}(x) & \cdots & c'_{mn}(x) \end{bmatrix}$$

Show that if the entries in A and B are differentiable functions of x and the sizes of the matrices are such that the stated operations can be performed, then

(a) $\dfrac{d}{dx}(kA) = k\dfrac{dA}{dx}$ 　　　　　(b) $\dfrac{d}{dx}(A + B) = \dfrac{dA}{dx} + \dfrac{dB}{dx}$

(c) $\dfrac{d}{dx}(AB) = \dfrac{dA}{dx}B + A\dfrac{dB}{dx}$

25. **(For Readers Who Have Studied Calculus)** Use part (c) of Exercise 24 to show that

$$\frac{dA^{-1}}{dx} = -A^{-1}\frac{dA}{dx}A^{-1}$$

State all the assumptions you make in obtaining this formula.

26. Find the values of a, b, and c that will make the equation

$$\frac{x^2 + x - 2}{(3x - 1)(x^2 + 1)} = \frac{a}{3x - 1} + \frac{bx + c}{x^2 + 1}$$

an identity.

Hint Multiply through by $(3x - 1)(x^2 + 1)$ and equate the corresponding coefficients of the polynomials on each side of the resulting equation.

27. If P is an $n \times 1$ matrix such that $P^T P = 1$, then $H = I - 2PP^T$ is called the corresponding *Householder matrix* (named after the American mathematician A. S. Householder).

(a) Verify that $P^T P = 1$ if $P^T = \begin{bmatrix} \frac{3}{4} & \frac{1}{6} & \frac{1}{4} & \frac{5}{12} & \frac{5}{12} \end{bmatrix}$ and compute the corresponding Householder matrix.

(b) Prove that if H is any Householder matrix, then $H = H^T$ and $H^T H = I$.

(c) Verify that the Householder matrix found in part (a) satisfies the conditions proved in part (b).

28. Assuming that the stated inverses exist, prove the following equalities.

(a) $(C^{-1} + D^{-1})^{-1} = C(C + D)^{-1}D$ 　　　　　(b) $(I + CD)^{-1}C = C(I + DC)^{-1}$

(c) $(C + DD^T)^{-1}D = C^{-1}D(I + D^T C^{-1}D)^{-1}$

29. (a) Show that if $a \neq b$, then

$$a^n + a^{n-1}b + a^{n-2}b^2 + \cdots + ab^{n-1} + b^n = \frac{a^{n+1} - b^{n+1}}{a - b}$$

(b) Use the result in part (a) to find A^n if

$$A = \begin{bmatrix} a & 0 & 0 \\ 0 & b & 0 \\ 1 & 0 & c \end{bmatrix}$$

Note This exercise is based on a problem by John M. Johnson, *The Mathematics Teacher*, Vol. 85, No. 9, 1992.

CHAPTER 1

Technology Exercises

The following exercises are designed to be solved using a technology utility. Typically, this will be MATLAB, *Mathematica*, Maple, Derive, or Mathcad, but it may also be some other type of linear algebra software or a scientific calculator with some linear algebra capabilities. For each exercise you will need to read the relevant documentation for the particular utility you are using. The goal of these exercises is to provide you with a basic proficiency with your technology utility. Once you have mastered the techniques in these exercises, you will be able to use your technology utility to solve many of the problems in the regular exercise sets.

Section 1.1   **T1.**   **Numbers and Numerical Operations** Read your documentation on entering and displaying numbers and performing the basic arithmetic operations of addition, subtraction, multiplication, division, raising numbers to powers, and extraction of roots. Determine how to control the number of digits in the screen display of a decimal number. If you are using a CAS, in which case you can compute with exact numbers rather than decimal approximations, then learn how to enter such numbers as π, $\sqrt{2}$, and $\frac{1}{3}$ exactly and convert them to decimal form. Experiment with numbers of your own choosing until you feel you have mastered the procedures and operations.

Section 1.2   **T1.**   **Matrices and Reduced Row-Echelon Form** Read your documentation on how to enter matrices and how to find the reduced row-echelon form of a matrix. Then use your utility to find the reduced row-echelon form of the augmented matrix in Example 4 of Section 1.2.

T2.   **Linear Systems With a Unique Solution** Read your documentation on how to solve a linear system, and then use your utility to solve the linear system in Example 3 of Section 1.1. Also, solve the system by reducing the augmented matrix to reduced row-echelon form.

T3.   **Linear Systems With Infinitely Many Solutions** Technology utilities vary on how they handle linear systems with infinitely many solutions. See how your utility handles the system in Example 4 of Section 1.2.

T4.   **Inconsistent Linear Systems** Technology utilities will often successfully identify inconsistent linear systems, but they can sometimes be fooled into reporting an inconsistent system as consistent, or vice versa. This typically happens when some of the numbers that occur in the computations are so small that roundoff error makes it difficult for the utility to determine whether or not they are equal to zero. Create some inconsistent linear systems and see how your utility handles them.

T5.   A polynomial whose graph passes through a given set of points is called an *__interpolating polynomial__* for those points. Some technology utilities have specific commands for finding interpolating polynomials. If your utility has this capability, read the documentation and then use this feature to solve Exercise 25 of Section 1.2.

Section 1.3   **T1.**   **Matrix Operations** Read your documentation on how to perform the basic operations on matrices—addition, subtraction, multiplication by scalars, and multiplication of matrices. Then perform the computations in Examples 3, 4, and 5. See what happens when you try to perform an operation on matrices with inconsistent sizes.

T2. Evaluate the expression $A^5 - 3A^3 + 7A - 4I$ for the matrix

$$A = \begin{bmatrix} 1 & -2 & 3 \\ -4 & 5 & -6 \\ 7 & -8 & 9 \end{bmatrix}$$

T3. **Extracting Rows and Columns** Read your documentation on how to extract rows and columns from a matrix, and then use your utility to extract various rows and columns from a matrix of your choice.

T4. **Transpose and Trace** Read your documentation on how to find the transpose and trace of a matrix, and then use your utility to find the transpose of the matrix A in Formula (12) and the trace of the matrix B in Example 12.

T5. **Constructing an Augmented Matrix** Read your documentation on how to create an augmented matrix $[A \mid \mathbf{b}]$ from matrices A and $\mathbf{b}$ that have previously been entered. Then use your utility to form the augmented matrix for the system $A\mathbf{x} = \mathbf{b}$ in Example 4 of Section 1.1 from the matrices A and $\mathbf{b}$.

Section 1.4 **T1.** **Zero and Identity Matrices** Typing in entries of a matrix can be tedious, so many technology utilities provide shortcuts for entering zero and identity matrices. Read your documentation on how to do this, and then enter some zero and identity matrices of various sizes.

T2. **Inverse** Read your documentation on how to find the inverse of a matrix, and then use your utility to perform the computations in Example 7.

T3. **Formula for the Inverse** If you are working with a CAS, use it to confirm Theorem 1.4.5.

T4. **Powers of a Matrix** Read your documentation on how to find powers of a matrix, and then use your utility to find various positive and negative powers of the matrix A in Example 8.

T5. Let

$$A = \begin{bmatrix} 1 & \frac{1}{2} & \frac{1}{3} \\ \frac{1}{4} & 1 & \frac{1}{5} \\ \frac{1}{6} & \frac{1}{7} & 1 \end{bmatrix}$$

Describe what happens to the matrix A^k when k is allowed to increase indefinitely (that is, as $k \to \infty$).

T6. By experimenting with different values of n, find an expression for the inverse of an $n \times n$ matrix of the form

$$A = \begin{bmatrix} 1 & 2 & 3 & 4 & \cdots & n-1 & n \\ 0 & 1 & 2 & 3 & \cdots & n-2 & n-1 \\ 0 & 0 & 1 & 2 & \cdots & n-3 & n-2 \\ \vdots & \vdots & \vdots & \vdots & & \vdots & \vdots \\ 0 & 0 & 0 & 0 & \cdots & 1 & 2 \\ 0 & 0 & 0 & 0 & \cdots & 0 & 1 \end{bmatrix}$$

Section 1.5 **T1.** Use your technology utility to verify Theorem 1.5.1 in several specific cases.

T2. **Singular Matrices** Find the inverse of the matrix in Example 4, and then see what your utility does when you try to invert the matrix in Example 5.

Section 1.6 **T1.** **Solving $A\mathbf{x} = \mathbf{b}$ by Inversion** Use the method of Example 4 to solve the system in Example 3 of Section 1.1.

T2. Compare the solution of $A\mathbf{x} = \mathbf{b}$ by Gaussian elimination and by inversion for several large matrices. Can you see the superiority of the former approach?

T3. Solve the linear system $A\mathbf{x} = 2\mathbf{x}$, given that

$$A = \begin{bmatrix} 0 & 0 & -2 \\ 1 & 2 & 1 \\ 1 & 0 & 3 \end{bmatrix}$$

Section 1.7

T1. Diagonal, Symmetric, and Triangular Matrices Many technology utilities provide shortcuts for entering diagonal, symmetric, and triangular matrices. Read your documentation on how to do this, and then experiment with entering various matrices of these types.

T2. Properties of Triangular Matrices Confirm the results in Theorem 1.7.1 using some triangular matrices of your choice.

T3. Confirm the results in Theorem 1.7.4. What happens if A is not square?

Determinants

CHAPTER CONTENTS

INTRODUCTION: We are all familiar with functions such as $f(x) = \sin x$ and $f(x) = x^2$, which associate a real number $f(x)$ with a real value of the variable x. Since both x and $f(x)$ assume only real values, such functions are described as real-valued functions of a real variable. In this section we shall study the "determinant function," which is a real-valued function of a matrix variable in the sense that it associates a real number $f(X)$ with a square matrix X. Our work on determinant functions will have important applications to the theory of systems of linear equations and will also lead us to an explicit formula for the inverse of an invertible matrix.

2.1
DETERMINANTS BY COFACTOR EXPANSION

As noted in the introduction to this chapter, a "determinant" is a certain kind of function that associates a real number with a square matrix. In this section we will define this function. As a consequence of our work here, we will obtain a formula for the inverse of an invertible matrix as well as a formula for the solution to certain systems of linear equations in terms of determinants.

Recall from Theorem 1.4.5 that the 2×2 matrix

$$A = \begin{bmatrix} a & b \\ c & d \end{bmatrix}$$

is invertible if $ad - bc \neq 0$. The expression $ad - bc$ occurs so frequently in mathematics that it has a name; it is called the **determinant** of the matrix A and is denoted by the symbol $\det(A)$ or $|A|$. With this notation, the formula for A^{-1} given in Theorem 1.4.5 is

$$A^{-1} = \frac{1}{\det(A)} \begin{bmatrix} d & -b \\ -c & a \end{bmatrix}$$

One of the goals of this chapter is to obtain analogs of this formula to square matrices of higher order. This will require that we extend the concept of a determinant to square matrices of all orders.

Minors and Cofactors

There are several ways in which we might proceed. The approach in this section is a recursive approach: It defines the determinant of an $n \times n$ matrix in terms of the determinants of certain $(n-1) \times (n-1)$ matrices. The $(n-1) \times (n-1)$ matrices that will appear in this definition are submatrices of the original matrix. These submatrices are given a special name:

> **DEFINITION**
>
> If A is a square matrix, then the **minor of entry a_{ij}** is denoted by M_{ij} and is defined to be the determinant of the submatrix that remains after the ith row and jth column are deleted from A. The number $(-1)^{i+j} M_{ij}$ is denoted by C_{ij} and is called the **cofactor of entry a_{ij}**.

EXAMPLE 1 Finding Minors and Cofactors

Let

$$A = \begin{bmatrix} 3 & 1 & -4 \\ 2 & 5 & 6 \\ 1 & 4 & 8 \end{bmatrix}$$

The minor of entry a_{11} is

$$M_{11} = \begin{vmatrix} 3 & 1 & -4 \\ 2 & 5 & 6 \\ 1 & 4 & 8 \end{vmatrix} = \begin{vmatrix} 5 & 6 \\ 4 & 8 \end{vmatrix} = 16$$

The cofactor of a_{11} is

$$C_{11} = (-1)^{1+1} M_{11} = M_{11} = 16$$

Similarly, the minor of entry a_{32} is

$$M_{32} = \begin{vmatrix} 3 & 1 & -4 \\ 2 & 5 & 6 \\ 1 & 4 & 8 \end{vmatrix} = \begin{vmatrix} 3 & -4 \\ 2 & 6 \end{vmatrix} = 26$$

The cofactor of a_{32} is

$$C_{32} - (-1)^{3+2}M_{32} = -M_{32} = -26 \quad \blacklozenge$$

Note that the cofactor and the minor of an element a_{ij} differ only in sign; that is, $C_{ij} = \pm M_{ij}$. A quick way to determine whether to use $+$ or $-$ is to use the fact that the sign relating C_{ij} and M_{ij} is in the ith row and jth column of the "checkerboard" array

$$\begin{bmatrix} + & - & + & - & + & \cdots \\ - & + & - & + & - & \cdots \\ + & - & + & - & + & \cdots \\ - & + & - & + & - & \cdots \\ \vdots & \vdots & \vdots & \vdots & \vdots & \end{bmatrix}$$

For example, $C_{11} = M_{11}$, $C_{21} = -M_{21}$, $C_{12} = -M_{12}$, $C_{22} = M_{22}$, and so on.

Strictly speaking, the determinant of a matrix is a number. However, it is common practice to "abuse" the terminology slightly and use the term *determinant* to refer to the matrix whose determinant is being computed. Thus we might refer to

$$\begin{vmatrix} 3 & 1 \\ 4 & -2 \end{vmatrix}$$

as a 2×2 determinant and call 3 the entry in the first row and first column of the determinant.

Cofactor Expansions　The definition of a 3×3 determinant in terms of minors and cofactors is

$$\det(A) = a_{11}M_{11} + a_{12}(-M_{12}) + a_{13}M_{13}$$
$$= a_{11}C_{11} + a_{12}C_{12} + a_{13}C_{13} \tag{1}$$

Equation (1) shows that the determinant of A can be computed by multiplying the entries in the first row of A by their corresponding cofactors and adding the resulting products. More generally, we define the determinant of an $n \times n$ matrix to be

$$\det(A) = a_{11}C_{11} + a_{12}C_{12} + \cdots + a_{1n}C_{1n}$$

This method of evaluating $\det(A)$ is called ***cofactor expansion*** along the first row of A.

EXAMPLE 2　Cofactor Expansion Along the First Row

Let $A = \begin{bmatrix} 3 & 1 & 0 \\ -2 & -4 & 3 \\ 5 & 4 & -2 \end{bmatrix}$. Evaluate $\det(A)$ by cofactor expansion along the first row of A.

Solution

From (1),

$$\det(A) = \begin{vmatrix} 3 & 1 & 0 \\ -2 & -4 & 3 \\ 5 & 4 & -2 \end{vmatrix} = 3 \begin{vmatrix} -4 & 3 \\ 4 & -2 \end{vmatrix} - 1 \begin{vmatrix} -2 & 3 \\ 5 & -2 \end{vmatrix} + 0 \begin{vmatrix} -2 & -4 \\ 5 & 4 \end{vmatrix}$$

$$= 3(-4) - (1)(-11) + 0 = -1 \quad \blacklozenge$$

If A is a 3×3 matrix, then its determinant is

$$\det(A) = \begin{vmatrix} a_{11} & a_{12} & a_{13} \\ a_{21} & a_{22} & a_{23} \\ a_{31} & a_{32} & a_{33} \end{vmatrix}$$

$$= a_{11} \begin{vmatrix} a_{22} & a_{23} \\ a_{32} & a_{33} \end{vmatrix} - a_{12} \begin{vmatrix} a_{21} & a_{23} \\ a_{31} & a_{33} \end{vmatrix} + a_{13} \begin{vmatrix} a_{21} & a_{22} \\ a_{31} & a_{32} \end{vmatrix}$$

$$= a_{11}(a_{22}a_{33} - a_{23}a_{32}) - a_{12}(a_{21}a_{33} - a_{23}a_{31}) + a_{13}(a_{21}a_{32} - a_{22}a_{31}) \qquad (2)$$

$$= a_{11}a_{22}a_{33} + a_{12}a_{23}a_{31} + a_{13}a_{21}a_{32} - a_{13}a_{22}a_{31} - a_{12}a_{21}a_{33} - a_{11}a_{23}a_{32} \qquad (3)$$

By rearranging the terms in (3) in various ways, it is possible to obtain other formulas like (2). There should be no trouble checking that all of the following are correct (see Exercise 28):

$$\det(A) = a_{11}C_{11} + a_{12}C_{12} + a_{13}C_{13}$$
$$= a_{11}C_{11} + a_{21}C_{21} + a_{31}C_{31}$$
$$= a_{21}C_{21} + a_{22}C_{22} + a_{23}C_{23}$$
$$= a_{12}C_{12} + a_{22}C_{22} + a_{32}C_{32}$$
$$= a_{31}C_{31} + a_{32}C_{32} + a_{33}C_{33}$$
$$= a_{13}C_{13} + a_{23}C_{23} + a_{33}C_{33} \qquad (4)$$

Note that in each equation, the entries and cofactors all come from the same row or column. These equations are called the *cofactor expansions* of $\det(A)$.

The results we have just given for 3×3 matrices form a special case of the following general theorem, which we state without proof.

THEOREM 2.1.1

> ## Expansions by Cofactors
>
> *The determinant of an $n \times n$ matrix A can be computed by multiplying the entries in any row (or column) by their cofactors and adding the resulting products; that is, for each $1 \leq i \leq n$ and $1 \leq j \leq n$,*
>
> $$\det(A) = a_{1j}C_{1j} + a_{2j}C_{2j} + \cdots + a_{nj}C_{nj}$$
> *(cofactor expansion along the jth column)*
>
> *and*
>
> $$\det(A) = a_{i1}C_{i1} + a_{i2}C_{i2} + \cdots + a_{in}C_{in}$$
> *(cofactor expansion along the ith row)*

Note that we may choose *any* row or *any* column.

EXAMPLE 3 Cofactor Expansion Along the First Column

Let A be the matrix in Example 2. Evaluate $\det(A)$ by cofactor expansion along the first column of A.

Solution

From (4)

$$\det(A) = \begin{vmatrix} 3 & 1 & 0 \\ -2 & -4 & 3 \\ 5 & 4 & -2 \end{vmatrix} = 3 \begin{vmatrix} -4 & 3 \\ 4 & -2 \end{vmatrix} - (-2) \begin{vmatrix} 1 & 0 \\ 4 & -2 \end{vmatrix} + 5 \begin{vmatrix} 1 & 0 \\ -4 & 3 \end{vmatrix}$$

$$= 3(-4) - (-2)(-2) + 5(3) = -1$$

This agrees with the result obtained in Example 2. ◆

REMARK In this example we had to compute three cofactors, but in Example 2 we only had to compute two of them, since the third was multiplied by zero. In general, the best strategy for evaluating a determinant by cofactor expansion is to expand along a row or column having the largest number of zeros.

EXAMPLE 4 Smart Choice of Row or Column

If A is the 4×4 matrix

$$A = \begin{bmatrix} 1 & 0 & 0 & -1 \\ 3 & 1 & 2 & 2 \\ 1 & 0 & -2 & 1 \\ 2 & 0 & 0 & 1 \end{bmatrix}$$

then to find $\det(A)$ it will be easiest to use cofactor expansion along the second column, since it has the most zeros:

$$\det(A) = 1 \cdot \begin{vmatrix} 1 & 0 & -1 \\ 1 & -2 & 1 \\ 2 & 0 & 1 \end{vmatrix}$$

For the 3×3 determinant, it will be easiest to use cofactor expansion along its second column, since it has the most zeros:

$$\det(A) = 1 \cdot -2 \cdot \begin{vmatrix} 1 & -1 \\ 2 & 1 \end{vmatrix}$$
$$= -2(1 + 2)$$
$$= -6$$

We would have found the same answer if we had used any other row or column. ◆

Adjoint of a Matrix

In a cofactor expansion we compute $\det(A)$ by multiplying the entries in a row or column by their cofactors and adding the resulting products. It turns out that if one multiplies the entries in any row by the corresponding cofactors from a *different* row, the sum of these products is always zero. (This result also holds for columns.) Although we omit the general proof, the next example illustrates the idea of the proof in a special case.

EXAMPLE 5 Entries and Cofactors from Different Rows

Let

$$A = \begin{bmatrix} a_{11} & a_{12} & a_{13} \\ a_{21} & a_{22} & a_{23} \\ a_{31} & a_{32} & a_{33} \end{bmatrix}$$

Consider the quantity

$$a_{11}C_{31} + a_{12}C_{32} + a_{13}C_{33}$$

that is formed by multiplying the entries in the first row by the cofactors of the corresponding entries in the third row and adding the resulting products. We now show that this quantity is equal to zero by the following trick. Construct a new matrix A' by replacing the third row of A with another copy of the first row. Thus

$$A' = \begin{bmatrix} a_{11} & a_{12} & a_{13} \\ a_{21} & a_{22} & a_{23} \\ a_{11} & a_{12} & a_{13} \end{bmatrix}$$

Let $C'_{31}, C'_{32}, C'_{33}$ be the cofactors of the entries in the third row of A'. Since the first two rows of A and A' are the same, and since the computations of $C_{31}, C_{32}, C_{33}, C'_{31}, C'_{32},$ and C'_{33} involve only entries from the first two rows of A and A', it follows that

$$C_{31} = C'_{31}, \qquad C_{32} = C'_{32}, \qquad C_{33} = C'_{33}$$

Since A' has two identical rows, it follows from (3) that

$$\det(A') = 0 \tag{5}$$

On the other hand, evaluating $\det(A')$ by cofactor expansion along the third row gives

$$\det(A') = a_{11}C'_{31} + a_{12}C'_{32} + a_{13}C'_{33} = a_{11}C_{31} + a_{12}C_{32} + a_{13}C_{33} \tag{6}$$

From (5) and (6) we obtain

$$a_{11}C_{31} + a_{12}C_{32} + a_{13}C_{33} = 0 \quad \blacklozenge$$

Now we'll use this fact to get a formula for A^{-1}.

DEFINITION

If A is any $n \times n$ matrix and C_{ij} is the cofactor of a_{ij}, then the matrix

$$\begin{bmatrix} C_{11} & C_{12} & \cdots & C_{1n} \\ C_{21} & C_{22} & \cdots & C_{2n} \\ \vdots & \vdots & & \vdots \\ C_{n1} & C_{n2} & \cdots & C_{nn} \end{bmatrix}$$

is called the ***matrix of cofactors from*** A. The transpose of this matrix is called the ***adjoint of*** A and is denoted by $\mathrm{adj}(A)$.

EXAMPLE 6 Adjoint of a 3 × 3 Matrix

Let

$$A = \begin{bmatrix} 3 & 2 & -1 \\ 1 & 6 & 3 \\ 2 & -4 & 0 \end{bmatrix}$$

The cofactors of A are

$$
\begin{array}{lll}
C_{11} = 12 & C_{12} = 6 & C_{13} = -16 \\
C_{21} = 4 & C_{22} = 2 & C_{23} = 16 \\
C_{31} = 12 & C_{32} = -10 & C_{33} = 16
\end{array}
$$

so the matrix of cofactors is

$$\begin{bmatrix} 12 & 6 & -16 \\ 4 & 2 & 16 \\ 12 & -10 & 16 \end{bmatrix}$$

and the adjoint of A is

$$\operatorname{adj}(A) = \begin{bmatrix} 12 & 4 & 12 \\ 6 & 2 & -10 \\ -16 & 16 & 16 \end{bmatrix} \quad \blacklozenge$$

We are now in a position to derive a formula for the inverse of an invertible matrix. We need to use an important fact that will be proved in Section 2.3: The square matrix A is invertible if and only if $\det(A)$ is not zero.

THEOREM 2.1.2

Inverse of a Matrix Using Its Adjoint

If A is an invertible matrix, then

$$A^{-1} = \frac{1}{\det(A)} \operatorname{adj}(A) \tag{7}$$

Proof We show first that

$$A \operatorname{adj}(A) = \det(A) I$$

Consider the product

$$A \operatorname{adj}(A) = \begin{bmatrix} a_{11} & a_{12} & \cdots & a_{1n} \\ a_{21} & a_{22} & \cdots & a_{2n} \\ \vdots & \vdots & & \vdots \\ a_{i1} & a_{i2} & \cdots & a_{in} \\ \vdots & \vdots & & \vdots \\ a_{n1} & a_{n2} & \cdots & a_{nn} \end{bmatrix} \begin{bmatrix} C_{11} & C_{21} & \cdots & C_{j1} & \cdots & C_{n1} \\ C_{12} & C_{22} & \cdots & C_{j2} & \cdots & C_{n2} \\ \vdots & \vdots & & \vdots & & \vdots \\ C_{1n} & C_{2n} & \cdots & C_{jn} & \cdots & C_{nn} \end{bmatrix}$$

The entry in the ith row and jth column of the product $A \operatorname{adj}(A)$ is

$$a_{i1}C_{j1} + a_{i2}C_{j2} + \cdots + a_{in}C_{jn} \tag{8}$$

(see the shaded lines above).

If $i = j$, then (8) is the cofactor expansion of $\det(A)$ along the ith row of A (Theorem 2.1.1), and if $i \neq j$, then the a's and the cofactors come from different rows of A, so the value of (8) is zero. Therefore,

$$A \operatorname{adj}(A) = \begin{bmatrix} \det(A) & 0 & \cdots & 0 \\ 0 & \det(A) & \cdots & 0 \\ \vdots & \vdots & & \vdots \\ 0 & 0 & \cdots & \det(A) \end{bmatrix} = \det(A)I \qquad (9)$$

Since A is invertible, $\det(A) \neq 0$. Therefore, Equation (9) can be rewritten as

$$\frac{1}{\det(A)}[A \operatorname{adj}(A)] = I \quad \text{or} \quad A\left[\frac{1}{\det(A)}\operatorname{adj}(A)\right] = I$$

Multiplying both sides on the left by A^{-1} yields

$$A^{-1} = \frac{1}{\det(A)}\operatorname{adj}(A) \qquad \blacksquare$$

EXAMPLE 7   Using the Adjoint to Find an Inverse Matrix

Use (7) to find the inverse of the matrix A in Example 6.

Solution

The reader can check that $\det(A) = 64$. Thus

$$A^{-1} = \frac{1}{\det(A)}\operatorname{adj}(A) = \frac{1}{64}\begin{bmatrix} 12 & 4 & 12 \\ 6 & 2 & -10 \\ -16 & 16 & 16 \end{bmatrix} = \begin{bmatrix} \frac{12}{64} & \frac{4}{64} & \frac{12}{64} \\ \frac{6}{64} & \frac{2}{64} & -\frac{10}{64} \\ -\frac{16}{64} & \frac{16}{64} & \frac{16}{64} \end{bmatrix} \blacklozenge$$

Applications of Formula (7)

Although the method in the preceding example is reasonable for inverting 3×3 matrices by hand, the inversion algorithm discussed in Section 1.5 is more efficient for larger matrices. It should be kept in mind, however, that the method of Section 1.5 is just a computational procedure, whereas Formula (7) is an actual formula for the inverse. As we shall now see, this formula is useful for deriving properties of the inverse.

In Section 1.7 we stated two results about inverses without proof.

- **Theorem 1.7.1c:** A triangular matrix is invertible if and only if its diagonal entries are all nonzero.

- **Theorem 1.7.1d:** The inverse of an invertible lower triangular matrix is lower triangular, and the inverse of an invertible upper triangular matrix is upper triangular.

We will now prove these results using the adjoint formula for the inverse. We need a preliminary result.

THEOREM 2.1.3

> *If A is an $n \times n$ triangular matrix (upper triangular, lower triangular, or diagonal), then $\det(A)$ is the product of the entries on the main diagonal of the matrix; that is,* $\det(A) = a_{11}a_{22}\cdots a_{nn}$.

For simplicity of notation, we will prove the result for a 4 × 4 lower triangular matrix

$$A = \begin{bmatrix} a_{11} & 0 & 0 & 0 \\ a_{21} & a_{22} & 0 & 0 \\ a_{31} & a_{32} & a_{33} & 0 \\ a_{41} & a_{42} & a_{43} & a_{44} \end{bmatrix}$$

The argument in the $n \times n$ case is similar, as is the case of upper triangular matrices.

Proof of Theorem 2.1.3 (4 × 4 lower triangular case) By Theorem 2.1.1, the determinant of A may be found by cofactor expansion along the first row:

$$\det(A) = \begin{vmatrix} a_{11} & 0 & 0 & 0 \\ a_{21} & a_{22} & 0 & 0 \\ a_{31} & a_{32} & a_{33} & 0 \\ a_{41} & a_{42} & a_{43} & a_{44} \end{vmatrix}$$

$$= a_{11} \begin{vmatrix} a_{22} & 0 & 0 \\ a_{32} & a_{33} & 0 \\ a_{42} & a_{43} & a_{44} \end{vmatrix}$$

Once again, it's easy to expand along the first row:

$$\det(A) = a_{11}a_{22} \begin{vmatrix} a_{33} & 0 \\ a_{43} & a_{44} \end{vmatrix}$$
$$= a_{11}a_{22}a_{33}|a_{44}|$$
$$= a_{11}a_{22}a_{33}a_{44}$$

where we have used the convention that the determinant of a 1×1 matrix $[a]$ is a. ∎

EXAMPLE 8   Determinant of an Upper Triangular Matrix

$$\begin{vmatrix} 2 & 7 & -3 & 8 & 3 \\ 0 & -3 & 7 & 5 & 1 \\ 0 & 0 & 6 & 7 & 6 \\ 0 & 0 & 0 & 9 & 8 \\ 0 & 0 & 0 & 0 & 4 \end{vmatrix} = (2)(-3)(6)(9)(4) = -1296 \quad \blacklozenge$$

Proof of Theorem 1.7.1c Let $A = [a_{ij}]$ be a triangular matrix, so that its diagonal entries are

$$a_{11}, a_{22}, \ldots, a_{nn}$$

From Theorem 2.1.3, the matrix A is invertible if and only if

$$\det(A) = a_{11}a_{22} \cdots a_{nn}$$

is nonzero, which is true if and only if the diagonal entries are all nonzero. ∎

We leave it as an exercise for the reader to use the adjoint formula for A^{-1} to show that if $A = [a_{ij}]$ is an invertible triangular matrix, then the successive diagonal entries of A^{-1} are

$$\frac{1}{a_{11}}, \frac{1}{a_{22}}, \ldots, \frac{1}{a_{nn}}$$

(See Example 3 of Section 1.7.)

Proof of Theorem 1.7.1d We will prove the result for upper triangular matrices and leave the lower triangular case as an exercise. Assume that A is upper triangular and invertible. Since

$$A^{-1} = \frac{1}{\det(A)} \text{adj}(A)$$

we can prove that A^{-1} is upper triangular by showing that $\text{adj}(A)$ is upper triangular, or, equivalently, that the matrix of cofactors is lower triangular. We can do this by showing that every cofactor C_{ij} with $i < j$ (i.e., above the main diagonal) is zero. Since

$$C_{ij} = (-1)^{i+j} M_{ij}$$

it suffices to show that each minor M_{ij} with $i < j$ is zero. For this purpose, let B_{ij} be the matrix that results when the ith row and jth column of A are deleted, so

$$M_{ij} = \det(B_{ij}) \tag{10}$$

From the assumption that $i < j$, it follows that B_{ij} is upper triangular (Exercise 32). Since A is upper triangular, its $(i + 1)$-st row begins with at least i zeros. But the ith row of B_{ij} is the $(i + 1)$-st row of A with the entry in the jth column removed. Since $i < j$, none of the first i zeros is removed by deleting the jth column; thus the ith row of B_{ij} starts with at least i zeros, which implies that this row has a zero on the main diagonal. It now follows from Theorem 2.1.3 that $\det(B_{ij}) = 0$ and from (10) that $M_{ij} = 0$. ∎

Cramer's Rule

The next theorem provides a formula for the solution of certain linear systems of n equations in n unknowns. This formula, known as **Cramer's rule**, is of marginal interest for computational purposes, but it is useful for studying the mathematical properties of a solution without the need for solving the system.

THEOREM 2.1.4

Cramer's Rule

If $A\mathbf{x} = \mathbf{b}$ is a system of n linear equations in n unknowns such that $\det(A) \neq 0$, then the system has a unique solution. This solution is

$$x_1 = \frac{\det(A_1)}{\det(A)}, \quad x_2 = \frac{\det(A_2)}{\det(A)}, \dots, \quad x_n = \frac{\det(A_n)}{\det(A)}$$

where A_j is the matrix obtained by replacing the entries in the jth column of A by the entries in the matrix

$$\mathbf{b} = \begin{bmatrix} b_1 \\ b_2 \\ \vdots \\ b_n \end{bmatrix}$$

Proof If $\det(A) \neq 0$, then A is invertible, and by Theorem 1.6.2, $\mathbf{x} = A^{-1}\mathbf{b}$ is the unique solution of $A\mathbf{x} = \mathbf{b}$. Therefore, by Theorem 2.1.2 we have

$$\mathbf{x} = A^{-1}\mathbf{b} = \frac{1}{\det(A)} \text{adj}(A)\mathbf{b} = \frac{1}{\det(A)} \begin{bmatrix} C_{11} & C_{21} & \cdots & C_{n1} \\ C_{12} & C_{22} & \cdots & C_{n2} \\ \vdots & \vdots & & \vdots \\ C_{1n} & C_{2n} & \cdots & C_{nn} \end{bmatrix} \begin{bmatrix} b_1 \\ b_2 \\ \vdots \\ b_n \end{bmatrix}$$

Multiplying the matrices out gives

$$\mathbf{x} = \frac{1}{\det(A)} \begin{bmatrix} b_1 C_{11} + b_2 C_{21} + \cdots + b_n C_{n1} \\ b_1 C_{12} + b_2 C_{22} + \cdots + b_n C_{n2} \\ \vdots \qquad \vdots \qquad \qquad \vdots \\ b_1 C_{1n} + b_2 C_{2n} + \cdots + b_n C_{nn} \end{bmatrix}$$

The entry in the jth row of $\mathbf{x}$ is therefore

$$x_j = \frac{b_1 C_{1j} + b_2 C_{2j} + \cdots + b_n C_{nj}}{\det(A)} \tag{11}$$

Now let

$$A_j = \begin{bmatrix} a_{11} & a_{12} & \cdots & a_{1j-1} & b_1 & a_{1j+1} & \cdots & a_{1n} \\ a_{21} & a_{22} & \cdots & a_{2j-1} & b_2 & a_{2j+1} & \cdots & a_{2n} \\ \vdots & \vdots & & \vdots & \vdots & \vdots & & \vdots \\ a_{n1} & a_{n2} & \cdots & a_{nj-1} & b_n & a_{nj+1} & \cdots & a_{nn} \end{bmatrix}$$

Since A_j differs from A only in the jth column, it follows that the cofactors of entries $b_1, b_2, \ldots, b_n$ in A_j are the same as the cofactors of the corresponding entries in the jth column of A. The cofactor expansion of $\det(A_j)$ along the jth column is therefore

$$\det(A_j) = b_1 C_{1j} + b_2 C_{2j} + \cdots + b_n C_{nj}$$

Substituting this result in (11) gives

$$x_j = \frac{\det(A_j)}{\det(A)} \qquad \blacksquare$$

Gabriel Cramer *(1704–1752)* was a Swiss mathematician. Although Cramer does not rank with the great mathematicians of his time, his contributions as a disseminator of mathematical ideas have earned him a well-deserved place in the history of mathematics. Cramer traveled extensively and met many of the leading mathematicians of his day.

Cramer's most widely known work, *Introduction à l'analyse des lignes courbes algébriques* (1750), was a study and classification of algebraic curves; Cramer's rule appeared in the appendix. Although the rule bears his name, variations of the idea were formulated earlier by various mathematicians. However, Cramer's superior notation helped clarify and popularize the technique.

Overwork combined with a fall from a carriage led to his death at the age of 48. Cramer was apparently a good-natured and pleasant person with broad interests. He wrote on philosophy of law and government and the history of mathematics. He served in public office, participated in artillery and fortifications activities for the government, instructed workers on techniques of cathedral repair, and undertook excavations of cathedral archives. Cramer received numerous honors for his activities.

EXAMPLE 9 Using Cramer's Rule to Solve a Linear System

Use Cramer's rule to solve

$$\begin{aligned} x_1 + \quad\ + 2x_3 &= 6 \\ -3x_1 + 4x_2 + 6x_3 &= 30 \\ -x_1 - 2x_2 + 3x_3 &= 8 \end{aligned}$$

Solution

$$A = \begin{bmatrix} 1 & 0 & 2 \\ -3 & 4 & 6 \\ -1 & -2 & 3 \end{bmatrix}, \qquad A_1 = \begin{bmatrix} 6 & 0 & 2 \\ 30 & 4 & 6 \\ 8 & -2 & 3 \end{bmatrix},$$

$$A_2 = \begin{bmatrix} 1 & 6 & 2 \\ -3 & 30 & 6 \\ -1 & 8 & 3 \end{bmatrix}, \qquad A_3 = \begin{bmatrix} 1 & 0 & 6 \\ -3 & 4 & 30 \\ -1 & -2 & 8 \end{bmatrix}$$

Therefore,

$$x_1 = \frac{\det(A_1)}{\det(A)} = \frac{-40}{44} = \frac{-10}{11}, \qquad x_2 = \frac{\det(A_2)}{\det(A)} = \frac{72}{44} = \frac{18}{11},$$

$$x_3 = \frac{\det(A_3)}{\det(A)} = \frac{152}{44} = \frac{38}{11} \qquad \blacklozenge$$

REMARK   To solve a system of n equations in n unknowns by Cramer's rule, it is necessary to evaluate $n + 1$ determinants of $n \times n$ matrices. For systems with more than three equations, Gaussian elimination is far more efficient. However, Cramer's rule does give a formula for the solution if the determinant of the coefficient matrix is nonzero.

EXERCISE SET 2.1

1. Let

$$A = \begin{bmatrix} 1 & -2 & 3 \\ 6 & 7 & -1 \\ -3 & 1 & 4 \end{bmatrix}$$

   (a) Find all the minors of A.   (b) Find all the cofactors.

2. Let

$$A = \begin{bmatrix} 4 & -1 & 1 & 6 \\ 0 & 0 & -3 & 3 \\ 4 & 1 & 0 & 14 \\ 4 & 1 & 3 & 2 \end{bmatrix}$$

   Find

   (a) M_{13} and C_{13} (b) M_{23} and C_{23} (c) M_{22} and C_{22} (d) M_{21} and C_{21}

3. Evaluate the determinant of the matrix in Exercise 1 by a cofactor expansion along

   (a)   the first row   (b)   the first column   (c)   the second row
   (d)   the second column   (e)   the third row   (f)   the third column

4. For the matrix in Exercise 1, find

   (a) adj(A) (b) A^{-1} using Theorem 2.1.2

In Exercises 5–10 evaluate det(A) by a cofactor expansion along a row or column of your choice.

5. $A = \begin{bmatrix} -3 & 0 & 7 \\ 2 & 5 & 1 \\ -1 & 0 & 5 \end{bmatrix}$ 6. $A = \begin{bmatrix} 3 & 3 & 1 \\ 1 & 0 & -4 \\ 1 & -3 & 5 \end{bmatrix}$ 7. $A = \begin{bmatrix} 1 & k & k^2 \\ 1 & k & k^2 \\ 1 & k & k^2 \end{bmatrix}$

8. $A = \begin{bmatrix} k+1 & k-1 & 7 \\ 2 & k-3 & 4 \\ 5 & k+1 & k \end{bmatrix}$ 9. $A = \begin{bmatrix} 3 & 3 & 0 & 5 \\ 2 & 2 & 0 & -2 \\ 4 & 1 & -3 & 0 \\ 2 & 10 & 3 & 2 \end{bmatrix}$

10. $A = \begin{bmatrix} 4 & 0 & 0 & 1 & 0 \\ 3 & 3 & 3 & -1 & 0 \\ 1 & 2 & 4 & 2 & 3 \\ 9 & 4 & 6 & 2 & 3 \\ 2 & 2 & 4 & 2 & 3 \end{bmatrix}$

In Exercises 11–14 find A^{-1} using Theorem 2.1.2.

11. $A = \begin{bmatrix} 2 & 5 & 5 \\ -1 & -1 & 0 \\ 2 & 4 & 3 \end{bmatrix}$ 12. $A = \begin{bmatrix} 2 & 0 & 3 \\ 0 & 3 & 2 \\ -2 & 0 & -4 \end{bmatrix}$

13. $A = \begin{bmatrix} 2 & -3 & 5 \\ 0 & 1 & -3 \\ 0 & 0 & 2 \end{bmatrix}$ 14. $A = \begin{bmatrix} 2 & 0 & 0 \\ 8 & 1 & 0 \\ -5 & 3 & 6 \end{bmatrix}$

15. Let

$$A = \begin{bmatrix} 1 & 3 & 1 & 1 \\ 2 & 5 & 2 & 2 \\ 1 & 3 & 8 & 9 \\ 1 & 3 & 2 & 2 \end{bmatrix}$$

(a) Evaluate A^{-1} using Theorem 2.1.2.

(b) Evaluate A^{-1} using the method of Example 4 in Section 1.5.

(c) Which method involves less computation?

In Exercises 16–21 solve by Cramer's rule, where it applies.

16. $7x_1 - 2x_2 = 3$
$3x_1 + x_2 = 5$

17. $4x + 5y \phantom{{} + 2z} = 2$
$11x + y + 2z = 3$
$x + 5y + 2z = 1$

18. $x - 4y + z = 6$
$4x - y + 2z = -1$
$2x + 2y - 3z = -20$

19. $x_1 - 3x_2 + x_3 = 4$
$2x_1 - x_2 \phantom{{} - 3x_3} = -2$
$4x_1 \phantom{{} - x_2} - 3x_3 = 0$

20. $-x_1 - 4x_2 + 2x_3 + x_4 = -32$
$2x_1 - x_2 + 7x_3 + 9x_4 = 14$
$-x_1 + x_2 + 3x_3 + x_4 = 11$
$x_1 - 2x_2 + x_3 - 4x_4 = {-4}$

21. $3x_1 - x_2 + x_3 = 4$
$-x_1 + 7x_2 - 2x_3 = 1$
$2x_1 + 6x_2 - x_3 = 5$

22. Show that the matrix

$$A = \begin{bmatrix} \cos\theta & \sin\theta & 0 \\ -\sin\theta & \cos\theta & 0 \\ 0 & 0 & 1 \end{bmatrix}$$

is invertible for all values of θ; then find A^{-1} using Theorem 2.1.2.

23. Use Cramer's rule to solve for y without solving for x, z, and w.

$$\begin{aligned} 4x + y + z + w &= 6 \\ 3x + 7y - z + w &= 1 \\ 7x + 3y - 5z + 8w &= -3 \\ x + y + z + 2w &= 3 \end{aligned}$$

24. Let $A\mathbf{x} = \mathbf{b}$ be the system in Exercise 23.

(a) Solve by Cramer's rule.　(b) Solve by Gauss–Jordan elimination.

(c) Which method involves fewer computations?

25. Prove that if $\det(A) = 1$ and all the entries in A are integers, then all the entries in A^{-1} are integers.

26. Let $A\mathbf{x} = \mathbf{b}$ be a system of n linear equations in n unknowns with integer coefficients and integer constants. Prove that if $\det(A) = 1$, the solution $\mathbf{x}$ has integer entries.

27. Prove that if A is an invertible lower triangular matrix, then A^{-1} is lower triangular.

28. Derive the last cofactor expansion listed in Formula (4).

29. Prove: The equation of the line through the distinct points (a_1, b_1) and (a_2, b_2) can be written as

$$\begin{vmatrix} x & y & 1 \\ a_1 & b_1 & 1 \\ a_2 & b_2 & 1 \end{vmatrix} = 0$$

30. Prove: (x_1, y_1), (x_2, y_2), and (x_3, y_3) are collinear points if and only if

$$\begin{vmatrix} x_1 & y_1 & 1 \\ x_2 & y_2 & 1 \\ x_3 & y_3 & 1 \end{vmatrix} = 0$$

31. (a) If $A = \begin{bmatrix} A_{11} & A_{12} \\ \hline 0 & A_{22} \end{bmatrix}$ is an "upper triangular" block matrix, where A_{11} and A_{22} are

square matrices, then $\det(A) = \det(A_{11})\det(A_{22})$. Use this result to evaluate $\det(A)$ for

$$\begin{bmatrix} 2 & -1 & 2 & 5 & 6 \\ 4 & 3 & -1 & 3 & 4 \\ \hline 0 & 0 & 1 & 3 & 5 \\ 0 & 0 & -2 & 6 & 2 \\ 0 & 0 & 3 & 5 & 2 \end{bmatrix}$$

(b) Verify your answer in part (a) by using a cofactor expansion to evaluate $\det(A)$.

32. Prove that if A is upper triangular and B_{ij} is the matrix that results when the ith row and jth column of A are deleted, then B_{ij} is upper triangular if $i < j$.

Discussion & Discovery

33. What is the maximum number of zeros that a 4×4 matrix can have without having a zero determinant? Explain your reasoning.

34. Let A be a matrix of the form

$$A = \begin{bmatrix} * & * & 0 & 0 & 0 \\ * & * & 0 & 0 & 0 \\ * & * & 0 & 0 & 0 \\ * & * & * & * & * \\ * & * & * & * & * \end{bmatrix}$$

How many different values can you obtain for $\det(A)$ by substituting numerical values (not necessarily all the same) for the $*$'s? Explain your reasoning.

35. Indicate whether the statement is always true or sometimes false. Justify your answer by giving a logical argument or a counterexample.

(a) $A\,\mathrm{adj}(A)$ is a diagonal matrix for every square matrix A.

(b) In theory, Cramer's rule can be used to solve any system of linear equations, although the amount of computation may be enormous.

(c) If A is invertible, then $\mathrm{adj}(A)$ must also be invertible.

(d) If A has a row of zeros, then so does $\mathrm{adj}(A)$.

2.2
EVALUATING DETERMINANTS BY ROW REDUCTION

In this section we shall show that the determinant of a square matrix can be evaluated by reducing the matrix to row-echelon form. This method is important since it is the most computationally efficient way to find the determinant of a general matrix.

A Basic Theorem

We begin with a fundamental theorem that will lead us to an efficient procedure for evaluating the determinant of a matrix of any order n.

THEOREM 2.2.1

Let A be a square matrix. If A has a row of zeros or a column of zeros, then $\det(A) = 0$.

Proof By Theorem 2.1.1, the determinant of A found by cofactor expansion along the row or column of all zeros is

$$\det(A) = 0 \cdot C_1 + 0 \cdot C_2 + \cdots + 0 \cdot C_n$$

where $C_1, \ldots, C_n$ are the cofactors for that row or column. Hence $\det(A)$ is zero. ∎

Here is another useful theorem:

THEOREM 2.2.2

> *Let A be a square matrix. Then* $\det(A) = \det(A^T)$.

Proof By Theorem 2.1.1, the determinant of A found by cofactor expansion along its first row is the same as the determinant of A^T found by cofactor expansion along its first column. ∎

REMARK Because of Theorem 2.2.2, nearly every theorem about determinants that contains the word *row* in its statement is also true when the word *column* is substituted for *row*. To prove a column statement, one need only transpose the matrix in question, to convert the column statement to a row statement, and then apply the corresponding known result for rows.

Elementary Row Operations

The next theorem shows how an elementary row operation on a matrix affects the value of its determinant.

THEOREM 2.2.3

> *Let A be an $n \times n$ matrix.*
>
> (a) *If B is the matrix that results when a single row or single column of A is multiplied by a scalar k, then $\det(B) = k \det(A)$.*
> (b) *If B is the matrix that results when two rows or two columns of A are interchanged, then $\det(B) = -\det(A)$.*
> (c) *If B is the matrix that results when a multiple of one row of A is added to another row or when a multiple of one column is added to another column, then $\det(B) = \det(A)$.*

We omit the proof but give the following example that illustrates the theorem for 3×3 determinants.

EXAMPLE 1   Theorem 2.2.3 Applied to 3 × 3 Determinants

We will verify the equation in the first row of Table 1 and leave the last two for the reader. By Theorem 2.1.1, the determinant of B may be found by cofactor expansion along the first row:

$$\det(B) = \begin{vmatrix} ka_{11} & ka_{12} & ka_{13} \\ a_{21} & a_{22} & a_{23} \\ a_{31} & a_{32} & a_{33} \end{vmatrix}$$

$$= ka_{11}C_{11} + ka_{12}C_{12} + ka_{33}C_{13}$$
$$= k(a_{11}C_{11} + a_{12}C_{12} + a_{33}C_{13})$$
$$= k \det(A)$$

Table 1

Relationship	Operation
$\begin{vmatrix} ka_{11} & ka_{12} & ka_{13} \\ a_{21} & a_{22} & a_{23} \\ a_{31} & a_{32} & a_{33} \end{vmatrix} = k \begin{vmatrix} a_{11} & a_{12} & a_{13} \\ a_{21} & a_{22} & a_{23} \\ a_{31} & a_{32} & a_{33} \end{vmatrix}$ $\det(B) = k\det(A)$	The first row of A is multiplied by k.
$\begin{vmatrix} a_{21} & a_{22} & a_{23} \\ a_{11} & a_{12} & a_{13} \\ a_{31} & a_{32} & a_{33} \end{vmatrix} = - \begin{vmatrix} a_{11} & a_{12} & a_{13} \\ a_{21} & a_{22} & a_{23} \\ a_{31} & a_{32} & a_{33} \end{vmatrix}$ $\det(B) = -\det(A)$	The first and second rows of A are interchanged.
$\begin{vmatrix} a_{11}+ka_{21} & a_{12}+ka_{22} & a_{13}+ka_{23} \\ a_{21} & a_{22} & a_{23} \\ a_{31} & a_{32} & a_{33} \end{vmatrix} = \begin{vmatrix} a_{11} & a_{12} & a_{13} \\ a_{21} & a_{22} & a_{23} \\ a_{31} & a_{32} & a_{33} \end{vmatrix}$ $\det(B) = \det(A)$	A multiple of the second row of A is added to the first row.

since C_{11}, C_{12}, and C_{13} do not depend on the first row of the matrix, and A and B differ only in their first rows. ◆

REMARK   As illustrated by the first equation in Table 1, part (*a*) of Theorem 2.2.3 enables us to bring a "common factor" from any row (or column) through the determinant sign.

Elementary Matrices   Recall that an elementary matrix results from performing a single elementary row operation on an identity matrix; thus, if we let $A = I_n$ in Theorem 2.2.3 [so that we have $\det(A) = \det(I_n) = 1$], then the matrix B is an elementary matrix, and the theorem yields the following result about determinants of elementary matrices.

THEOREM 2.2.4

Let E be an $n \times n$ elementary matrix.

(a) *If E results from multiplying a row of I_n by k, then $\det(E) = k$.*
(b) *If E results from interchanging two rows of I_n, then $\det(E) = -1$.*
(c) *If E results from adding a multiple of one row of I_n to another, then $\det(E) = 1$.*

EXAMPLE 2   Determinants of Elementary Matrices

The following determinants of elementary matrices, which are evaluated by inspection, illustrate Theorem 2.2.4.

$$\begin{vmatrix} 1 & 0 & 0 & 0 \\ 0 & 3 & 0 & 0 \\ 0 & 0 & 1 & 0 \\ 0 & 0 & 0 & 1 \end{vmatrix} = 3, \qquad \begin{vmatrix} 0 & 0 & 0 & 1 \\ 0 & 1 & 0 & 0 \\ 0 & 0 & 1 & 0 \\ 1 & 0 & 0 & 0 \end{vmatrix} = -1, \qquad \begin{vmatrix} 1 & 0 & 0 & 7 \\ 0 & 1 & 0 & 0 \\ 0 & 0 & 1 & 0 \\ 0 & 0 & 0 & 1 \end{vmatrix} = 1 \quad ◆$$

The second row of I_4 was multiplied by 3.   The first and last rows of I_4 were interchanged.   7 times the last row of I_4 was added to the first row.

Matrices with Proportional Rows or Columns

If a square matrix A has two proportional rows, then a row of zeros can be introduced by adding a suitable multiple of one of the rows to the other. Similarly for columns. But adding a multiple of one row or column to another does not change the determinant, so from Theorem 2.2.1, we must have $\det(A) = 0$. This proves the following theorem.

THEOREM 2.2.5

> *If A is a square matrix with two proportional rows or two proportional columns, then* $\det(A) = 0.$

EXAMPLE 3   Introducing Zero Rows

The following computation illustrates the introduction of a row of zeros when there are two proportional rows:

$$\begin{vmatrix} 1 & 3 & -2 & 4 \\ 2 & 6 & -4 & 8 \\ 3 & 9 & 1 & 5 \\ 1 & 1 & 4 & 8 \end{vmatrix} = \begin{vmatrix} 1 & 3 & -2 & 4 \\ 0 & 0 & 0 & 0 \\ 3 & 9 & 1 & 5 \\ 1 & 1 & 4 & 8 \end{vmatrix} = 0$$

← The second row is 2 times the first, so we added -2 times the first row to the second to introduce a row of zeros.

Each of the following matrices has two proportional rows or columns; thus, each has a determinant of zero.

$$\begin{bmatrix} -1 & 4 \\ -2 & 8 \end{bmatrix}, \quad \begin{bmatrix} 1 & -2 & 7 \\ -4 & 8 & 5 \\ 2 & -4 & 3 \end{bmatrix}, \quad \begin{bmatrix} 3 & -1 & 4 & -5 \\ 6 & -2 & 5 & 2 \\ 5 & 8 & 1 & 4 \\ -9 & 3 & -12 & 15 \end{bmatrix} \blacklozenge$$

Evaluating Determinants by Row Reduction

We shall now give a method for evaluating determinants that involves substantially less computation than the cofactor expansion method. The idea of the method is to reduce the given matrix to upper triangular form by elementary row operations, then compute the determinant of the upper triangular matrix (an easy computation), and then relate that determinant to that of the original matrix. Here is an example:

EXAMPLE 4   Using Row Reduction to Evaluate a Determinant

Evaluate $\det(A)$ where

$$A = \begin{bmatrix} 0 & 1 & 5 \\ 3 & -6 & 9 \\ 2 & 6 & 1 \end{bmatrix}$$

Solution

We will reduce A to row-echelon form (which is upper triangular) and apply Theorem 2.2.3:

$$\det(A) = \begin{vmatrix} 0 & 1 & 5 \\ 3 & -6 & 9 \\ 2 & 6 & 1 \end{vmatrix} = -\begin{vmatrix} 3 & -6 & 9 \\ 0 & 1 & 5 \\ 2 & 6 & 1 \end{vmatrix}$$

← The first and second rows of A were interchanged.

$$= -3\begin{vmatrix} 1 & -2 & 3 \\ 0 & 1 & 5 \\ 2 & 6 & 1 \end{vmatrix}$$

← A common factor of 3 from the first row was taken through the determinant sign.

$$= -3 \begin{vmatrix} 1 & -2 & 3 \\ 0 & 1 & 5 \\ 0 & 10 & -5 \end{vmatrix} \qquad \longleftarrow \quad \begin{array}{l} -2 \text{ times the first row was} \\ \text{added to the third row.} \end{array}$$

$$= -3 \begin{vmatrix} 1 & -2 & 3 \\ 0 & 1 & 5 \\ 0 & 0 & -55 \end{vmatrix} \qquad \longleftarrow \quad \begin{array}{l} -10 \text{ times the second row} \\ \text{was added to the third row.} \end{array}$$

$$= (-3)(-55) \begin{vmatrix} 1 & -2 & 3 \\ 0 & 1 & 5 \\ 0 & 0 & 1 \end{vmatrix} \qquad \longleftarrow \quad \begin{array}{l} \text{A common factor of } -55 \\ \text{from the last row was taken} \\ \text{through the determinant sign.} \end{array}$$

$$= (-3)(-55)(1) = 165 \quad \blacklozenge$$

REMARK The method of row reduction is well suited for computer evaluation of determinants because it is computationally efficient and easily programmed. However, cofactor expansion is often easier for hand computation.

EXAMPLE 5 Using Column Operations to Evaluate a Determinant

Compute the determinant of

$$A = \begin{bmatrix} 1 & 0 & 0 & 3 \\ 2 & 7 & 0 & 6 \\ 0 & 6 & 3 & 0 \\ 7 & 3 & 1 & -5 \end{bmatrix}$$

Solution

This determinant could be computed as above by using elementary row operations to reduce A to row-echelon form, but we can put A in lower triangular form in one step by adding -3 times the first column to the fourth to obtain

$$\det(A) = \det \begin{bmatrix} 1 & 0 & 0 & 0 \\ 2 & 7 & 0 & 0 \\ 0 & 6 & 3 & 0 \\ 7 & 3 & 1 & -26 \end{bmatrix} = (1)(7)(3)(-26) = -546$$

This example points out the utility of keeping an eye open for column operations that can shorten computations. $\blacklozenge$

Cofactor expansion and row or column operations can sometimes be used in combination to provide an effective method for evaluating determinants. The following example illustrates this idea.

EXAMPLE 6 Row Operations and Cofactor Expansion

Evaluate $\det(A)$ where

$$A = \begin{bmatrix} 3 & 5 & -2 & 6 \\ 1 & 2 & -1 & 1 \\ 2 & 4 & 1 & 5 \\ 3 & 7 & 5 & 3 \end{bmatrix}$$

Solution

By adding suitable multiples of the second row to the remaining rows, we obtain

$$\det(A) = \begin{vmatrix} 0 & -1 & 1 & 3 \\ 1 & 2 & -1 & 1 \\ 0 & 0 & 3 & 3 \\ 0 & 1 & 8 & 0 \end{vmatrix}$$

$$= - \begin{vmatrix} -1 & 1 & 3 \\ 0 & 3 & 3 \\ 1 & 8 & 0 \end{vmatrix}$$ ⟵ Cofactor expansion along the first column

$$= - \begin{vmatrix} -1 & 1 & 3 \\ 0 & 3 & 3 \\ 0 & 9 & 3 \end{vmatrix}$$ ⟵ We added the first row to the third row.

$$= -(-1) \begin{vmatrix} 3 & 3 \\ 9 & 3 \end{vmatrix}$$ ⟵ Cofactor expansion along the first column

$$= -18 \ \blacklozenge$$

EXERCISE SET 2.2

1. Verify that $\det(A) = \det(A^T)$ for

(a) $A = \begin{bmatrix} -2 & 3 \\ 1 & 4 \end{bmatrix}$ (b) $A = \begin{bmatrix} 2 & -1 & 3 \\ 1 & 2 & 4 \\ 5 & -3 & 6 \end{bmatrix}$

2. Evaluate the following determinants by inspection.

(a) $\begin{vmatrix} 3 & -17 & 4 \\ 0 & 5 & 1 \\ 0 & 0 & -2 \end{vmatrix}$ (b) $\begin{vmatrix} \sqrt{2} & 0 & 0 & 0 \\ -8 & \sqrt{2} & 0 & 0 \\ 7 & 0 & -1 & 0 \\ 9 & 5 & 6 & 1 \end{vmatrix}$

(c) $\begin{vmatrix} -2 & 1 & 3 \\ 1 & -7 & 4 \\ -2 & 1 & 3 \end{vmatrix}$ (d) $\begin{vmatrix} 1 & -2 & 3 \\ 2 & -4 & 6 \\ 5 & -8 & 1 \end{vmatrix}$

3. Find the determinants of the following elementary matrices by inspection.

(a) $\begin{bmatrix} 1 & 0 & 0 & 0 \\ 0 & 1 & 0 & 0 \\ 0 & 0 & -5 & 0 \\ 0 & 0 & 0 & 1 \end{bmatrix}$ (b) $\begin{bmatrix} 1 & 0 & 0 & 0 \\ 0 & 0 & 1 & 0 \\ 0 & 1 & 0 & 0 \\ 0 & 0 & 0 & 1 \end{bmatrix}$ (c) $\begin{bmatrix} 1 & 0 & 0 & 0 \\ 0 & 1 & 0 & -9 \\ 0 & 0 & 1 & 0 \\ 0 & 0 & 0 & 1 \end{bmatrix}$

In Exercises 4–11 evaluate the determinant of the given matrix by reducing the matrix to row-echelon form.

4. $\begin{bmatrix} 3 & 6 & -9 \\ 0 & 0 & -2 \\ -2 & 1 & 5 \end{bmatrix}$ 5. $\begin{bmatrix} 0 & 3 & 1 \\ 1 & 1 & 2 \\ 3 & 2 & 4 \end{bmatrix}$ 6. $\begin{bmatrix} 1 & -3 & 0 \\ -2 & 4 & 1 \\ 5 & -2 & 2 \end{bmatrix}$

7. $\begin{bmatrix} 3 & -6 & 9 \\ -2 & 7 & -2 \\ 0 & 1 & 5 \end{bmatrix}$ 8. $\begin{bmatrix} 1 & -2 & 3 & 1 \\ 5 & -9 & 6 & 3 \\ -1 & 2 & -6 & -2 \\ 2 & 8 & 6 & 1 \end{bmatrix}$ 9. $\begin{bmatrix} 2 & 1 & 3 & 1 \\ 1 & 0 & 1 & 1 \\ 0 & 2 & 1 & 0 \\ 0 & 1 & 2 & 3 \end{bmatrix}$

10. $\begin{bmatrix} 0 & 1 & 1 & 1 \\ \frac{1}{2} & \frac{1}{2} & 1 & \frac{1}{2} \\ \frac{2}{3} & \frac{1}{3} & \frac{1}{3} & 0 \\ -\frac{1}{3} & \frac{2}{3} & 0 & 0 \end{bmatrix}$ **11.** $\begin{bmatrix} 1 & 3 & 1 & 5 & 3 \\ -2 & -7 & 0 & -4 & 2 \\ 0 & 0 & 1 & 0 & 1 \\ 0 & 0 & 2 & 1 & 1 \\ 0 & 0 & 0 & 1 & 1 \end{bmatrix}$

12. Given that $\begin{vmatrix} a & b & c \\ d & e & f \\ g & h & i \end{vmatrix} = -6$, find

(a) $\begin{vmatrix} d & e & f \\ g & h & i \\ a & b & c \end{vmatrix}$ (b) $\begin{vmatrix} 3a & 3b & 3c \\ -d & -e & -f \\ 4g & 4h & 4i \end{vmatrix}$

(c) $\begin{vmatrix} a+g & b+h & c+i \\ d & e & f \\ g & h & i \end{vmatrix}$ (d) $\begin{vmatrix} -3a & -3b & -3c \\ d & e & f \\ g-4d & h-4e & i-4f \end{vmatrix}$

13. Use row reduction to show that

$$\begin{vmatrix} 1 & 1 & 1 \\ a & b & c \\ a^2 & b^2 & c^2 \end{vmatrix} = (b-a)(c-a)(c-b)$$

14. Use an argument like that in the proof of Theorem 2.1.3 to show that

(a) $\det \begin{bmatrix} 0 & 0 & a_{13} \\ 0 & a_{22} & a_{23} \\ a_{31} & a_{32} & a_{33} \end{bmatrix} = -a_{13}a_{22}a_{31}$

(b) $\det \begin{bmatrix} 0 & 0 & 0 & a_{14} \\ 0 & 0 & a_{23} & a_{24} \\ 0 & a_{32} & a_{33} & a_{34} \\ a_{41} & a_{42} & a_{43} & a_{44} \end{bmatrix} = a_{14}a_{23}a_{32}a_{41}$

15. Prove the following special cases of Theorem 2.2.3.

(a) $\begin{vmatrix} a_{21} & a_{22} & a_{23} \\ a_{11} & a_{12} & a_{13} \\ a_{31} & a_{32} & a_{33} \end{vmatrix} = - \begin{vmatrix} a_{11} & a_{12} & a_{13} \\ a_{21} & a_{22} & a_{23} \\ a_{31} & a_{32} & a_{33} \end{vmatrix}$

(b) $\begin{vmatrix} a_{11}+ka_{21} & a_{12}+ka_{22} & a_{13}+ka_{23} \\ a_{21} & a_{22} & a_{23} \\ a_{31} & a_{32} & a_{33} \end{vmatrix} = \begin{vmatrix} a_{11} & a_{12} & a_{13} \\ a_{21} & a_{22} & a_{23} \\ a_{31} & a_{32} & a_{33} \end{vmatrix}$

16. Repeat Exercises 4–7 using a combination of row reduction and cofactor expansion, as in Example 6.

17. Repeat Exercises 8–11 using a combination of row reduction and cofactor expansion, as in Example 6.

Discussion
Discovery

18. In each part, find $\det(A)$ by inspection, and explain your reasoning.

(a) $A = \begin{bmatrix} 0 & 0 & 1 \\ 0 & 1 & 0 \\ 1 & 0 & 0 \end{bmatrix}$ (b) $A = \begin{bmatrix} 0 & 0 & 0 & 1 \\ 0 & 0 & 1 & 0 \\ 0 & 1 & 0 & 0 \\ 1 & 0 & 0 & 0 \end{bmatrix}$

19. By inspection, solve the equation

$$\begin{vmatrix} x & 5 & 7 \\ 0 & x+1 & 6 \\ 0 & 0 & 2x-1 \end{vmatrix} = 0$$

Explain your reasoning.

20. (a) By inspection, find two solutions of the equation

$$\begin{vmatrix} 1 & x & x^2 \\ 1 & 1 & 1 \\ 1 & -3 & 9 \end{vmatrix} = 0$$

(b) Is it possible that there are other solutions? Justify your answer.

21. How many arithmetic operations are needed, in general, to find $\det(A)$ by row reduction? By cofactor expansion?

2.3
PROPERTIES OF THE DETERMINANT FUNCTION

In this section we shall develop some of the fundamental properties of the determinant function. Our work here will give us some further insight into the relationship between a square matrix and its determinant. One of the immediate consequences of this material will be the determinant test for the invertibility of a matrix.

Basic Properties of Determinants

Suppose that A and B are $n \times n$ matrices and k is any scalar. We begin by considering possible relationships between $\det(A)$, $\det(B)$, and

$$\det(kA), \quad \det(A+B), \quad \text{and} \quad \det(AB)$$

Since a common factor of any row of a matrix can be moved through the det sign, and since each of the n rows in kA has a common factor of k, we obtain

$$\det(kA) = k^n \det(A) \tag{1}$$

For example,

$$\begin{vmatrix} ka_{11} & ka_{12} & ka_{13} \\ ka_{21} & ka_{22} & ka_{23} \\ ka_{31} & ka_{32} & ka_{33} \end{vmatrix} = k^3 \begin{vmatrix} a_{11} & a_{12} & a_{13} \\ a_{21} & a_{22} & a_{23} \\ a_{31} & a_{32} & a_{33} \end{vmatrix}$$

Unfortunately, no simple relationship exists among $\det(A)$, $\det(B)$, and $\det(A+B)$. In particular, we emphasize that $\det(A+B)$ will usually *not* be equal to $\det(A) + \det(B)$. The following example illustrates this fact.

EXAMPLE 1 det(A + B) ≠ det(A) + det(B)

Consider

$$A = \begin{bmatrix} 1 & 2 \\ 2 & 5 \end{bmatrix}, \qquad B = \begin{bmatrix} 3 & 1 \\ 1 & 3 \end{bmatrix}, \qquad A+B = \begin{bmatrix} 4 & 3 \\ 3 & 8 \end{bmatrix}$$

We have $\det(A) = 1$, $\det(B) = 8$, and $\det(A+B) = 23$; thus

$$\det(A+B) \neq \det(A) + \det(B) \quad \blacklozenge$$

In spite of the negative tone of the preceding example, there is one important relationship concerning sums of determinants that is often useful. To obtain it, consider two 2×2 matrices that differ only in the second row:

$$A = \begin{bmatrix} a_{11} & a_{12} \\ a_{21} & a_{22} \end{bmatrix} \quad \text{and} \quad B = \begin{bmatrix} a_{11} & a_{12} \\ b_{21} & b_{22} \end{bmatrix}$$

We have

$$\det(A) + \det(B) = (a_{11}a_{22} - a_{12}a_{21}) + (a_{11}b_{22} - a_{12}b_{21})$$
$$= a_{11}(a_{22} + b_{22}) - a_{12}(a_{21} + b_{21})$$
$$= \det \begin{bmatrix} a_{11} & a_{12} \\ a_{21} + b_{21} & a_{22} + b_{22} \end{bmatrix}$$

Thus

$$\det \begin{bmatrix} a_{11} & a_{12} \\ a_{21} & a_{22} \end{bmatrix} + \det \begin{bmatrix} a_{11} & a_{12} \\ b_{21} & b_{22} \end{bmatrix} = \det \begin{bmatrix} a_{11} & a_{12} \\ a_{21} + b_{21} & a_{22} + b_{22} \end{bmatrix}$$

This is a special case of the following general result.

THEOREM 2.3.1

Let A, B, and C be $n \times n$ matrices that differ only in a single row, say the rth, and assume that the rth row of C can be obtained by adding corresponding entries in the rth rows of A and B. Then

$$\det(C) = \det(A) + \det(B)$$

The same result holds for columns.

EXAMPLE 2 Using Theorem 2.3.1

By evaluating the determinants, the reader can check that

$$\det \begin{bmatrix} 1 & 7 & 5 \\ 2 & 0 & 3 \\ 1+0 & 4+1 & 7+(-1) \end{bmatrix} = \det \begin{bmatrix} 1 & 7 & 5 \\ 2 & 0 & 3 \\ 1 & 4 & 7 \end{bmatrix} + \det \begin{bmatrix} 1 & 7 & 5 \\ 2 & 0 & 3 \\ 0 & 1 & -1 \end{bmatrix} \quad \blacklozenge$$

Determinant of a Matrix Product

When one considers the complexity of the definitions of matrix multiplication and determinants, it would seem unlikely that any simple relationship should exist between them. This is what makes the elegant simplicity of the following result so surprising: We will show that if A and B are square matrices of the same size, then

$$\det(AB) = \det(A)\det(B) \tag{2}$$

The proof of this theorem is fairly intricate, so we will have to develop some preliminary results first. We begin with the special case of (2) in which A is an elementary matrix. Because this special case is only a prelude to (2), we call it a lemma.

LEMMA 2.3.2

If B is an $n \times n$ matrix and E is an $n \times n$ elementary matrix, then

$$\det(EB) = \det(E)\det(B)$$

Proof We shall consider three cases, each depending on the row operation that produces matrix E.

Case 1. If E results from multiplying a row of I_n by k, then by Theorem 1.5.1, EB results from B by multiplying a row by k; so from Theorem 2.2.3a we have

$$\det(EB) = k \det(B)$$

But from Theorem 2.2.4a we have $\det(E) = k$, so

$$\det(EB) = \det(E) \det(B)$$

Cases 2 and 3. The proofs of the cases where E results from interchanging two rows of I_n or from adding a multiple of one row to another follow the same pattern as Case 1 and are left as exercises. ■

Remark It follows by repeated applications of Lemma 2.3.2 that if B is an $n \times n$ matrix and $E_1, E_2, \ldots, E_r$ are $n \times n$ elementary matrices, then

$$\det(E_1 E_2 \cdots E_r B) = \det(E_1) \det(E_2) \cdots \det(E_r) \det(B) \tag{3}$$

For example,

$$\det(E_1 E_2 B) = \det(E_1) \det(E_2 B) = \det(E_1) \det(E_2) \det(B)$$

Determinant Test for Invertibility

The next theorem provides an important criterion for invertibility in terms of determinants, and it will be used in proving (2).

THEOREM 2.3.3

> *A square matrix A is invertible if and only if $\det(A) \neq 0$.*

Proof Let R be the reduced row-echelon form of A. As a preliminary step, we will show that $\det(A)$ and $\det(R)$ are both zero or both nonzero: Let $E_1, E_2, \ldots, E_r$ be the elementary matrices that correspond to the elementary row operations that produce R from A. Thus

$$R = E_r \cdots E_2 E_1 A$$

and from (3),

$$\det(R) = \det(E_r) \cdots \det(E_2) \det(E_1) \det(A) \tag{4}$$

But from Theorem 2.2.4 the determinants of the elementary matrices are all nonzero. (Keep in mind that multiplying a row by zero is *not* an allowable elementary row operation, so $k \neq 0$ in this application of Theorem 2.2.4.) Thus, it follows from (4) that $\det(A)$ and $\det(R)$ are both zero or both nonzero. Now to the main body of the proof.

If A is invertible, then by Theorem 1.6.4 we have $R = I$, so $\det(R) = 1 \neq 0$ and consequently $\det(A) \neq 0$. Conversely, if $\det(A) \neq 0$, then $\det(R) \neq 0$, so R cannot have a row of zeros. It follows from Theorem 1.4.3 that $R = I$, so A is invertible by Theorem 1.6.4. ■

It follows from Theorems 2.3.3 and 2.2.5 that a square matrix with two proportional rows or columns is not invertible.

EXAMPLE 3   Determinant Test for Invertibility

Since the first and third rows of

$$A = \begin{bmatrix} 1 & 2 & 3 \\ 1 & 0 & 1 \\ 2 & 4 & 6 \end{bmatrix}$$

are proportional, $\det(A) = 0$. Thus A is not invertible. ◆

We are now ready for the result concerning products of matrices.

THEOREM 2.3.4

> *If A and B are square matrices of the same size, then*
>
> $$\det(AB) = \det(A)\det(B)$$

Proof   We divide the proof into two cases that depend on whether or not A is invertible. If the matrix A is not invertible, then by Theorem 1.6.5 neither is the product AB. Thus, from Theorem 2.3.3, we have $\det(AB) = 0$ and $\det(A) = 0$, so it follows that $\det(AB) = \det(A)\det(B)$.

Now assume that A is invertible. By Theorem 1.6.4, the matrix A is expressible as a product of elementary matrices, say

$$A = E_1 E_2 \cdots E_r \tag{5}$$

so

$$AB = E_1 E_2 \cdots E_r B$$

Applying (3) to this equation yields

$$\det(AB) = \det(E_1)\det(E_2) \cdots \det(E_r)\det(B)$$

and applying (3) again yields

$$\det(AB) = \det(E_1 E_2 \cdots E_r)\det(B)$$

which, from (5), can be written as $\det(AB) = \det(A)\det(B)$. ∎

EXAMPLE 4   Verifying That det(AB) = det(A) det(B)

Consider the matrices

$$A = \begin{bmatrix} 3 & 1 \\ 2 & 1 \end{bmatrix}, \qquad B = \begin{bmatrix} -1 & 3 \\ 5 & 8 \end{bmatrix}, \qquad AB = \begin{bmatrix} 2 & 17 \\ 3 & 14 \end{bmatrix}$$

We leave it for the reader to verify that

$$\det(A) = 1, \quad \det(B) = -23, \quad \text{and} \quad \det(AB) = -23$$

Thus $\det(AB) = \det(A)\det(B)$, as guaranteed by Theorem 2.3.4. ◆

The following theorem gives a useful relationship between the determinant of an invertible matrix and the determinant of its inverse.

THEOREM 2.3.5

If A is invertible, then

$$\det(A^{-1}) = \frac{1}{\det(A)}$$

Proof   Since $A^{-1}A = I$, it follows that $\det(A^{-1}A) = \det(I)$. Therefore, we must have $\det(A^{-1})\det(A) = 1$. Since $\det(A) \neq 0$, the proof can be completed by dividing through by $\det(A)$. ∎

Linear Systems of the Form $A\mathbf{x} = \lambda\mathbf{x}$

Many applications of linear algebra are concerned with systems of n linear equations in n unknowns that are expressed in the form

$$A\mathbf{x} = \lambda\mathbf{x} \tag{6}$$

where λ is a scalar. Such systems are really homogeneous linear systems in disguise, since (6) can be rewritten as $\lambda\mathbf{x} - A\mathbf{x} = \mathbf{0}$ or, by inserting an identity matrix and factoring, as

$$(\lambda I - A)\mathbf{x} = \mathbf{0} \tag{7}$$

Here is an example:

EXAMPLE 5   Finding $\lambda I - A$

The linear system

$$x_1 + 3x_2 = \lambda x_1$$
$$4x_1 + 2x_2 = \lambda x_2$$

can be written in matrix form as

$$\begin{bmatrix} 1 & 3 \\ 4 & 2 \end{bmatrix} \begin{bmatrix} x_1 \\ x_2 \end{bmatrix} = \lambda \begin{bmatrix} x_1 \\ x_2 \end{bmatrix}$$

which is of form (6) with

$$A = \begin{bmatrix} 1 & 3 \\ 4 & 2 \end{bmatrix} \quad \text{and} \quad \mathbf{x} = \begin{bmatrix} x_1 \\ x_2 \end{bmatrix}$$

This system can be rewritten as

$$\lambda \begin{bmatrix} x_1 \\ x_2 \end{bmatrix} - \begin{bmatrix} 1 & 3 \\ 4 & 2 \end{bmatrix} \begin{bmatrix} x_1 \\ x_2 \end{bmatrix} = \begin{bmatrix} 0 \\ 0 \end{bmatrix}$$

or

$$\lambda \begin{bmatrix} 1 & 0 \\ 0 & 1 \end{bmatrix} \begin{bmatrix} x_1 \\ x_2 \end{bmatrix} - \begin{bmatrix} 1 & 3 \\ 4 & 2 \end{bmatrix} \begin{bmatrix} x_1 \\ x_2 \end{bmatrix} = \begin{bmatrix} 0 \\ 0 \end{bmatrix}$$

or

$$\begin{bmatrix} \lambda - 1 & -3 \\ -4 & \lambda - 2 \end{bmatrix} \begin{bmatrix} x_1 \\ x_2 \end{bmatrix} = \begin{bmatrix} 0 \\ 0 \end{bmatrix}$$

which is of form (7) with

$$\lambda I - A = \begin{bmatrix} \lambda - 1 & -3 \\ -4 & \lambda - 2 \end{bmatrix} \quad \blacklozenge$$

The primary problem of interest for linear systems of the form (7) is to determine those values of λ for which the system has a nontrivial solution; such a value of λ is called a ***characteristic value*** or an ***eigenvalue***[†] of A. If λ is an eigenvalue of A, then the nontrivial solutions of (7) are called the ***eigenvectors*** of A corresponding to λ.

It follows from Theorem 2.3.3 that the system $(\lambda I - A)\mathbf{x} = \mathbf{0}$ has a nontrivial solution if and only if

$$\det(\lambda I - A) = 0 \tag{8}$$

This is called the ***characteristic equation*** of A; the eigenvalues of A can be found by solving this equation for λ.

Eigenvalues and eigenvectors will be studied again in subsequent chapters, where we will discuss their geometric interpretation and develop their properties in more depth.

EXAMPLE 6 Eigenvalues and Eigenvectors

Find the eigenvalues and corresponding eigenvectors of the matrix A in Example 5.

Solution

The characteristic equation of A is

$$\det(\lambda I - A) = \begin{vmatrix} \lambda - 1 & -3 \\ -4 & \lambda - 2 \end{vmatrix} = 0 \quad \text{or} \quad \lambda^2 - 3\lambda - 10 = 0$$

The factored form of this equation is $(\lambda + 2)(\lambda - 5) = 0$, so the eigenvalues of A are $\lambda = -2$ and $\lambda = 5$.

By definition,

$$\mathbf{x} = \begin{bmatrix} x_1 \\ x_2 \end{bmatrix}$$

is an eigenvector of A if and only if $\mathbf{x}$ is a nontrivial solution of $(\lambda I - A)\mathbf{x} = \mathbf{0}$; that is,

$$\begin{bmatrix} \lambda - 1 & -3 \\ -4 & \lambda - 2 \end{bmatrix} \begin{bmatrix} x_1 \\ x_2 \end{bmatrix} = \begin{bmatrix} 0 \\ 0 \end{bmatrix} \tag{9}$$

If $\lambda = -2$, then (9) becomes

$$\begin{bmatrix} -3 & -3 \\ -4 & -4 \end{bmatrix} \begin{bmatrix} x_1 \\ x_2 \end{bmatrix} = \begin{bmatrix} 0 \\ 0 \end{bmatrix}$$

Solving this system yields (verify) $x_1 = -t$, $x_2 = t$, so the eigenvectors corresponding to $\lambda = -2$ are the nonzero solutions of the form

$$\mathbf{x} = \begin{bmatrix} x_1 \\ x_2 \end{bmatrix} = \begin{bmatrix} -t \\ t \end{bmatrix}$$

Again from (9), the eigenvectors of A corresponding to $\lambda = 5$ are the nontrivial solutions of

$$\begin{bmatrix} 4 & -3 \\ -4 & 3 \end{bmatrix} \begin{bmatrix} x_1 \\ x_2 \end{bmatrix} = \begin{bmatrix} 0 \\ 0 \end{bmatrix}$$

[†]The word *eigenvalue* is a mixture of German and English. The German prefix *eigen* can be translated as "proper," which stems from the older literature where eigenvalues were known as *proper values*; they were also called *latent roots*.

We leave it for the reader to solve this system and show that the eigenvectors of A corresponding to $\lambda = 5$ are the nonzero solutions of the form

$$\mathbf{x} = \begin{bmatrix} \frac{3}{4}t \\ t \end{bmatrix} \; \blacklozenge$$

Summary

In Theorem 1.6.4 we listed five results that are equivalent to the invertibility of a matrix A. We conclude this section by merging Theorem 2.3.3 with that list to produce the following theorem that relates all of the major topics we have studied thus far.

THEOREM 2.3.6

Equivalent Statements

If A is an $n \times n$ matrix, then the following statements are equivalent.

(a) *A is invertible.*
(b) *$A\mathbf{x} = \mathbf{0}$ has only the trivial solution.*
(c) *The reduced row-echelon form of A is I_n.*
(d) *A can be expressed as a product of elementary matrices.*
(e) *$A\mathbf{x} = \mathbf{b}$ is consistent for every $n \times 1$ matrix $\mathbf{b}$.*
(f) *$A\mathbf{x} = \mathbf{b}$ has exactly one solution for every $n \times 1$ matrix $\mathbf{b}$.*
(g) $\det(A) \neq 0$.

EXERCISE SET 2.3

1. Verify that $\det(kA) = k^n \det(A)$ for

    (a) $A = \begin{bmatrix} -1 & 2 \\ 3 & 4 \end{bmatrix}$; $k = 2$ (b) $A = \begin{bmatrix} 2 & -1 & 3 \\ 3 & 2 & 1 \\ 1 & 4 & 5 \end{bmatrix}$; $k = -2$

2. Verify that $\det(AB) = \det(A)\det(B)$ for

    $$A = \begin{bmatrix} 2 & 1 & 0 \\ 3 & 4 & 0 \\ 0 & 0 & 2 \end{bmatrix} \quad \text{and} \quad B = \begin{bmatrix} 1 & -1 & 3 \\ 7 & 1 & 2 \\ 5 & 0 & 1 \end{bmatrix}$$

    Is $\det(A + B) = \det(A) + \det(B)$?

3. By inspection, explain why $\det(A) = 0$.

    $$A = \begin{bmatrix} -2 & 8 & 1 & 4 \\ 3 & 2 & 5 & 1 \\ 1 & 10 & 6 & 5 \\ 4 & -6 & 4 & -3 \end{bmatrix}$$

4. Use Theorem 2.3.3 to determine which of the following matrices are invertible.

    (a) $\begin{bmatrix} 1 & 0 & -1 \\ 9 & -1 & 4 \\ 8 & 9 & -1 \end{bmatrix}$ (b) $\begin{bmatrix} 4 & 2 & 8 \\ -2 & 1 & -4 \\ 3 & 1 & 6 \end{bmatrix}$

    (c) $\begin{bmatrix} \sqrt{2} & -\sqrt{7} & 0 \\ 3\sqrt{2} & -3\sqrt{7} & 0 \\ 5 & -9 & 0 \end{bmatrix}$ (d) $\begin{bmatrix} -3 & 0 & 1 \\ 5 & 0 & 6 \\ 8 & 0 & 3 \end{bmatrix}$

5. Let

$$A = \begin{bmatrix} a & b & c \\ d & e & f \\ g & h & i \end{bmatrix}$$

Assuming that $\det(A) = -7$, find

(a) $\det(3A)$ (b) $\det(A^{-1})$ (c) $\det(2A^{-1})$

(d) $\det((2A)^{-1})$ (e) $\det \begin{bmatrix} a & g & d \\ b & h & e \\ c & i & f \end{bmatrix}$

6. Without directly evaluating, show that $x = 0$ and $x = 2$ satisfy

$$\begin{vmatrix} x^2 & x & 2 \\ 2 & 1 & 1 \\ 0 & 0 & -5 \end{vmatrix} = 0$$

7. Without directly evaluating, show that

$$\det \begin{bmatrix} b+c & c+a & b+a \\ a & b & c \\ 1 & 1 & 1 \end{bmatrix} = 0$$

In Exercises 8–11 prove the identity without evaluating the determinants.

8. $\begin{vmatrix} a_1 & b_1 & a_1 + b_1 + c_1 \\ a_2 & b_2 & a_2 + b_2 + c_2 \\ a_3 & b_3 & a_3 + b_3 + c_3 \end{vmatrix} = \begin{vmatrix} a_1 & b_1 & c_1 \\ a_2 & b_2 & c_2 \\ a_3 & b_3 & c_3 \end{vmatrix}$

9. $\begin{vmatrix} a_1 + b_1 & a_1 - b_1 & c_1 \\ a_2 + b_2 & a_2 - b_2 & c_2 \\ a_3 + b_3 & a_3 - b_3 & c_3 \end{vmatrix} = -2 \begin{vmatrix} a_1 & b_1 & c_1 \\ a_2 & b_2 & c_2 \\ a_3 & b_3 & c_3 \end{vmatrix}$

10. $\begin{vmatrix} a_1 + b_1 t & a_2 + b_2 t & a_3 + b_3 t \\ a_1 t + b_1 & a_2 t + b_2 & a_3 t + b_3 \\ c_1 & c_2 & c_3 \end{vmatrix} = (1 - t^2) \begin{vmatrix} a_1 & a_2 & a_3 \\ b_1 & b_2 & b_3 \\ c_1 & c_2 & c_3 \end{vmatrix}$

11. $\begin{vmatrix} a_1 & b_1 + ta_1 & c_1 + rb_1 + sa_1 \\ a_2 & b_2 + ta_2 & c_2 + rb_2 + sa_2 \\ a_3 & b_3 + ta_3 & c_3 + rb_3 + sa_3 \end{vmatrix} = \begin{vmatrix} a_1 & a_2 & a_3 \\ b_1 & b_2 & b_3 \\ c_1 & c_2 & c_3 \end{vmatrix}$

12. For which value(s) of k does A fail to be invertible?

(a) $A = \begin{bmatrix} k-3 & -2 \\ -2 & k-2 \end{bmatrix}$ (b) $A = \begin{bmatrix} 1 & 2 & 4 \\ 3 & 1 & 6 \\ k & 3 & 2 \end{bmatrix}$

13. Use Theorem 2.3.3 to show that

$$\begin{bmatrix} \sin^2 \alpha & \sin^2 \beta & \sin^2 \gamma \\ \cos^2 \alpha & \cos^2 \beta & \cos^2 \gamma \\ 1 & 1 & 1 \end{bmatrix}$$

is not invertible for any values of α, β, and γ.

14. Express the following linear systems in the form $(\lambda I - A)\mathbf{x} = \mathbf{0}$.

(a) $\begin{aligned} x_1 + 2x_2 &= \lambda x_1 \\ 2x_1 + x_2 &= \lambda x_2 \end{aligned}$ (b) $\begin{aligned} 2x_1 + 3x_2 &= \lambda x_1 \\ 4x_1 + 3x_2 &= \lambda x_2 \end{aligned}$ (c) $\begin{aligned} 3x_1 + x_2 &= \lambda x_1 \\ -5x_1 - 3x_2 &= \lambda x_2 \end{aligned}$

15. For each of the systems in Exercise 14, find

    (i) the characteristic equation;

    (ii) the eigenvalues;

    (iii) the eigenvectors corresponding to each of the eigenvalues.

16. Let A and B be $n \times n$ matrices. Show that if A is invertible, then $\det(B) = \det(A^{-1}BA)$.

17. (a) Express

$$\begin{vmatrix} a_1 + b_1 & c_1 + d_1 \\ a_2 + b_2 & c_2 + d_2 \end{vmatrix}$$

    as a sum of four determinants whose entries contain no sums.

    (b) Express

$$\begin{vmatrix} a_1 + b_1 & c_1 + d_1 & e_1 + f_1 \\ a_2 + b_2 & c_2 + d_2 & e_2 + f_2 \\ a_3 + b_3 & c_3 + d_3 & e_3 + f_3 \end{vmatrix}$$

    as a sum of eight determinants whose entries contain no sums.

18. Prove that a square matrix A is invertible if and only if $A^T A$ is invertible.

19. Prove Cases 2 and 3 of Lemma 2.3.2.

Discussion & Discovery

20. Let A and B be $n \times n$ matrices. You know from earlier work that AB and BA need not be equal. Is the same true for $\det(AB)$ and $\det(BA)$? Explain your reasoning.

21. Let A and B be $n \times n$ matrices. You know from earlier work that AB is invertible if A and B are invertible. What can you say about the invertibility of AB if one or both of the factors are singular? Explain your reasoning.

22. Indicate whether the statement is always true or sometimes false. Justify each answer by giving a logical argument or a counterexample.

    (a) $\det(2A) = 2\det(A)$

    (b) $|A^2| = |A|^2$

    (c) $\det(I + A) = 1 + \det(A)$

    (d) If $\det(A) = 0$, then the homogeneous system $A\mathbf{x} = \mathbf{0}$ has infinitely many solutions.

23. Indicate whether the statement is always true or sometimes false. Justify your answer by giving a logical argument or a counterexample.

    (a) If $\det(A) = 0$, then A is not expressible as a product of elementary matrices.

    (b) If the reduced row-echelon form of A has a row of zeros, then $\det(A) = 0$.

    (c) The determinant of a matrix is unchanged if the columns are written in reverse order.

    (d) There is no square matrix A such that $\det(AA^T) = -1$.

2.4
A COMBINATORIAL APPROACH TO DETERMINANTS

There is a combinatorial view of determinants that actually predates matrices. In this section we explore this connection.

    There is another way to approach determinants that complements the cofactor expansion approach. It is based on permutations.

> **DEFINITION**
>
> A *permutation* of the set of integers $\{1, 2, \ldots, n\}$ is an arrangement of these integers in some order without omissions or repetitions.

EXAMPLE 1 Permutations of Three Integers

There are six different permutations of the set of integers $\{1, 2, 3\}$. These are

$$(1, 2, 3) \qquad (2, 1, 3) \qquad (3, 1, 2)$$
$$(1, 3, 2) \qquad (2, 3, 1) \qquad (3, 2, 1) \ \blacklozenge$$

One convenient method of systematically listing permutations is to use a *permutation tree*. This method is illustrated in our next example.

EXAMPLE 2 Permutations of Four Integers

List all permutations of the set of integers $\{1, 2, 3, 4\}$.

Solution

Consider Figure 2.4.1. The four dots labeled 1, 2, 3, 4 at the top of the figure represent the possible choices for the first number in the permutation. The three branches emanating from these dots represent the possible choices for the second position in the permutation. Thus, if the permutation begins $(2, -, -, -)$, the three possibilities for the second position are 1, 3, and 4. The two branches emanating from each dot in the second position represent the possible choices for the third position. Thus, if the permutation begins $(2, 3, -, -)$, the two possible choices for the third position are 1 and 4. Finally, the single branch emanating from each dot in the third position represents the only possible choice for the fourth position. Thus, if the permutation begins with $(2, 3, 4, -)$, the only choice for the fourth position is 1. The different permutations can now be listed by tracing out all the possible paths through the "tree" from the first position to the last position. We obtain the following list by this process.

$$
\begin{array}{cccc}
(1, 2, 3, 4) & (2, 1, 3, 4) & (3, 1, 2, 4) & (4, 1, 2, 3) \\
(1, 2, 4, 3) & (2, 1, 4, 3) & (3, 1, 4, 2) & (4, 1, 3, 2) \\
(1, 3, 2, 4) & (2, 3, 1, 4) & (3, 2, 1, 4) & (4, 2, 1, 3) \\
(1, 3, 4, 2) & (2, 3, 4, 1) & (3, 2, 4, 1) & (4, 2, 3, 1) \\
(1, 4, 2, 3) & (2, 4, 1, 3) & (3, 4, 1, 2) & (4, 3, 1, 2) \\
(1, 4, 3, 2) & (2, 4, 3, 1) & (3, 4, 2, 1) & (4, 3, 2, 1) \ \blacklozenge
\end{array}
$$

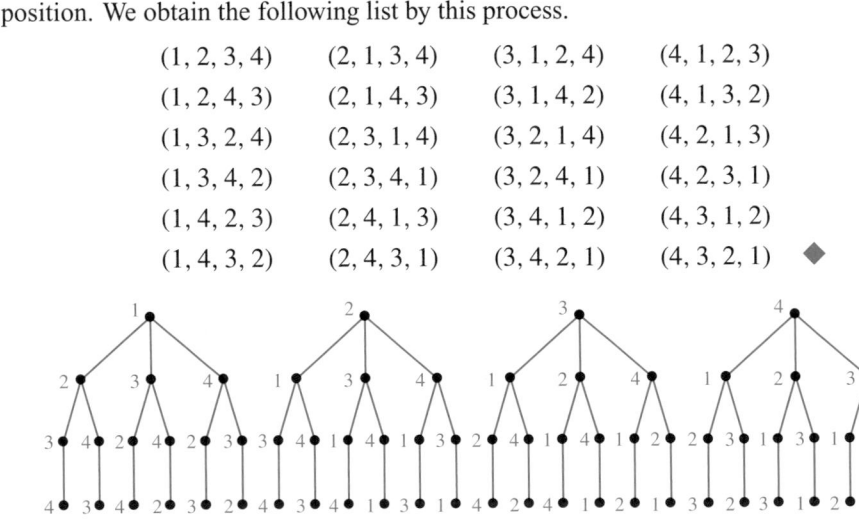

Figure 2.4.1

From this example we see that there are 24 permutations of $\{1, 2, 3, 4\}$. This result could have been anticipated without actually listing the permutations by arguing as follows. Since the first position can be filled in four ways and then the second position in three ways, there are $4 \cdot 3$ ways of filling the first two positions. Since the third position can then be filled in two ways, there are $4 \cdot 3 \cdot 2$ ways of filling the first three positions. Finally, since the last position can then be filled in only one way, there are $4 \cdot 3 \cdot 2 \cdot 1 = 24$ ways of filling all four positions. In general, the set $\{1, 2, \ldots, n\}$ will have $n(n-1)(n-2) \cdots 2 \cdot 1 = n!$ different permutations.

We will denote a general permutation of the set $\{1, 2, \ldots, n\}$ by $(j_1, j_2, \ldots, j_n)$. Here, j_1 is the first integer in the permutation, j_2 is the second, and so on. An ***inversion*** is said to occur in a permutation $(j_1, j_2, \ldots, j_n)$ whenever a larger integer precedes a smaller one. The total number of inversions occurring in a permutation can be obtained as follows: (1) find the number of integers that are less than j_1 and that follow j_1 in the permutation; (2) find the number of integers that are less than j_2 and that follow j_2 in the permutation. Continue this counting process for $j_3, \ldots, j_{n-1}$. The sum of these numbers will be the total number of inversions in the permutation.

EXAMPLE 3 Counting Inversions

Determine the number of inversions in the following permutations:

$$\text{(a)} \ (6, 1, 3, 4, 5, 2) \qquad \text{(b)} \ (2, 4, 1, 3) \qquad \text{(c)} \ (1, 2, 3, 4)$$

Solution

(a) The number of inversions is $5 + 0 + 1 + 1 + 1 = 8$.

(b) The number of inversions is $1 + 2 + 0 = 3$.

(c) There are zero inversions in this permutation. ◆

DEFINITION

A permutation is called ***even*** if the total number of inversions is an even integer and is called ***odd*** if the total number of inversions is an odd integer.

EXAMPLE 4 Classifying Permutations

The following table classifies the various permutations of $\{1, 2, 3\}$ as even or odd.

Permutation	Number of Inversions	Classification
(1, 2, 3)	0	even
(1, 3, 2)	1	odd
(2, 1, 3)	1	odd
(2, 3, 1)	2	even
(3, 1, 2)	2	even
(3, 2, 1)	3	odd

◆

Combinatorial Definition of the Determinant

By an *elementary product* from an $n \times n$ matrix A we shall mean any product of n entries from A, no two of which come from the same row or the same column.

EXAMPLE 5 Elementary Products

List all elementary products from the matrices

$$\text{(a)} \begin{bmatrix} a_{11} & a_{12} \\ a_{21} & a_{22} \end{bmatrix} \qquad \text{(b)} \begin{bmatrix} a_{11} & a_{12} & a_{13} \\ a_{21} & a_{22} & a_{23} \\ a_{31} & a_{32} & a_{33} \end{bmatrix}$$

Solution (a)

Since each elementary product has two factors, and since each factor comes from a different row, an elementary product can be written in the form

$$a_{1\text{-}} a_{2\text{-}}$$

where the blanks designate column numbers. Since no two factors in the product come from the same column, the column numbers must be $\underline{1\,2}$ or $\underline{2\,1}$. Thus the only elementary products are $a_{11}a_{22}$ and $a_{12}a_{21}$.

Solution (b)

Since each elementary product has three factors, each of which comes from a different row, an elementary product can be written in the form

$$a_{1\text{-}} a_{2\text{-}} a_{3\text{-}}$$

Since no two factors in the product come from the same column, the column numbers have no repetitions; consequently, they must form a permutation of the set $\{1, 2, 3\}$. These $3! = 6$ permutations yield the following list of elementary products.

$$\begin{array}{ccc} a_{11}a_{22}a_{33} & a_{12}a_{21}a_{33} & a_{13}a_{21}a_{32} \\ a_{11}a_{23}a_{32} & a_{12}a_{23}a_{31} & a_{13}a_{22}a_{31} \end{array} \quad \blacklozenge$$

As this example points out, an $n \times n$ matrix A has $n!$ elementary products. They are the products of the form $a_{1j_1} a_{2j_2} \cdots a_{nj_n}$, where $(j_1, j_2, \ldots, j_n)$ is a permutation of the set $\{1, 2, \ldots, n\}$. By a *signed elementary product from A* we shall mean an elementary product $a_{1j_1} a_{2j_2} \cdots a_{nj_n}$ multiplied by $+1$ or -1. We use the $+$ if $(j_1, j_2, \ldots, j_n)$ is an even permutation and the $-$ if $(j_1, j_2, \ldots, j_n)$ is an odd permutation.

EXAMPLE 6 Signed Elementary Products

List all signed elementary products from the matrices

$$\text{(a)} \begin{bmatrix} a_{11} & a_{12} \\ a_{21} & a_{22} \end{bmatrix} \qquad \text{(b)} \begin{bmatrix} a_{11} & a_{12} & a_{13} \\ a_{21} & a_{22} & a_{23} \\ a_{31} & a_{32} & a_{33} \end{bmatrix}$$

Solution

(a)

Elementary Product	Associated Permutation	Even or Odd	Signed Elementary Product
$a_{11}a_{22}$	$(1, 2)$	even	$a_{11}a_{22}$
$a_{12}a_{21}$	$(2, 1)$	odd	$-a_{12}a_{21}$

(b)

Elementary Product	Associated Permutation	Even or Odd	Signed Elementary Product
$a_{11}a_{22}a_{33}$	$(1, 2, 3)$	even	$a_{11}a_{22}a_{33}$
$a_{11}a_{23}a_{32}$	$(1, 3, 2)$	odd	$-a_{11}a_{23}a_{32}$
$a_{12}a_{21}a_{33}$	$(2, 1, 3)$	odd	$-a_{12}a_{21}a_{33}$
$a_{12}a_{23}a_{31}$	$(2, 3, 1)$	even	$a_{12}a_{23}a_{31}$
$a_{13}a_{21}a_{32}$	$(3, 1, 2)$	even	$a_{13}a_{21}a_{32}$
$a_{13}a_{22}a_{31}$	$(3, 2, 1)$	odd	$-a_{13}a_{22}a_{31}$

◆

We are now in a position to give the combinatorial definition of the determinant function.

DEFINITION

Let A be a square matrix. We define $\det(A)$ to be the sum of all signed elementary products from A.

EXAMPLE 7 Determinants of 2 × 2 and 3 × 3 Matrices

Referring to Example 6, we obtain

(a) $\det \begin{bmatrix} a_{11} & a_{12} \\ a_{21} & a_{22} \end{bmatrix} = a_{11}a_{22} - a_{12}a_{21}$

(b) $\det \begin{bmatrix} a_{11} & a_{12} & a_{13} \\ a_{21} & a_{22} & a_{23} \\ a_{31} & a_{32} & a_{33} \end{bmatrix} = a_{11}a_{22}a_{33} + a_{12}a_{23}a_{31} + a_{13}a_{21}a_{32}$
$$- a_{13}a_{22}a_{31} - a_{12}a_{21}a_{33} - a_{11}a_{23}a_{32} \quad ◆$$

Of course, this definition of $\det(A)$ agrees with the definition in Section 2.1, although we will not prove this.

These expressions suggest the mnemonic devices given in Figure 2.4.2. The formula in part (a) of Example 7 is obtained from Figure 2.4.2a by multiplying the entries on the rightward arrow and subtracting the product of the entries on the leftward arrow. The formula in part (b) of Example 7 is obtained by recopying the first and second columns

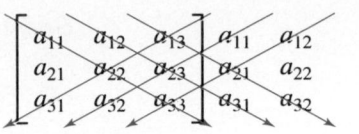

(a) Determinant of a 2 × 2 matrix   (b) Determinant of a 3 × 3 matrix

Figure 2.4.2

as shown in Figure 2.4.2b. The determinant is then computed by summing the products on the rightward arrows and subtracting the products on the leftward arrows.

WARNING   We emphasize that the methods shown in Figure 2.4.2 do not work for determinants of 4 × 4 matrices or higher.

EXAMPLE 8 Evaluating Determinants

Evaluate the determinants of

$$A = \begin{bmatrix} 3 & 1 \\ 4 & -2 \end{bmatrix} \quad \text{and} \quad B = \begin{bmatrix} 1 & 2 & 3 \\ -4 & 5 & 6 \\ 7 & -8 & 9 \end{bmatrix}$$

Solution

Using the method of Figure 2.4.2a gives

$$\det(A) = (3)(-2) - (1)(4) = -10$$

Using the method of Figure 2.4.2b gives

$$\det(B) = (45) + (84) + (96) - (105) - (-48) - (-72) = 240$$

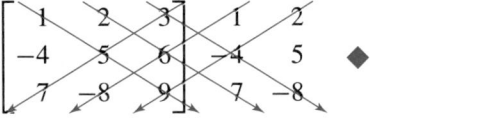

The determinant of A may be written as

$$\det(A) = \sum \pm a_{1j_1} a_{2j_2} \cdots a_{nj_n} \tag{1}$$

where $\sum$ indicates that the terms are to be summed over all permutations $(j_1, j_2, \ldots, j_n)$ and the + or − is selected in each term according to whether the permutation is even or odd. This notation is useful when the combinatorial definition of a determinant needs to be emphasized.

REMARK   Evaluating determinants directly from this definition leads to computational difficulties. Indeed, evaluating a 4 × 4 determinant directly would involve computing 4! = 24 signed elementary products, and a 10 × 10 determinant would require the computation of 10! = 3,628,800 signed elementary products. Even the fastest of digital computers cannot handle the computation of a 25×25 determinant by this method in a practical amount of time.

1. Find the number of inversions in each of the following permutations of $\{1, 2, 3, 4, 5\}$.
   (a) $(4\ 1\ 3\ 5\ 2)$ (b) $(5\ 3\ 4\ 2\ 1)$ (c) $(3\ 2\ 5\ 4\ 1)$
   (d) $(5\ 4\ 3\ 2\ 1)$ (e) $(1\ 2\ 3\ 4\ 5)$ (f) $(1\ 4\ 2\ 3\ 5)$

2. Classify each of the permutations in Exercise 1 as even or odd.

In Exercises 3–12 evaluate the determinant using the method of this section.

3. $\begin{vmatrix} 3 & 5 \\ -2 & 4 \end{vmatrix}$ 4. $\begin{vmatrix} 4 & 1 \\ 8 & 2 \end{vmatrix}$ 5. $\begin{vmatrix} -5 & 6 \\ -7 & -2 \end{vmatrix}$

6. $\begin{vmatrix} \sqrt{2} & \sqrt{6} \\ 4 & \sqrt{3} \end{vmatrix}$ 7. $\begin{vmatrix} a-3 & 5 \\ -3 & a-2 \end{vmatrix}$ 8. $\begin{vmatrix} -2 & 7 & 6 \\ 5 & 1 & -2 \\ 3 & 8 & 4 \end{vmatrix}$

9. $\begin{vmatrix} -2 & 1 & 4 \\ 3 & 5 & -7 \\ 1 & 6 & 2 \end{vmatrix}$ 10. $\begin{vmatrix} -1 & 1 & 2 \\ 3 & 0 & -5 \\ 1 & 7 & 2 \end{vmatrix}$

11. $\begin{vmatrix} 3 & 0 & 0 \\ 2 & -1 & 5 \\ 1 & 9 & -4 \end{vmatrix}$ 12. $\begin{vmatrix} c & -4 & 3 \\ 2 & 1 & c^2 \\ 4 & c-1 & 2 \end{vmatrix}$

13. Find all values of λ for which $\det(A) = 0$, using the method of this section.

    (a) $\begin{bmatrix} \lambda-2 & 1 \\ -5 & \lambda+4 \end{bmatrix}$ (b) $\begin{bmatrix} \lambda-4 & 0 & 0 \\ 0 & \lambda & 2 \\ 0 & 3 & \lambda-1 \end{bmatrix}$

14. Classify each permutation of $\{1, 2, 3, 4\}$ as even or odd.

15. (a) Use the results in Exercise 14 to construct a formula for the determinant of a 4×4 matrix.

    (b) Why do the mnemonics of Figure 2.4.2 fail for a 4×4 matrix?

16. Use the formula obtained in Exercise 15 to evaluate

$$\begin{vmatrix} 4 & -9 & 9 & 2 \\ -2 & 5 & 6 & 4 \\ 1 & 2 & -5 & -3 \\ 1 & -2 & 0 & -2 \end{vmatrix}$$

17. Use the combinatorial definition of the determinant to evaluate

    (a) $\begin{vmatrix} 0 & 0 & 0 & 0 & -3 \\ 0 & 0 & 0 & -4 & 0 \\ 0 & 0 & -1 & 0 & 0 \\ 0 & 2 & 0 & 0 & 0 \\ 5 & 0 & 0 & 0 & 0 \end{vmatrix}$ (b) $\begin{vmatrix} 5 & 0 & 0 & 0 & 0 \\ 0 & 0 & 0 & 0 & -4 \\ 0 & 0 & 3 & 0 & 0 \\ 0 & 0 & 0 & 1 & 0 \\ 0 & -2 & 0 & 0 & 0 \end{vmatrix}$

18. Solve for x.

$$\begin{vmatrix} x & -1 \\ 3 & 1-x \end{vmatrix} = \begin{vmatrix} 1 & 0 & -3 \\ 2 & x & -6 \\ 1 & 3 & x-5 \end{vmatrix}$$

19. Show that the value of the determinant

$$\begin{vmatrix} \sin\theta & \cos\theta & 0 \\ -\cos\theta & \sin\theta & 0 \\ \sin\theta - \cos\theta & \sin\theta + \cos\theta & 1 \end{vmatrix}$$

does not depend on θ, using the method of this section.

20. Prove that the matrices

$$A = \begin{bmatrix} a & b \\ 0 & c \end{bmatrix} \quad \text{and} \quad B = \begin{bmatrix} d & e \\ 0 & f \end{bmatrix}$$

commute if and only if

$$\begin{vmatrix} b & a-c \\ e & d-f \end{vmatrix} = 0$$

Discussion & Discovery

21. Explain why the determinant of an $n \times n$ matrix with integer entries must be an integer, using the method of this section.

22. What can you say about the determinant of an $n \times n$ matrix all of whose entries are 1? Explain your reasoning, using the method of this section.

23. (a) Explain why the determinant of an $n \times n$ matrix with a row of zeros must have a zero determinant, using the method of this section.

(b) Explain why the determinant of an $n \times n$ matrix with a column of zeros must have a zero determinant.

24. Use Formula (1) to discover a formula for the determinant of an $n \times n$ diagonal matrix. Express the formula in words.

25. Use Formula (1) to discover a formula for the determinant of an $n \times n$ upper triangular matrix. Express the formula in words. Do the same for a lower triangular matrix.

CHAPTER 2
Supplementary Exercises

1. Use Cramer's rule to solve for x' and y' in terms of x and y.

$$x = \tfrac{3}{5}x' - \tfrac{4}{5}y'$$
$$y = \tfrac{4}{5}x' + \tfrac{3}{5}y'$$

2. Use Cramer's rule to solve for x' and y' in terms of x and y.

$$x = x'\cos\theta - y'\sin\theta$$
$$y = x'\sin\theta + y'\cos\theta$$

3. By examining the determinant of the coefficient matrix, show that the following system has a nontrivial solution if and only if $\alpha = \beta$.

$$\begin{aligned} x + \ y + \alpha z &= 0 \\ x + \ y + \beta z &= 0 \\ \alpha x + \beta y + \ z &= 0 \end{aligned}$$

4. Let A be a 3×3 matrix, each of whose entries is 1 or 0. What is the largest possible value for $\det(A)$?

5. (a) For the triangle in the accompanying figure, use trigonometry to show that

$$b\cos\gamma + c\cos\beta = a$$
$$c\cos\alpha + a\cos\gamma = b$$
$$a\cos\beta + b\cos\alpha = c$$

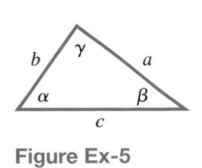

Figure Ex-5

and then apply Cramer's rule to show that

$$\cos \alpha = \frac{b^2 + c^2 - a^2}{2bc}$$

(b)   Use Cramer's rule to obtain similar formulas for $\cos \beta$ and $\cos \gamma$.

6.   Use determinants to show that for all real values of λ, the only solution of

$$x - 2y = \lambda x$$
$$x - y = \lambda y$$

is $x = 0$, $y = 0$.

7.   Prove: If A is invertible, then $\text{adj}(A)$ is invertible and

$$[\text{adj}(A)]^{-1} = \frac{1}{\det(A)} A = \text{adj}(A^{-1})$$

8.   Prove: If A is an $n \times n$ matrix, then $\det[\text{adj}(A)] = [\det(A)]^{n-1}$.

9.   **(For Readers Who Have Studied Calculus)** Show that if $f_1(x)$, $f_2(x)$, $g_1(x)$, and $g_2(x)$ are differentiable functions, and if

$$W = \begin{vmatrix} f_1(x) & f_2(x) \\ g_1(x) & g_2(x) \end{vmatrix}, \quad \text{then} \quad \frac{dW}{dx} = \begin{vmatrix} f_1'(x) & f_2'(x) \\ g_1(x) & g_2(x) \end{vmatrix} + \begin{vmatrix} f_1(x) & f_2(x) \\ g_1'(x) & g_2'(x) \end{vmatrix}$$

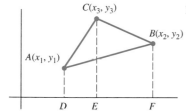

Figure Ex-10

10.   (a)   In the accompanying figure, the area of the triangle ABC can be expressed as

$$\text{area } ABC = \text{ area } ADEC + \text{ area } CEFB - \text{ area } ADFB$$

Use this and the fact that the area of a trapezoid equals $\frac{1}{2}$ the altitude times the sum of the parallel sides to show that

$$\text{area } ABC = \frac{1}{2} \begin{vmatrix} x_1 & y_1 & 1 \\ x_2 & y_2 & 1 \\ x_3 & y_3 & 1 \end{vmatrix}$$

Note   In the derivation of this formula, the vertices are labeled such that the triangle is traced counterclockwise proceeding from (x_1, y_1) to (x_2, y_2) to (x_3, y_3). For a clockwise orientation, the determinant above yields the *negative* of the area.

(b)   Use the result in (a) to find the area of the triangle with vertices $(3, 3)$, $(4, 0)$, $(-2, -1)$.

11.   Prove: If the entries in each row of an $n \times n$ matrix A add up to zero, then the determinant of A is zero.

Hint   Consider the product AX, where X is the $n \times 1$ matrix, each of whose entries is one.

12.   Let A be an $n \times n$ matrix and B the matrix that results when the rows of A are written in reverse order (last row becomes the first, and so forth). How are $\det(A)$ and $\det(B)$ related?

13.   Indicate how A^{-1} will be affected if

(a)   the ith and jth rows of A are interchanged.

(b)   the ith row of A is multiplied by a nonzero scalar, c.

(c) c times the ith row of A is added to the jth row.

14.   Let A be an $n \times n$ matrix. Suppose that B_1 is obtained by adding the same number t to each entry in the ith row of A and that B_2 is obtained by subtracting t from each entry in the ith row of A. Show that $\det(A) = \frac{1}{2}[\det(B_1) + \det(B_2)]$.

15.   Let

$$A = \begin{bmatrix} a_{11} & a_{12} & a_{13} \\ a_{21} & a_{22} & a_{23} \\ a_{31} & a_{32} & a_{33} \end{bmatrix}$$

(a)   Express $\det(\lambda I - A)$ as a polynomial $p(\lambda) = \lambda^3 + b\lambda^2 + c\lambda + d$.

(b)   Express the coefficients b and d in terms of determinants and traces.

16. Without directly evaluating the determinant, show that

$$\begin{vmatrix} \sin\alpha & \cos\alpha & \sin(\alpha+\delta) \\ \sin\beta & \cos\beta & \sin(\beta+\delta) \\ \sin\gamma & \cos\gamma & \sin(\gamma+\delta) \end{vmatrix} = 0$$

17. Use the fact that 21,375, 38,798, 34,162, 40,223, and 79,154 are all divisible by 19 to show that

$$\begin{vmatrix} 2 & 1 & 3 & 7 & 5 \\ 3 & 8 & 7 & 9 & 8 \\ 3 & 4 & 1 & 6 & 2 \\ 4 & 0 & 2 & 2 & 3 \\ 7 & 9 & 1 & 5 & 4 \end{vmatrix}$$

is divisible by 19 without directly evaluating the determinant.

18. Find the eigenvalues and corresponding eigenvectors for each of the following systems.

(a)
$$\begin{aligned} x_2 + 9x_3 &= \lambda x_1 \\ x_1 + 4x_2 - 7x_3 &= \lambda x_2 \\ x_1 \qquad\quad - 3x_3 &= \lambda x_3 \end{aligned}$$

(b)
$$\begin{aligned} x_2 + x_3 &= \lambda x_1 \\ x_1 \qquad - x_3 &= \lambda x_2 \\ x_1 + 5x_2 + 3x_3 &= \lambda x_3 \end{aligned}$$

CHAPTER 2

Technology Exercises

The following exercises are designed to be solved using a technology utility. Typically, this will be MATLAB, *Mathematica*, Maple, Derive, or Mathcad, but it may also be some other type of linear algebra software or a scientific calculator with some linear algebra capabilities. For each exercise you will need to read the relevant documentation for the particular utility you are using. The goal of these exercises is to provide you with a basic proficiency with your technology utility. Once you have mastered the techniques in these exercises, you will be able to use your technology utility to solve many of the problems in the regular exercise sets.

Section 2.1

T1. **(Determinants)** Read your documentation on how to compute determinants, and then compute several determinants.

T2. **(Minors, Cofactors, and Adjoints)** Technology utilities vary widely in their treatment of minors, cofactors, and adjoints. For example, some utilities have commands for computing minors but not cofactors, and some provide direct commands for finding adjoints, whereas others do not. Thus, depending on your utility, you may have to piece together commands or do some sign adjustment by hand to find cofactors and adjoints. Read your documentation, and then find the adjoint of the matrix *A* in Example 6.

T3. Use Cramer's rule to find a polynomial of degree 3 that passes through the points $(0, 1)$, $(1, -1)$, $(2, -1)$, and $(3, 7)$. Verify your results by plotting the points and the curve on one graph.

Section 2.2

T1. **(Determinant of a Transpose)** Confirm part (*b*) of Theorem 2.2.1 using some matrices of your choice.

Section 2.3

T1. **(Determinant of a Product)** Confirm Theorem 2.3.4 for some matrices of your choice.

T2. **(Determinant of an Inverse)** Confirm Theorem 2.3.5 for some matrices of your choice.

T3. **(Characteristic Equation)** If you are working with a CAS, use it to find the characteristic equation of the matrix *A* in Example 6. Also, read your documentation on how to solve equations, and then solve the equation $\det(\lambda I - A) = 0$ for the eigenvalues of *A*.

Section 2.4

T1. **(Determinant Formulas)** If you are working with a CAS, use it to confirm the formulas in Example 7. Also, use it to obtain the formula requested in Exercise 15 of Section 2.4.

T2. **(Simplification)** If you are working with a CAS, read the documentation on simplifying algebraic expressions, and then use the determinant and simplification commands in combination to show that

$$\begin{vmatrix} a & b & c & d \\ -b & a & d & -c \\ -c & -d & a & b \\ -d & c & -b & a \end{vmatrix} = (a^2 + b^2 + c^2 + d^2)^2$$

T3. Use the method of Exercise T2 to find a simple formula for the determinant

$$\begin{vmatrix} (a+b)^2 & c^2 & c^2 \\ a^2 & (b+c)^2 & a^2 \\ b^2 & b^2 & (c+a)^2 \end{vmatrix}$$

Vectors in 2-Space and 3-Space

CHAPTER CONTENTS

I NTRODUCTION: Many physical quantities, such as area, length, mass, and temperature, are completely described once the magnitude of the quantity is given. Such quantities are called **scalars**. Other physical quantities are not completely determined until both a magnitude and a direction are specified. These quantities are called **vectors**. For example, wind movement is usually described by giving the speed and direction, say 20 mph northeast. The wind speed and wind direction form a vector called the wind **velocity**. Other examples of vectors are **force** and **displacement**. In this chapter our goal is to review some of the basic theory of vectors in two and three dimensions.

Note. *Readers already familiar with the contents of this chapter can go to Chapter 4 with no loss of continuity.*

3.1

INTRODUCTION TO VECTORS (GEOMETRIC)

In this section, vectors in 2-space and 3-space will be introduced geometrically, arithmetic operations on vectors will be defined, and some basic properties of these arithmetic operations will be established.

Geometric Vectors

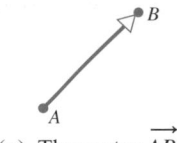

(*a*) The vector $\overrightarrow{AB}$

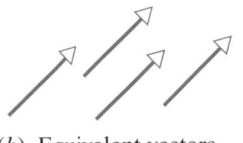

(*b*) Equivalent vectors

Figure 3.1.1

Vectors can be represented geometrically as directed line segments or arrows in 2-space or 3-space. The direction of the arrow specifies the direction of the vector, and the length of the arrow describes its magnitude. The tail of the arrow is called the ***initial point*** of the vector, and the tip of the arrow the ***terminal point***. Symbolically, we shall denote vectors in lowercase boldface type (for instance, **a**, **k**, **v**, **w**, and **x**). When discussing vectors, we shall refer to numbers as ***scalars***. For now, all our scalars will be real numbers and will be denoted in lowercase italic type (for instance, *a*, *k*, *v*, *w*, and *x*).

If, as in Figure 3.1.1*a*, the initial point of a vector **v** is *A* and the terminal point is *B*, we write

$$\mathbf{v} = \overrightarrow{AB}$$

Vectors with the same length and same direction, such as those in Figure 3.1.1*b*, are called ***equivalent***. Since we want a vector to be determined solely by its length and direction, equivalent vectors are regarded as ***equal*** even though they may be located in different positions. If **v** and **w** are equivalent, we write

$$\mathbf{v} = \mathbf{w}$$

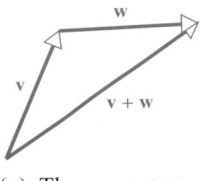

(*a*) The sum **v** + **w**

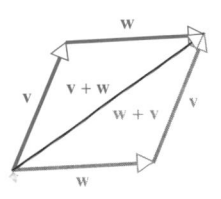

(*b*) **v** + **w** = **w** + **v**

Figure 3.1.2

DEFINITION

If **v** and **w** are any two vectors, then the ***sum*** **v** + **w** is the vector determined as follows: Position the vector **w** so that its initial point coincides with the terminal point of **v**. The vector **v** + **w** is represented by the arrow from the initial point of **v** to the terminal point of **w** (Figure 3.1.2*a*).

In Figure 3.1.2*b* we have constructed two sums, **v** + **w** (color arrows) and **w** + **v** (gray arrows). It is evident that

$$\mathbf{v} + \mathbf{w} = \mathbf{w} + \mathbf{v}$$

and that the sum coincides with the diagonal of the parallelogram determined by **v** and **w** when these vectors are positioned so that they have the same initial point.

The vector of length zero is called the ***zero vector*** and is denoted by **0**. We define

$$\mathbf{0} + \mathbf{v} = \mathbf{v} + \mathbf{0} = \mathbf{v}$$

for every vector **v**. Since there is no natural direction for the zero vector, we shall agree that it can be assigned any direction that is convenient for the problem being considered. If **v** is any nonzero vector, then −**v**, the ***negative*** of **v**, is defined to be the vector that has the same magnitude as **v** but is oppositely directed (Figure 3.1.3). This vector has the property

$$\mathbf{v} + (-\mathbf{v}) = \mathbf{0}$$

(Why?) In addition, we define −**0** = **0**. Subtraction of vectors is defined as follows:

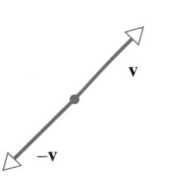

Figure 3.1.3 The negative of **v** has the same length as **v** but is oppositely directed.

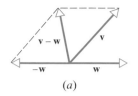

(a)

(b)

Figure 3.1.4

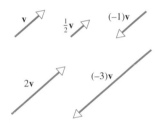

Figure 3.1.5

Vectors in Coordinate Systems

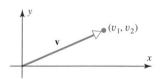

Figure 3.1.6 v_1 and v_2 are the components of **v**.

> **DEFINITION**
>
> If **v** and **w** are any two vectors, then the *difference* of **w** from **v** is defined by
> $$\mathbf{v} - \mathbf{w} = \mathbf{v} + (-\mathbf{w})$$
> (Figure 3.1.4a).

To obtain the difference $\mathbf{v} - \mathbf{w}$ without constructing $-\mathbf{w}$, position **v** and **w** so that their initial points coincide; the vector from the terminal point of **w** to the terminal point of **v** is then the vector $\mathbf{v} - \mathbf{w}$ (Figure 3.1.4b).

> **DEFINITION**
>
> If **v** is a nonzero vector and k is a nonzero real number (scalar), then the *product* $k\mathbf{v}$ is defined to be the vector whose length is $|k|$ times the length of **v** and whose direction is the same as that of **v** if $k > 0$ and opposite to that of **v** if $k < 0$. We define $k\mathbf{v} = \mathbf{0}$ if $k = 0$ or $\mathbf{v} = \mathbf{0}$.

Figure 3.1.5 illustrates the relation between a vector **v** and the vectors $\frac{1}{2}\mathbf{v}$, $(-1)\mathbf{v}$, $2\mathbf{v}$, and $(-3)\mathbf{v}$. Note that the vector $(-1)\mathbf{v}$ has the same length as **v** but is oppositely directed. Thus $(-1)\mathbf{v}$ is just the negative of **v**; that is,

$$(-1)\mathbf{v} = -\mathbf{v}$$

A vector of the form $k\mathbf{v}$ is called a *scalar multiple* of **v**. As evidenced by Figure 3.1.5, vectors that are scalar multiples of each other are parallel. Conversely, it can be shown that nonzero parallel vectors are scalar multiples of each other. We omit the proof.

Problems involving vectors can often be simplified by introducing a rectangular coordinate system. For the moment we shall restrict the discussion to vectors in 2-space (the plane). Let **v** be any vector in the plane, and assume, as in Figure 3.1.6, that **v** has been positioned so that its initial point is at the origin of a rectangular coordinate system. The coordinates (v_1, v_2) of the terminal point of **v** are called the *components of* **v**, and we write

$$\mathbf{v} = (v_1, v_2)$$

If equivalent vectors, **v** and **w**, are located so that their initial points fall at the origin, then it is obvious that their terminal points must coincide (since the vectors have the same length and direction); thus the vectors have the same components. Conversely, vectors with the same components are equivalent since they have the same length and the same direction. In summary, two vectors

$$\mathbf{v} = (v_1, v_2) \quad \text{and} \quad \mathbf{w} = (w_1, w_2)$$

are equivalent if and only if

$$v_1 = w_1 \quad \text{and} \quad v_2 = w_2$$

The operations of vector addition and multiplication by scalars are easy to carry out in terms of components. As illustrated in Figure 3.1.7, if

$$\mathbf{v} = (v_1, v_2) \quad \text{and} \quad \mathbf{w} = (w_1, w_2)$$

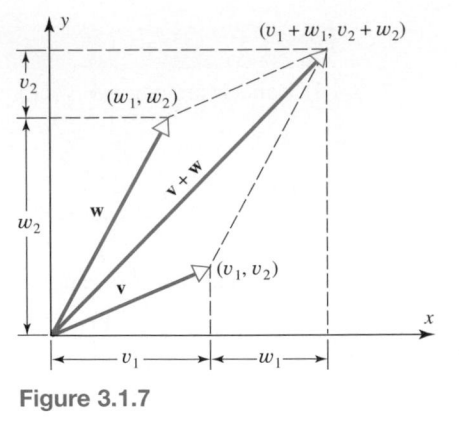

Figure 3.1.7

then

$$\mathbf{v} + \mathbf{w} = (v_1 + w_1, v_2 + w_2) \tag{1}$$

If $\mathbf{v} = (v_1, v_2)$ and k is any scalar, then by using a geometric argument involving similar triangles, it can be shown (Exercise 16) that

$$k\mathbf{v} = (kv_1, kv_2) \tag{2}$$

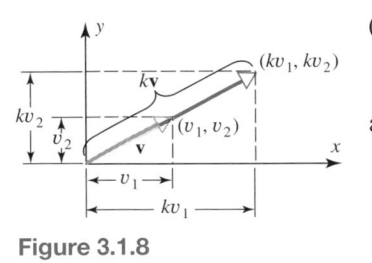

Figure 3.1.8

(Figure 3.1.8). Thus, for example, if $\mathbf{v} = (1, -2)$ and $\mathbf{w} = (7, 6)$, then

$$\mathbf{v} + \mathbf{w} = (1, -2) + (7, 6) = (1 + 7, -2 + 6) = (8, 4)$$

and

$$4\mathbf{v} = 4(1, -2) = (4(1), 4(-2)) = (4, -8)$$

Since $\mathbf{v} - \mathbf{w} = \mathbf{v} + (-1)\mathbf{w}$, it follows from Formulas (1) and (2) that

$$\mathbf{v} - \mathbf{w} = (v_1 - w_1, v_2 - w_2)$$

(Verify.)

Vectors in 3-Space   Just as vectors in the plane can be described by pairs of real numbers, vectors in 3-space can be described by triples of real numbers by introducing a ***rectangular coordinate*** system. To construct such a coordinate system, select a point O, called the ***origin***, and choose three mutually perpendicular lines, called ***coordinate axes***, passing through the origin. Label these axes x, y, and z, and select a positive direction for each coordinate

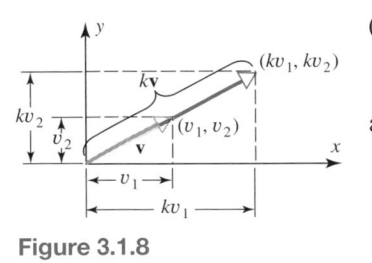

(a) (b)

Figure 3.1.9

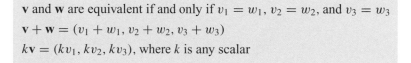

Figure 3.1.10

axis as well as a unit of length for measuring distances (Figure 3.1.9*a*). Each pair of coordinate axes determines a plane called a ***coordinate plane***. These are referred to as the ***xy-plane***, the ***xz-plane***, and the ***yz-plane***. To each point P in 3-space we assign a triple of numbers (x, y, z), called the ***coordinates of P***, as follows: Pass three planes through P parallel to the coordinate planes, and denote the points of intersection of these planes with the three coordinate axes by X, Y, and Z (Figure 3.1.9*b*). The coordinates of P are defined to be the signed lengths

$$x = OX, \qquad y = OY, \qquad z = OZ$$

In Figure 3.1.10*a* we have constructed the point whose coordinates are $(4, 5, 6)$ and in Figure 3.1.10*b* the point whose coordinates are $(-3, 2, -4)$.

Rectangular coordinate systems in 3-space fall into two categories, ***left-handed*** and ***right-handed***. A right-handed system has the property that an ordinary screw pointed in the positive direction on the z-axis would be advanced if the positive x-axis were rotated 90° toward the positive y-axis (Figure 3.1.11*a*); the system is left-handed if the screw would be retracted (Figure 3.1.11*b*).

REMARK In this book we shall use only right-handed coordinate systems.

If, as in Figure 3.1.12, a vector **v** in 3-space is positioned so its initial point is at the origin of a rectangular coordinate system, then the coordinates of the terminal point are called the ***components*** of **v**, and we write

$$\mathbf{v} = (v_1, v_2, v_3)$$

If $\mathbf{v} = (v_1, v_2, v_3)$ and $\mathbf{w} = (w_1, w_2, w_3)$ are two vectors in 3-space, then arguments similar to those used for vectors in a plane can be used to establish the following results.

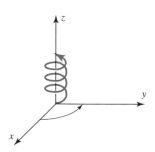

(*a*) Right-handed

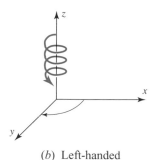

(*b*) Left-handed

Figure 3.1.11

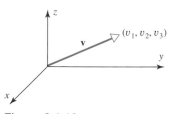

Figure 3.1.12

> **v** and **w** are equivalent if and only if $v_1 = w_1$, $v_2 = w_2$, and $v_3 = w_3$
>
> $\mathbf{v} + \mathbf{w} = (v_1 + w_1, v_2 + w_2, v_3 + w_3)$
>
> $k\mathbf{v} = (kv_1, kv_2, kv_3)$, where k is any scalar

EXAMPLE 1 Vector Computations with Components

If $\mathbf{v} = (1, -3, 2)$ and $\mathbf{w} = (4, 2, 1)$, then

$$\mathbf{v} + \mathbf{w} = (5, -1, 3), \qquad 2\mathbf{v} = (2, -6, 4), \qquad -\mathbf{w} = (-4, -2, -1),$$
$$\mathbf{v} - \mathbf{w} = \mathbf{v} + (-\mathbf{w}) = (-3, -5, 1) \quad \blacklozenge$$

Application to Computer Color Models

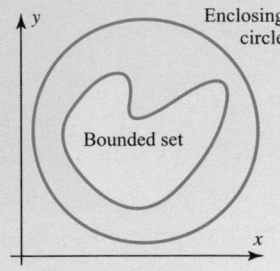

(*a*) Set enclosed by a circle

Colors on computer monitors are commonly based on what is called the **RGB** *color model*. Colors in this system are created by adding together percentages of the primary colors red (R), green (G), and blue (B). One way to do this is to identify the primary colors with the vectors

$$\mathbf{r} = (1, 0, 0) \quad \text{(pure red)},$$

$$\mathbf{g} = (0, 1, 0) \quad \text{(pure green)},$$

$$\mathbf{b} = (0, 0, 1) \quad \text{(pure blue)}$$

in R^3 and to create all other colors by forming linear combinations of $\mathbf{r}$, $\mathbf{g}$, and $\mathbf{b}$ using coefficients between 0 and 1, inclusive; these coefficients represent the percentage of each pure color in the mix. The set of all such color vectors is called **RGB** *space* or the **RGB** *color cube*. Thus, each color vector $\mathbf{c}$ in this cube is expressible as a linear combination of the form

$$\mathbf{c} = c_1 \mathbf{r} + c_2 \mathbf{g} + c_3 \mathbf{b}$$

$$= c_1(1, 0, 0) + c_2(0, 1, 0) + c_3(0, 0, 1)$$

$$= (c_1, c_2, c_3)$$

where $0 \leq c_i \leq 1$. As indicated in the figure, the corners of the cube represent the pure primary colors together with the colors, black, white, magenta, cyan, and yellow. The vectors along the diagonal running from black to white correspond to shades of gray.

(*b*) This set cannot be enclosed by any circle.

Sometimes a vector is positioned so that its initial point is not at the origin. If the vector $\overrightarrow{P_1 P_2}$ has initial point $P_1(x_1, y_1, z_1)$ and terminal point $P_2(x_2, y_2, z_2)$, then

$$\overrightarrow{P_1 P_2} = (x_2 - x_1, y_2 - y_1, z_2 - z_1)$$

That is, the components of $\overrightarrow{P_1 P_2}$ are obtained by subtracting the coordinates of the initial point from the coordinates of the terminal point. This may be seen using Figure 3.1.13: The vector $\overrightarrow{P_1 P_2}$ is the difference of vectors $\overrightarrow{OP_2}$ and $\overrightarrow{OP_1}$, so

$$\overrightarrow{P_1 P_2} = \overrightarrow{OP_2} - \overrightarrow{OP_1} = (x_2, y_2, z_2) - (x_1, y_1, z_1) = (x_2 - x_1, y_2 - y_1, z_2 - z_1)$$

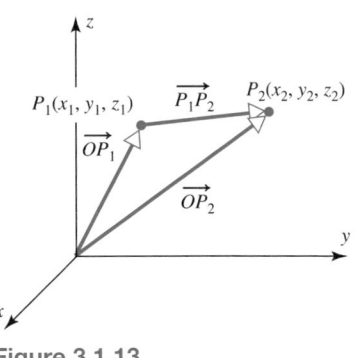

Figure 3.1.13

EXAMPLE 2 **Finding the Components of a Vector**

The components of the vector $\mathbf{v} = \overrightarrow{P_1 P_2}$ with initial point $P_1(2, -1, 4)$ and terminal point $P_2(7, 5, -8)$ are

$$\mathbf{v} = (7 - 2, 5 - (-1), (-8) - 4) = (5, 6, -12) \quad \blacklozenge$$

In 2-space the vector with initial point $P_1(x_1, y_1)$ and terminal point $P_2(x_2, y_2)$ is

$$\overrightarrow{P_1 P_2} = (x_2 - x_1, y_2 - y_1)$$

Translation of Axes

The solutions to many problems can be simplified by translating the coordinate axes to obtain new axes parallel to the original ones.

In Figure 3.1.14*a* we have translated the axes of an xy-coordinate system to obtain an $x'y'$-coordinate system whose origin O' is at the point $(x, y) = (k, l)$. A point P in 2-space now has both (x, y) coordinates and (x', y') coordinates. To see how the two are related, consider the vector $\overrightarrow{O'P}$ (Figure 3.1.14*b*). In the xy-system its initial point is at (k, l) and its terminal point is at (x, y), so $\overrightarrow{O'P} = (x - k, y - l)$. In the $x'y'$-system its initial point is at $(0, 0)$ and its terminal point is at (x', y'), so $\overrightarrow{O'P} = (x', y')$. Therefore,

$$x' = x - k, \qquad y' = y - l$$

These formulas are called the ***translation equations***.

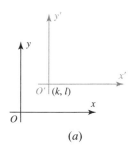

(a)

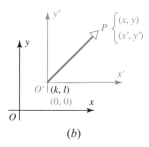

(b)

Figure 3.1.14

EXAMPLE 3   Using the Translation Equations

Suppose that an xy-coordinate system is translated to obtain an $x'y'$-coordinate system whose origin has xy-coordinates $(k, l) = (4, 1)$.

(a) Find the $x'y'$-coordinates of the point with the xy-coordinates $P(2, 0)$.

(b) Find the xy-coordinates of the point with $x'y'$-coordinates $Q(-1, 5)$.

Solution (a)

The translation equations are

$$x' = x - 4, \qquad y' = y - 1$$

so the $x'y'$-coordinates of $P(2, 0)$ are $x' = 2 - 4 = -2$ and $y' = 0 - 1 = -1$.

Solution (b)

The translation equations in (a) can be rewritten as

$$x = x' + 4, \qquad y = y' + 1$$

so the xy-coordinates of Q are $x = -1 + 4 = 3$ and $y = 5 + 1 = 6$. ◆

In 3-space the translation equations are

$$x' = x - k, \qquad y' = y - l, \qquad z' = z - m$$

where (k, l, m) are the xyz-coordinates of the $x'y'z'$-origin.

EXERCISE SET
3.1

1. Draw a right-handed coordinate system and locate the points whose coordinates are
   - (a) $(3, 4, 5)$
   - (b) $(-3, 4, 5)$
   - (c) $(3, -4, 5)$
   - (d) $(3, 4, -5)$
   - (e) $(-3, -4, 5)$
   - (f) $(-3, 4, -5)$
   - (g) $(3, -4, -5)$
   - (h) $(-3, -4, -5)$
   - (i) $(-3, 0, 0)$
   - (j) $(3, 0, 3)$
   - (k) $(0, 0, -3)$
   - (l) $(0, 3, 0)$

2. Sketch the following vectors with the initial points located at the origin:
   - (a) $\mathbf{v}_1 = (3, 6)$
   - (b) $\mathbf{v}_2 = (-4, -8)$
   - (c) $\mathbf{v}_3 = (-4, -3)$
   - (d) $\mathbf{v}_4 = (5, -4)$
   - (e) $\mathbf{v}_5 = (3, 0)$
   - (f) $\mathbf{v}_6 = (0, -7)$
   - (g) $\mathbf{v}_7 = (3, 4, 5)$
   - (h) $\mathbf{v}_8 = (3, 3, 0)$
   - (i) $\mathbf{v}_9 = (0, 0, -3)$

3. Find the components of the vector having initial point P_1 and terminal point P_2.
   - (a) $P_1(4, 8)$, $P_2(3, 7)$
   - (b) $P_1(3, -5)$, $P_2(-4, -7)$
   - (c) $P_1(-5, 0)$, $P_2(-3, 1)$
   - (d) $P_1(0, 0)$, $P_2(a, b)$
   - (e) $P_1(3, -7, 2)$, $P_2(-2, 5, -4)$
   - (f) $P_1(-1, 0, 2)$, $P_2(0, -1, 0)$
   - (g) $P_1(a, b, c)$, $P_2(0, 0, 0)$
   - (h) $P_1(0, 0, 0)$, $P_2(a, b, c)$

4. Find a nonzero vector $\mathbf{u}$ with initial point $P(-1, 3, -5)$ such that
   - (a) $\mathbf{u}$ has the same direction as $\mathbf{v} = (6, 7, -3)$
   - (b) $\mathbf{u}$ is oppositely directed to $\mathbf{v} = (6, 7, -3)$

5. Find a nonzero vector $\mathbf{u}$ with terminal point $Q(3, 0, -5)$ such that
   - (a) $\mathbf{u}$ has the same direction as $\mathbf{v} = (4, -2, -1)$
   - (b) $\mathbf{u}$ is oppositely directed to $\mathbf{v} = (4, -2, -1)$

6. Let $\mathbf{u} = (-3, 1, 2)$, $\mathbf{v} = (4, 0, -8)$, and $\mathbf{w} = (6, -1, -4)$. Find the components of
   - (a) $\mathbf{v} - \mathbf{w}$
   - (b) $6\mathbf{u} + 2\mathbf{v}$
   - (c) $-\mathbf{v} + \mathbf{u}$
   - (d) $5(\mathbf{v} - 4\mathbf{u})$
   - (e) $-3(\mathbf{v} - 8\mathbf{w})$
   - (f) $(2\mathbf{u} - 7\mathbf{w}) - (8\mathbf{v} + \mathbf{u})$

7. Let $\mathbf{u}$, $\mathbf{v}$, and $\mathbf{w}$ be the vectors in Exercise 6. Find the components of the vector $\mathbf{x}$ that satisfies $2\mathbf{u} - \mathbf{v} + \mathbf{x} = 7\mathbf{x} + \mathbf{w}$.

8. Let $\mathbf{u}$, $\mathbf{v}$, and $\mathbf{w}$ be the vectors in Exercise 6. Find scalars c_1, c_2, and c_3 such that

$$c_1\mathbf{u} + c_2\mathbf{v} + c_3\mathbf{w} = (2, 0, 4)$$

9. Show that there do not exist scalars c_1, c_2, and c_3 such that

$$c_1(-2, 9, 6) + c_2(-3, 2, 1) + c_3(1, 7, 5) = (0, 5, 4)$$

10. Find all scalars c_1, c_2, and c_3 such that

$$c_1(1, 2, 0) + c_2(2, 1, 1) + c_3(0, 3, 1) = (0, 0, 0)$$

11. Let P be the point $(2, 3, -2)$ and Q the point $(7, -4, 1)$.
    - (a) Find the midpoint of the line segment connecting P and Q.
    - (b) Find the point on the line segment connecting P and Q that is $\frac{3}{4}$ of the way from P to Q.

12. Suppose an xy-coordinate system is translated to obtain an $x'y'$-coordinate system whose origin O' has xy-coordinates $(2, -3)$.
    - (a) Find the $x'y'$-coordinates of the point P whose xy-coordinates are $(7, 5)$.
    - (b) Find the xy-coordinates of the point Q whose $x'y'$-coordinates are $(-3, 6)$.
    - (c) Draw the xy and $x'y'$-coordinate axes and locate the points P and Q.
    - (d) If $\mathbf{v} = (3, 7)$ is a vector in the xy-coordinate system, what are the components of $\mathbf{v}$ in the $x'y'$-coordinate system?
    - (e) If $\mathbf{v} = (v_1, v_2)$ is a vector in the xy-coordinate system, what are the components of $\mathbf{v}$ in the $x'y'$-coordinate system?

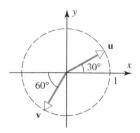

Figure Ex-15

13. Let P be the point $(1, 3, 7)$. If the point $(4, 0, -6)$ is the midpoint of the line segment connecting P and Q, what is Q?

14. Suppose that an xyz-coordinate system is translated to obtain an $x'y'z'$-coordinate system. Let $\mathbf{v}$ be a vector whose components are $\mathbf{v} = (v_1, v_2, v_3)$ in the xyz-system. Show that $\mathbf{v}$ has the same components in the $x'y'z'$-system.

15. Find the components of $\mathbf{u}$, $\mathbf{v}$, $\mathbf{u} + \mathbf{v}$, and $\mathbf{u} - \mathbf{v}$ for the vectors shown in the accompanying figure.

16. Prove geometrically that if $\mathbf{v} = (v_1, v_2)$, then $k\mathbf{v} = (kv_1, kv_2)$. (Restrict the proof to the case $k > 0$ illustrated in Figure 3.1.8. The complete proof would involve various cases that depend on the sign of k and the quadrant in which the vector falls.)

Discussion & Discovery

17. Consider Figure 3.1.13. Discuss a geometric interpretation of the vector
$$\mathbf{u} = \overrightarrow{OP_1} + \tfrac{1}{2}(\overrightarrow{OP_2} - \overrightarrow{OP_1})$$

18. Draw a picture that shows four nonzero vectors whose sum is zero.

19. If you were given four nonzero vectors, how would you construct geometrically a fifth vector that is equal to the sum of the first four? Draw a picture to illustrate your method.

20. Consider a clock with vectors drawn from the center to each hour as shown in the accompanying figure.

   (a) What is the sum of the 12 vectors that result if the vector terminating at 12 is doubled in length and the other vectors are left alone?

   (b) What is the sum of the 12 vectors that result if the vectors terminating at 3 and 9 are each tripled and the others are left alone?

   (c) What is the sum of the 9 vectors that remain if the vectors terminating at 5, 11, and 8 are removed?

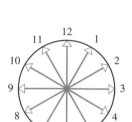

Figure Ex-20

21. Indicate whether the statement is true (T) or false (F). Justify your answer.

   (a) If $\mathbf{x} + \mathbf{y} = \mathbf{x} + \mathbf{z}$, then $\mathbf{y} = \mathbf{x}$.

   (b) If $\mathbf{u} + \mathbf{v} = \mathbf{0}$, then $a\mathbf{u} + b\mathbf{v} = \mathbf{0}$ for all a and b.

   (c) Parallel vectors with the same length are equal.

   (d) If $a\mathbf{x} = \mathbf{0}$, then either $a = 0$ or $\mathbf{x} = \mathbf{0}$.

   (e) If $a\mathbf{u} + b\mathbf{v} = \mathbf{0}$, then $\mathbf{u}$ and $\mathbf{v}$ are parallel vectors.

   (f) The vectors $\mathbf{u} = (\sqrt{2}, \sqrt{3})$ and $\mathbf{v} = \left(\frac{1}{\sqrt{2}}, \frac{1}{2}\sqrt{3}\right)$ are equivalent.

3.2
NORM OF A VECTOR; VECTOR ARITHMETIC

In this section we shall establish the basic rules of vector arithmetic.

Properties of Vector Operations

The following theorem lists the most important properties of vectors in 2-space and 3-space.

THEOREM 3.2.1

> **Properties of Vector Arithmetic**
>
> *If $\mathbf{u}$, $\mathbf{v}$, and $\mathbf{w}$ are vectors in 2- or 3-space and k and l are scalars, then the following relationships hold.*
>
> (*a*) $\mathbf{u} + \mathbf{v} = \mathbf{v} + \mathbf{u}$ (*b*) $(\mathbf{u} + \mathbf{v}) + \mathbf{w} = \mathbf{u} + (\mathbf{v} + \mathbf{w})$
>
> (*c*) $\mathbf{u} + \mathbf{0} = \mathbf{0} + \mathbf{u} = \mathbf{u}$ (*d*) $\mathbf{u} + (-\mathbf{u}) = \mathbf{0}$

> (e) $k(l\mathbf{u}) = (kl)\mathbf{u}$ (f) $k(\mathbf{u} + \mathbf{v}) = k\mathbf{u} + k\mathbf{v}$
>
> (g) $(k + l)\mathbf{u} = k\mathbf{u} + l\mathbf{u}$ (h) $1\mathbf{u} = \mathbf{u}$

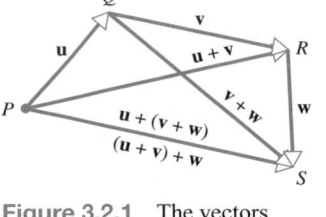

Figure 3.2.1 The vectors
$\mathbf{u} + (\mathbf{v} + \mathbf{w})$ and $(\mathbf{u} + \mathbf{v}) + \mathbf{w}$
are equal.

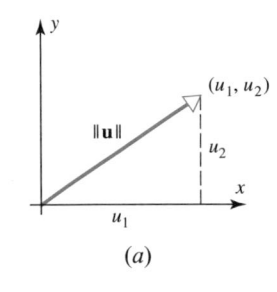

(a)

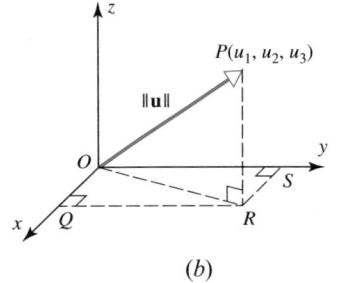

(b)

Figure 3.2.2

Norm of a Vector

Before discussing the proof, we note that we have developed two approaches to vectors: *geometric*, in which vectors are represented by arrows or directed line segments, and *analytic*, in which vectors are represented by pairs or triples of numbers called components. As a consequence, the equations in Theorem 1 can be proved either geometrically or analytically. To illustrate, we shall prove part (b) both ways. The remaining proofs are left as exercises.

Proof of part (b) (analytic) We shall give the proof for vectors in 3-space; the proof for 2-space is similar. If $\mathbf{u} = (u_1, u_2, u_3)$, $\mathbf{v} = (v_1, v_2, v_3)$, and $\mathbf{w} = (w_1, w_2, w_3)$, then

$$
\begin{aligned}
(\mathbf{u} + \mathbf{v}) + \mathbf{w} &= [(u_1, u_2, u_3) + (v_1, v_2, v_3)] + (w_1, w_2, w_3) \\
&= (u_1 + v_1, u_2 + v_2, u_3 + v_3) + (w_1, w_2, w_3) \\
&= ([u_1 + v_1] + w_1, [u_2 + v_2] + w_2, [u_3 + v_3] + w_3) \\
&= (u_1 + [v_1 + w_1], u_2 + [v_2 + w_2], u_3 + [v_3 + w_3]) \\
&= (u_1, u_2, u_3) + (v_1 + w_1, v_2 + w_2, v_3 + w_3) \\
&= \mathbf{u} + (\mathbf{v} + \mathbf{w})
\end{aligned}
$$

Proof of part (b) (geometric) Let $\mathbf{u}$, $\mathbf{v}$, and $\mathbf{w}$ be represented by $\overrightarrow{PQ}$, $\overrightarrow{QR}$, and $\overrightarrow{RS}$ as shown in Figure 3.2.1. Then

$$\mathbf{v} + \mathbf{w} = \overrightarrow{QS} \quad \text{and} \quad \mathbf{u} + (\mathbf{v} + \mathbf{w}) = \overrightarrow{PS}$$

Also,

$$\mathbf{u} + \mathbf{v} = \overrightarrow{PR} \quad \text{and} \quad (\mathbf{u} + \mathbf{v}) + \mathbf{w} = \overrightarrow{PS}$$

Therefore,

$$\mathbf{u} + (\mathbf{v} + \mathbf{w}) = (\mathbf{u} + \mathbf{v}) + \mathbf{w} \qquad\blacksquare$$

REMARK In light of part (b) of this theorem, the symbol $\mathbf{u} + \mathbf{v} + \mathbf{w}$ is unambiguous since the same sum is obtained no matter where parentheses are inserted. Moreover, if the vectors $\mathbf{u}$, $\mathbf{v}$, and $\mathbf{w}$ are placed "tip to tail," then the sum $\mathbf{u} + \mathbf{v} + \mathbf{w}$ is the vector from the initial point of $\mathbf{u}$ to the terminal point of $\mathbf{w}$ (Figure 3.2.1).

The ***length*** of a vector $\mathbf{u}$ is often called the ***norm*** of $\mathbf{u}$ and is denoted by $\|\mathbf{u}\|$. It follows from the Theorem of Pythagoras that the norm of a vector $\mathbf{u} = (u_1, u_2)$ in 2-space is

$$\|\mathbf{u}\| = \sqrt{u_1^2 + u_2^2} \tag{1}$$

(Figure 3.2.2a). Let $\mathbf{u} = (u_1, u_2, u_3)$ be a vector in 3-space. Using Figure 3.2.2b and two applications of the Theorem of Pythagoras, we obtain

$$\|\mathbf{u}\|^2 = (OR)^2 + (RP)^2 = (OQ)^2 + (OS)^2 + (RP)^2 = u_1^2 + u_2^2 + u_3^2$$

Thus

$$\|\mathbf{u}\| = \sqrt{u_1^2 + u_2^2 + u_3^2} \tag{2}$$

A vector of norm 1 is called a ***unit vector***.

Global Positioning

GPS (*Global Positioning System*) is the system used by the military, ships, airplane pilots, surveyors, utility companies, automobiles, and hikers to locate current positions by communicating with a system of satellites. The system, which is operated by the U.S. Department of Defense, nominally uses 24 satellites that orbit the Earth every 12 hours at a height of about 11,000 miles. These satellites move in six orbital planes that have been chosen to make between five and eight satellites visible from any point on Earth.

To explain how the system works, assume that the Earth is a sphere, and suppose that there is an xyz-coordinate system with its origin at the Earth's center and its z-axis through the North Pole. Let us assume that relative to this coordinate system a ship is at an unknown point (x, y, z) at some time t. For simplicity, assume that distances are measured in units equal to the Earth's radius, so that the coordinates of the ship always satisfy the equation

$$x^2 + y^2 + z^2 = 1$$

The GPS identifies the ship's coordinates (x, y, z) at a time t using a triangulation system and computed distances from four satellites. These distances are computed using the speed of light (approximately 0.469 Earth radii per hundredth of a second) and the time it takes for the signal to travel from the satellite to the ship. For example, if the ship receives the signal

at time t and the satellite indicates that it transmitted the signal at time t_0, then the distance d traveled by the signal will be

$$d = 0.469(t - t_0)$$

In theory, knowing three ship-to-satellite distances would suffice to determine the three unknown coordinates of the ship. However, the problem is that the ships (or other GPS users) do not generally have clocks that can compute t with sufficient accuracy for global positioning. Thus, the variable t must be regarded as a fourth unknown, and hence the need for the distance to a fourth satellite. Suppose that in addition to transmitting the time t_0, each satellite also transmits its coordinates (x_0, y_0, z_0) at that time, thereby allowing d to be computed as

$$d = \sqrt{(x - x_0)^2 + (y - y_0)^2 + (z - z_0)^2}$$

If we now equate the squares of d from both equations and round off to three decimal places, then we obtain the second-degree equation

$$(x - x_0)^2 + (y - y_0)^2 + (z - z_0)^2$$
$$= 0.22(t - t_0)^2$$

Since there are four different satellites, and we can get an equation like this for each one, we can produce four equations in the unknowns x, y, z, and t_0. Although these are second-degree equations, it is possible to use these equations and some algebra to produce a system of linear equations that can be solved for the unknowns.

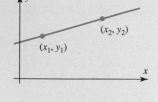

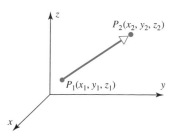

Figure 3.2.3 The distance between P_1 and P_2 is the norm of the vector $\overrightarrow{P_1 P_2}$.

If $P_1(x_1, y_1, z_1)$ and $P_2(x_2, y_2, z_2)$ are two points in 3-space, then the **distance** d between them is the norm of the vector $\overrightarrow{P_1 P_2}$ (Figure 3.2.3). Since

$$\overrightarrow{P_1 P_2} = (x_2 - x_1, y_2 - y_1, z_2 - z_1)$$

it follows from (2) that

$$d = \sqrt{(x_2 - x_1)^2 + (y_2 - y_1)^2 + (z_2 - z_1)^2} \tag{3}$$

Similarly, if $P_1(x_1, y_1)$ and $P_2(x_2, y_2)$ are points in 2-space, then the distance between them is given by

$$d = \sqrt{(x_2 - x_1)^2 + (y_2 - y_1)^2} \tag{4}$$

EXAMPLE 1 Finding Norm and Distance

The norm of the vector $\mathbf{u} = (-3, 2, 1)$ is

$$\|\mathbf{u}\| = \sqrt{(-3)^2 + (2)^2 + (1)^2} = \sqrt{14}$$

The distance d between the points $P_1(2, -1, -5)$ and $P_2(4, -3, 1)$ is

$$d = \sqrt{(4 - 2)^2 + (-3 + 1)^2 + (1 + 5)^2} = \sqrt{44} = 2\sqrt{11} \; \blacklozenge$$

From the definition of the product $k\mathbf{u}$, the length of the vector $k\mathbf{u}$ is $|k|$ times the length of $\mathbf{u}$. Expressed as an equation, this statement says that

$$\|k\mathbf{u}\| = |k|\|\mathbf{u}\| \tag{5}$$

This useful formula is applicable in both 2-space and 3-space.

EXERCISE SET 3.2

1. Find the norm of $\mathbf{v}$.
   (a) $\mathbf{v} = (4, -3)$
   (b) $\mathbf{v} = (2, 3)$
   (c) $\mathbf{v} = (-5, 0)$
   (d) $\mathbf{v} = (2, 2, 2)$
   (e) $\mathbf{v} = (-7, 2, -1)$
   (f) $\mathbf{v} = (0, 6, 0)$

2. Find the distance between P_1 and P_2.
   (a) $P_1(3, 4),\ P_2(5, 7)$
   (b) $P_1(-3, 6),\ P_2(-1, -4)$
   (c) $P_1(7, -5, 1),\ P_2(-7, -2, -1)$
   (d) $P_1(3, 3, 3),\ P_2(6, 0, 3)$

3. Let $\mathbf{u} = (2, -2, 3)$, $\mathbf{v} = (1, -3, 4)$, $\mathbf{w} = (3, 6, -4)$. In each part, evaluate the expression.
   (a) $\|\mathbf{u} + \mathbf{v}\|$
   (b) $\|\mathbf{u}\| + \|\mathbf{v}\|$
   (c) $\|-2\mathbf{u}\| + 2\|\mathbf{u}\|$
   (d) $\|3\mathbf{u} - 5\mathbf{v} + \mathbf{w}\|$
   (e) $\dfrac{1}{\|\mathbf{w}\|}\mathbf{w}$
   (f) $\left\|\dfrac{1}{\|\mathbf{w}\|}\mathbf{w}\right\|$

4. If $\|\mathbf{v}\| = 2$ and $\|\mathbf{w}\| = 3$, what are the largest and smallest values possible for $\|\mathbf{v} - \mathbf{w}\|$? Give a geometric explanation of your results.

5. Let $\mathbf{u} = (2, 0, 4)$ and $\mathbf{v} = (1, 3, -6)$. In each of the following, determine, if possible, scalars k, l such that
   (a) $k\mathbf{u} + l\mathbf{v} = (5, 9, -14)$
   (b) $k\mathbf{u} + l\mathbf{v} = (9, 15, -21)$

6. Let $\mathbf{u} = (2, 6, -7)$, $\mathbf{v} = (-1, -1, 8)$, and $k = 3$. If $(2, 14, 11) = k\mathbf{u} + l\mathbf{v}$, what is the value of l?

7. Let $\mathbf{v} = (-1, 2, 5)$. Find all scalars k such that $\|k\mathbf{v}\| = 4$.

8. Let $\mathbf{u} = (7, -3, 1)$, $\mathbf{v} = (9, 6, 6)$, $\mathbf{w} = (2, 1, -8)$, $k = -2$, and $l = 5$. Verify that these vectors and scalars satisfy the stated equalities from Theorem 1.
   (a) part (b)
   (b) part (e)
   (c) part (f)
   (d) part (g)

9. (a) Show that if $\mathbf{v}$ is any nonzero vector, then $\dfrac{1}{\|\mathbf{v}\|}\mathbf{v}$ is a unit vector.

   (b) Use the result in part (a) to find a unit vector that has the same direction as the vector $\mathbf{v} = (3, 4)$.

   (c) Use the result in part (a) to find a unit vector that is oppositely directed to the vector $\mathbf{v} = (-2, 3, -6)$.

10. (a) Show that the components of the vector $\mathbf{v} = (v_1, v_2)$ in Figure Ex-10a are $v_1 = \|\mathbf{v}\| \cos \theta$ and $v_2 = \|\mathbf{v}\| \sin \theta$.

(b) Let $\mathbf{u}$ and $\mathbf{v}$ be the vectors in Figure Ex-10b. Use the result in part (a) to find the components of $4\mathbf{u} - 5\mathbf{v}$.

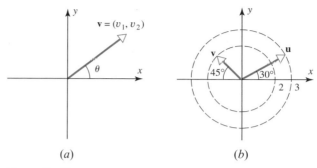

(a) (b)

Figure Ex-10

11. Let $\mathbf{p}_0 = (x_0, y_0, z_0)$ and $\mathbf{p} = (x, y, z)$. Describe the set of all points (x, y, z) for which $\|\mathbf{p} - \mathbf{p}_0\| = 1$.

12. Prove geometrically that if $\mathbf{u}$ and $\mathbf{v}$ are vectors in 2- or 3-space, then $\|\mathbf{u} + \mathbf{v}\| \le \|\mathbf{u}\| + \|\mathbf{v}\|$.

13. Prove parts (a), (c), and (e) of Theorem 1 analytically.

14. Prove parts (d), (g), and (h) of Theorem 1 analytically.

Discussion
&
Discovery

15. For the inequality stated in Exercise 9, is it possible to have $\|\mathbf{u} + \mathbf{v}\| = \|\mathbf{u}\| + \|\mathbf{v}\|$? Explain your reasoning.

16. (a) What relationship must hold for the point $\mathbf{p} = (a, b, c)$ to be equidistant from the origin and the xz-plane? Make sure that the relationship you state is valid for positive and negative values of a, b, and c.

(b) What relationship must hold for the point $\mathbf{p} = (a, b, c)$ to be farther from the origin than from the xz-plane? Make sure that the relationship you state is valid for positive and negative values of a, b, and c.

17. (a) What does the inequality $\|\mathbf{x}\| < 1$ tell you about the location of the point $\mathbf{x}$ in the plane?

(b) Write down an inequality that describes the set of points that lie outside the circle of radius 1, centered at the point $\mathbf{x}_0$.

18. The triangles in the accompanying figure should suggest a geometric proof of Theorem 3.2.1(f) for the case where $k > 0$. Give the proof.

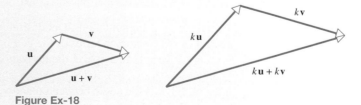

Figure Ex-18

3.3
DOT PRODUCT; PROJECTIONS

In this section we shall discuss an important way of multiplying vectors in 2-space or 3-space. We shall then give some applications of this multiplication to geometry.

Dot Product of Vectors

Let **u** and **v** be two nonzero vectors in 2-space or 3-space, and assume these vectors have been positioned so that their initial points coincide. By the *angle between* **u** *and* **v**, we shall mean the angle θ determined by **u** and **v** that satisfies $0 \leq \theta \leq \pi$ (Figure 3.3.1).

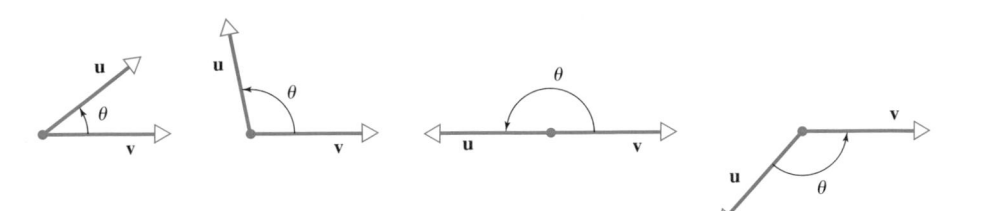

Figure 3.3.1 The angle θ between **u** and **v** satisfies $0 \leq \theta \leq \pi$.

> **DEFINITION**
>
> If **u** and **v** are vectors in 2-space or 3-space and θ is the angle between **u** and **v**, then the *dot product* or *Euclidean inner product* **u** · **v** is defined by
>
> $$\mathbf{u} \cdot \mathbf{v} = \begin{cases} \|\mathbf{u}\|\|\mathbf{v}\| \cos\theta & \text{if } \mathbf{u} \neq \mathbf{0} \text{ and } \mathbf{v} \neq \mathbf{0} \\ 0 & \text{if } \mathbf{u} = \mathbf{0} \text{ or } \mathbf{v} = \mathbf{0} \end{cases} \qquad (1)$$

EXAMPLE 1 Dot Product

As shown in Figure 3.3.2, the angle between the vectors **u** = $(0, 0, 1)$ and **v** = $(0, 2, 2)$ is 45°. Thus

$$\mathbf{u} \cdot \mathbf{v} = \|\mathbf{u}\|\|\mathbf{v}\| \cos\theta = \left(\sqrt{0^2 + 0^2 + 1^2}\right)\left(\sqrt{0^2 + 2^2 + 2^2}\right)\left(\frac{1}{\sqrt{2}}\right) = 2 \quad \blacklozenge$$

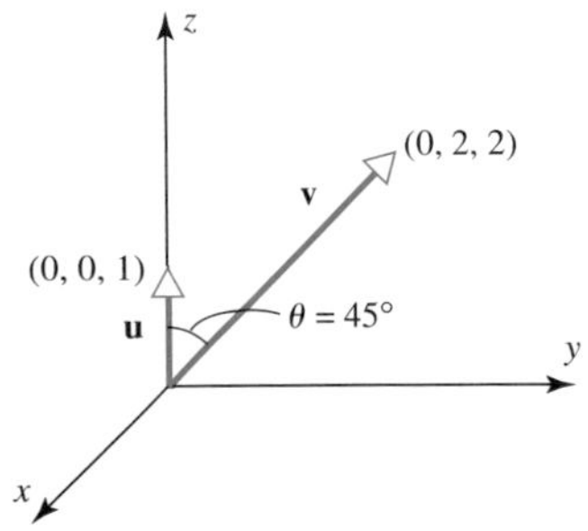

Figure 3.3.2

Component Form of the Dot Product

For purposes of computation, it is desirable to have a formula that expresses the dot product of two vectors in terms of the components of the vectors. We will derive such a formula for vectors in 3-space; the derivation for vectors in 2-space is similar.

Let **u** = (u_1, u_2, u_3) and **v** = (v_1, v_2, v_3) be two nonzero vectors. If, as shown in Figure 3.3.3, θ is the angle between **u** and **v**, then the law of cosines yields

$$\|\overrightarrow{PQ}\|^2 = \|\mathbf{u}\|^2 + \|\mathbf{v}\|^2 - 2\|\mathbf{u}\|\|\mathbf{v}\| \cos\theta \qquad (2)$$

Since $\overrightarrow{PQ} = \mathbf{v} - \mathbf{u}$, we can rewrite (2) as

$$\|\mathbf{u}\|\|\mathbf{v}\| \cos\theta = \tfrac{1}{2}(\|\mathbf{u}\|^2 + \|\mathbf{v}\|^2 - \|\mathbf{v} - \mathbf{u}\|^2)$$

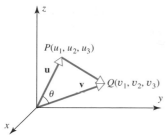

Figure 3.3.3

or

$$\mathbf{u} \cdot \mathbf{v} = \tfrac{1}{2}(\|\mathbf{u}\|^2 + \|\mathbf{v}\|^2 - \|\mathbf{v} - \mathbf{u}\|^2)$$

Substituting

$$\|\mathbf{u}\|^2 = u_1^2 + u_2^2 + u_3^2, \qquad \|\mathbf{v}\|^2 = v_1^2 + v_2^2 + v_3^2,$$

and

$$\|\mathbf{v} - \mathbf{u}\|^2 = (v_1 - u_1)^2 + (v_2 - u_2)^2 + (v_3 - u_3)^2$$

we obtain, after simplifying,

$$\mathbf{u} \cdot \mathbf{v} = u_1 v_1 + u_2 v_2 + u_3 v_3 \tag{3}$$

Although we derived this formula under the assumption that $\mathbf{u}$ and $\mathbf{v}$ are nonzero, the formula is also valid if $\mathbf{u} = \mathbf{0}$ or $\mathbf{v} = \mathbf{0}$ (verify).

If $\mathbf{u} = (u_1, u_2)$ and $\mathbf{v} = (v_1, v_2)$ are two vectors in 2-space, then the formula corresponding to (3) is

$$\mathbf{u} \cdot \mathbf{v} = u_1 v_1 + u_2 v_2 \tag{4}$$

Finding the Angle Between Vectors

If $\mathbf{u}$ and $\mathbf{v}$ are nonzero vectors, then Formula (1) can be written as

$$\cos \theta = \frac{\mathbf{u} \cdot \mathbf{v}}{\|\mathbf{u}\| \|\mathbf{v}\|} \tag{5}$$

EXAMPLE 2 Dot Product Using (3)

Consider the vectors $\mathbf{u} = (2, -1, 1)$ and $\mathbf{v} = (1, 1, 2)$. Find $\mathbf{u} \cdot \mathbf{v}$ and determine the angle θ between $\mathbf{u}$ and $\mathbf{v}$.

Solution

$$\mathbf{u} \cdot \mathbf{v} = u_1 v_1 + u_2 v_2 + u_3 v_3 = (2)(1) + (-1)(1) + (1)(2) = 3$$

For the given vectors we have $\|\mathbf{u}\| = \|\mathbf{v}\| = \sqrt{6}$, so from (5),

$$\cos \theta = \frac{\mathbf{u} \cdot \mathbf{v}}{\|\mathbf{u}\| \|\mathbf{v}\|} = \frac{3}{\sqrt{6}\sqrt{6}} = \frac{1}{2}$$

Thus, $\theta = 60°$. ◆

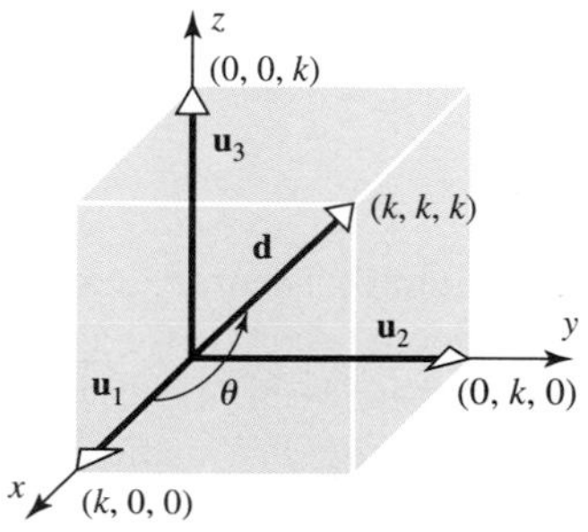

EXAMPLE 3 A Geometric Problem

Find the angle between a diagonal of a cube and one of its edges.

Solution

Let k be the length of an edge and introduce a coordinate system as shown in Figure 3.3.4. If we let $\mathbf{u}_1 = (k, 0, 0)$, $\mathbf{u}_2 = (0, k, 0)$, and $\mathbf{u}_3 = (0, 0, k)$, then the vector

$$\mathbf{d} = (k, k, k) = \mathbf{u}_1 + \mathbf{u}_2 + \mathbf{u}_3$$

Figure 3.3.4

is a diagonal of the cube. The angle θ between **d** and the edge $\mathbf{u}_1$ satisfies

$$\cos\theta = \frac{\mathbf{u}_1 \cdot \mathbf{d}}{\|\mathbf{u}_1\|\|\mathbf{d}\|} = \frac{k^2}{(k)(\sqrt{3k^2})} = \frac{1}{\sqrt{3}}$$

Thus

$$\theta = \cos^{-1}\left(\frac{1}{\sqrt{3}}\right) \approx 54.74°$$

Note that this is independent of k, as expected. ◆

The following theorem shows how the dot product can be used to obtain information about the angle between two vectors; it also establishes an important relationship between the norm and the dot product.

THEOREM 3.3.1

Let **u** and **v** be vectors in 2- or 3-space.

(a) $\mathbf{v} \cdot \mathbf{v} = \|\mathbf{v}\|^2$; that is, $\|\mathbf{v}\| = (\mathbf{v} \cdot \mathbf{v})^{1/2}$
(b) If the vectors **u** and **v** are nonzero and θ is the angle between them, then

θ is acute	if and only if	$\mathbf{u} \cdot \mathbf{v} > 0$
θ is obtuse	if and only if	$\mathbf{u} \cdot \mathbf{v} < 0$
$\theta = \pi/2$	if and only if	$\mathbf{u} \cdot \mathbf{v} = 0$

Proof (a) Since the angle θ between **v** and **v** is 0, we have

$$\mathbf{v} \cdot \mathbf{v} = \|\mathbf{v}\|\|\mathbf{v}\|\cos\theta = \|\mathbf{v}\|^2 \cos 0 = \|\mathbf{v}\|^2$$

Proof (b) Since θ satisfies $0 \le \theta \le \pi$, it follows that θ is acute if and only if $\cos\theta > 0$, that θ is obtuse if and only if $\cos\theta < 0$, and that $\theta = \pi/2$ if and only if $\cos\theta = 0$. But $\cos\theta$ has the same sign as $\mathbf{u} \cdot \mathbf{v}$ since $\mathbf{u} \cdot \mathbf{v} = \|\mathbf{u}\|\|\mathbf{v}\|\cos\theta$, $\|\mathbf{u}\| > 0$, and $\|\mathbf{v}\| > 0$. Thus, the result follows. ∎

EXAMPLE 4 Finding Dot Products from Components

If $\mathbf{u} = (1, -2, 3)$, $\mathbf{v} = (-3, 4, 2)$, and $\mathbf{w} = (3, 6, 3)$, then

$$\mathbf{u} \cdot \mathbf{v} = (1)(-3) + (-2)(4) + (3)(2) = -5$$
$$\mathbf{v} \cdot \mathbf{w} = (-3)(3) + (4)(6) + (2)(3) = 21$$
$$\mathbf{u} \cdot \mathbf{w} = (1)(3) + (-2)(6) + (3)(3) = 0$$

Therefore, **u** and **v** make an obtuse angle, **v** and **w** make an acute angle, and **u** and **w** are perpendicular. ◆

Orthogonal Vectors

Perpendicular vectors are also called **orthogonal** vectors. In light of Theorem 3.3.1*b*, two *nonzero* vectors are orthogonal if and only if their dot product is zero. If we agree to consider **u** and **v** to be perpendicular when either or both of these vectors is **0**, then we can state without exception that *two vectors* **u** *and* **v** *are orthogonal* (*perpendicular*) *if and only if* $\mathbf{u} \cdot \mathbf{v} = 0$. To indicate that **u** and **v** are orthogonal vectors, we write $\mathbf{u} \perp \mathbf{v}$.

EXAMPLE 5   A Vector Perpendicular to a Line

Show that in 2-space the nonzero vector $\mathbf{n} = (a, b)$ is perpendicular to the line $ax + by + c = 0$.

Solution

Let $P_1(x_1, y_1)$ and $P_2(x_2, y_2)$ be distinct points on the line, so that

$$ax_1 + by_1 + c = 0$$
$$ax_2 + by_2 + c = 0 \tag{6}$$

Since the vector $\overrightarrow{P_1 P_2} = (x_2 - x_1, y_2 - y_1)$ runs along the line (Figure 3.3.5), we need only show that $\mathbf{n}$ and $\overrightarrow{P_1 P_2}$ are perpendicular. But on subtracting the equations in (6), we obtain

$$a(x_2 - x_1) + b(y_2 - y_1) = 0$$

which can be expressed in the form

$$(a, b) \cdot (x_2 - x_1, y_2 - y_1) = 0 \quad \text{or} \quad \mathbf{n} \cdot \overrightarrow{P_1 P_2} = 0$$

Thus $\mathbf{n}$ and $\overrightarrow{P_1 P_2}$ are perpendicular. ◆

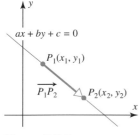

Figure 3.3.5

The following theorem lists the most important properties of the dot product. They are useful in calculations involving vectors.

THEOREM 3.3.2

> ### Properties of the Dot Product
>
> *If* $\mathbf{u}$, $\mathbf{v}$, *and* $\mathbf{w}$ *are vectors in 2- or 3-space and k is a scalar, then*
>
> (a) $\mathbf{u} \cdot \mathbf{v} = \mathbf{v} \cdot \mathbf{u}$
> (b) $\mathbf{u} \cdot (\mathbf{v} + \mathbf{w}) = \mathbf{u} \cdot \mathbf{v} + \mathbf{u} \cdot \mathbf{w}$
> (c) $k(\mathbf{u} \cdot \mathbf{v}) = (k\mathbf{u}) \cdot \mathbf{v} = \mathbf{u} \cdot (k\mathbf{v})$
> (d) $\mathbf{v} \cdot \mathbf{v} > 0$ *if* $\mathbf{v} \neq \mathbf{0}$, *and* $\mathbf{v} \cdot \mathbf{v} = 0$ *if* $\mathbf{v} = \mathbf{0}$

Proof   We shall prove (*c*) for vectors in 3-space and leave the remaining proofs as exercises. Let $\mathbf{u} = (u_1, u_2, u_3)$ and $\mathbf{v} = (v_1, v_2, v_3)$; then

$$
\begin{aligned}
k(\mathbf{u} \cdot \mathbf{v}) &= k(u_1 v_1 + u_2 v_2 + u_3 v_3) \\
&= (ku_1)v_1 + (ku_2)v_2 + (ku_3)v_3 \\
&= (k\mathbf{u}) \cdot \mathbf{v}
\end{aligned}
$$

Similarly,

$$k(\mathbf{u} \cdot \mathbf{v}) = \mathbf{u} \cdot (k\mathbf{v}) \qquad \blacksquare$$

An Orthogonal Projection

In many applications it is of interest to "decompose" a vector $\mathbf{u}$ into a sum of two terms, one parallel to a specified nonzero vector $\mathbf{a}$ and the other perpendicular to $\mathbf{a}$. If $\mathbf{u}$ and $\mathbf{a}$ are positioned so that their initial points coincide at a point Q, we can decompose the vector $\mathbf{u}$ as follows (Figure 3.3.6): Drop a perpendicular from the tip of $\mathbf{u}$ to the line through $\mathbf{a}$, and construct the vector $\mathbf{w}_1$ from Q to the foot of this perpendicular. Next form the difference

$$\mathbf{w}_2 = \mathbf{u} - \mathbf{w}_1$$

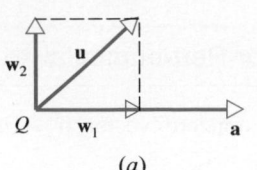
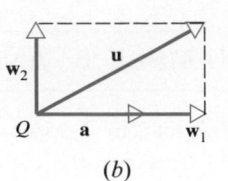
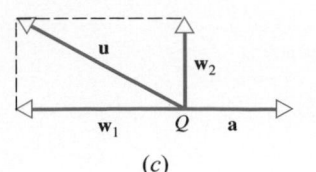

(a) (b) (c)

Figure 3.3.6 The vector $\mathbf{u}$ is the sum of $\mathbf{w}_1$ and $\mathbf{w}_2$, where $\mathbf{w}_1$ is parallel to $\mathbf{a}$ and $\mathbf{w}_2$ is perpendicular to $\mathbf{a}$.

As indicated in Figure 3.3.6, the vector $\mathbf{w}_1$ is parallel to $\mathbf{a}$, the vector $\mathbf{w}_2$ is perpendicular to $\mathbf{a}$, and

$$\mathbf{w}_1 + \mathbf{w}_2 = \mathbf{w}_1 + (\mathbf{u} - \mathbf{w}_1) = \mathbf{u}$$

The vector $\mathbf{w}_1$ is called the ***orthogonal projection of* u *on* a** or sometimes the ***vector component of* u *along* a**. It is denoted by

$$\text{proj}_{\mathbf{a}}\mathbf{u} \tag{7}$$

The vector $\mathbf{w}_2$ is called the ***vector component of* u *orthogonal to* a**. Since we have $\mathbf{w}_2 = \mathbf{u} - \mathbf{w}_1$, this vector can be written in notation (7) as

$$\mathbf{w}_2 = \mathbf{u} - \text{proj}_{\mathbf{a}}\mathbf{u}$$

The following theorem gives formulas for calculating $\text{proj}_{\mathbf{a}}\mathbf{u}$ and $\mathbf{u} - \text{proj}_{\mathbf{a}}\mathbf{u}$.

THEOREM 3.3.3

If $\mathbf{u}$ *and* $\mathbf{a}$ *are vectors in 2-space or 3-space and if* $\mathbf{a} \neq \mathbf{0}$, *then*

$$\text{proj}_{\mathbf{a}}\mathbf{u} = \frac{\mathbf{u} \cdot \mathbf{a}}{\|\mathbf{a}\|^2}\mathbf{a} \quad \textit{(vector component of } \mathbf{u} \textit{ along } \mathbf{a})$$

$$\mathbf{u} - \text{proj}_{\mathbf{a}}\mathbf{u} = \mathbf{u} - \frac{\mathbf{u} \cdot \mathbf{a}}{\|\mathbf{a}\|^2}\mathbf{a} \quad \textit{(vector component of } \mathbf{u} \textit{ orthogonal to } \mathbf{a})$$

Proof Let $\mathbf{w}_1 = \text{proj}_{\mathbf{a}}\mathbf{u}$ and $\mathbf{w}_2 = \mathbf{u} - \text{proj}_{\mathbf{a}}\mathbf{u}$. Since $\mathbf{w}_1$ is parallel to $\mathbf{a}$, it must be a scalar multiple of $\mathbf{a}$, so it can be written in the form $\mathbf{w}_1 = k\mathbf{a}$. Thus

$$\mathbf{u} = \mathbf{w}_1 + \mathbf{w}_2 = k\mathbf{a} + \mathbf{w}_2 \tag{8}$$

Taking the dot product of both sides of (8) with $\mathbf{a}$ and using Theorems 3.3.1*a* and 3.3.2 yields

$$\mathbf{u} \cdot \mathbf{a} = (k\mathbf{a} + \mathbf{w}_2) \cdot \mathbf{a} = k\|\mathbf{a}\|^2 + \mathbf{w}_2 \cdot \mathbf{a} \tag{9}$$

But $\mathbf{w}_2 \cdot \mathbf{a} = 0$ since $\mathbf{w}_2$ is perpendicular to $\mathbf{a}$; so (9) yields

$$k = \frac{\mathbf{u} \cdot \mathbf{a}}{\|\mathbf{a}\|^2}$$

Since $\text{proj}_{\mathbf{a}}\mathbf{u} = \mathbf{w}_1 = k\mathbf{a}$, we obtain

$$\text{proj}_{\mathbf{a}}\mathbf{u} = \frac{\mathbf{u} \cdot \mathbf{a}}{\|\mathbf{a}\|^2}\mathbf{a} \qquad \blacksquare$$

EXAMPLE 6 Vector Component of u Along a

Let $\mathbf{u} = (2, -1, 3)$ and $\mathbf{a} = (4, -1, 2)$. Find the vector component of $\mathbf{u}$ along $\mathbf{a}$ and the vector component of $\mathbf{u}$ orthogonal to $\mathbf{a}$.

Solution

$$\mathbf{u} \cdot \mathbf{a} = (2)(4) + (-1)(-1) + (3)(2) = 15$$
$$\|\mathbf{a}\|^2 = 4^2 + (-1)^2 + 2^2 = 21$$

Thus the vector component of $\mathbf{u}$ along $\mathbf{a}$ is

$$\text{proj}_{\mathbf{a}}\mathbf{u} = \frac{\mathbf{u} \cdot \mathbf{a}}{\|\mathbf{a}\|^2}\mathbf{a} = \tfrac{15}{21}(4, -1, 2) = \left(\tfrac{20}{7}, -\tfrac{5}{7}, \tfrac{10}{7}\right)$$

and the vector component of $\mathbf{u}$ orthogonal to $\mathbf{a}$ is

$$\mathbf{u} - \text{proj}_{\mathbf{a}}\mathbf{u} = (2, -1, 3) - \left(\tfrac{20}{7}, -\tfrac{5}{7}, \tfrac{10}{7}\right) = \left(-\tfrac{6}{7}, -\tfrac{2}{7}, \tfrac{11}{7}\right)$$

As a check, the reader may wish to verify that the vectors $\mathbf{u} - \text{proj}_{\mathbf{a}}\mathbf{u}$ and $\mathbf{a}$ are perpendicular by showing that their dot product is zero. ◆

A formula for the length of the vector component of $\mathbf{u}$ along $\mathbf{a}$ can be obtained by writing

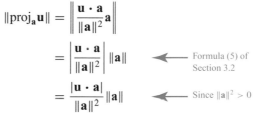

$$\|\text{proj}_{\mathbf{a}}\mathbf{u}\| = \left\|\frac{\mathbf{u} \cdot \mathbf{a}}{\|\mathbf{a}\|^2}\mathbf{a}\right\|$$

$$= \left|\frac{\mathbf{u} \cdot \mathbf{a}}{\|\mathbf{a}\|^2}\right| \|\mathbf{a}\| \quad \longleftarrow \quad \text{Formula (5) of Section 3.2}$$

$$= \frac{|\mathbf{u} \cdot \mathbf{a}|}{\|\mathbf{a}\|^2} \|\mathbf{a}\| \quad \longleftarrow \quad \text{Since } \|\mathbf{a}\|^2 > 0$$

which yields

$$\|\text{proj}_{\mathbf{a}}\mathbf{u}\| = \frac{|\mathbf{u} \cdot \mathbf{a}|}{\|\mathbf{a}\|} \tag{10}$$

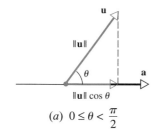

(a) $0 \le \theta < \dfrac{\pi}{2}$

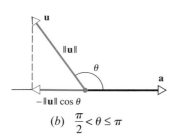

(b) $\dfrac{\pi}{2} < \theta \le \pi$

Figure 3.3.7

If θ denotes the angle between $\mathbf{u}$ and $\mathbf{a}$, then $\mathbf{u} \cdot \mathbf{a} = \|\mathbf{u}\|\|\mathbf{a}\| \cos\theta$, so (10) can also be written as

$$\|\text{proj}_{\mathbf{a}}\mathbf{u}\| = \|\mathbf{u}\| |\cos\theta| \tag{11}$$

(Verify.) A geometric interpretation of this result is given in Figure 3.3.7.

As an example, we will use vector methods to derive a formula for the distance from a point in the plane to a line.

EXAMPLE 7 Distance Between a Point and a Line

Find a formula for the distance D between point $P_0(x_0, y_0)$ and the line $ax + by + c = 0$.

Solution

Let $Q(x_1, y_1)$ be any point on the line, and position the vector $\mathbf{n} = (a, b)$ so that its initial point is at Q.

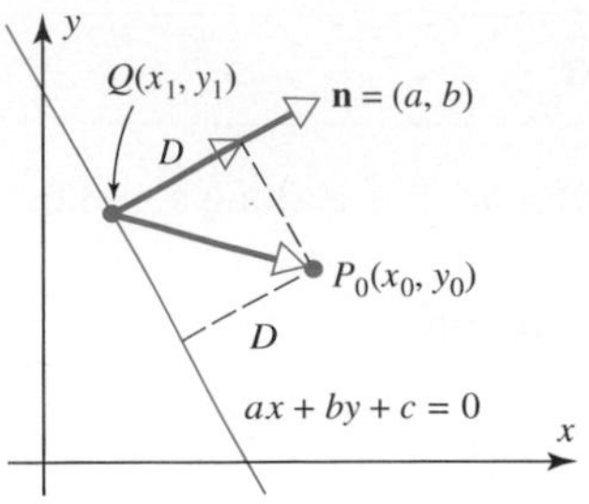

Figure 3.3.8

By virtue of Example 5, the vector **n** is perpendicular to the line (Figure 3.3.8). As indicated in the figure, the distance D is equal to the length of the orthogonal projection of $\overrightarrow{QP_0}$ on **n**; thus, from (10),

$$D = \|\text{proj}_{\mathbf{n}} \overrightarrow{QP_0}\| = \frac{|\overrightarrow{QP_0} \cdot \mathbf{n}|}{\|\mathbf{n}\|}$$

But

$$\overrightarrow{QP_0} = (x_0 - x_1, y_0 - y_1)$$
$$\overrightarrow{QP_0} \cdot \mathbf{n} = a(x_0 - x_1) + b(y_0 - y_1)$$
$$\|\mathbf{n}\| = \sqrt{a^2 + b^2}$$

so

$$D = \frac{|a(x_0 - x_1) + b(y_0 - y_1)|}{\sqrt{a^2 + b^2}} \tag{12}$$

Since the point $Q(x_1, y_1)$ lies on the line, its coordinates satisfy the equation of the line, so

$$ax_1 + by_1 + c = 0 \quad \text{or} \quad c = -ax_1 - by_1$$

Substituting this expression in (12) yields the formula

$$D = \frac{|ax_0 + by_0 + c|}{\sqrt{a^2 + b^2}} \quad \blacklozenge \tag{13}$$

EXAMPLE 8   Using the Distance Formula

It follows from Formula (13) that the distance D from the point $(1, -2)$ to the line $3x + 4y - 6 = 0$ is

$$D = \frac{|(3)(1) + 4(-2) - 6|}{\sqrt{3^2 + 4^2}} = \frac{|-11|}{\sqrt{25}} = \frac{11}{5} \quad \blacklozenge$$

EXERCISE SET 3.3

1.   Find $\mathbf{u} \cdot \mathbf{v}$.
   (a) $\mathbf{u} = (2, 3)$, $\mathbf{v} = (5, -7)$ (b) $\mathbf{u} = (-6, -2)$, $\mathbf{v} = (4, 0)$
   (c) $\mathbf{u} = (1, -5, 4)$, $\mathbf{v} = (3, 3, 3)$ (d) $\mathbf{u} = (-2, 2, 3)$, $\mathbf{v} = (1, 7, -4)$

2.   In each part of Exercise 1, find the cosine of the angle θ between $\mathbf{u}$ and $\mathbf{v}$.

3.   Determine whether $\mathbf{u}$ and $\mathbf{v}$ make an acute angle, make an obtuse angle, or are orthogonal.
   (a) $\mathbf{u} = (6, 1, 4)$, $\mathbf{v} = (2, 0, -3)$ (b) $\mathbf{u} = (0, 0, -1)$, $\mathbf{v} = (1, 1, 1)$
   (c) $\mathbf{u} = (-6, 0, 4)$, $\mathbf{v} = (3, 1, 6)$ (d) $\mathbf{u} = (2, 4, -8)$, $\mathbf{v} = (5, 3, 7)$

4.   Find the orthogonal projection of $\mathbf{u}$ on $\mathbf{a}$.
   (a) $\mathbf{u} = (6, 2)$, $\mathbf{a} = (3, -9)$ (b) $\mathbf{u} = (-1, -2)$, $\mathbf{a} = (-2, 3)$
   (c) $\mathbf{u} = (3, 1, -7)$, $\mathbf{a} = (1, 0, 5)$ (d) $\mathbf{u} = (1, 0, 0)$, $\mathbf{a} = (4, 3, 8)$

5.   In each part of Exercise 4, find the vector component of $\mathbf{u}$ orthogonal to $\mathbf{a}$.

6.   In each part, find $\|\text{proj}_{\mathbf{a}} \mathbf{u}\|$.
   (a) $\mathbf{u} = (1, -2)$, $\mathbf{a} = (-4, -3)$ (b) $\mathbf{u} = (5, 6)$, $\mathbf{a} = (2, -1)$
   (c) $\mathbf{u} = (3, 0, 4)$, $\mathbf{a} = (2, 3, 3)$ (d) $\mathbf{u} = (3, -2, 6)$, $\mathbf{a} = (1, 2, -7)$

7. Let $\mathbf{u} = (5, -2, 1)$, $\mathbf{v} = (1, 6, 3)$, and $k = -4$. Verify Theorem 3.3.2 for these quantities.

8. (a) Show that $\mathbf{v} = (a, b)$ and $\mathbf{w} = (-b, a)$ are orthogonal vectors.

   (b) Use the result in part (a) to find two vectors that are orthogonal to $\mathbf{v} = (2, -3)$.

   (c) Find two unit vectors that are orthogonal to $(-3, 4)$.

9. Let $\mathbf{u} = (3, 4)$, $\mathbf{v} = (5, -1)$, and $\mathbf{w} = (7, 1)$. Evaluate the expressions.

   (a) $\mathbf{u} \cdot (7\mathbf{v} + \mathbf{w})$ (b) $\|(\mathbf{u} \cdot \mathbf{w})\mathbf{w}\|$ (c) $\|\mathbf{u}\|(\mathbf{v} \cdot \mathbf{w})$ (d) $(\|\mathbf{u}\|\mathbf{v}) \cdot \mathbf{w}$

10. Find five different nonzero vectors that are orthogonal to $\mathbf{u} = (5, -2, 3)$.

11. Use vectors to find the cosines of the interior angles of the triangle with vertices $(0, -1)$, $(1, -2)$, and $(4, 1)$.

12. Show that $A(3, 0, 2)$, $B(4, 3, 0)$, and $C(8, 1, -1)$ are vertices of a right triangle. At which vertex is the right angle?

13. Find a unit vector that is orthogonal to both $\mathbf{u} = (1, 0, 1)$ and $\mathbf{v} = (0, 1, 1)$.

14. A vector $\mathbf{a}$ in the xy-plane has a length of 9 units and points in a direction that is $120°$ counterclockwise from the positive x-axis, and a vector $\mathbf{b}$ in that plane has a length of 5 units and points in the positive y-direction. Find $\mathbf{a} \cdot \mathbf{b}$.

15. A vector $\mathbf{a}$ in the xy-plane points in a direction that is $47°$ counterclockwise from the positive x-axis, and a vector $\mathbf{b}$ in that plane points in a direction that is $43°$ clockwise from the positive x-axis. What can you say about the value of $\mathbf{a} \cdot \mathbf{b}$?

16. Let $\mathbf{p} = (2, k)$ and $\mathbf{q} = (3, 5)$. Find k such that

   (a) $\mathbf{p}$ and $\mathbf{q}$ are parallel   (b) $\mathbf{p}$ and $\mathbf{q}$ are orthogonal

   (c) the angle between $\mathbf{p}$ and $\mathbf{q}$ is $\pi/3$ (d) the angle between $\mathbf{p}$ and $\mathbf{q}$ is $\pi/4$

17. Use Formula (13) to calculate the distance between the point and the line.

   (a) $4x + 3y + 4 = 0$; $(-3, 1)$ (b) $y = -4x + 2$; $(2, -5)$ (c) $3x + y = 5$; $(1, 8)$

18. Establish the identity $\|\mathbf{u} + \mathbf{v}\|^2 + \|\mathbf{u} - \mathbf{v}\|^2 = 2\|\mathbf{u}\|^2 + 2\|\mathbf{v}\|^2$.

19. Establish the identity $\mathbf{u} \cdot \mathbf{v} = \frac{1}{4}\|\mathbf{u} + \mathbf{v}\|^2 - \frac{1}{4}\|\mathbf{u} - \mathbf{v}\|^2$.

20. Find the angle between a diagonal of a cube and one of its faces.

21. Let $\mathbf{i}$, $\mathbf{j}$, and $\mathbf{k}$ be unit vectors along the positive x, y, and z axes of a rectangular coordinate system in 3-space. If $\mathbf{v} = (a, b, c)$ is a nonzero vector, then the angles α, β, and γ between $\mathbf{v}$ and the vectors $\mathbf{i}$, $\mathbf{j}$, and $\mathbf{k}$, respectively, are called the *direction angles* of $\mathbf{v}$ (see accompanying figure), and the numbers $\cos\alpha$, $\cos\beta$, and $\cos\gamma$ are called the *direction cosines* of $\mathbf{v}$.

   (a) Show that $\cos\alpha = a/\|\mathbf{v}\|$.

   (b) Find $\cos\beta$ and $\cos\gamma$.

   (c) Show that $\mathbf{v}/\|\mathbf{v}\| = (\cos\alpha, \cos\beta, \cos\gamma)$.

   (d) Show that $\cos^2\alpha + \cos^2\beta + \cos^2\gamma = 1$.

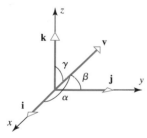

Figure Ex-21

22. Use the result in Exercise 21 to estimate, to the nearest degree, the angles that a diagonal of a box with dimensions 10 cm × 15 cm × 25 cm makes with the edges of the box.

   Note   A calculator is needed.

23. Referring to Exercise 21, show that two nonzero vectors, $\mathbf{v}_1$ and $\mathbf{v}_2$, in 3-space are perpendicular if and only if their direction cosines satisfy

$$\cos\alpha_1 \cos\alpha_2 + \cos\beta_1 \cos\beta_2 + \cos\gamma_1 \cos\gamma_2 = 0$$

24. (a) Find the area of the triangle with vertices $A(2, 3)$, $C(4, 7)$, and $D(-5, 8)$.

   (b) Find the coordinates of the point B such that the quadrilateral $ABCD$ is a parallelogram. What is the area of this parallelogram?

25. Show that if $\mathbf{v}$ is orthogonal to both $\mathbf{w}_1$ and $\mathbf{w}_2$, then $\mathbf{v}$ is orthogonal to $k_1\mathbf{w}_1 + k_2\mathbf{w}_2$ for all scalars k_1 and k_2.

26. Let **u** and **v** be nonzero vectors in 2- or 3-space, and let $k = \|\mathbf{u}\|$ and $l = \|\mathbf{v}\|$. Show that the vector $\mathbf{w} = l\mathbf{u} + k\mathbf{v}$ bisects the angle between **u** and **v**.

Discussion & Discovery

27. In each part, something is wrong with the expression. What?

(a) $\mathbf{u} \cdot (\mathbf{v} \cdot \mathbf{w})$ (b) $(\mathbf{u} \cdot \mathbf{v}) + \mathbf{w}$ (c) $\|\mathbf{u} \cdot \mathbf{v}\|$ (d) $k \cdot (\mathbf{u} + \mathbf{v})$

28. Is it possible to have $\text{proj}_{\mathbf{a}}\mathbf{u} = \text{proj}_{\mathbf{u}}\mathbf{a}$? Explain your reasoning.

29. If $\mathbf{u} \neq \mathbf{0}$, is it valid to cancel **u** from both sides of the equation $\mathbf{u} \cdot \mathbf{v} = \mathbf{u} \cdot \mathbf{w}$ and conclude that $\mathbf{v} = \mathbf{w}$? Explain your reasoning.

30. Suppose that **u**, **v**, and **w** are mutually orthogonal nonzero vectors in 3-space, and suppose that you know the dot products of these vectors with a vector **r** in 3-space. Find an expression for **r** in terms of **u**, **v**, **w**, and the dot products.

Hint Look for an expression of the form $\mathbf{r} = c_1\mathbf{u} + c_2\mathbf{v} + c_3\mathbf{w}$.

31. Suppose that **u** and **v** are orthogonal vectors in 2-space or 3-space. What famous theorem is described by the equation $\|\mathbf{u} + \mathbf{v}\|^2 = \|\mathbf{u}\|^2 + \|\mathbf{v}\|^2$? Draw a picture to support your answer.

3.4 CROSS PRODUCT

In many applications of vectors to problems in geometry, physics, and engineering, it is of interest to construct a vector in 3-space that is perpendicular to two given vectors. In this section we shall show how to do this.

Cross Product of Vectors

Recall from Section 3.3 that the dot product of two vectors in 2-space or 3-space produces a scalar. We will now define a type of vector multiplication that produces a vector as the product but that is applicable only in 3-space.

> **DEFINITION**
>
> If $\mathbf{u} = (u_1, u_2, u_3)$ and $\mathbf{v} = (v_1, v_2, v_3)$ are vectors in 3-space, then the **cross product** $\mathbf{u} \times \mathbf{v}$ is the vector defined by
>
> $$\mathbf{u} \times \mathbf{v} = (u_2v_3 - u_3v_2, u_3v_1 - u_1v_3, u_1v_2 - u_2v_1)$$
>
> or, in determinant notation,
>
> $$\mathbf{u} \times \mathbf{v} = \left(\begin{vmatrix} u_2 & u_3 \\ v_2 & v_3 \end{vmatrix}, -\begin{vmatrix} u_1 & u_3 \\ v_1 & v_3 \end{vmatrix}, \begin{vmatrix} u_1 & u_2 \\ v_1 & v_2 \end{vmatrix} \right) \tag{1}$$

REMARK Instead of memorizing (1), you can obtain the components of $\mathbf{u} \times \mathbf{v}$ as follows:

- Form the 2×3 matrix $\begin{bmatrix} u_1 & u_2 & u_3 \\ v_1 & v_2 & v_3 \end{bmatrix}$ whose first row contains the components of **u** and whose second row contains the components of **v**.

- To find the first component of $\mathbf{u} \times \mathbf{v}$, delete the first column and take the determinant; to find the second component, delete the second column and take the negative of the determinant; and to find the third component, delete the third column and take the determinant.

EXAMPLE 1 Calculating a Cross Product

Find $\mathbf{u} \times \mathbf{v}$, where $\mathbf{u} = (1, 2, -2)$ and $\mathbf{v} = (3, 0, 1)$.

Solution

From either (1) or the mnemonic in the preceding remark, we have

$$\mathbf{u} \times \mathbf{v} = \left(\begin{vmatrix} 2 & -2 \\ 0 & 1 \end{vmatrix}, - \begin{vmatrix} 1 & -2 \\ 3 & 1 \end{vmatrix}, \begin{vmatrix} 1 & 2 \\ 3 & 0 \end{vmatrix} \right)$$

$$= (2, -7, -6) \quad \blacklozenge$$

There is an important difference between the dot product and cross product of two vectors—the dot product is a scalar and the cross product is a vector. The following theorem gives some important relationships between the dot product and cross product and also shows that $\mathbf{u} \times \mathbf{v}$ is orthogonal to both $\mathbf{u}$ and $\mathbf{v}$.

THEOREM 3.4.1

> ### Relationships Involving Cross Product and Dot Product
>
> *If* $\mathbf{u}$, $\mathbf{v}$, *and* $\mathbf{w}$ *are vectors in 3-space, then*
>
> (*a*) $\mathbf{u} \cdot (\mathbf{u} \times \mathbf{v}) = 0$ $\qquad\qquad$ ($\mathbf{u} \times \mathbf{v}$ *is orthogonal to* $\mathbf{u}$)
> (*b*) $\mathbf{v} \cdot (\mathbf{u} \times \mathbf{v}) = 0$ $\qquad\qquad$ ($\mathbf{u} \times \mathbf{v}$ *is orthogonal to* $\mathbf{v}$)
> (*c*) $\|\mathbf{u} \times \mathbf{v}\|^2 = \|\mathbf{u}\|^2 \|\mathbf{v}\|^2 - (\mathbf{u} \cdot \mathbf{v})^2$ $\quad$ (*Lagrange's identity*)
> (*d*) $\mathbf{u} \times (\mathbf{v} \times \mathbf{w}) = (\mathbf{u} \cdot \mathbf{w})\mathbf{v} - (\mathbf{u} \cdot \mathbf{v})\mathbf{w}$ (*relationship between cross and dot products*)
> (*e*) $(\mathbf{u} \times \mathbf{v}) \times \mathbf{w} = (\mathbf{u} \cdot \mathbf{w})\mathbf{v} - (\mathbf{v} \cdot \mathbf{w})\mathbf{u}$ (*relationship between cross and dot products*)

Proof (a) Let $\mathbf{u} = (u_1, u_2, u_3)$ and $\mathbf{v} = (v_1, v_2, v_3)$. Then

$$\mathbf{u} \cdot (\mathbf{u} \times \mathbf{v}) = (u_1, u_2, u_3) \cdot (u_2 v_3 - u_3 v_2, u_3 v_1 - u_1 v_3, u_1 v_2 - u_2 v_1)$$

$$= u_1(u_2 v_3 - u_3 v_2) + u_2(u_3 v_1 - u_1 v_3) + u_3(u_1 v_2 - u_2 v_1) = 0$$

Proof (b) Similar to (*a*).

Proof (c) Since

$$\|\mathbf{u} \times \mathbf{v}\|^2 = (u_2 v_3 - u_3 v_2)^2 + (u_3 v_1 - u_1 v_3)^2 + (u_1 v_2 - u_2 v_1)^2 \qquad (2)$$

and

$$\|\mathbf{u}\|^2 \|\mathbf{v}\|^2 - (\mathbf{u} \cdot \mathbf{v})^2 = (u_1^2 + u_2^2 + u_3^2)(v_1^2 + v_2^2 + v_3^2) - (u_1 v_1 + u_2 v_2 + u_3 v_3)^2 \qquad (3)$$

the proof can be completed by "multiplying out" the right sides of (2) and (3) and verifying their equality.

Proof (d) and (e) See Exercises 26 and 27. $\qquad\qquad\qquad\qquad\qquad$ ■

EXAMPLE 2 u × v Is Perpendicular to u and to v

Consider the vectors

$$\mathbf{u} = (1, 2, -2) \quad \text{and} \quad \mathbf{v} = (3, 0, 1)$$

Joseph Louis Lagrange *(1736–1813)* was a French-Italian mathematician and astronomer. Although his father wanted him to become a lawyer, Lagrange was attracted to mathematics and astronomy after reading a memoir by the astronomer Halley. At age 16 he began to study mathematics on his own and by age 19 was appointed to a professorship at the Royal Artillery School in Turin. The following year he solved some famous problems using new methods that eventually blossomed into a branch of mathematics called the *calculus of variations*. These methods and Lagrange's applications of them to problems in celestial mechanics were so monumental that by age 25 he was regarded by many of his contemporaries as the greatest living mathematician. One of Lagrange's most famous works is a memoir, *Mécanique Analytique*, in which he reduced the theory of mechanics to a few general formulas from which all other necessary equations could be derived.

Napoleon was a great admirer of Lagrange and showered him with many honors. In spite of his fame, Lagrange was a shy and modest man. On his death, he was buried with honor in the Pantheon.

In Example 1 we showed that

$$\mathbf{u} \times \mathbf{v} = (2, -7, -6)$$

Since

$$\mathbf{u} \cdot (\mathbf{u} \times \mathbf{v}) = (1)(2) + (2)(-7) + (-2)(-6) = 0$$

and

$$\mathbf{v} \cdot (\mathbf{u} \times \mathbf{v}) = (3)(2) + (0)(-7) + (1)(-6) = 0$$

$\mathbf{u} \times \mathbf{v}$ is orthogonal to both $\mathbf{u}$ and $\mathbf{v}$, as guaranteed by Theorem 3.4.1. ◆

The main arithmetic properties of the cross product are listed in the next theorem.

THEOREM 3.4.2

Properties of Cross Product

If $\mathbf{u}$, $\mathbf{v}$, *and* $\mathbf{w}$ *are any vectors in 3-space and k is any scalar, then*

(*a*) $\mathbf{u} \times \mathbf{v} = -(\mathbf{v} \times \mathbf{u})$
(*b*) $\mathbf{u} \times (\mathbf{v} + \mathbf{w}) = (\mathbf{u} \times \mathbf{v}) + (\mathbf{u} \times \mathbf{w})$
(*c*) $(\mathbf{u} + \mathbf{v}) \times \mathbf{w} = (\mathbf{u} \times \mathbf{w}) + (\mathbf{v} \times \mathbf{w})$
(*d*) $k(\mathbf{u} \times \mathbf{v}) = (k\mathbf{u}) \times \mathbf{v} = \mathbf{u} \times (k\mathbf{v})$
(*e*) $\mathbf{u} \times \mathbf{0} = \mathbf{0} \times \mathbf{u} = \mathbf{0}$
(*f*) $\mathbf{u} \times \mathbf{u} = \mathbf{0}$

The proofs follow immediately from Formula (1) and properties of determinants; for example, (*a*) can be proved as follows:

Proof (a)   Interchanging $\mathbf{u}$ and $\mathbf{v}$ in (1) interchanges the rows of the three determinants on the right side of (1) and hence changes the sign of each component in the cross product. Thus $\mathbf{u} \times \mathbf{v} = -(\mathbf{v} \times \mathbf{u})$. ■

The proofs of the remaining parts are left as exercises.

EXAMPLE 3 Standard Unit Vectors

Consider the vectors

$$\mathbf{i} = (1, 0, 0), \qquad \mathbf{j} = (0, 1, 0), \qquad \mathbf{k} = (0, 0, 1)$$

These vectors each have length 1 and lie along the coordinate axes (Figure 3.4.1). They are called the ***standard unit vectors*** in 3-space. Every vector $\mathbf{v} = (v_1, v_2, v_3)$ in 3-space is expressible in terms of $\mathbf{i}, \mathbf{j}$, and $\mathbf{k}$ since we can write

$$\mathbf{v} = (v_1, v_2, v_3) = v_1(1, 0, 0) + v_2(0, 1, 0) + v_3(0, 0, 1) = v_1\mathbf{i} + v_2\mathbf{j} + v_3\mathbf{k}$$

For example,

$$(2, -3, 4) = 2\mathbf{i} - 3\mathbf{j} + 4\mathbf{k}$$

From (1) we obtain

$$\mathbf{i} \times \mathbf{j} = \left(\begin{vmatrix} 0 & 0 \\ 1 & 0 \end{vmatrix}, -\begin{vmatrix} 1 & 0 \\ 0 & 0 \end{vmatrix}, \begin{vmatrix} 1 & 0 \\ 0 & 1 \end{vmatrix} \right) = (0, 0, 1) = \mathbf{k} \quad \blacklozenge$$

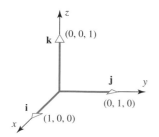

Figure 3.4.1 The standard unit vectors.

The reader should have no trouble obtaining the following results:

$$\begin{array}{lll} \mathbf{i} \times \mathbf{i} = \mathbf{0} & \mathbf{j} \times \mathbf{j} = \mathbf{0} & \mathbf{k} \times \mathbf{k} = \mathbf{0} \\ \mathbf{i} \times \mathbf{j} = \mathbf{k} & \mathbf{j} \times \mathbf{k} = \mathbf{i} & \mathbf{k} \times \mathbf{i} = \mathbf{j} \\ \mathbf{j} \times \mathbf{i} = -\mathbf{k} & \mathbf{k} \times \mathbf{j} = -\mathbf{i} & \mathbf{i} \times \mathbf{k} = -\mathbf{j} \end{array}$$

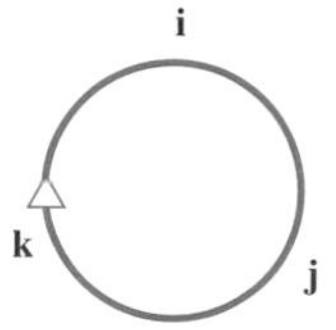

Figure 3.4.2

Figure 3.4.2 is helpful for remembering these results. Referring to this diagram, the cross product of two consecutive vectors going clockwise is the next vector around, and the cross product of two consecutive vectors going counterclockwise is the negative of the next vector around.

Determinant Form of Cross Product

It is also worth noting that a cross product can be represented symbolically in the form of a formal 3×3 determinant:

$$\mathbf{u} \times \mathbf{v} = \begin{vmatrix} \mathbf{i} & \mathbf{j} & \mathbf{k} \\ u_1 & u_2 & u_3 \\ v_1 & v_2 & v_3 \end{vmatrix} = \begin{vmatrix} u_2 & u_3 \\ v_2 & v_3 \end{vmatrix} \mathbf{i} - \begin{vmatrix} u_1 & u_3 \\ v_1 & v_3 \end{vmatrix} \mathbf{j} + \begin{vmatrix} u_1 & u_2 \\ v_1 & v_2 \end{vmatrix} \mathbf{k} \qquad (4)$$

For example, if $\mathbf{u} = (1, 2, -2)$ and $\mathbf{v} = (3, 0, 1)$, then

$$\mathbf{u} \times \mathbf{v} = \begin{vmatrix} \mathbf{i} & \mathbf{j} & \mathbf{k} \\ 1 & 2 & -2 \\ 3 & 0 & 1 \end{vmatrix} = 2\mathbf{i} - 7\mathbf{j} - 6\mathbf{k}$$

which agrees with the result obtained in Example 1.

WARNING   It is not true in general that $\mathbf{u} \times (\mathbf{v} \times \mathbf{w}) = (\mathbf{u} \times \mathbf{v}) \times \mathbf{w}$. For example,

$$\mathbf{i} \times (\mathbf{j} \times \mathbf{j}) = \mathbf{i} \times \mathbf{0} = \mathbf{0}$$

and

$$(\mathbf{i} \times \mathbf{j}) \times \mathbf{j} = \mathbf{k} \times \mathbf{j} = -\mathbf{i}$$

so

$$\mathbf{i} \times (\mathbf{j} \times \mathbf{j}) \neq (\mathbf{i} \times \mathbf{j}) \times \mathbf{j}$$

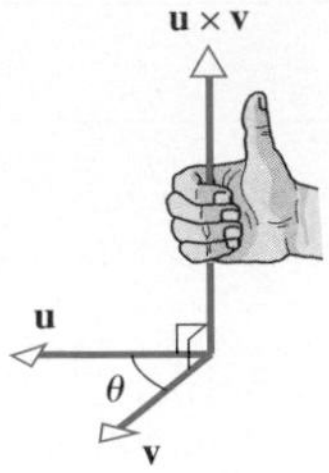

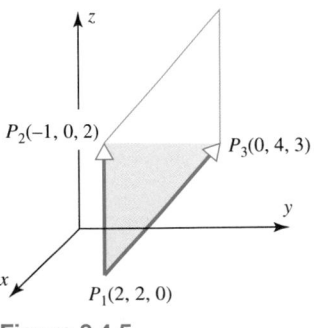

Figure 3.4.3

Geometric Interpretation of Cross Product

We know from Theorem 3.4.1 that $\mathbf{u} \times \mathbf{v}$ is orthogonal to both $\mathbf{u}$ and $\mathbf{v}$. If $\mathbf{u}$ and $\mathbf{v}$ are nonzero vectors, it can be shown that the direction of $\mathbf{u} \times \mathbf{v}$ can be determined using the following "right-hand rule"[†] (Figure 3.4.3): Let θ be the angle between $\mathbf{u}$ and $\mathbf{v}$, and suppose $\mathbf{u}$ is rotated through the angle θ until it coincides with $\mathbf{v}$. If the fingers of the right hand are cupped so that they point in the direction of rotation, then the thumb indicates (roughly) the direction of $\mathbf{u} \times \mathbf{v}$.

The reader may find it instructive to practice this rule with the products

$$\mathbf{i} \times \mathbf{j} = \mathbf{k}, \qquad \mathbf{j} \times \mathbf{k} = \mathbf{i}, \qquad \mathbf{k} \times \mathbf{i} = \mathbf{j}$$

If $\mathbf{u}$ and $\mathbf{v}$ are vectors in 3-space, then the norm of $\mathbf{u} \times \mathbf{v}$ has a useful geometric interpretation. Lagrange's identity, given in Theorem 3.4.1, states that

$$\|\mathbf{u} \times \mathbf{v}\|^2 = \|\mathbf{u}\|^2 \|\mathbf{v}\|^2 - (\mathbf{u} \cdot \mathbf{v})^2 \tag{5}$$

If θ denotes the angle between $\mathbf{u}$ and $\mathbf{v}$, then $\mathbf{u} \cdot \mathbf{v} = \|\mathbf{u}\| \|\mathbf{v}\| \cos \theta$, so (5) can be rewritten as

$$\|\mathbf{u} \times \mathbf{v}\|^2 = \|\mathbf{u}\|^2 \|\mathbf{v}\|^2 - \|\mathbf{u}\|^2 \|\mathbf{v}\|^2 \cos^2 \theta$$
$$= \|\mathbf{u}\|^2 \|\mathbf{v}\|^2 (1 - \cos^2 \theta)$$
$$= \|\mathbf{u}\|^2 \|\mathbf{v}\|^2 \sin^2 \theta$$

Since $0 \leq \theta \leq \pi$, it follows that $\sin \theta \geq 0$, so this can be rewritten as

$$\|\mathbf{u} \times \mathbf{v}\| = \|\mathbf{u}\| \|\mathbf{v}\| \sin \theta \tag{6}$$

But $\|\mathbf{v}\| \sin \theta$ is the altitude of the parallelogram determined by $\mathbf{u}$ and $\mathbf{v}$ (Figure 3.4.4). Thus, from (6), the area A of this parallelogram is given by

$$A = (\text{base})(\text{altitude}) = \|\mathbf{u}\| \|\mathbf{v}\| \sin \theta = \|\mathbf{u} \times \mathbf{v}\|$$

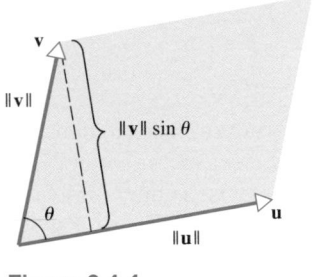

Figure 3.4.4

This result is even correct if $\mathbf{u}$ and $\mathbf{v}$ are collinear, since the parallelogram determined by $\mathbf{u}$ and $\mathbf{v}$ has zero area and from (6) we have $\mathbf{u} \times \mathbf{v} = \mathbf{0}$ because $\theta = 0$ in this case. Thus we have the following theorem.

THEOREM 3.4.3

Area of a Parallelogram

If $\mathbf{u}$ and $\mathbf{v}$ are vectors in 3-space, then $\|\mathbf{u} \times \mathbf{v}\|$ is equal to the area of the parallelogram determined by $\mathbf{u}$ and $\mathbf{v}$.

EXAMPLE 4 Area of a Triangle

Find the area of the triangle determined by the points $P_1(2, 2, 0)$, $P_2(-1, 0, 2)$, and $P_3(0, 4, 3)$.

Solution

The area A of the triangle is $\frac{1}{2}$ the area of the parallelogram determined by the vectors $\overrightarrow{P_1 P_2}$ and $\overrightarrow{P_1 P_3}$ (Figure 3.4.5). Using the method discussed in Example 2 of Section 3.1,

Figure 3.4.5

[†]Recall that we agreed to consider only right-handed coordinate systems in this text. Had we used left-handed systems instead, a "left-hand rule" would apply here.

$\overrightarrow{P_1P_2} = (-3, -2, 2)$ and $\overrightarrow{P_1P_3} = (-2, 2, 3)$. It follows that

$$\overrightarrow{P_1P_2} \times \overrightarrow{P_1P_3} = (-10, 5, -10)$$

and consequently,

$$A = \tfrac{1}{2}\|\overrightarrow{P_1P_2} \times \overrightarrow{P_1P_3}\| = \tfrac{1}{2}(15) = \tfrac{15}{2} \quad \blacklozenge$$

DEFINITION

If **u**, **v**, and **w** are vectors in 3-space, then

$$\mathbf{u} \cdot (\mathbf{v} \times \mathbf{w})$$

is called the *scalar triple product* of **u**, **v**, and **w**.

The scalar triple product of $\mathbf{u} = (u_1, u_2, u_3)$, $\mathbf{v} = (v_1, v_2, v_3)$, and $\mathbf{w} = (w_1, w_2, w_3)$ can be calculated from the formula

$$\mathbf{u} \cdot (\mathbf{v} \times \mathbf{w}) = \begin{vmatrix} u_1 & u_2 & u_3 \\ v_1 & v_2 & v_3 \\ w_1 & w_2 & w_3 \end{vmatrix} \tag{7}$$

This follows from Formula (4) since

$$\mathbf{u} \cdot (\mathbf{v} \times \mathbf{w}) = \mathbf{u} \cdot \left(\begin{vmatrix} v_2 & v_3 \\ w_2 & w_3 \end{vmatrix} \mathbf{i} - \begin{vmatrix} v_1 & v_3 \\ w_1 & w_3 \end{vmatrix} \mathbf{j} + \begin{vmatrix} v_1 & v_2 \\ w_1 & w_2 \end{vmatrix} \mathbf{k} \right)$$

$$= \begin{vmatrix} v_2 & v_3 \\ w_2 & w_3 \end{vmatrix} u_1 - \begin{vmatrix} v_1 & v_3 \\ w_1 & w_3 \end{vmatrix} u_2 + \begin{vmatrix} v_1 & v_2 \\ w_1 & w_2 \end{vmatrix} u_3$$

$$= \begin{vmatrix} u_1 & u_2 & u_3 \\ v_1 & v_2 & v_3 \\ w_1 & w_2 & w_3 \end{vmatrix}$$

EXAMPLE 5 Calculating a Scalar Triple Product

Calculate the scalar triple product $\mathbf{u} \cdot (\mathbf{v} \times \mathbf{w})$ of the vectors

$$\mathbf{u} = 3\mathbf{i} - 2\mathbf{j} - 5\mathbf{k}, \qquad \mathbf{v} = \mathbf{i} + 4\mathbf{j} - 4\mathbf{k}, \qquad \mathbf{w} = 3\mathbf{j} + 2\mathbf{k}$$

Solution

From (7),

$$\mathbf{u} \cdot (\mathbf{v} \times \mathbf{w}) = \begin{vmatrix} 3 & -2 & -5 \\ 1 & 4 & -4 \\ 0 & 3 & 2 \end{vmatrix}$$

$$= 3 \begin{vmatrix} 4 & -4 \\ 3 & 2 \end{vmatrix} - (-2) \begin{vmatrix} 1 & -4 \\ 0 & 2 \end{vmatrix} + (-5) \begin{vmatrix} 1 & 4 \\ 0 & 3 \end{vmatrix}$$

$$= 60 + 4 - 15 = 49 \quad \blacklozenge$$

REMARK The symbol $(\mathbf{u} \cdot \mathbf{v}) \times \mathbf{w}$ makes no sense because we cannot form the cross product of a scalar and a vector. Thus no ambiguity arises if we write $\mathbf{u} \cdot \mathbf{v} \times \mathbf{w}$ rather than $\mathbf{u} \cdot (\mathbf{v} \times \mathbf{w})$. However, for clarity we shall usually keep the parentheses.

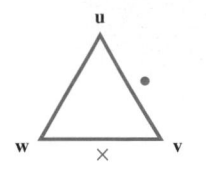

Figure 3.4.6

It follows from (7) that

$$\mathbf{u} \cdot (\mathbf{v} \times \mathbf{w}) = \mathbf{w} \cdot (\mathbf{u} \times \mathbf{v}) = \mathbf{v} \cdot (\mathbf{w} \times \mathbf{u})$$

since the 3×3 determinants that represent these products can be obtained from one another by *two* row interchanges. (Verify.) These relationships can be remembered by moving the vectors $\mathbf{u}$, $\mathbf{v}$, and $\mathbf{w}$ clockwise around the vertices of the triangle in Figure 3.4.6.

Geometric Interpretation of Determinants

The next theorem provides a useful geometric interpretation of 2×2 and 3×3 determinants.

THEOREM 3.4.4

(a) *The absolute value of the determinant*

$$\det \begin{bmatrix} u_1 & u_2 \\ v_1 & v_2 \end{bmatrix}$$

is equal to the area of the parallelogram in 2-space determined by the vectors $\mathbf{u} = (u_1, u_2)$ *and* $\mathbf{v} = (v_1, v_2)$. *(See Figure 3.4.7a.)*

(b) *The absolute value of the determinant*

$$\det \begin{bmatrix} u_1 & u_2 & u_3 \\ v_1 & v_2 & v_3 \\ w_1 & w_2 & w_3 \end{bmatrix}$$

is equal to the volume of the parallelepiped in 3-space determined by the vectors $\mathbf{u} = (u_1, u_2, u_3)$, $\mathbf{v} = (v_1, v_2, v_3)$, *and* $\mathbf{w} = (w_1, w_2, w_3)$. *(See Figure 3.4.7b.)*

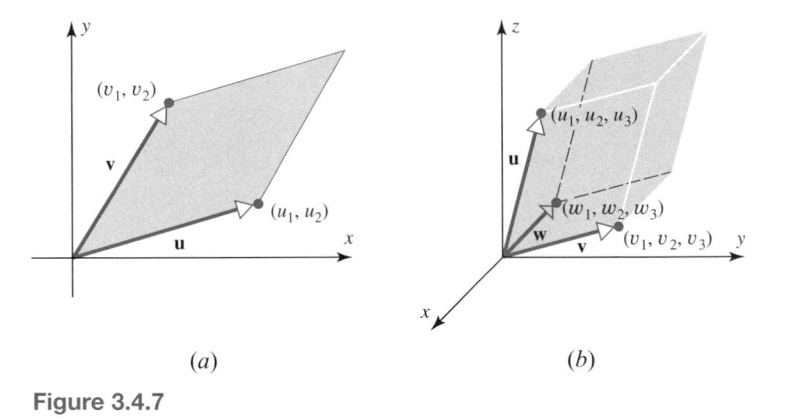

(a) (b)

Figure 3.4.7

Proof (a) The key to the proof is to use Theorem 3.4.3. However, that theorem applies to vectors in 3-space, whereas $\mathbf{u} = (u_1, u_2)$ and $\mathbf{v} = (v_1, v_2)$ are vectors in 2-space. To circumvent this "dimension problem," we shall view $\mathbf{u}$ and $\mathbf{v}$ as vectors in the xy-plane of an xyz-coordinate system (Figure 3.4.8a), in which case these vectors are expressed

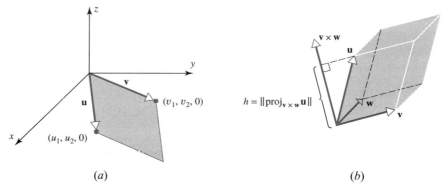

Figure 3.4.8

as $\mathbf{u} = (u_1, u_2, 0)$ and $\mathbf{v} = (v_1, v_2, 0)$. Thus

$$\mathbf{u} \times \mathbf{v} = \begin{vmatrix} \mathbf{i} & \mathbf{j} & \mathbf{k} \\ u_1 & u_2 & 0 \\ v_1 & v_2 & 0 \end{vmatrix} = \begin{vmatrix} u_1 & u_2 \\ v_1 & v_2 \end{vmatrix} \mathbf{k} = \det \begin{bmatrix} u_1 & u_2 \\ v_1 & v_2 \end{bmatrix} \mathbf{k}$$

It now follows from Theorem 3.4.3 and the fact that $\|\mathbf{k}\| = 1$ that the area A of the parallelogram determined by $\mathbf{u}$ and $\mathbf{v}$ is

$$A = \|\mathbf{u} \times \mathbf{v}\| = \left\| \det \begin{bmatrix} u_1 & u_2 \\ v_1 & v_2 \end{bmatrix} \mathbf{k} \right\| = \left| \det \begin{bmatrix} u_1 & u_2 \\ v_1 & v_2 \end{bmatrix} \right| \|\mathbf{k}\| = \left| \det \begin{bmatrix} u_1 & u_2 \\ v_1 & v_2 \end{bmatrix} \right|$$

which completes the proof.

Proof (b) As shown in Figure 3.4.8*b*, take the base of the parallelepiped determined by $\mathbf{u}$, $\mathbf{v}$, and $\mathbf{w}$ to be the parallelogram determined by $\mathbf{v}$ and $\mathbf{w}$. It follows from Theorem 3.4.3 that the area of the base is $\|\mathbf{v} \times \mathbf{w}\|$ and, as illustrated in Figure 3.4.8*b*, the height h of the parallelepiped is the length of the orthogonal projection of $\mathbf{u}$ on $\mathbf{v} \times \mathbf{w}$. Therefore, by Formula (10) of Section 3.3,

$$h = \|\text{proj}_{\mathbf{v} \times \mathbf{w}} \mathbf{u}\| = \frac{|\mathbf{u} \cdot (\mathbf{v} \times \mathbf{w})|}{\|\mathbf{v} \times \mathbf{w}\|}$$

It follows that the volume V of the parallelepiped is

$$V = (\text{area of base}) \cdot \text{height} = \|\mathbf{v} \times \mathbf{w}\| \frac{|\mathbf{u} \cdot (\mathbf{v} \times \mathbf{w})|}{\|\mathbf{v} \times \mathbf{w}\|} = |\mathbf{u} \cdot (\mathbf{v} \times \mathbf{w})|$$

so from (7),

$$V = \left| \det \begin{bmatrix} u_1 & u_2 & u_3 \\ v_1 & v_2 & v_3 \\ w_1 & w_2 & w_3 \end{bmatrix} \right|$$

which completes the proof. ■

Remark If V denotes the volume of the parallelepiped determined by vectors $\mathbf{u}$, $\mathbf{v}$, and $\mathbf{w}$, then it follows from Theorem 3.3 and Formula (7) that

$$V = \begin{bmatrix} \text{volume of parallelepiped} \\ \text{determined by } \mathbf{u}, \mathbf{v}, \text{ and } \mathbf{w} \end{bmatrix} = |\mathbf{u} \cdot (\mathbf{v} \times \mathbf{w})| \tag{8}$$

From this and Theorem 3.3.1*b*, we can conclude that

$$\mathbf{u} \cdot (\mathbf{v} \times \mathbf{w}) = \pm V$$

where the $+$ or $-$ results depending on whether $\mathbf{u}$ makes an acute or an obtuse angle with $\mathbf{v} \times \mathbf{w}$.

Formula (8) leads to a useful test for ascertaining whether three given vectors lie in the same plane. Since three vectors not in the same plane determine a parallelepiped of positive volume, it follows from (8) that $|\mathbf{u} \cdot (\mathbf{v} \times \mathbf{w})| = 0$ if and only if the vectors $\mathbf{u}$, $\mathbf{v}$, and $\mathbf{w}$ lie in the same plane. Thus we have the following result.

THEOREM 3.4.5

> *If the vectors* $\mathbf{u} = (u_1, u_2, u_3)$, $\mathbf{v} = (v_1, v_2, v_3)$, *and* $\mathbf{w} = (w_1, w_2, w_3)$ *have the same initial point, then they lie in the same plane if and only if*
>
> $$\mathbf{u} \cdot (\mathbf{v} \times \mathbf{w}) = \begin{vmatrix} u_1 & u_2 & u_3 \\ v_1 & v_2 & v_3 \\ w_1 & w_2 & w_3 \end{vmatrix} = 0$$

Independence of Cross Product and Coordinates

Initially, we defined a vector to be a directed line segment or arrow in 2-space or 3-space; coordinate systems and components were introduced later in order to simplify computations with vectors. Thus, a vector has a "mathematical existence" regardless of whether a coordinate system has been introduced. Further, the components of a vector are not determined by the vector alone; they depend as well on the coordinate system chosen. For example, in Figure 3.4.9 we have indicated a fixed vector $\mathbf{v}$ in the plane and two different coordinate systems. In the xy-coordinate system the components of $\mathbf{v}$ are $(1, 1)$, and in the $x'y'$-system they are $(\sqrt{2}, 0)$.

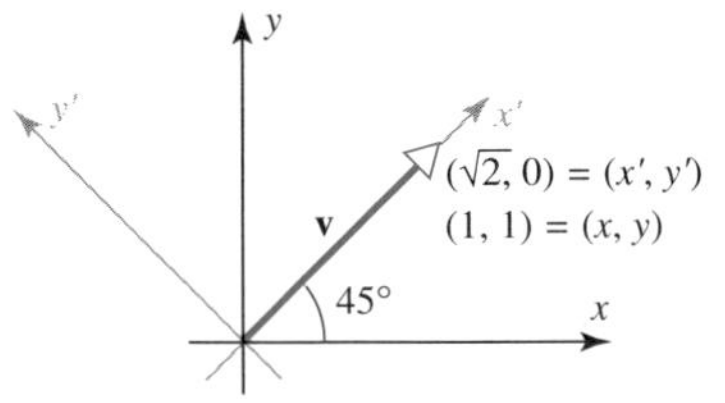

Figure 3.4.9

This raises an important question about our definition of cross product. Since we defined the cross product $\mathbf{u} \times \mathbf{v}$ in terms of the components of $\mathbf{u}$ and $\mathbf{v}$, and since these components depend on the coordinate system chosen, it seems possible that two *fixed* vectors $\mathbf{u}$ and $\mathbf{v}$ might have different cross products in different coordinate systems. Fortunately, this is not the case. To see that this is so, we need only recall that

- $\mathbf{u} \times \mathbf{v}$ is perpendicular to both $\mathbf{u}$ and $\mathbf{v}$.
- The orientation of $\mathbf{u} \times \mathbf{v}$ is determined by the right-hand rule.
- $\|\mathbf{u} \times \mathbf{v}\| = \|\mathbf{u}\|\|\mathbf{v}\| \sin \theta$.

These three properties completely determine the vector $\mathbf{u} \times \mathbf{v}$: the first and second properties determine the direction, and the third property determines the length. Since these properties of $\mathbf{u} \times \mathbf{v}$ depend only on the lengths and relative positions of $\mathbf{u}$ and $\mathbf{v}$ and not on the particular right-hand coordinate system being used, the vector $\mathbf{u} \times \mathbf{v}$ will remain unchanged if a different right-hand coordinate system is introduced. We say that the definition of $\mathbf{u} \times \mathbf{v}$ is ***coordinate free***. This result is of importance to physicists and engineers who often work with many coordinate systems in the same problem.

EXAMPLE 6 $\mathbf{u} \times \mathbf{v}$ Is Independent of the Coordinate System
---

Consider two perpendicular vectors $\mathbf{u}$ and $\mathbf{v}$, each of length 1 (Figure 3.4.10*a*). If we introduce an xyz-coordinate system as shown in Figure 3.4.10*b*, then

$$\mathbf{u} = (1, 0, 0) = \mathbf{i} \quad \text{and} \quad \mathbf{v} = (0, 1, 0) = \mathbf{j}$$

so that

$$\mathbf{u} \times \mathbf{v} = \mathbf{i} \times \mathbf{j} = \mathbf{k} = (0, 0, 1)$$

However, if we introduce an $x'y'z'$-coordinate system as shown in Figure 3.4.10*c*, then

$$\mathbf{u} = (0, 0, 1) = \mathbf{k} \quad \text{and} \quad \mathbf{v} = (1, 0, 0) = \mathbf{i}$$

so that

$$\mathbf{u} \times \mathbf{v} = \mathbf{k} \times \mathbf{i} = \mathbf{j} = (0, 1, 0)$$

But it is clear from Figures 3.4.10*b* and 3.4.10*c* that the vector $(0, 0, 1)$ in the xyz-system is the same as the vector $(0, 1, 0)$ in the $x'y'z'$-system. Thus we obtain the same vector $\mathbf{u} \times \mathbf{v}$ whether we compute with coordinates from the xyz-system or with coordinates from the $x'y'z'$-system. ◆

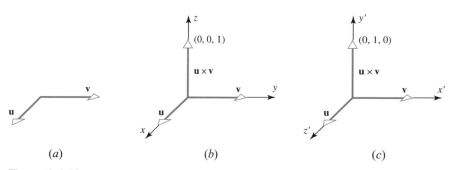

(a)   *(b)*   *(c)*

Figure 3.4.10

EXERCISE SET
3.4

1. Let $\mathbf{u} = (3, 2, -1)$, $\mathbf{v} = (0, 2, -3)$, and $\mathbf{w} = (2, 6, 7)$. Compute
    (a) $\mathbf{v} \times \mathbf{w}$ (b) $\mathbf{u} \times (\mathbf{v} \times \mathbf{w})$ (c) $(\mathbf{u} \times \mathbf{v}) \times \mathbf{w}$
    (d) $(\mathbf{u} \times \mathbf{v}) \times (\mathbf{v} \times \mathbf{w})$ (e) $\mathbf{u} \times (\mathbf{v} - 2\mathbf{w})$ (f) $(\mathbf{u} \times \mathbf{v}) - 2\mathbf{w}$

2. Find a vector that is orthogonal to both $\mathbf{u}$ and $\mathbf{v}$.
    (a) $\mathbf{u} = (-6, 4, 2)$, $\mathbf{v} = (3, 1, 5)$ (b) $\mathbf{u} = (-2, 1, 5)$, $\mathbf{v} = (3, 0, -3)$

3. Find the area of the parallelogram determined by $\mathbf{u}$ and $\mathbf{v}$.
    (a) $\mathbf{u} = (1, -1, 2)$, $\mathbf{v} = (0, 3, 1)$ (b) $\mathbf{u} = (2, 3, 0)$, $\mathbf{v} = (-1, 2, -2)$
    (c) $\mathbf{u} = (3, -1, 4)$, $\mathbf{v} = (6, -2, 8)$

4. Find the area of the triangle having vertices P, Q, and R.
    (a) $P(2, 6, -1)$, $Q(1, 1, 1)$, $R(4, 6, 2)$ (b) $P(1, -1, 2)$, $Q(0, 3, 4)$, $R(6, 1, 8)$

5. Verify parts (a), (b), and (c) of Theorem 3.4.1 for the vectors $\mathbf{u} = (4, 2, 1)$ and $\mathbf{v} = (-3, 2, 7)$.

6. Verify parts (a), (b), and (c) of Theorem 3.4.2 for $\mathbf{u} = (5, -1, 2)$, $\mathbf{v} = (6, 0, -2)$, and $\mathbf{w} = (1, 2, -1)$.

7. Find a vector $\mathbf{v}$ that is orthogonal to the vector $\mathbf{u} = (2, -3, 5)$.

8. Find the scalar triple product $\mathbf{u} \cdot (\mathbf{v} \times \mathbf{w})$.

   (a) $\mathbf{u} = (-1, 2, 4)$, $\mathbf{v} = (3, 4, -2)$, $\mathbf{w} = (-1, 2, 5)$

   (b) $\mathbf{u} = (3, -1, 6)$, $\mathbf{v} = (2, 4, 3)$, $\mathbf{w} = (5, -1, 2)$

9. Suppose that $\mathbf{u} \cdot (\mathbf{v} \times \mathbf{w}) = 3$. Find

   (a) $\mathbf{u} \cdot (\mathbf{w} \times \mathbf{v})$ (b) $(\mathbf{v} \times \mathbf{w}) \cdot \mathbf{u}$ (c) $\mathbf{w} \cdot (\mathbf{u} \times \mathbf{v})$

   (d) $\mathbf{v} \cdot (\mathbf{u} \times \mathbf{w})$ (e) $(\mathbf{u} \times \mathbf{w}) \cdot \mathbf{v}$ (f) $\mathbf{v} \cdot (\mathbf{w} \times \mathbf{w})$

10. Find the volume of the parallelepiped with sides $\mathbf{u}$, $\mathbf{v}$, and $\mathbf{w}$.

    (a) $\mathbf{u} = (2, -6, 2)$, $\mathbf{v} = (0, 4, -2)$, $\mathbf{w} = (2, 2, -4)$

    (b) $\mathbf{u} = (3, 1, 2)$, $\mathbf{v} = (4, 5, 1)$, $\mathbf{w} = (1, 2, 4)$

11. Determine whether $\mathbf{u}$, $\mathbf{v}$, and $\mathbf{w}$ lie in the same plane when positioned so that their initial points coincide.

    (a) $\mathbf{u} = (-1, -2, 1)$, $\mathbf{v} = (3, 0, -2)$, $\mathbf{w} = (5, -4, 0)$

    (b) $\mathbf{u} = (5, -2, 1)$, $\mathbf{v} = (4, -1, 1)$, $\mathbf{w} = (1, -1, 0)$

    (c) $\mathbf{u} = (4, -8, 1)$, $\mathbf{v} = (2, 1, -2)$, $\mathbf{w} = (3, -4, 12)$

12. Find all unit vectors parallel to the yz-plane that are perpendicular to the vector $(3, -1, 2)$.

13. Find all unit vectors in the plane determined by $\mathbf{u} = (3, 0, 1)$ and $\mathbf{v} = (1, -1, 1)$ that are perpendicular to the vector $\mathbf{w} = (1, 2, 0)$.

14. Let $\mathbf{a} = (a_1, a_2, a_3)$, $\mathbf{b} = (b_1, b_2, b_3)$, $\mathbf{c} = (c_1, c_2, c_3)$, and $\mathbf{d} = (d_1, d_2, d_3)$. Show that

$$(\mathbf{a} + \mathbf{d}) \cdot (\mathbf{b} \times \mathbf{c}) = \mathbf{a} \cdot (\mathbf{b} \times \mathbf{c}) + \mathbf{d} \cdot (\mathbf{b} \times \mathbf{c})$$

15. Simplify $(\mathbf{u} + \mathbf{v}) \times (\mathbf{u} - \mathbf{v})$.

16. Use the cross product to find the sine of the angle between the vectors $\mathbf{u} = (2, 3, -6)$ and $\mathbf{v} = (2, 3, 6)$.

17. (a) Find the area of the triangle having vertices $A(1, 0, 1)$, $B(0, 2, 3)$, and $C(2, 1, 0)$.

    (b) Use the result of part (a) to find the length of the altitude from vertex C to side AB.

18. Show that if $\mathbf{u}$ is a vector from any point on a line to a point P not on the line, and $\mathbf{v}$ is a vector parallel to the line, then the distance between P and the line is given by $\|\mathbf{u} \times \mathbf{v}\|/\|\mathbf{v}\|$.

19. Use the result of Exercise 18 to find the distance between the point P and the line through the points A and B.

    (a) $P(-3, 1, 2)$, $A(1, 1, 0)$, $B(-2, 3, -4)$ (b) $P(4, 3, 0)$, $A(2, 1, -3)$, $B(0, 2, -1)$

20. Prove: If θ is the angle between $\mathbf{u}$ and $\mathbf{v}$ and $\mathbf{u} \cdot \mathbf{v} \neq 0$, then $\tan \theta = \|\mathbf{u} \times \mathbf{v}\|/(\mathbf{u} \cdot \mathbf{v})$.

21. Consider the parallelepiped with sides $\mathbf{u} = (3, 2, 1)$, $\mathbf{v} = (1, 1, 2)$, and $\mathbf{w} = (1, 3, 3)$.

    (a) Find the area of the face determined by $\mathbf{u}$ and $\mathbf{w}$.

    (b) Find the angle between $\mathbf{u}$ and the plane containing the face determined by $\mathbf{v}$ and $\mathbf{w}$.

    Note   The *angle between a vector and a plane* is defined to be the complement of the angle θ between the vector and that normal to the plane for which $0 \leq \theta \leq \pi/2$.

22. Find a vector $\mathbf{n}$ that is perpendicular to the plane determined by the points $A(0, -2, 1)$, $B(1, -1, -2)$, and $C(-1, 1, 0)$. [See the note in Exercise 21.]

23. Let $\mathbf{m}$ and $\mathbf{n}$ be vectors whose components in the xyz-system of Figure 3.4.10 are $\mathbf{m} = (0, 0, 1)$ and $\mathbf{n} = (0, 1, 0)$.

    (a)   Find the components of $\mathbf{m}$ and $\mathbf{n}$ in the $x'y'z'$-system of Figure 3.4.10.

    (b)   Compute $\mathbf{m} \times \mathbf{n}$ using the components in the xyz-system.

    (c)   Compute $\mathbf{m} \times \mathbf{n}$ using the components in the $x'y'z'$-system.

    (d)   Show that the vectors obtained in (b) and (c) are the same.

24. Prove the following identities.

    (a) $(\mathbf{u} + k\mathbf{v}) \times \mathbf{v} = \mathbf{u} \times \mathbf{v}$ (b) $\mathbf{u} \cdot (\mathbf{v} \times \mathbf{z}) = -(\mathbf{u} \times \mathbf{z}) \cdot \mathbf{v}$

25. Let $\mathbf{u}$, $\mathbf{v}$, and $\mathbf{w}$ be nonzero vectors in 3-space with the same initial point, but such that no two of them are collinear. Show that

    (a) $\mathbf{u} \times (\mathbf{v} \times \mathbf{w})$ lies in the plane determined by $\mathbf{v}$ and $\mathbf{w}$

    (b) $(\mathbf{u} \times \mathbf{v}) \times \mathbf{w}$ lies in the plane determined by $\mathbf{u}$ and $\mathbf{v}$

26. Prove part (d) of Theorem 3.4.1.

    Hint   First prove the result in the case where $\mathbf{w} = \mathbf{i} = (1, 0, 0)$, then when $\mathbf{w} = \mathbf{j} = (0, 1, 0)$, and then when $\mathbf{w} = \mathbf{k} = (0, 0, 1)$. Finally, prove it for an arbitrary vector $\mathbf{w} = (w_1, w_2, w_3)$ by writing $\mathbf{w} = w_1\mathbf{i} + w_2\mathbf{j} + w_3\mathbf{k}$.

27. Prove part (e) of Theorem 3.4.1.

    Hint   Apply part (a) of Theorem 3.4.2 to the result in part (d) of Theorem 3.4.1.

28. Let $\mathbf{u} = (1, 3, -1)$, $\mathbf{v} = (1, 1, 2)$, and $\mathbf{w} = (3, -1, 2)$. Calculate $\mathbf{u} \times (\mathbf{v} \times \mathbf{w})$ using the technique of Exercise 26; then check your result by calculating directly.

29. Prove: If $\mathbf{a}$, $\mathbf{b}$, $\mathbf{c}$, and $\mathbf{d}$ lie in the same plane, then $(\mathbf{a} \times \mathbf{b}) \times (\mathbf{c} \times \mathbf{d}) = \mathbf{0}$.

30. It is a theorem of solid geometry that the volume of a tetrahedron is $\frac{1}{3}$ (area of base) · (height). Use this result to prove that the volume of a tetrahedron whose sides are the vectors $\mathbf{a}$, $\mathbf{b}$, and $\mathbf{c}$ is $\frac{1}{6}|\mathbf{a} \cdot (\mathbf{b} \times \mathbf{c})|$ (see the accompanying figure).

31. Use the result of Exercise 30 to find the volume of the tetrahedron with vertices P, Q, R, S.

    (a) $P(-1, 2, 0)$, $Q(2, 1, -3)$, $R(1, 0, 1)$, $S(3, -2, 3)$

    (b) $P(0, 0, 0)$, $Q(1, 2, -1)$, $R(3, 4, 0)$, $S(-1, -3, 4)$

32. Prove part (b) of Theorem 3.4.2.

33. Prove parts (c) and (d) of Theorem 3.4.2.

34. Prove parts (e) and (f) of Theorem 3.4.2.

Figure Ex-30

Discussion & Discovery

35. (a)   Suppose that $\mathbf{u}$ and $\mathbf{v}$ are noncollinear vectors with their initial points at the origin in 3-space. Make a sketch that illustrates how $\mathbf{w} = \mathbf{v} \times (\mathbf{u} \times \mathbf{v})$ is oriented in relation to $\mathbf{u}$ and $\mathbf{v}$.

    (b)   For $\mathbf{w}$ as in part (a), what can you say about the values of $\mathbf{u} \cdot \mathbf{w}$ and $\mathbf{v} \cdot \mathbf{w}$? Explain your reasoning.

36. If $\mathbf{u} \neq \mathbf{0}$, is it valid to cancel $\mathbf{u}$ from both sides of the equation $\mathbf{u} \times \mathbf{v} = \mathbf{u} \times \mathbf{w}$ and conclude that $\mathbf{v} = \mathbf{w}$? Explain your reasoning.

37. Something is wrong with one of the following expressions. Which one is it and what is wrong?

$$\mathbf{u} \cdot (\mathbf{v} \times \mathbf{w}), \qquad \mathbf{u} \times \mathbf{v} \times \mathbf{w}, \qquad (\mathbf{u} \cdot \mathbf{v}) \times \mathbf{w}$$

38. What can you say about the vectors $\mathbf{u}$ and $\mathbf{v}$ if $\mathbf{u} \times \mathbf{v} = \mathbf{0}$?

39. Give some examples of algebraic rules that hold for multiplication of real numbers but not for the cross product of vectors.

3.5
LINES AND PLANES IN 3-SPACE

In this section we shall use vectors to derive equations of lines and planes in 3-space. We shall then use these equations to solve some basic geometric problems.

Planes in 3-Space

In analytic geometry a line in 2-space can be specified by giving its slope and one of its points. Similarly, one can specify a plane in 3-space by giving its inclination and specifying one of its points. A convenient method for describing the inclination of a plane is to specify a nonzero vector, called a *normal*, that is perpendicular to the plane.

Suppose that we want to find the equation of the plane passing through the point $P_0(x_0, y_0, z_0)$ and having the nonzero vector $\mathbf{n} = (a, b, c)$ as a normal. It is evident from Figure 3.5.1 that the plane consists precisely of those points $P(x, y, z)$ for which the vector $\overrightarrow{P_0 P}$ is orthogonal to $\mathbf{n}$; that is,

$$\mathbf{n} \cdot \overrightarrow{P_0 P} = 0 \tag{1}$$

Since $\overrightarrow{P_0 P} = (x - x_0, y - y_0, z - z_0)$, Equation (1) can be written as

$$a(x - x_0) + b(y - y_0) + c(z - z_0) = 0 \tag{2}$$

We call this the *point-normal* form of the equation of a plane.

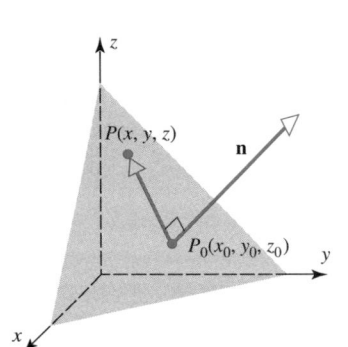

Figure 3.5.1 Plane with normal vector.

EXAMPLE 1   Finding the Point-Normal Equation of a Plane

Find an equation of the plane passing through the point $(3, -1, 7)$ and perpendicular to the vector $\mathbf{n} = (4, 2, -5)$.

Solution

From (2) a point-normal form is

$$4(x - 3) + 2(y + 1) - 5(z - 7) = 0 \quad \blacklozenge$$

By multiplying out and collecting terms, we can rewrite (2) in the form

$$ax + by + cz + d = 0$$

where a, b, c, and d are constants, and a, b, and c are not all zero. For example, the equation in Example 1 can be rewritten as

$$4x + 2y - 5z + 25 = 0$$

As the next theorem shows, planes in 3-space are represented by equations of the form $ax + by + cz + d = 0$.

THEOREM 3.5.1

If a, b, c, and d are constants and a, b, and c are not all zero, then the graph of the equation

$$ax + by + cz + d = 0 \tag{3}$$

is a plane having the vector $\mathbf{n} = (a, b, c)$ as a normal.

Equation (3) is a linear equation in x, y, and z; it is called the ***general form*** of the equation of a plane.

Proof By hypothesis, the coefficients a, b, and c are not all zero. Assume, for the moment, that $a \neq 0$. Then the equation $ax + by + cz + d = 0$ can be rewritten in the form $a(x + (d/a)) + by + cz = 0$. But this is a point-normal form of the plane passing through the point $(-d/a, 0, 0)$ and having $\mathbf{n} = (a, b, c)$ as a normal.

If $a = 0$, then either $b \neq 0$ or $c \neq 0$. A straightforward modification of the above argument will handle these other cases. ∎

Just as the solutions of a system of linear equations

$$ax + by = k_1$$
$$cx + dy = k_2$$

correspond to points of intersection of the lines $ax + by = k_1$ and $cx + dy = k_2$ in the xy-plane, so the solutions of a system

$$ax + by + cz = k_1$$
$$dx + ey + fz = k_2 \tag{4}$$
$$gx + hy + iz = k_3$$

correspond to the points of intersection of the three planes $ax + by + cz = k_1$, $dx + ey + fz = k_2$, and $gx + hy + iz = k_3$.

In Figure 3.5.2 we have illustrated the geometric possibilities that occur when (4) has zero, one, or infinitely many solutions.

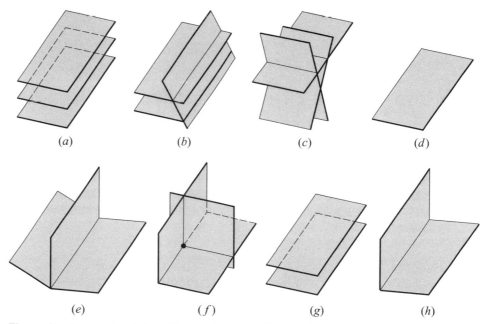

(a) $\qquad$ (b) $\qquad$ (c) $\qquad$ (d)

(e) $\qquad$ (f) $\qquad$ (g) $\qquad$ (h)

Figure 3.5.2 (a) No solutions (3 parallel planes). (b) No solutions (2 parallel planes). (c) No solutions (3 planes with no common intersection). (d) Infinitely many solutions (3 coincident planes). (e) Infinitely many solutions (3 planes intersecting in a line). (f) One solution (3 planes intersecting at a point). (g) No solutions (2 coincident planes parallel to a third plane). (h) Infinitely many solutions (2 coincident planes intersecting a third plane).

EXAMPLE 2 Equation of a Plane Through Three Points

Find the equation of the plane passing through the points $P_1(1, 2, -1)$, $P_2(2, 3, 1)$, and $P_3(3, -1, 2)$.

Solution

Since the three points lie in the plane, their coordinates must satisfy the general equation $ax + by + cz + d = 0$ of the plane. Thus

$$a + 2b - c + d = 0$$
$$2a + 3b + c + d = 0$$
$$3a - b + 2c + d = 0$$

Solving this system gives $a = -\frac{9}{16}t$, $b = -\frac{1}{16}t$, $c = \frac{5}{16}t$, $d = t$. Letting $t = -16$, for example, yields the desired equation

$$9x + y - 5z - 16 = 0$$

We note that any other choice of t gives a multiple of this equation, so that any value of $t \neq 0$ would also give a valid equation of the plane.

Alternative Solution

Since the points $P_1(1, 2, -1)$, $P_2(2, 3, 1)$, and $P_3(3, -1, 2)$ lie in the plane, the vectors $\overrightarrow{P_1 P_2} = (1, 1, 2)$ and $\overrightarrow{P_1 P_3} = (2, -3, 3)$ are parallel to the plane. Therefore, the equation $\overrightarrow{P_1 P_2} \times \overrightarrow{P_1 P_3} = (9, 1, -5)$ is normal to the plane, since it is perpendicular to both $\overrightarrow{P_1 P_2}$ and $\overrightarrow{P_1 P_3}$. From this and the fact that P_1 lies in the plane, a point-normal form for the equation of the plane is

$$9(x - 1) + (y - 2) - 5(z + 1) = 0 \quad \text{or} \quad 9x + y - 5z - 16 = 0 \quad \blacklozenge$$

Vector Form of Equation of a Plane

Vector notation provides a useful alternative way of writing the point-normal form of the equation of a plane: Referring to Figure 3.5.3, let $\mathbf{r} = (x, y, z)$ be the vector from the origin to the point $P(x, y, z)$, let $\mathbf{r}_0 = (x_0, y_0, z_0)$ be the vector from the origin to the point $P_0(x_0, y_0, z_0)$, and let $\mathbf{n} = (a, b, c)$ be a vector normal to the plane. Then $\overrightarrow{P_0 P} = \mathbf{r} - \mathbf{r}_0$, so Formula (1) can be rewritten as

$$\mathbf{n} \cdot (\mathbf{r} - \mathbf{r}_0) = 0 \tag{5}$$

This is called the **vector form of the equation of a plane**.

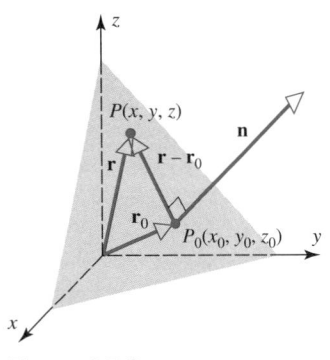

Figure 3.5.3

EXAMPLE 3 Vector Equation of a Plane Using (5)

The equation

$$(-1, 2, 5) \cdot (x - 6, y - 3, z + 4) = 0$$

is the vector equation of the plane that passes through the point $(6, 3, -4)$ and is perpendicular to the vector $\mathbf{n} = (-1, 2, 5)$. $\blacklozenge$

Lines in 3-Space

We shall now show how to obtain equations for lines in 3-space. Suppose that l is the line in 3-space through the point $P_0(x_0, y_0, z_0)$ and parallel to the nonzero vector

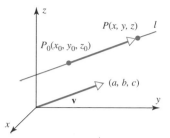

Figure 3.5.4 $\overrightarrow{P_0 P}$ is parallel to **v**.

$\mathbf{v} = (a, b, c)$. It is clear (Figure 3.5.4) that l consists precisely of those points $P(x, y, z)$ for which the vector $\overrightarrow{P_0 P}$ is parallel to **v**—that is, for which there is a scalar t such that

$$\overrightarrow{P_0 P} = t\mathbf{v} \qquad (6)$$

In terms of components, (6) can be written as

$$(x - x_0, y - y_0, z - z_0) = (ta, tb, tc)$$

from which it follows that $x - x_0 = ta$, $y - y_0 = tb$, and $z - z_0 = tc$, so

$$x = x_0 + ta, \qquad y = y_0 + tb, \qquad z = z_0 + tc$$

As the parameter t varies from $-\infty$ to $+\infty$, the point $P(x, y, z)$ traces out the line l. The equations

$$x = x_0 + ta, \quad y = y_0 + tb, \quad z = z_0 + tc \qquad (-\infty < t < +\infty) \qquad (7)$$

are called ***parametric equations*** for l.

EXAMPLE 4 Parametric Equations of a Line

The line through the point $(1, 2, -3)$ and parallel to the vector $\mathbf{v} = (4, 5, -7)$ has parametric equations

$$x = 1 + 4t, \quad y = 2 + 5t, \quad z = -3 - 7t \qquad (-\infty < t < +\infty) \blacklozenge$$

EXAMPLE 5 Intersection of a Line and the *xy*-Plane

(a) Find parametric equations for the line l passing through the points $P_1(2, 4, -1)$ and $P_2(5, 0, 7)$.

(b) Where does the line intersect the xy-plane?

Solution (a)

Since the vector $\overrightarrow{P_1 P_2} = (3, -4, 8)$ is parallel to l and $P_1(2, 4, -1)$ lies on l, the line l is given by

$$x = 2 + 3t, \quad y = 4 - 4t, \quad z = -1 + 8t \qquad (-\infty < t < +\infty)$$

Solution (b)

The line intersects the xy-plane at the point where $z = -1 + 8t = 0$, that is, where $t = \frac{1}{8}$. Substituting this value of t in the parametric equations for l yields, as the point of intersection,

$$(x, y, z) = \left(\frac{19}{8}, \frac{7}{2}, 0\right) \qquad \blacksquare$$

EXAMPLE 6 Line of Intersection of Two Planes

Find parametric equations for the line of intersection of the planes

$$3x + 2y - 4z - 6 = 0 \quad \text{and} \quad x - 3y - 2z - 4 = 0$$

Solution

The line of intersection consists of all points (x, y, z) that satisfy the two equations in the system

$$3x + 2y - 4z = 6$$
$$x - 3y - 2z = 4$$

Solving this system by Gaussian elimination gives $x = \frac{26}{11} + \frac{16}{11}t$, $y = -\frac{6}{11} - \frac{2}{11}t$, $z = t$. Therefore, the line of intersection can be represented by the parametric equations

$$x = \tfrac{26}{11} + \tfrac{16}{11}t, \quad y = -\tfrac{6}{11} - \tfrac{2}{11}t, \quad z = t \qquad (-\infty < t < +\infty) \qquad \blacksquare$$

Vector Form of Equation of a Line

Vector notation provides a useful alternative way of writing the parametric equations of a line: Referring to Figure 3.5.5, let $\mathbf{r} = (x, y, z)$ be the vector from the origin to the point $P(x, y, z)$, let $\mathbf{r}_0 = (x_0, y_0, z_0)$ be the vector from the origin to the point $P_0(x_0, y_0, z_0)$, and let $\mathbf{v} = (a, b, c)$ be a vector parallel to the line. Then $\overrightarrow{P_0 P} = \mathbf{r} - \mathbf{r}_0$, so Formula (6) can be rewritten as

$$\mathbf{r} - \mathbf{r}_0 = t\mathbf{v}$$

Taking into account the range of t-values, this can be rewritten as

$$\mathbf{r} = \mathbf{r}_0 + t\mathbf{v} \qquad (-\infty < t < +\infty) \tag{8}$$

This is called the ***vector form of the equation of a line*** in 3-space.

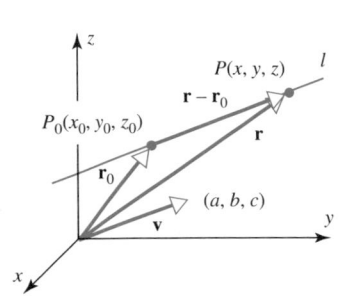

Figure 3.5.5 Vector interpretation of a line in 3-space.

EXAMPLE 7 A Line Parallel to a Given Vector

The equation

$$(x, y, z) = (-2, 0, 3) + t(4, -7, 1) \qquad (-\infty < t < +\infty)$$

is the vector equation of the line through the point $(-2, 0, 3)$ that is parallel to the vector $\mathbf{v} = (4, -7, 1)$. ◆

Problems Involving Distance

We conclude this section by discussing two basic "distance problems" in 3-space:

Problems:

(a) Find the distance between a point and a plane.

(b) Find the distance between two parallel planes.

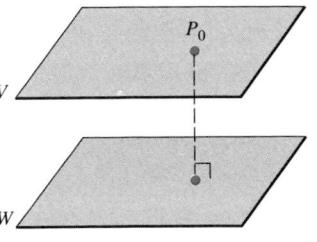

Figure 3.5.6 The distance between the parallel planes V and W is equal to the distance between P_0 and W.

The two problems are related. If we can find the distance between a point and a plane, then we can find the distance between parallel planes by computing the distance between either one of the planes and an arbitrary point P_0 in the other (Figure 3.5.6).

THEOREM 3.5.2

Distance Between a Point and a Plane

The distance D between a point $P_0(x_0, y_0, z_0)$ and the plane $ax + by + cz + d = 0$ is

$$D = \frac{|ax_0 + by_0 + cz_0 + d|}{\sqrt{a^2 + b^2 + c^2}} \tag{9}$$

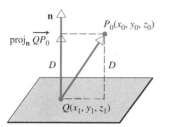

Figure 3.5.7 Distance from P_0 to plane.

Proof Let $Q(x_1, y_1, z_1)$ be any point in the plane. Position the normal $\mathbf{n} = (a, b, c)$ so that its initial point is at Q. As illustrated in Figure 3.5.7, the distance D is equal to the length of the orthogonal projection of $\overrightarrow{QP_0}$ on $\mathbf{n}$. Thus, from (10) of Section 3.3,

$$D = \|\text{proj}_{\mathbf{n}} \overrightarrow{QP_0}\| = \frac{|\overrightarrow{QP_0} \cdot \mathbf{n}|}{\|\mathbf{n}\|}$$

But

$$\overrightarrow{QP_0} = (x_0 - x_1, y_0 - y_1, z_0 - z_1)$$

$$\overrightarrow{QP_0} \cdot \mathbf{n} = a(x_0 - x_1) + b(y_0 - y_1) + c(z_0 - z_1)$$

$$\|\mathbf{n}\| = \sqrt{a^2 + b^2 + c^2}$$

Thus

$$D = \frac{|a(x_0 - x_1) + b(y_0 - y_1) + c(z_0 - z_1)|}{\sqrt{a^2 + b^2 + c^2}} \tag{10}$$

Since the point $Q(x_1, y_1, z_1)$ lies in the plane, its coordinates satisfy the equation of the plane; thus

$$ax_1 + by_1 + cz_1 + d = 0$$

or

$$d = -ax_1 - by_1 - cz_1$$

Substituting this expression in (10) yields (9). ∎

REMARK Note the similarity between (9) and the formula for the distance between a point and a line in 2-space [(13) of Section 3.3].

EXAMPLE 8 Distance Between a Point and a Plane

Find the distance D between the point $(1, -4, -3)$ and the plane $2x - 3y + 6z = -1$.

Solution

To apply (9), we first rewrite the equation of the plane in the form

$$2x - 3y + 6z + 1 = 0$$

Then

$$D = \frac{|2(1) + (-3)(-4) + 6(-3) + 1|}{\sqrt{2^2 + (-3)^2 + 6^2}} = \frac{|-3|}{7} = \frac{3}{7} \quad \blacklozenge$$

Given two planes, either they intersect, in which case we can ask for their line of intersection, as in Example 6, or they are parallel, in which case we can ask for the distance between them. The following example illustrates the latter problem.

EXAMPLE 9 Distance Between Parallel Planes

The planes

$$x + 2y - 2z = 3 \quad \text{and} \quad 2x + 4y - 4z = 7$$

are parallel since their normals, $(1, 2, -2)$ and $(2, 4, -4)$, are parallel vectors. Find the distance between these planes.

Solution

To find the distance D between the planes, we may select an arbitrary point in one of the planes and compute its distance to the other plane. By setting $y = z = 0$ in the equation $x + 2y - 2z = 3$, we obtain the point $P_0(3, 0, 0)$ in this plane. From (9), the distance between P_0 and the plane $2x + 4y - 4z = 7$ is

$$D = \frac{|2(3) + 4(0) + (-4)(0) - 7|}{\sqrt{2^2 + 4^2 + (-4)^2}} = \frac{1}{6} \qquad \blacksquare$$

EXERCISE SET 3.5

1. Find a point-normal form of the equation of the plane passing through P and having $\mathbf{n}$ as a normal.
   (a) $P(-1, 3, -2)$; $\mathbf{n} = (-2, 1, -1)$ (b) $P(1, 1, 4)$; $\mathbf{n} = (1, 9, 8)$
   (c) $P(2, 0, 0)$; $\mathbf{n} = (0, 0, 2)$ (d) $P(0, 0, 0)$; $\mathbf{n} = (1, 2, 3)$

2. Write the equations of the planes in Exercise 1 in general form.

3. Find a point-normal form of the equations of the following planes.
   (a) $-3x + 7y + 2z = 10$ (b) $x - 4z = 0$

4. Find an equation for the plane passing through the given points.
   (a) $P(-4, -1, -1)$, $Q(-2, 0, 1)$, $R(-1, -2, -3)$
   (b) $P(5, 4, 3)$, $Q(4, 3, 1)$, $R(1, 5, 4)$

5. Determine whether the planes are parallel.
   (a) $4x - y + 2z = 5$ and $7x - 3y + 4z = 8$
   (b) $x - 4y - 3z - 2 = 0$ and $3x - 12y - 9z - 7 = 0$
   (c) $2y = 8x - 4z + 5$ and $x = \frac{1}{2}z + \frac{1}{4}y$

6. Determine whether the line and plane are parallel.
   (a) $x = -5 - 4t$, $y = 1 - t$, $z = 3 + 2t$; $x + 2y + 3z - 9 = 0$
   (b) $x = 3t$, $y = 1 + 2t$, $z = 2 - t$; $4x - y + 2z = 1$

7. Determine whether the planes are perpendicular.
   (a) $3x - y + z - 4 = 0$, $x + 2z = -1$ (b) $x - 2y + 3z = 4$, $-2x + 5y + 4z = -1$

8. Determine whether the line and plane are perpendicular.
   (a) $x = -2 - 4t$, $y = 3 - 2t$, $z = 1 + 2t$; $2x + y - z = 5$
   (b) $x = 2 + t$, $y = 1 - t$, $z = 5 + 3t$; $6x + 6y - 7 = 0$

9. Find parametric equations for the line passing through P and parallel to $\mathbf{n}$.
   (a) $P(3, -1, 2)$; $\mathbf{n} = (2, 1, 3)$ (b) $P(-2, 3, -3)$; $\mathbf{n} = (6, -6, -2)$
   (c) $P(2, 2, 6)$; $\mathbf{n} = (0, 1, 0)$ (d) $P(0, 0, 0)$; $\mathbf{n} = (1, -2, 3)$

10. Find parametric equations for the line passing through the given points.

    (a) $(5, -2, 4)$, $(7, 2, -4)$ (b) $(0, 0, 0)$, $(2, -1, -3)$

11. Find parametric equations for the line of intersection of the given planes.

    (a) $7x - 2y + 3z = -2$ and $-3x + y + 2z + 5 = 0$

    (b) $2x + 3y - 5z = 0$ and $y = 0$

12. Find the vector form of the equation of the plane that passes through P_0 and has normal $\mathbf{n}$.

    (a) $P_0(-1, 2, 4)$; $\mathbf{n} = (-2, 4, 1)$ (b) $P_0(2, 0, -5)$; $\mathbf{n} = (-1, 4, 3)$

    (c) $P_0(5, -2, 1)$; $\mathbf{n} = (-1, 0, 0)$ (d) $P_0(0, 0, 0)$; $\mathbf{n} = (a, b, c)$

13. Determine whether the planes are parallel.

    (a) $(-1, 2, 4) \cdot (x - 5, y + 3, z - 7) = 0$; $(2, -4, -8) \cdot (x + 3, y + 5, z - 9) = 0$

    (b) $(3, 0, -1) \cdot (x + 1, y - 2, z - 3) = 0$; $(-1, 0, 3) \cdot (x + 1, y - z, z - 3) = 0$

14. Determine whether the planes are perpendicular.

    (a) $(-2, 1, 4) \cdot (x - 1, y, z + 3) = 0$; $(1, -2, 1) \cdot (x + 3, y - 5, z) = 0$

    (b) $(3, 0, -2) \cdot (x + 4, y - 7, z + 1) = 0$; $(1, 1, 1) \cdot (x, y, z) = 0$

15. Find the vector form of the equation of the line through P_0 and parallel to $\mathbf{v}$.

    (a) $P_0(-1, 2, 3)$; $\mathbf{v} = (7, -1, 5)$ (b) $P_0(2, 0, -1)$; $\mathbf{v} = (1, 1, 1)$

    (c) $P_0(2, -4, 1)$; $\mathbf{v} = (0, 0, -2)$ (d) $P_0(0, 0, 0)$; $\mathbf{v} = (a, b, c)$

16. Show that the line

$$x = 0, \quad y = t, \quad z = t \qquad (-\infty < t < +\infty)$$

    (a) lies in the plane $6x + 4y - 4z = 0$

    (b) is parallel to and below the plane $5x - 3y + 3z = 1$

    (c) is parallel to and above the plane $6x + 2y - 2z = 3$

17. Find an equation for the plane through $(-2, 1, 7)$ that is perpendicular to the line $x - 4 = 2t$, $y + 2 = 3t$, $z = -5t$.

18. Find an equation of

    (a) the xy-plane   (b) the xz-plane   (c) the yz-plane

19. Find an equation of the plane that contains the point (x_0, y_0, z_0) and is

    (a) parallel to the xy-plane

    (b) parallel to the yz-plane

    (c) parallel to the xz-plane

20. Find an equation for the plane that passes through the origin and is parallel to the plane $7x + 4y - 2z + 3 = 0$.

21. Find an equation for the plane that passes through the point $(3, -6, 7)$ and is parallel to the plane $5x - 2y + z - 5 = 0$.

22. Find the point of intersection of the line

$$x - 9 = -5t, \quad y + 1 = -t, \quad z - 3 = t \qquad (-\infty < t < +\infty)$$

    and the plane $2x - 3y + 4z + 7 = 0$.

23. Find an equation for the plane that contains the line $x = -1 + 3t$, $y = 5 + 2t$, $z = 2 - t$ and is perpendicular to the plane $2x - 4y + 2z = 9$.

24. Find an equation for the plane that passes through $(2, 4, -1)$ and contains the line of intersection of the planes $x - y - 4z = 2$ and $-2x + y + 2z = 3$.

25. Show that the points $(-1, -2, -3)$, $(-2, 0, 1)$, $(-4, -1, -1)$, and $(2, 0, 1)$ lie in the same plane.

26. Find parametric equations for the line through $(-2, 5, 0)$ that is parallel to the planes $2x + y - 4z = 0$ and $-x + 2y + 3z + 1 = 0$.

27. Find an equation for the plane through $(-2, 1, 5)$ that is perpendicular to the planes $4x - 2y + 2z = -1$ and $3x + 3y - 6z = 5$.

28. Find an equation for the plane through $(2, -1, 4)$ that is perpendicular to the line of intersection of the planes $4x + 2y + 2z = -1$ and $3x + 6y + 3z = 7$.

29. Find an equation for the plane that is perpendicular to the plane $8x - 2y + 6z = 1$ and passes through the points $P_1(-1, 2, 5)$ and $P_2(2, 1, 4)$.

30. Show that the lines

$$x = 3 - 2t, \quad y = 4 + t, \quad z = 1 - t \quad (-\infty < t < +\infty)$$

and

$$x = 5 + 2t, \quad y = 1 - t, \quad z = 7 + t \quad (-\infty < t < +\infty)$$

are parallel, and find an equation for the plane they determine.

31. Find an equation for the plane that contains the point $(1, -1, 2)$ and the line $x = t, y = t + 1$, $z = -3 + 2t$.

32. Find an equation for the plane that contains the line $x = 1 + t, y = 3t, z = 2t$ and is parallel to the line of intersection of the planes $-x + 2y + z = 0$ and $x + z + 1 = 0$.

33. Find an equation for the plane, each of whose points is equidistant from $(-1, -4, -2)$ and $(0, -2, 2)$.

34. Show that the line

$$x - 5 = -t, \quad y + 3 = 2t, \quad z + 1 = -5t \quad (-\infty < t < +\infty)$$

is parallel to the plane $-3x + y + z - 9 = 0$.

35. Show that the lines

$$x - 3 = 4t, \quad y - 4 = t, \quad z - 1 = 0 \quad (-\infty < t < +\infty)$$

and

$$x + 1 = 12t, \quad y - 7 = 6t, \quad z - 5 = 3t \quad (-\infty < t < +\infty)$$

intersect, and find the point of intersection.

36. Find an equation for the plane containing the lines in Exercise 35.

37. Find parametric equations for the line of intersection of the planes

(a) $-3x + 2y + z = -5$ and $7x + 3y - 2z = -2$

(b) $5x - 7y + 2z = 0$ and $y = 0$

38. Show that the plane whose intercepts with the coordinate axes are $x = a$, $y = b$, and $z = c$ has equation

$$\frac{x}{a} + \frac{y}{b} + \frac{z}{c} = 1$$

provided that a, b, and c are nonzero.

39. Find the distance between the point and the plane.

(a) $(3, 1, -2)$; $x + 2y - 2z = 4$

(b) $(-1, 2, 1)$; $2x + 3y - 4z = 1$

(c) $(0, 3, -2)$; $x - y - z = 3$

40. Find the distance between the given parallel planes.

(a) $3x - 4y + z = 1$ and $6x - 8y + 2z = 3$

(b) $-4x + y - 3z = 0$ and $8x - 2y + 6z = 0$

(c) $2x - y + z = 1$ and $2x - y + z = -1$

41. Find the distance between the line $x = 3t - 1$, $y = 2 - t$, $z = t$ and each of the following points.

    (a) $(0, 0, 0)$ (b) $(2, 0, -5)$ (c) $(2, 1, 1)$

42. Show that if a, b, and c are nonzero, then the line

$$x = x_0 + at, \quad y = y_0 + bt, \quad z = z_0 + ct \qquad (-\infty < t < +\infty)$$

consists of all points (x, y, z) that satisfy

$$\frac{x - x_0}{a} = \frac{y - y_0}{b} = \frac{z - z_0}{c}$$

These are called ***symmetric equations*** for the line.

43. Find symmetric equations for the lines in parts (a) and (b) of Exercise 9.

    Note   See Exercise 42 for terminology.

44. In each part, find equations for two planes whose intersection is the given line.

    (a) $x = 7 - 4t$, $y = -5 - 2t$, $z = 5 + t$ $(-\infty < t < +\infty)$

    (b) $x = 4t$, $y = 2t$, $z = 7t$ $(-\infty < t < +\infty)$

    Hint   Each equality in the symmetric equations of a line represents a plane containing the line. See Exercise 42 for terminology.

45. Two intersecting planes in 3-space determine two angles of intersection: an acute angle $(0 \leq \theta \leq 90°)$ and its supplement $180° - \theta$ (see the accompanying figure). If $\mathbf{n}_1$ and $\mathbf{n}_2$ are nonzero normals to the planes, then the angle between $\mathbf{n}_1$ and $\mathbf{n}_2$ is θ or $180° - \theta$, depending on the directions of the normals (see the accompanying figure). In each part, find the acute angle of intersection of the planes to the nearest degree.

    (a) $x = 0$ and $2x - y + z - 4 = 0$

    (b) $x + 2y - 2z = 5$ and $6x - 3y + 2z = 8$

    Note   A calculator is needed.

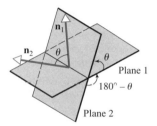

Figure Ex-45

46. Find the acute angle between the plane $x - y - 3z = 5$ and the line $x = 2 - t$, $y = 2t$, $z = 3t - 1$ to the nearest degree.

    Hint   See Exercise 45.

Discussion & Discovery

47. What do the lines $\mathbf{r} = \mathbf{r}_0 + t\mathbf{v}$ and $\mathbf{r} = \mathbf{r}_0 - t\mathbf{v}$ have in common? Explain.

48. What is the relationship between the line $x = x_0 + at$, $y = y_0 + bt$, $z = z_0 + ct$ and the plane $ax + by + cz = 0$? Explain your reasoning.

49. Let $\mathbf{r}_1$ and $\mathbf{r}_2$ be vectors from the origin to the points $P_1(x_1, y_1, z_1)$ and $P_2(x_2, y_2, z_2)$, respectively. What does the equation

$$\mathbf{r} = (1 - t)\mathbf{r}_1 + t\mathbf{r}_2 \qquad (0 \leq t \leq 1)$$

represent geometrically? Explain your reasoning.

50. Write parametric equations for two perpendicular lines through the point (x_0, y_0, z_0).

51. How can you tell whether the line $\mathbf{x} = \mathbf{x}_0 + t\mathbf{v}$ in 3-space is parallel to the plane $\mathbf{x} = \mathbf{x}_0 + t_1\mathbf{v}_1 + t_2\mathbf{v}_2$?

52. Indicate whether the statement is true (T) or false (F). Justify your answer.

    (a) If a, b, and c are not all zero, then the line $x = at$, $y = bt$, $z = ct$ is perpendicular to the plane $ax + by + cz = 0$.

    (b) Two nonparallel lines in 3-space must intersect in at least one point.

    (c) If $\mathbf{u}$, $\mathbf{v}$, and $\mathbf{w}$ are vectors in 3-space such that $\mathbf{u} + \mathbf{v} + \mathbf{w} = \mathbf{0}$, then the three vectors lie in some plane.

    (d) The equation $\mathbf{x} = t\mathbf{v}$ represents a line for every vector $\mathbf{v}$ in 2-space.

CHAPTER 3

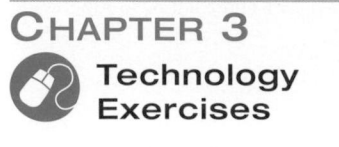

Technology Exercises

The following exercises are designed to be solved using a technology utility. Typically, this will be MATLAB, *Mathematica*, Maple, Derive, or Mathcad, but it may also be some other type of linear algebra software or a scientific calculator with some linear algebra capabilities. For each exercise you will need to read the relevant documentation for the particular utility you are using. The goal of these exercises is to provide you with a basic proficiency with your technology utility. Once you have mastered the techniques in these exercises, you will be able to use your technology utility to solve many of the problems in the regular exercise sets.

Section 3.1

T1. **(Vectors)** Read your documentation on how to enter vectors and how to add, subtract, and multiply them by scalars. Then perform the computations in Example 1.

T2. **(Drawing Vectors)** If you are using a technology utility that can draw line segments in two- or three-dimensional space, try drawing some line segments with initial and terminal points of your choice. You may also want to see if your utility allows you to create arrowheads, in which case you can make your line segments look like geometric vectors.

Section 3.3

T1. **(Dot Product and Norm)** Some technology utilities provide commands for calculating dot products and norms, whereas others provide only a command for the dot product. In the latter case, norms can be computed from the formula $\|\mathbf{v}\| = \sqrt{\mathbf{v} \cdot \mathbf{v}}$. Read your documentation on how to find dot products (and norms, if available), and then perform the computations in Example 2.

T2. **(Projections)** See if you can program your utility to calculate $\text{proj}_{\mathbf{a}}\mathbf{u}$ when the user enters the vectors $\mathbf{a}$ and $\mathbf{u}$. Check your work by having your program perform the computations in Example 6.

Section 3.4

T1. **(Cross Product)** Read your documentation on how to find cross products, and then perform the computation in Example 1.

T2. **(Cross Product Formula)** If you are working with a CAS, use it to confirm Formula (1a).

T3. **(Cross Product Properties)** If you are working with a CAS, use it to prove the results in Theorem 3.4.1.

T4. **(Area of a Triangle)** See if you can program your technology utility to find the area of the triangle in 3-space determined by three points when the user enters their coordinates. Check your work by calculating the area of the triangle in Example 4.

T5. **(Triple Scalar Product Formula)** If you are working with a CAS, use it to prove Formula (7) by showing that the difference between the two sides is zero.

T6. **(Volume of a Parallelepiped)** See if you can program your technology utility to find the volume of the parallelepiped in 3-space determined by vectors $\mathbf{u}$, $\mathbf{v}$, and $\mathbf{w}$ when the user enters the vectors. Check your work by solving Exercise 10 in Exercise Set 3.4.

Euclidean Vector Spaces

CHAPTER CONTENTS

INTRODUCTION: The idea of using pairs of numbers to locate points in the plane and triples of numbers to locate points in 3-space was first clearly spelled out in the mid-seventeenth century. By the latter part of the eighteenth century, mathematicians and physicists began to realize that there was no need to stop with triples. It was recognized that quadruples of numbers (a_1, a_2, a_3, a_4) could be regarded as points in "four-dimensional" space, quintuples $(a_1, a_2, a_3, a_4, a_5)$ as points in "five-dimensional" space, and so on, an n-tuple of numbers being a point in "n-dimensional" space. Our goal in this chapter is to study the properties of operations on vectors in this kind of space.

4.1
EUCLIDEAN
n-SPACE

Although our geometric visualization does not extend beyond 3-space, it is nevertheless possible to extend many familiar ideas beyond 3-space by working with analytic or numerical properties of points and vectors rather than the geometric properties. In this section we shall make these ideas more precise.

Vectors in *n*-Space

We begin with a definition.

> **DEFINITION**
>
> If *n* is a positive integer, then an ***ordered n-tuple*** is a sequence of *n* real numbers $(a_1, a_2, \ldots, a_n)$. The set of all ordered *n*-tuples is called ***n-space*** and is denoted by R^n.

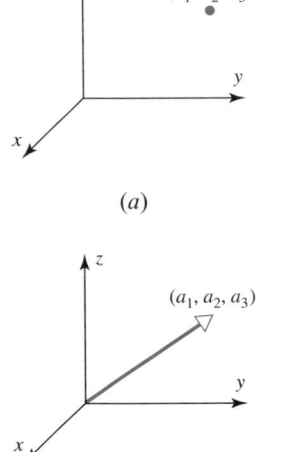

(a)

(b)

Figure 4.1.1 The ordered triple (a_1, a_2, a_3) can be interpreted geometrically as a point or as a vector.

When $n = 2$ or 3, it is customary to use the terms ***ordered pair*** and ***ordered triple***, respectively, rather than *ordered 2-tuple* and *ordered 3-tuple*. When $n = 1$, each ordered *n*-tuple consists of one real number, so R^1 may be viewed as the set of real numbers. It is usual to write R rather than R^1 for this set.

It might have occurred to you in the study of 3-space that the symbol (a_1, a_2, a_3) has two different geometric interpretations: it can be interpreted as a point, in which case a_1, a_2, and a_3 are the coordinates (Figure 4.1.1*a*), or it can be interpreted as a vector, in which case a_1, a_2, and a_3 are the components (Figure 4.1.1*b*). It follows, therefore, that an ordered *n*-tuple $(a_1, a_2, \ldots, a_n)$ can be viewed either as a "generalized point" or as a "generalized vector"—the distinction is mathematically unimportant. Thus we can describe the 5-tuple $(-2, 4, 0, 1, 6)$ either as a point in R^5 or as a vector in R^5.

> **DEFINITION**
>
> Two vectors $\mathbf{u} = (u_1, u_2, \ldots, u_n)$ and $\mathbf{v} = (v_1, v_2, \ldots, v_n)$ in R^n are called ***equal*** if
> $$u_1 = v_1, \quad u_2 = v_2, \ldots, \quad u_n = v_n$$
> The ***sum* u + v** is defined by
> $$\mathbf{u} + \mathbf{v} = (u_1 + v_1, u_2 + v_2, \ldots, u_n + v_n)$$
> and if k is any scalar, the ***scalar multiple* $k\mathbf{u}$** is defined by
> $$k\mathbf{u} = (ku_1, ku_2, \ldots, ku_n)$$

The operations of addition and scalar multiplication in this definition are called the ***standard operations*** on R^n.

The ***zero vector*** in R^n is denoted by **0** and is defined to be the vector
$$\mathbf{0} = (0, 0, \ldots, 0)$$

If $\mathbf{u} = (u_1, u_2, \ldots, u_n)$ is any vector in R^n, then the ***negative*** (or ***additive inverse***) of **u** is denoted by $-\mathbf{u}$ and is defined by
$$-\mathbf{u} = (-u_1, -u_2, \ldots, -u_n)$$

The ***difference*** of vectors in R^n is defined by
$$\mathbf{v} - \mathbf{u} = \mathbf{v} + (-\mathbf{u})$$

Some Examples of Vectors in Higher-Dimensional Spaces

- **Experimental Data**—A scientist performs an experiment and makes n numerical measurements each time the experiment is performed. The result of each experiment can be regarded as a vector $\mathbf{y} = (y_1, y_2, \ldots, y_n)$ in R^n in which $y_1, y_2, \ldots, y_n$ are the measured values.

- **Storage and Warehousing**—A national trucking company has 15 depots for storing and servicing its trucks. At each point in time the distribution of trucks in the service depots can be described by a 15-tuple $\mathbf{x} = (x_1, x_2, \ldots, x_{15})$ in which x_1 is the number of trucks in the first depot, x_2 is the number in the second depot, and so forth.

- **Electrical Circuits**—A certain kind of processing chip is designed to receive four input voltages and produces three output voltages in response. The input voltages can be regarded as vectors in R^4 and the output voltages as vectors in R^3. Thus, the chip can be viewed as a device that transforms each input vector $\mathbf{v} = (v_1, v_2, v_3, v_4)$ in R^4 into some output vector $\mathbf{w} = (w_1, w_2, w_3)$ in R^3.

- **Graphical Images**—One way in which color images are created on computer screens is by assigning each pixel (an addressable point on the screen) three numbers that describe the *hue*, *saturation*, and *brightness* of the pixel. Thus, a complete color image can be viewed as a set of 5-tuples of the form $\mathbf{v} = (x, y, h, s, b)$ in which x and y are the screen coordinates of a pixel and h, s, and b are its hue, saturation, and brightness.

- **Economics**—Our approach to economic analysis is to divide an economy into sectors (manufacturing, services, utilities, and so forth) and to measure the output of each sector by a dollar value. Thus, in an economy with 10 sectors the economic output of the entire economy can be represented by a 10-tuple $\mathbf{s} = (s_1, s_2, \ldots, s_{10})$ in which the numbers $s_1, s_2, \ldots, s_{10}$ are the outputs of the individual sectors.

- **Mechanical Systems**—Suppose that six particles move along the same coordinate line so that at time t their coordinates are $x_1, x_2, \ldots, x_6$ and their velocities are $v_1, v_2, \ldots, v_6$, respectively. This information can be represented by the vector

$$\mathbf{v} = (x_1, x_2, x_3, x_4, x_5, x_6,$$
$$v_1, v_2, v_3, v_4, v_5, v_6, t)$$

in R^{13}. This vector is called the *state* of the particle system at time t.

- **Physics**—In string theory the smallest, indivisible components of the Universe are not particles but loops that behave like vibrating strings. Whereas Einstein's space-time universe was four-dimensional, strings reside in an 11-dimensional world.

or, in terms of components,

$$\mathbf{v} - \mathbf{u} = (v_1 - u_1, v_2 - u_2, \ldots, v_n - u_n)$$

Properties of Vector Operations in n-Space

The most important arithmetic properties of addition and scalar multiplication of vectors in R^n are listed in the following theorem. The proofs are all easy and are left as exercises.

THEOREM 4.1.1

Properties of Vectors in R^n

If $\mathbf{u} = (u_1, u_2, \ldots, u_n)$, $\mathbf{v} = (v_1, v_2, \ldots, v_n)$, *and* $\mathbf{w} = (w_1, w_2, \ldots, w_n)$ *are vectors in* R^n *and* k *and* m *are scalars, then:*

(a) $\mathbf{u} + \mathbf{v} = \mathbf{v} + \mathbf{u}$

(b) $\mathbf{u} + (\mathbf{v} + \mathbf{w}) = (\mathbf{u} + \mathbf{v}) + \mathbf{w}$

(c) $\mathbf{u} + \mathbf{0} = \mathbf{0} + \mathbf{u} = \mathbf{u}$

(d) $\mathbf{u} + (-\mathbf{u}) = \mathbf{0}$; *that is,* $\mathbf{u} - \mathbf{u} = \mathbf{0}$

(e) $k(m\mathbf{u}) = (km)\mathbf{u}$

(f) $k(\mathbf{u} + \mathbf{v}) = k\mathbf{u} + k\mathbf{v}$

(g) $(k + m)\mathbf{u} = k\mathbf{u} + m\mathbf{u}$

(h) $1\mathbf{u} = \mathbf{u}$

This theorem enables us to manipulate vectors in R^n without expressing the vectors in terms of components. For example, to solve the vector equation $\mathbf{x} + \mathbf{u} = \mathbf{v}$ for $\mathbf{x}$, we can add $-\mathbf{u}$ to both sides and proceed as follows:

$$(\mathbf{x} + \mathbf{u}) + (-\mathbf{u}) = \mathbf{v} + (-\mathbf{u})$$
$$\mathbf{x} + (\mathbf{u} - \mathbf{u}) = \mathbf{v} - \mathbf{u}$$
$$\mathbf{x} + \mathbf{0} = \mathbf{v} - \mathbf{u}$$
$$\mathbf{x} = \mathbf{v} - \mathbf{u}$$

The reader will find it instructive to name the parts of Theorem 4.1.1 that justify the last three steps in this computation.

Euclidean *n*-Space

To extend the notions of distance, norm, and angle to R^n, we begin with the following generalization of the dot product on R^2 and R^3 [Formulas (3) and (4) of Section 3.3].

DEFINITION

If $\mathbf{u} = (u_1, u_2, \ldots, u_n)$ and $\mathbf{v} = (v_1, v_2, \ldots, v_n)$ are any vectors in R^n, then the *Euclidean inner product* $\mathbf{u} \cdot \mathbf{v}$ is defined by

$$\mathbf{u} \cdot \mathbf{v} = u_1 v_1 + u_2 v_2 + \cdots + u_n v_n$$

Observe that when $n = 2$ or 3, the Euclidean inner product is the ordinary dot product.

EXAMPLE 1 Inner Product of Vectors in R^4

The Euclidean inner product of the vectors

$$\mathbf{u} = (-1, 3, 5, 7) \quad \text{and} \quad \mathbf{v} = (5, -4, 7, 0)$$

in R^4 is

$$\mathbf{u} \cdot \mathbf{v} = (-1)(5) + (3)(-4) + (5)(7) + (7)(0) = 18 \quad \blacklozenge$$

Since so many of the familiar ideas from 2-space and 3-space carry over to *n*-space, it is common to refer to R^n, with the operations of addition, scalar multiplication, and the Euclidean inner product, as *Euclidean n-space*.

The four main arithmetic properties of the Euclidean inner product are listed in the next theorem.

THEOREM 4.1.2

Properties of Euclidean Inner Product

If $\mathbf{u}$, $\mathbf{v}$, *and* $\mathbf{w}$ *are vectors in* R^n *and k is any scalar, then:*

(*a*) $\mathbf{u} \cdot \mathbf{v} = \mathbf{v} \cdot \mathbf{u}$ (*b*) $(\mathbf{u} + \mathbf{v}) \cdot \mathbf{w} = \mathbf{u} \cdot \mathbf{w} + \mathbf{v} \cdot \mathbf{w}$

(*c*) $(k\mathbf{u}) \cdot \mathbf{v} = k(\mathbf{u} \cdot \mathbf{v})$ (*d*) $\mathbf{v} \cdot \mathbf{v} \geq 0$. *Further,* $\mathbf{v} \cdot \mathbf{v} = 0$ *if and only if* $\mathbf{v} = \mathbf{0}$.

We shall prove parts (*b*) and (*d*) and leave proofs of the rest as exercises.

Application of Dot Products to ISBNs

Most books published in the last 25 years have been assigned a unique 10-digit number called an *International Standard Book Number* or ISBN. The first nine digits of this number are split into three groups—the first group representing the country or group of countries in which the book originates, the second identifying the publisher, and the third assigned to the book title itself. The tenth and final digit, called a *check digit*, is computed from the first nine digits and is used to ensure that an electronic transmission of the ISBN, say over the Internet, occurs without error.

To explain how this is done, regard the first nine digits of the ISBN as a vector **b** in R^9, and let **a** be the vector

$$\mathbf{a} = (1, 2, 3, 4, 5, 6, 7, 8, 9)$$

Then the check digit c is computed using the following procedure:

1. Form the dot product $\mathbf{a} \cdot \mathbf{b}$.
2. Divide $\mathbf{a} \cdot \mathbf{b}$ by 11, thereby producing a remainder c that is an integer between 0 and 10, inclusive. The check digit is taken to be c, with the proviso that $c = 10$ is written as X to avoid double digits.

For example, the ISBN of the brief edition of *Calculus*, sixth edition, by Howard Anton is

$$0\text{-}471\text{-}15307\text{-}9$$

which has a check digit of 9. This is consistent with the first nine digits of the ISBN, since

$$\mathbf{a} \cdot \mathbf{b} = (1, 2, 3, 4, 5, 6, 7, 8, 9)$$
$$\cdot (0, 4, 7, 1, 1, 5, 3, 0, 7) = 152$$

Dividing 152 by 11 produces a quotient of 13 and a remainder of 9, so the check digit is $c = 9$. If an electronic order is placed for a book with a certain ISBN, then the warehouse can use the above procedure to verify that the check digit is consistent with the first nine digits, thereby reducing the possibility of a costly shipping error.

Proof (b) Let $\mathbf{u} = (u_1, u_2, \ldots, u_n), \mathbf{v} = (v_1, v_2, \ldots, v_n)$, and $\mathbf{w} = (w_1, w_2, \ldots, w_n)$. Then

$$(\mathbf{u} + \mathbf{v}) \cdot \mathbf{w} = (u_1 + v_1, u_2 + v_2, \ldots, u_n + v_n) \cdot (w_1, w_2, \ldots, w_n)$$
$$= (u_1 + v_1)w_1 + (u_2 + v_2)w_2 + \cdots + (u_n + v_n)w_n$$
$$= (u_1 w_1 + u_2 w_2 + \cdots + u_n w_n) + (v_1 w_1 + v_2 w_2 + \cdots + v_n w_n)$$
$$= \mathbf{u} \cdot \mathbf{w} + \mathbf{v} \cdot \mathbf{w}$$

Proof (d) We have $\mathbf{v} \cdot \mathbf{v} = v_1^2 + v_2^2 + \cdots + v_n^2 \geq 0$. Further, equality holds if and only if $v_1 = v_2 = \cdots = v_n = 0$—that is, if and only if $\mathbf{v} = \mathbf{0}$. ∎

EXAMPLE 2 Length and Distance in R^4

Theorem 4.1.2 allows us to perform computations with Euclidean inner products in much the same way as we perform them with ordinary arithmetic products. For example,

$$(3\mathbf{u} + 2\mathbf{v}) \cdot (4\mathbf{u} + \mathbf{v}) = (3\mathbf{u}) \cdot (4\mathbf{u} + \mathbf{v}) + (2\mathbf{v}) \cdot (4\mathbf{u} + \mathbf{v})$$
$$= (3\mathbf{u}) \cdot (4\mathbf{u}) + (3\mathbf{u}) \cdot \mathbf{v} + (2\mathbf{v}) \cdot (4\mathbf{u}) + (2\mathbf{v}) \cdot \mathbf{v}$$
$$= 12(\mathbf{u} \cdot \mathbf{u}) + 3(\mathbf{u} \cdot \mathbf{v}) + 8(\mathbf{v} \cdot \mathbf{u}) + 2(\mathbf{v} \cdot \mathbf{v})$$
$$= 12(\mathbf{u} \cdot \mathbf{u}) + 11(\mathbf{u} \cdot \mathbf{v}) + 2(\mathbf{v} \cdot \mathbf{v})$$

The reader should determine which parts of Theorem 4.1.2 were used in each step. ◆

Norm and Distance in Euclidean *n*-Space

By analogy with the familiar formulas in R^2 and R^3, we define the *Euclidean norm* (or *Euclidean length*) of a vector $\mathbf{u} = (u_1, u_2, \ldots, u_n)$ in R^n by

$$\|\mathbf{u}\| = (\mathbf{u} \cdot \mathbf{u})^{1/2} = \sqrt{u_1^2 + u_2^2 + \cdots + u_n^2} \tag{1}$$

[Compare this formula to Formulas (1) and (2) in Section 3.2.]

Similarly, the *Euclidean distance* between the points $\mathbf{u} = (u_1, u_2, \ldots, u_n)$ and $\mathbf{v} = (v_1, v_2, \ldots, v_n)$ in R^n is defined by

$$d(\mathbf{u}, \mathbf{v}) = \|\mathbf{u} - \mathbf{v}\| = \sqrt{(u_1 - v_1)^2 + (u_2 - v_2)^2 + \cdots + (u_n - v_n)^2} \tag{2}$$

[See Formulas (3) and (4) of Section 3.2.]

EXAMPLE 3 Finding Norm and Distance

If $\mathbf{u} = (1, 3, -2, 7)$ and $\mathbf{v} = (0, 7, 2, 2)$, then in the Euclidean space R^4,

$$\|\mathbf{u}\| = \sqrt{(1)^2 + (3)^2 + (-2)^2 + (7)^2} = \sqrt{63} = 3\sqrt{7}$$

and

$$d(\mathbf{u}, \mathbf{v}) = \sqrt{(1 - 0)^2 + (3 - 7)^2 + (-2 - 2)^2 + (7 - 2)^2} = \sqrt{58} \quad \blacklozenge$$

The following theorem provides one of the most important inequalities in linear algebra: the *Cauchy–Schwarz* inequality.

THEOREM 4.1.3

> **Cauchy–Schwarz Inequality in R^n**
>
> *If $\mathbf{u} = (u_1, u_2, \ldots, u_n)$ and $\mathbf{v} = (v_1, v_2, \ldots, v_n)$ are vectors in R^n, then*
>
> $$|\mathbf{u} \cdot \mathbf{v}| \leq \|\mathbf{u}\| \|\mathbf{v}\| \tag{3}$$

In terms of components, (3) is the same as

$$|u_1 v_1 + u_2 v_2 + \cdots + u_n v_n| \leq (u_1^2 + u_2^2 + \cdots + u_n^2)^{1/2} (v_1^2 + v_2^2 + \cdots + v_n^2)^{1/2} \tag{4}$$

We omit the proof at this time, since a more general version of this theorem will be proved later in the text. However, for vectors in R^2 and R^3, this result is a simple consequence of Formula (1) of Section 3.3: If $\mathbf{u}$ and $\mathbf{v}$ are nonzero vectors in R^2 or R^3, then

$$|\mathbf{u} \cdot \mathbf{v}| = |\|\mathbf{u}\| \|\mathbf{v}\| \cos \theta| = \|\mathbf{u}\| \|\mathbf{v}\| |\cos \theta| \leq \|\mathbf{u}\| \|\mathbf{v}\| \tag{5}$$

and if either $\mathbf{u} = \mathbf{0}$ or $\mathbf{v} = \mathbf{0}$, then both sides of (3) are zero, so the inequality holds in this case as well.

The next two theorems list the basic properties of length and distance in Euclidean *n*-space.

Augustin Louis (Baron de) Cauchy

Herman Amandus Schwarz

Augustin Louis (Baron de) Cauchy
(1789–1857), French mathematician. Cauchy's early education was acquired from his father, a barrister and master of the classics. Cauchy entered L'Ecole Polytechnique in 1805 to study engineering, but because of poor health, he was advised to concentrate on mathematics. His major mathematical work began in 1811 with a series of brilliant solutions to some difficult outstanding problems.

Cauchy's mathematical contributions for the next 35 years were brilliant and staggering in quantity: over 700 papers filling 26 modern volumes. Cauchy's work initiated the era of modern analysis; he brought to mathematics standards of precision and rigor undreamed of by earlier mathematicians.

Cauchy's life was inextricably tied to the political upheavals of the time. A strong partisan of the Bourbons, he left his wife and children in 1830 to follow the Bourbon king Charles X into exile. For his loyalty he was made a baron by the ex-king. Cauchy eventually returned to France but refused to accept a university position until the government waived its requirement that he take a loyalty oath.

It is difficult to get a clear picture of the man. Devoutly Catholic, he sponsored charitable work for unwed mothers and criminals and relief for Ireland. Yet other aspects of his life cast him in an unfavorable light. The Norwegian mathematician Abel described him as "mad, infinitely Catholic, and bigoted." Some writers praise his teaching, yet others say he rambled incoherently and, according to a report of the day, he once devoted an entire lecture to extracting the square root of seventeen to ten decimal places by a method well known to his students. In any event, Cauchy is undeniably one of the greatest minds in the history of science.

Herman Amandus Schwarz *(1843–1921)*, German mathematician. Schwarz was the leading mathematician in Berlin in the first part of the twentieth century. Because of a devotion to his teaching duties at the University of Berlin and a propensity for treating both important and trivial facts with equal thoroughness, he did not publish in great volume. He tended to focus on narrow concrete problems, but his techniques were often extremely clever and influenced the work of other mathematicians. A version of the inequality that bears his name appeared in a paper about surfaces of minimal area published in 1885.

THEOREM 4.1.4

> ## Properties of Length in R^n
>
> *If* **u** *and* **v** *are vectors in* R^n *and* k *is any scalar, then:*
>
> *(a)* $\|\mathbf{u}\| \geq 0$ *(b)* $\|\mathbf{u}\| = 0$ *if and only if* $\mathbf{u} = \mathbf{0}$
> *(c)* $\|k\mathbf{u}\| = |k|\|\mathbf{u}\|$ *(d)* $\|\mathbf{u} + \mathbf{v}\| \leq \|\mathbf{u}\| + \|\mathbf{v}\|$ *(Triangle inequality)*

We shall prove (*c*) and (*d*) and leave (*a*) and (*b*) as exercises.

Proof (c)   If $\mathbf{u} = (u_1, u_2, \ldots, u_n)$, then $k\mathbf{u} = (ku_1, ku_2, \ldots, ku_n)$, so

$$\|k\mathbf{u}\| = \sqrt{(ku_1)^2 + (ku_2)^2 + \cdots + (ku_n)^2}$$

$$= |k|\sqrt{u_1^2 + u_2^2 + \cdots + u_n^2}$$

$$= |k|\|\mathbf{u}\|$$

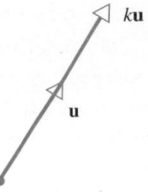

(a) $\|k\mathbf{u}\| = |k|\,\|\mathbf{u}\|$

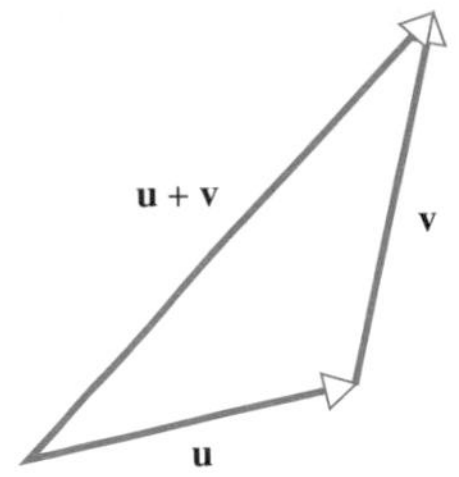

(b) $\|\mathbf{u}+\mathbf{v}\| \le \|\mathbf{u}\| + \|\mathbf{v}\|$

Figure 4.1.2

Proof (d)

$$\begin{aligned}\|\mathbf{u}+\mathbf{v}\|^2 &= (\mathbf{u}+\mathbf{v})\cdot(\mathbf{u}+\mathbf{v}) = (\mathbf{u}\cdot\mathbf{u}) + 2(\mathbf{u}\cdot\mathbf{v}) + (\mathbf{v}\cdot\mathbf{v})\\ &= \|\mathbf{u}\|^2 + 2(\mathbf{u}\cdot\mathbf{v}) + \|\mathbf{v}\|^2\\ &\le \|\mathbf{u}\|^2 + 2|\mathbf{u}\cdot\mathbf{v}| + \|\mathbf{v}\|^2 \quad \longleftarrow \text{Property of absolute value}\\ &\le \|\mathbf{u}\|^2 + 2\|\mathbf{u}\|\|\mathbf{v}\| + \|\mathbf{v}\|^2 \quad \longleftarrow \text{Cauchy–Schwarz inequality}\\ &= (\|\mathbf{u}\| + \|\mathbf{v}\|)^2\end{aligned}$$

The result now follows on taking square roots of both sides. ■

Part (*c*) of this theorem states that multiplying a vector by a scalar k multiplies the length of that vector by a factor of $|k|$ (Figure 4.1.2*a*). Part (*d*) of this theorem is known as the ***triangle inequality*** because it generalizes the familiar result from Euclidean geometry that states that the sum of the lengths of any two sides of a triangle is at least as large as the length of the third side (Figure 4.1.2*b*).

The results in the next theorem are immediate consequences of those in Theorem 4.1.4, as applied to the distance function $d(\mathbf{u}, \mathbf{v})$ on R^n. They generalize the familiar results for R^2 and R^3.

THEOREM 4.1.5

Properties of Distance in R^n

If **u**, **v**, *and* **w** *are vectors in* R^n *and k is any scalar, then:*

(*a*) $d(\mathbf{u}, \mathbf{v}) \ge 0$ (*b*) $d(\mathbf{u}, \mathbf{v}) = 0$ *if and only if* $\mathbf{u} = \mathbf{v}$

(*c*) $d(\mathbf{u}, \mathbf{v}) = d(\mathbf{v}, \mathbf{u})$ (*d*) $d(\mathbf{u}, \mathbf{v}) \le d(\mathbf{u}, \mathbf{w}) + d(\mathbf{w}, \mathbf{v})$ (*Triangle inequality*)

We shall prove part (*d*) and leave the remaining parts as exercises.

Proof (d) From (2) and part (*d*) of Theorem 4.1.4, we have

$$\begin{aligned}d(\mathbf{u}, \mathbf{v}) &= \|\mathbf{u}-\mathbf{v}\| = \|(\mathbf{u}-\mathbf{w}) + (\mathbf{w}-\mathbf{v})\|\\ &\le \|\mathbf{u}-\mathbf{w}\| + \|\mathbf{w}-\mathbf{v}\| = d(\mathbf{u}, \mathbf{w}) + d(\mathbf{w}, \mathbf{v})\end{aligned}$$ ■

$d(\mathbf{u}, \mathbf{w}) \le d(\mathbf{u}, \mathbf{v}) + d(\mathbf{v}, \mathbf{w})$

Figure 4.1.3

Part (*d*) of this theorem, which is also called the *triangle inequality*, generalizes the familiar result from Euclidean geometry that states that the shortest distance between two points is along a straight line (Figure 4.1.3).

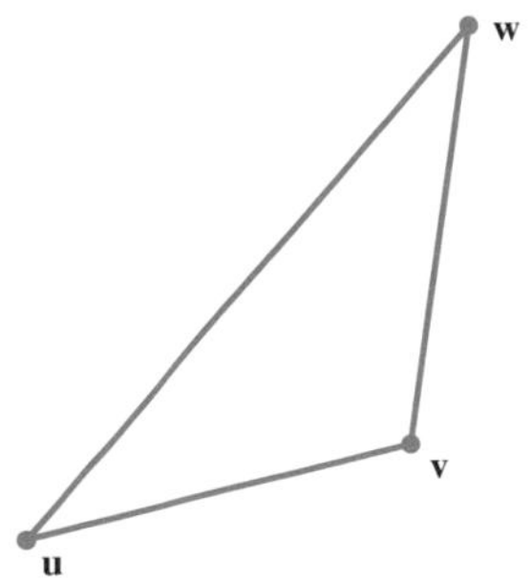

Formula (1) expresses the norm of a vector in terms of a dot product. The following useful theorem expresses the dot product in terms of norms.

THEOREM 4.1.6

If **u** *and* **v** *are vectors in* R^n *with the Euclidean inner product, then*

$$\mathbf{u}\cdot\mathbf{v} = \tfrac{1}{4}\|\mathbf{u}+\mathbf{v}\|^2 - \tfrac{1}{4}\|\mathbf{u}-\mathbf{v}\|^2 \tag{6}$$

Proof

$$\begin{aligned}\|\mathbf{u}+\mathbf{v}\|^2 &= (\mathbf{u}+\mathbf{v})\cdot(\mathbf{u}+\mathbf{v}) = \|\mathbf{u}\|^2 + 2(\mathbf{u}\cdot\mathbf{v}) + \|\mathbf{v}\|^2\\ \|\mathbf{u}-\mathbf{v}\|^2 &= (\mathbf{u}-\mathbf{v})\cdot(\mathbf{u}-\mathbf{v}) = \|\mathbf{u}\|^2 - 2(\mathbf{u}\cdot\mathbf{v}) + \|\mathbf{v}\|^2\end{aligned}$$

from which (6) follows by simple algebra. ■

Some problems that use this theorem are given in the exercises.

Orthogonality

Recall that in the Euclidean spaces R^2 and R^3, two vectors **u** and **v** are defined to be *orthogonal* (perpendicular) if $\mathbf{u} \cdot \mathbf{v} = 0$ (Section 3.3). Motivated by this, we make the following definition.

> **DEFINITION**
>
> Two vectors **u** and **v** in R^n are called *orthogonal* if $\mathbf{u} \cdot \mathbf{v} = 0$.

EXAMPLE 4 Orthogonal Vectors in R^4

In the Euclidean space R^4 the vectors

$$\mathbf{u} = (-2, 3, 1, 4) \quad \text{and} \quad \mathbf{v} = (1, 2, 0, -1)$$

are orthogonal, since

$$\mathbf{u} \cdot \mathbf{v} = (-2)(1) + (3)(2) + (1)(0) + (4)(-1) = 0 \quad \blacklozenge$$

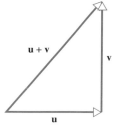

Figure 4.1.4

Properties of orthogonal vectors will be discussed in more detail later in the text, but we note at this point that many of the familiar properties of orthogonal vectors in the Euclidean spaces R^2 and R^3 continue to hold in the Euclidean space R^n. For example, if **u** and **v** are orthogonal vectors in R^2 or R^3, then **u**, **v**, and $\mathbf{u} + \mathbf{v}$ form the sides of a right triangle (Figure 4.1.4); thus, by the Theorem of Pythagoras,

$$\|\mathbf{u} + \mathbf{v}\|^2 = \|\mathbf{u}\|^2 + \|\mathbf{v}\|^2$$

The following theorem shows that this result extends to R^n.

THEOREM 4.1.7

> **Pythagorean Theorem in R^n**
>
> *If* **u** *and* **v** *are orthogonal vectors in R^n with the Euclidean inner product, then*
>
> $$\|\mathbf{u} + \mathbf{v}\|^2 = \|\mathbf{u}\|^2 + \|\mathbf{v}\|^2$$

Proof

$$\|\mathbf{u} + \mathbf{v}\|^2 = (\mathbf{u} + \mathbf{v}) \cdot (\mathbf{u} + \mathbf{v}) = \|\mathbf{u}\|^2 + 2(\mathbf{u} \cdot \mathbf{v}) + \|\mathbf{v}\|^2 = \|\mathbf{u}\|^2 + \|\mathbf{v}\|^2 \quad \blacksquare$$

Alternative Notations for Vectors in R^n

It is often useful to write a vector $\mathbf{u} = (u_1, u_2, \ldots, u_n)$ in R^n in matrix notation as a row matrix or a column matrix:

$$\mathbf{u} = \begin{bmatrix} u_1 \\ u_2 \\ \vdots \\ u_n \end{bmatrix} \quad \text{or} \quad \mathbf{u} = \begin{bmatrix} u_1 & u_2 & \cdots & u_n \end{bmatrix}$$

This is justified because the matrix operations

$$\mathbf{u} + \mathbf{v} = \begin{bmatrix} u_1 \\ u_2 \\ \vdots \\ u_n \end{bmatrix} + \begin{bmatrix} v_1 \\ v_2 \\ \vdots \\ v_n \end{bmatrix} = \begin{bmatrix} u_1 + v_1 \\ u_2 + v_2 \\ \vdots \\ u_n + v_n \end{bmatrix}, \qquad k\mathbf{u} = k \begin{bmatrix} u_1 \\ u_2 \\ \vdots \\ u_n \end{bmatrix} = \begin{bmatrix} ku_1 \\ ku_2 \\ \vdots \\ ku_n \end{bmatrix}$$

or

$$\mathbf{u} + \mathbf{v} = [u_1 \quad u_2 \quad \cdots \quad u_n] + [v_1 \quad v_2 \quad \cdots \quad v_n]$$
$$= [u_1 + v_1 \quad u_2 + v_2 \quad \cdots \quad u_n + v_n]$$
$$k\mathbf{u} = k[u_1 \quad u_2 \quad \cdots \quad u_n] = [ku_1 \quad ku_2 \quad \cdots \quad ku_n]$$

produce the same results as the vector operations

$$\mathbf{u} + \mathbf{v} = (u_1, u_2, \ldots, u_n) + (v_1, v_2, \ldots, v_n) = (u_1 + v_1, u_2 + v_2, \ldots, u_n + v_n)$$
$$k\mathbf{u} = k(u_1, u_2, \ldots, u_n) = (ku_1, ku_2, \ldots, ku_n)$$

The only difference is the form in which the vectors are written.

A Matrix Formula for the Dot Product

If we use column matrix notation for the vectors

$$\mathbf{u} = \begin{bmatrix} u_1 \\ u_2 \\ \vdots \\ u_n \end{bmatrix} \quad \text{and} \quad \mathbf{v} = \begin{bmatrix} v_1 \\ v_2 \\ \vdots \\ v_n \end{bmatrix}$$

and omit the brackets on 1×1 matrices, then it follows that

$$\mathbf{v}^T \mathbf{u} = [v_1 \quad v_2 \quad \cdots \quad v_n] \begin{bmatrix} u_1 \\ u_2 \\ \vdots \\ u_n \end{bmatrix} = [u_1 v_1 + u_2 v_2 + \cdots + u_n v_n] = [\mathbf{u} \cdot \mathbf{v}] = \mathbf{u} \cdot \mathbf{v}$$

Thus, for vectors in column matrix notation, we have the following formula for the Euclidean inner product:

$$\mathbf{u} \cdot \mathbf{v} = \mathbf{v}^T \mathbf{u} \tag{7}$$

For example, if

$$\mathbf{u} = \begin{bmatrix} -1 \\ 3 \\ 5 \\ 7 \end{bmatrix} \quad \text{and} \quad \mathbf{v} = \begin{bmatrix} 5 \\ -4 \\ 7 \\ 0 \end{bmatrix}$$

then

$$\mathbf{u} \cdot \mathbf{v} = \mathbf{v}^T \mathbf{u} = [5 \quad -4 \quad 7 \quad 0] \begin{bmatrix} -1 \\ 3 \\ 5 \\ 7 \end{bmatrix} = [18] = 18$$

If A is an $n \times n$ matrix, then it follows from Formula (7) and properties of the transpose that

$$A\mathbf{u} \cdot \mathbf{v} = \mathbf{v}^T(A\mathbf{u}) = (\mathbf{v}^T A)\mathbf{u} = (A^T \mathbf{v})^T \mathbf{u} = \mathbf{u} \cdot A^T \mathbf{v}$$
$$\mathbf{u} \cdot A\mathbf{v} = (A\mathbf{v})^T \mathbf{u} = (\mathbf{v}^T A^T)\mathbf{u} = \mathbf{v}^T(A^T \mathbf{u}) = A^T \mathbf{u} \cdot \mathbf{v}$$

The resulting formulas

$$A\mathbf{u} \cdot \mathbf{v} = \mathbf{u} \cdot A^T \mathbf{v} \tag{8}$$

$$\mathbf{u} \cdot A\mathbf{v} = A^T \mathbf{u} \cdot \mathbf{v} \tag{9}$$

provide an important link between multiplication by an $n \times n$ matrix A and multiplication by A^T.

EXAMPLE 5 Verifying That $A\mathbf{u} \cdot \mathbf{v} = \mathbf{u} \cdot A^T\mathbf{v}$

Suppose that

$$A = \begin{bmatrix} 1 & -2 & 3 \\ 2 & 4 & 1 \\ -1 & 0 & 1 \end{bmatrix}, \qquad \mathbf{u} = \begin{bmatrix} -1 \\ 2 \\ 4 \end{bmatrix}, \qquad \mathbf{v} = \begin{bmatrix} -2 \\ 0 \\ 5 \end{bmatrix}$$

Then

$$A\mathbf{u} = \begin{bmatrix} 1 & -2 & 3 \\ 2 & 4 & 1 \\ -1 & 0 & 1 \end{bmatrix} \begin{bmatrix} -1 \\ 2 \\ 4 \end{bmatrix} = \begin{bmatrix} 7 \\ 10 \\ 5 \end{bmatrix}$$

$$A^T\mathbf{v} = \begin{bmatrix} 1 & 2 & -1 \\ -2 & 4 & 0 \\ 3 & 1 & 1 \end{bmatrix} \begin{bmatrix} -2 \\ 0 \\ 5 \end{bmatrix} = \begin{bmatrix} -7 \\ 4 \\ -1 \end{bmatrix}$$

from which we obtain

$$A\mathbf{u} \cdot \mathbf{v} = 7(-2) + 10(0) + 5(5) = 11$$
$$\mathbf{u} \cdot A^T\mathbf{v} = (-1)(-7) + 2(4) + 4(-1) = 11$$

Thus $A\mathbf{u} \cdot \mathbf{v} = \mathbf{u} \cdot A^T\mathbf{v}$ as guaranteed by Formula (8). We leave it for the reader to verify that (9) also holds. ◆

A Dot Product View of Matrix Multiplication

Dot products provide another way of thinking about matrix multiplication. Recall that if $A = [a_{ij}]$ is an $m \times r$ matrix and $B = [b_{ij}]$ is an $r \times n$ matrix, then the ijth entry of AB is

$$a_{i1}b_{1j} + a_{i2}b_{2j} + \cdots + a_{ir}b_{rj}$$

which is the dot product of the ith row vector of A

$$[a_{i1} \quad a_{i2} \quad \cdots \quad a_{ir}]$$

and the jth column vector of B

$$\begin{bmatrix} b_{1j} \\ b_{2j} \\ \vdots \\ b_{rj} \end{bmatrix}$$

Thus, if the row vectors of A are $\mathbf{r}_1, \mathbf{r}_2, \ldots, \mathbf{r}_m$ and the column vectors of B are $\mathbf{c}_1, \mathbf{c}_2, \ldots, \mathbf{c}_n$, then the matrix product AB can be expressed as

$$AB = \begin{bmatrix} \mathbf{r}_1 \cdot \mathbf{c}_1 & \mathbf{r}_1 \cdot \mathbf{c}_2 & \cdots & \mathbf{r}_1 \cdot \mathbf{c}_n \\ \mathbf{r}_2 \cdot \mathbf{c}_1 & \mathbf{r}_2 \cdot \mathbf{c}_2 & \cdots & \mathbf{r}_2 \cdot \mathbf{c}_n \\ \vdots & \vdots & & \vdots \\ \mathbf{r}_m \cdot \mathbf{c}_1 & \mathbf{r}_m \cdot \mathbf{c}_2 & \cdots & \mathbf{r}_m \cdot \mathbf{c}_n \end{bmatrix} \tag{10}$$

In particular, a linear system $A\mathbf{x} = \mathbf{b}$ can be expressed in dot product form as

$$\begin{bmatrix} \mathbf{r}_1 \cdot \mathbf{x} \\ \mathbf{r}_2 \cdot \mathbf{x} \\ \vdots \\ \mathbf{r}_m \cdot \mathbf{x} \end{bmatrix} = \begin{bmatrix} b_1 \\ b_2 \\ \vdots \\ b_m \end{bmatrix} \tag{11}$$

where $\mathbf{r}_1, \mathbf{r}_2, \ldots, \mathbf{r}_m$ are the row vectors of A, and $b_1, b_2, \ldots, b_m$ are the entries of $\mathbf{b}$.

EXAMPLE 6 A Linear System Written in Dot Product Form

The following is an example of a linear system expressed in dot product form (11).

System	Dot Product Form

$$3x_1 - 4x_2 + x_3 = 1$$
$$2x_1 - 7x_2 - 4x_3 = 5$$
$$x_1 + 5x_2 - 8x_3 = 0$$

$$\begin{bmatrix} (3, -4, 1) \cdot (x_1, x_2, x_3) \\ (2, -7, -4) \cdot (x_1, x_2, x_3) \\ (1, 5, -8) \cdot (x_1, x_2, x_3) \end{bmatrix} = \begin{bmatrix} 1 \\ 5 \\ 0 \end{bmatrix} \quad \blacklozenge$$

EXERCISE SET
4.1

1. Let $\mathbf{u} = (-3, 2, 1, 0)$, $\mathbf{v} = (4, 7, -3, 2)$, and $\mathbf{w} = (5, -2, 8, 1)$. Find

   (a) $\mathbf{v} - \mathbf{w}$ (b) $2\mathbf{u} + 7\mathbf{v}$ (c) $-\mathbf{u} + (\mathbf{v} - 4\mathbf{w})$

   (d) $6(\mathbf{u} - 3\mathbf{v})$ (e) $-\mathbf{v} - \mathbf{w}$ (f) $(6\mathbf{v} - \mathbf{w}) - (4\mathbf{u} + \mathbf{v})$

2. Let $\mathbf{u}$, $\mathbf{v}$, and $\mathbf{w}$ be the vectors in Exercise 1. Find the vector $\mathbf{x}$ that satisfies $5\mathbf{x} - 2\mathbf{v} = 2(\mathbf{w} - 5\mathbf{x})$.

3. Let $\mathbf{u}_1 = (-1, 3, 2, 0)$, $\mathbf{u}_2 = (2, 0, 4, -1)$, $\mathbf{u}_3 = (7, 1, 1, 4)$, and $\mathbf{u}_4 = (6, 3, 1, 2)$. Find scalars c_1, c_2, c_3, and c_4 such that $c_1\mathbf{u}_1 + c_2\mathbf{u}_2 + c_3\mathbf{u}_3 + c_4\mathbf{u}_4 = (0, 5, 6, -3)$.

4. Show that there do not exist scalars c_1, c_2, and c_3 such that

$$c_1(1, 0, 1, 0) + c_2(1, 0, -2, 1) + c_3(2, 0, 1, 2) = (1, -2, 2, 3)$$

5. In each part, compute the Euclidean norm of the vector.

   (a) $(-2, 5)$ (b) $(1, 2, -2)$ (c) $(3, 4, 0, -12)$ (d) $(-2, 1, 1, -3, 4)$

6. Let $\mathbf{u} = (4, 1, 2, 3)$, $\mathbf{v} = (0, 3, 8, -2)$, and $\mathbf{w} = (3, 1, 2, 2)$. Evaluate each expression.

   (a) $\|\mathbf{u} + \mathbf{v}\|$ (b) $\|\mathbf{u}\| + \|\mathbf{v}\|$ (c) $\|-2\mathbf{u}\| + 2\|\mathbf{u}\|$

   (d) $\|3\mathbf{u} - 5\mathbf{v} + \mathbf{w}\|$ (e) $\dfrac{1}{\|\mathbf{w}\|}\mathbf{w}$ (f) $\left\|\dfrac{1}{\|\mathbf{w}\|}\mathbf{w}\right\|$

7. Show that if $\mathbf{v}$ is a nonzero vector in R^n, then $(1/\|\mathbf{v}\|)\mathbf{v}$ has Euclidean norm 1.

8. Let $\mathbf{v} = (-2, 3, 0, 6)$. Find all scalars k such that $\|k\mathbf{v}\| = 5$.

9. Find the Euclidean inner product $\mathbf{u} \cdot \mathbf{v}$.

   (a) $\mathbf{u} = (2, 5)$, $\mathbf{v} = (-4, 3)$

   (b) $\mathbf{u} = (4, 8, 2)$, $\mathbf{v} = (0, 1, 3)$

   (c) $\mathbf{u} = (3, 1, 4, -5)$, $\mathbf{v} = (2, 2, -4, -3)$

   (d) $\mathbf{u} = (-1, 1, 0, 4, -3)$, $\mathbf{v} = (-2, -2, 0, 2, -1)$

10. (a) Find two vectors in R^2 with Euclidean norm 1 whose Euclidean inner product with $(3, -1)$ is zero.

    (b) Show that there are infinitely many vectors in R^3 with Euclidean norm 1 whose Euclidean inner product with $(1, -3, 5)$ is zero.

11. Find the Euclidean distance between $\mathbf{u}$ and $\mathbf{v}$.

    (a) $\mathbf{u} = (1, -2)$, $\mathbf{v} = (2, 1)$

    (b) $\mathbf{u} = (2, -2, 2)$, $\mathbf{v} = (0, 4, -2)$

    (c) $\mathbf{u} = (0, -2, -1, 1)$, $\mathbf{v} = (-3, 2, 4, 4)$

    (d) $\mathbf{u} = (3, -3, -2, 0, -3)$, $\mathbf{v} = (-4, 1, -1, 5, 0)$

12. Verify parts (b), (e), (f), and (g) of Theorem 4.1.1 for $\mathbf{u} = (2, 0, -3, 1)$, $\mathbf{v} = (4, 0, 3, 5)$, $\mathbf{w} = (1, 6, 2, -1)$, $k = 5$, and $l = -3$.

13. Verify parts (b) and (c) of Theorem 4.1.2 for the values of **u**, **v**, **w**, and k in Exercise 12.

14. In each part, determine whether the given vectors are orthogonal.
    (a) $\mathbf{u} = (-1, 3, 2)$, $\mathbf{v} = (4, 2, -1)$ (b) $\mathbf{u} = (-2, -2, -2)$, $\mathbf{v} = (1, 1, 1)$
    (c) $\mathbf{u} = (u_1, u_2, u_3)$, $\mathbf{v} = (0, 0, 0)$ (d) $\mathbf{u} = (-4, 6, -10, 1)$, $\mathbf{v} = (2, 1, -2, 9)$
    (c) $\mathbf{u} = (0, 3, -2, 1)$, $\mathbf{v} - (5, 2, -1, 0)$ (f) $\mathbf{u} = (a, b)$, $\mathbf{v} = (-b, a)$

15. For which values of k are **u** and **v** orthogonal?
    (a) $\mathbf{u} = (2, 1, 3)$, $\mathbf{v} = (1, 7, k)$ (b) $\mathbf{u} = (k, k, 1)$, $\mathbf{v} = (k, 5, 6)$

16. Find two vectors of norm 1 that are orthogonal to the three vectors $\mathbf{u} = (2, 1, \quad 4, 0)$, $\mathbf{v} = (-1, -1, 2, 2)$, and $\mathbf{w} = (3, 2, 5, 4)$.

17. In each part, verify that the Cauchy–Schwarz inequality holds.
    (a) $\mathbf{u} = (3, 2)$, $\mathbf{v} = (4, -1)$ (b) $\mathbf{u} = (-3, 1, 0)$, $\mathbf{v} = (2, -1, 3)$
    (c) $\mathbf{u} = (-4, 2, 1)$, $\mathbf{v} = (8, -4, -2)$ (d) $\mathbf{u} = (0, -2, 2, 1)$, $\mathbf{v} = (-1, -1, 1, 1)$

18. In each part, verify that Formulas (8) and (9) hold.

    (a) $A = \begin{bmatrix} 2 & -1 \\ 3 & 4 \end{bmatrix}$, $\mathbf{u} = \begin{bmatrix} 3 \\ 1 \end{bmatrix}$, $\mathbf{v} = \begin{bmatrix} -2 \\ 6 \end{bmatrix}$

    (b) $A = \begin{bmatrix} -1 & 2 & 4 \\ 3 & 1 & 0 \\ 5 & -2 & 3 \end{bmatrix}$, $\mathbf{u} = \begin{bmatrix} -1 \\ 2 \\ 5 \end{bmatrix}$, $\mathbf{v} = \begin{bmatrix} 0 \\ 2 \\ -4 \end{bmatrix}$

19. Solve the following linear system for x_1, x_2, and x_3.

$$(1, -1, 4) \cdot (x_1, x_2, x_3) = 10$$
$$(3, 2, 0) \cdot (x_1, x_2, x_3) = 1$$
$$(4, -5, -1) \cdot (x_1, x_2, x_3) = 7$$

20. Find $\mathbf{u} \cdot \mathbf{v}$ given that $\|\mathbf{u} + \mathbf{v}\| = 1$ and $\|\mathbf{u} \quad \mathbf{v}\| = 5$.

21. Use Theorem 4.1.6 to show that **u** and **v** are orthogonal vectors in R^n if $\|\mathbf{u} + \mathbf{v}\| = \|\mathbf{u} - \mathbf{v}\|$. Interpret this result geometrically in R^2.

22. The formulas for the vector components in Theorem 3.3.3 hold in R^n as well. Given that $\mathbf{a} = (-1, 1, 2, 3)$ and $\mathbf{u} = (2, 1, 4, -1)$, find the vector component of **u** along **a** and the vector component of **u** orthogonal to **a**.

23. Determine whether the two lines

$$\mathbf{r} = (3, 2, 3, -1) + t(4, 6, 4, -2) \quad \text{and} \quad \mathbf{r} = (0, 3, 5, 4) + s(1, -3, -4, -2)$$

intersect in R^4.

24. Prove the following generalization of Theorem 4.1.7. If $\mathbf{v}_1, \mathbf{v}_2, \ldots, \mathbf{v}_r$ are pairwise orthogonal vectors in R^n, then

$$\|\mathbf{v}_1 + \mathbf{v}_2 + \cdots + \mathbf{v}_r\|^2 = \|\mathbf{v}_1\|^2 + \|\mathbf{v}_2\|^2 + \cdots + \|\mathbf{v}_r\|^2$$

25. Prove: If **u** and **v** are $n \times 1$ matrices and A is an $n \times n$ matrix, then

$$(\mathbf{v}^T A^T A \mathbf{u})^2 \leq (\mathbf{u}^T A^T A \mathbf{u})(\mathbf{v}^T A^T A \mathbf{v})$$

26. Use the Cauchy–Schwarz inequality to prove that for all real values of a, b, and θ,

$$(a \cos \theta + b \sin \theta)^2 \leq a^2 + b^2$$

27. Prove: If **u**, **v**, and **w** are vectors in R^n and k is any scalar, then
    (a) $\mathbf{u} \cdot (k\mathbf{v}) = k(\mathbf{u} \cdot \mathbf{v})$ (b) $\mathbf{u} \cdot (\mathbf{v} + \mathbf{w}) = \mathbf{u} \cdot \mathbf{v} + \mathbf{u} \cdot \mathbf{w}$

28. Prove parts (a) through (d) of Theorem 4.1.1.

29. Prove parts (e) through (h) of Theorem 4.1.1.

30. Prove parts (*a*) and (*c*) of Theorem 4.1.2.

31. Prove parts (*a*) and (*b*) of Theorem 4.1.4.

32. Prove parts (*a*), (*b*), and (*c*) of Theorem 4.1.5.

33. Suppose that $a_1, a_2, \ldots, a_n$ are positive real numbers. In R^2, the vectors $\mathbf{v}_1 = (a_1, 0)$ and $\mathbf{v}_2 = (0, a_2)$ determine a rectangle of area $A = a_1 a_2$ (see the accompanying figure), and in R^3, the vectors $\mathbf{v}_1 = (a_1, 0, 0)$, $\mathbf{v}_2 = (0, a_2, 0)$, and $\mathbf{v}_3 = (0, 0, a_3)$ determine a box of volume $V = a_1 a_2 a_3$ (see the accompanying figure). The area A and the volume V are sometimes called the **Euclidean measure** of the rectangle and box, respectively.

(a) How would you define the Euclidean measure of the "box" in R^n that is determined by the vectors

$$\mathbf{v}_1 = (a_1, 0, 0, \ldots, 0), \quad \mathbf{v}_2 = (0, a_2, 0, \ldots, 0), \ldots, \quad \mathbf{v}_n = (0, 0, 0, \ldots, a_n)?$$

(b) How would you define the Euclidean length of the "diagonal" of the box in part (a)?

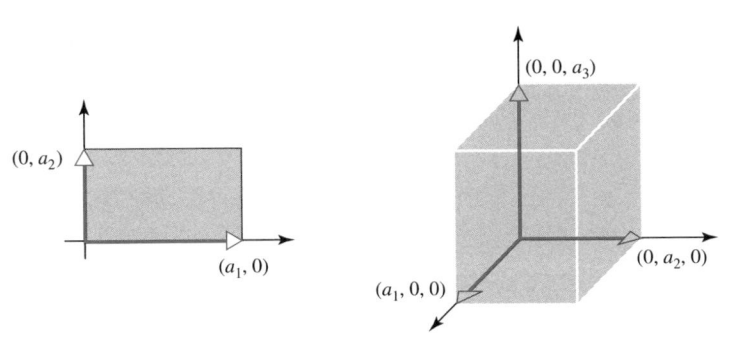

Area $A = a_1 a_2$ Volume $V = a_1 a_2 a_3$

Figure Ex-33

Discussion & Discovery

34. (a) Suppose that $\mathbf{u}$ and $\mathbf{v}$ are vectors in R^n. Show that

$$\|\mathbf{u} + \mathbf{v}\|^2 + \|\mathbf{u} - \mathbf{v}\|^2 = 2(\|\mathbf{u}\|^2 + \|\mathbf{v}\|^2)$$

(b) The result in part (a) states a theorem about parallelograms in R^2. What is the theorem?

35. (a) If $\mathbf{u}$ and $\mathbf{v}$ are orthogonal vectors in R^n such that $\|\mathbf{u}\| = 1$ and $\|\mathbf{v}\| = 1$, then $d(\mathbf{u}, \mathbf{v}) = $ ____.

(b) Draw a picture to illustrate this result.

36. In the accompanying figure the vectors $\mathbf{u}, \mathbf{v}$, and $\mathbf{u} - \mathbf{v}$ form a triangle in R^2, and θ denotes the angle between $\mathbf{u}$ and $\mathbf{v}$. It follows from the law of cosines in trigonometry that

$$\|\mathbf{u} - \mathbf{v}\|^2 = \|\mathbf{u}\|^2 + \|\mathbf{v}\|^2 - 2\|\mathbf{u}\|\|\mathbf{v}\| \cos\theta$$

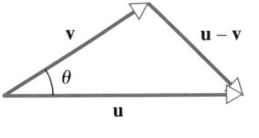

Figure Ex-36

Do you think that this formula still holds if $\mathbf{u}$ and $\mathbf{v}$ are vectors in R^n? Justify your answer.

37. Indicate whether each statement is always true or sometimes false. Justify your answer by giving a logical argument or a counterexample.

(a) If $\|\mathbf{u} + \mathbf{v}\|^2 = \|\mathbf{u}\|^2 + \|\mathbf{v}\|^2$, then $\mathbf{u}$ and $\mathbf{v}$ are orthogonal.

(b) If $\mathbf{u}$ is orthogonal to $\mathbf{v}$ and $\mathbf{w}$, then $\mathbf{u}$ is orthogonal to $\mathbf{v} + \mathbf{w}$.

(c) If $\mathbf{u}$ is orthogonal to $\mathbf{v} + \mathbf{w}$, then $\mathbf{u}$ is orthogonal to $\mathbf{v}$ and $\mathbf{w}$.

(d) If $\|\mathbf{u} - \mathbf{v}\| = 0$, then $\mathbf{u} = \mathbf{v}$.

(e) If $\|k\mathbf{u}\| = k\|\mathbf{u}\|$, then $k \geq 0$.

4.2

LINEAR TRANS-FORMATIONS FROM R^n TO R^m

In this section we shall begin the study of functions of the form $\mathbf{w} = F(\mathbf{x})$, *where the independent variable* $\mathbf{x}$ *is a vector in* R^n *and the dependent variable* $\mathbf{w}$ *is a vector in* R^m. *We shall concentrate on a special class of such functions called "linear transformations." Linear transformations are fundamental in the study of linear algebra and have many important applications in physics, engineering, social sciences, and various branches of mathematics.*

Functions from R^n to R

Recall that a ***function*** is a rule f that associates with each element in a set A one and only one element in a set B. If f associates the element b with the element a, then we write $b = f(a)$ and say that b is the ***image*** of a under f or that $f(a)$ is the ***value*** of f at a. The set A is called the ***domain*** of f and the set B is called the ***codomain*** of f. The subset of B consisting of all possible values for f as a varies over A is called the ***range*** of f. For the most common functions, A and B are sets of real numbers, in which case f is called a ***real-valued function of a real variable***. Other common functions occur when B is a set of real numbers and A is a set of vectors in R^2, R^3, or, more generally, R^n. Some examples are shown in Table 1. Two functions f_1 and f_2 are regarded as ***equal***, written $f_1 = f_2$, if they have the same domain and $f_1(a) = f_2(a)$ for all a in the domain.

Table 1

Formula	Example	Classification	Description
$f(x)$	$f(x) = x^2$	Real-valued function of a real variable	Function from R to R
$f(x, y)$	$f(x, y) = x^2 + y^2$	Real-valued function of two real variables	Function from R^2 to R
$f(x, y, z)$	$f(x, y, z) = x^2 + y^2 + z^2$	Real-valued function of three real variables	Function from R^3 to R
$f(x_1, x_2, \ldots, x_n)$	$f(x_1, x_2, \ldots, x_n) = x_1^2 + x_2^2 + \cdots + x_n^2$	Real-valued function of n real variables	Function from R^n to R

Functions from R^n to R^m

If the domain of a function f is R^n and the codomain is R^m (m and n possibly the same), then f is called a ***map*** or ***transformation*** from R^n to R^m, and we say that the function f ***maps*** R^n into R^m. We denote this by writing $f: R^n \to R^m$. The functions in Table 1 are transformations for which $m = 1$. In the case where $m = n$, the transformation $f: R^n \to R^n$ is called an ***operator*** on R^n. The first entry in Table 1 is an operator on R.

To illustrate one important way in which transformations can arise, suppose that $f_1, f_2, \ldots, f_m$ are real-valued functions of n real variables, say

$$\begin{aligned} w_1 &= f_1(x_1, x_2, \ldots, x_n) \\ w_2 &= f_2(x_1, x_2, \ldots, x_n) \\ &\vdots \qquad\qquad \vdots \\ w_m &= f_m(x_1, x_2, \ldots, x_n) \end{aligned} \tag{1}$$

These m equations assign a unique point $(w_1, w_2, \ldots, w_m)$ in R^m to each point $(x_1, x_2, \ldots, x_n)$ in R^n and thus define a transformation from R^n to R^m. If we denote this transformation by T, then $T: R^n \rightarrow R^m$ and

$$T(x_1, x_2, \ldots, x_n) = (w_1, w_2, \ldots, w_m)$$

EXAMPLE 1 A Transformation from R^2 to R^3

The equations

$$w_1 = x_1 + x_2$$
$$w_2 = 3x_1x_2$$
$$w_3 = x_1^2 - x_2^2$$

define a transformation $T: R^2 \rightarrow R^3$. With this transformation, the image of the point (x_1, x_2) is

$$T(x_1, x_2) = (x_1 + x_2, 3x_1x_2, x_1^2 - x_2^2)$$

Thus, for example, $T(1, -2) = (-1, -6, -3)$. ◆

Linear Transformations from R^n to R^m

In the special case where the equations in (1) are linear, the transformation $T: R^n \rightarrow R^m$ defined by those equations is called a **linear transformation** (or a **linear operator** if $m = n$). Thus a linear transformation $T: R^n \rightarrow R^m$ is defined by equations of the form

$$
\begin{aligned}
w_1 &= a_{11}x_1 + a_{12}x_2 + \cdots + a_{1n}x_n \\
w_2 &= a_{21}x_1 + a_{22}x_2 + \cdots + a_{2n}x_n \\
&\;\;\vdots \qquad\quad \vdots \qquad\quad \vdots \qquad\qquad \vdots \\
w_m &= a_{m1}x_1 + a_{m2}x_2 + \cdots + a_{mn}x_n
\end{aligned}
\tag{2}
$$

or, in matrix notation,

$$
\begin{bmatrix} w_1 \\ w_2 \\ \vdots \\ w_m \end{bmatrix}
=
\begin{bmatrix}
a_{11} & a_{12} & \cdots & a_{1n} \\
a_{21} & a_{22} & \cdots & a_{2n} \\
\vdots & \vdots & & \vdots \\
a_{m1} & a_{m2} & \cdots & a_{mn}
\end{bmatrix}
\begin{bmatrix} x_1 \\ x_2 \\ \vdots \\ x_n \end{bmatrix}
\tag{3}
$$

or more briefly by

$$\mathbf{w} = A\mathbf{x} \tag{4}$$

The matrix $A = [a_{ij}]$ is called the **standard matrix** for the linear transformation T, and T is called **multiplication by A**.

EXAMPLE 2 A Linear Transformation from R^4 to R^3

The linear transformation $T: R^4 \rightarrow R^3$ defined by the equations

$$
\begin{aligned}
w_1 &= 2x_1 - 3x_2 + \;\;x_3 - 5x_4 \\
w_2 &= 4x_1 + \;\;x_2 - 2x_3 + \;\;x_4 \\
w_3 &= 5x_1 - \;\;x_2 + 4x_3
\end{aligned}
\tag{5}
$$

can be expressed in matrix form as

$$
\begin{bmatrix} w_1 \\ w_2 \\ w_3 \end{bmatrix} = \begin{bmatrix} 2 & -3 & 1 & -5 \\ 4 & 1 & -2 & 1 \\ 5 & -1 & 4 & 0 \end{bmatrix} \begin{bmatrix} x_1 \\ x_2 \\ x_3 \\ x_4 \end{bmatrix} \tag{6}
$$

so the standard matrix for T is

$$
A = \begin{bmatrix} 2 & -3 & 1 & -5 \\ 4 & 1 & -2 & 1 \\ 5 & -1 & 4 & 0 \end{bmatrix}
$$

The image of a point (x_1, x_2, x_3, x_4) can be computed directly from the defining equations (5) or from (6) by matrix multiplication. For example, if $(x_1, x_2, x_3, x_4) = (1, -3, 0, 2)$, then substituting in (5) yields

$$ w_1 = 1, \qquad w_2 = 3, \qquad w_3 = 8 $$

(verify) or alternatively from (6),

$$
\begin{bmatrix} w_1 \\ w_2 \\ w_3 \end{bmatrix} = \begin{bmatrix} 2 & -3 & 1 & -5 \\ 4 & 1 & -2 & 1 \\ 5 & -1 & 4 & 0 \end{bmatrix} \begin{bmatrix} 1 \\ -3 \\ 0 \\ 2 \end{bmatrix} = \begin{bmatrix} 1 \\ 3 \\ 8 \end{bmatrix} \blacklozenge
$$

Some Notational Matters

If $T: R^n \to R^m$ is multiplication by A, and if it is important to emphasize that A is the standard matrix for T, we shall denote the linear transformation $T: R^n \to R^m$ by $T_A: R^n \to R^m$. Thus

$$ T_A(\mathbf{x}) = A\mathbf{x} \tag{7} $$

It is understood in this equation that the vector $\mathbf{x}$ in R^n is expressed as a column matrix.

Sometimes it is awkward to introduce a new letter to denote the standard matrix for a linear transformation $T: R^n \to R^m$. In such cases we will denote the standard matrix for T by the symbol $[T]$. With this notation, equation (7) would take the form

$$ T(\mathbf{x}) = [T]\mathbf{x} \tag{8} $$

Occasionally, the two notations for a standard matrix will be mixed, in which case we have the relationship

$$ [T_A] = A \tag{9} $$

REMARK Amidst all of this notation, it is important to keep in mind that we have established a correspondence between $m \times n$ matrices and linear transformations from R^n to R^m: To each matrix A there corresponds a linear transformation T_A (multiplication by A), and to each linear transformation $T: R^n \to R^m$, there corresponds an $m \times n$ matrix $[T]$ (the standard matrix for T).

Geometry of Linear Transformations

Depending on whether n-tuples are regarded as points or vectors, the geometric effect of an operator $T: R^n \to R^n$ is to transform each point (or vector) in R^n into some new point (or vector) (Figure 4.2.1).

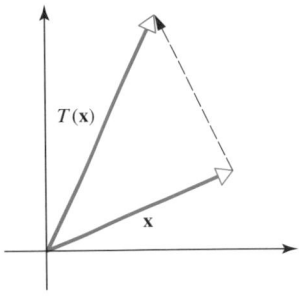

(a) T maps points to points

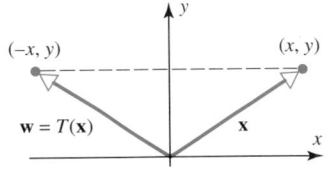

(b) T maps vectors to vectors

Figure 4.2.1

EXAMPLE 3 Zero Transformation from R^n to R^m

If 0 is the $m \times n$ zero matrix and $\mathbf{0}$ is the zero vector in R^n, then for every vector $\mathbf{x}$ in R^n,

$$T_0(\mathbf{x}) = 0\mathbf{x} = \mathbf{0}$$

so multiplication by zero maps every vector in R^n into the zero vector in R^m. We call T_0 the ***zero transformation*** from R^n to R^m. Sometimes the zero transformation is denoted by 0. Although this is the same notation used for the zero matrix, the appropriate interpretation will usually be clear from the context. ◆

EXAMPLE 4 Identity Operator on R^n

If I is the $n \times n$ identity matrix, then for every vector $\mathbf{x}$ in R^n,

$$T_I(\mathbf{x}) = I\mathbf{x} = \mathbf{x}$$

so multiplication by I maps every vector in R^n into itself. We call T_I the ***identity operator*** on R^n. Sometimes the identity operator is denoted by I. Although this is the same notation used for the identity matrix, the appropriate interpretation will usually be clear from the context. ◆

Among the most important linear operators on R^2 and R^3 are those that produce reflections, projections, and rotations. We shall now discuss such operators.

Reflection Operators

Consider the operator $T: R^2 \to R^2$ that maps each vector into its symmetric image about the y-axis (Figure 4.2.2).

If we let $\mathbf{w} = T(\mathbf{x})$, then the equations relating the components of $\mathbf{x}$ and $\mathbf{w}$ are

$$\begin{aligned} w_1 &= -x = -x + 0y \\ w_2 &= y = 0x + y \end{aligned} \tag{10}$$

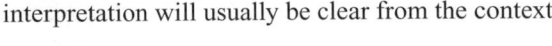

or, in matrix form,

$$\begin{bmatrix} w_1 \\ w_2 \end{bmatrix} = \begin{bmatrix} -1 & 0 \\ 0 & 1 \end{bmatrix} \begin{bmatrix} x \\ y \end{bmatrix} \tag{11}$$

Figure 4.2.2

Since the equations in (10) are linear, T is a linear operator, and from (11) the standard matrix for T is

$$[T] = \begin{bmatrix} -1 & 0 \\ 0 & 1 \end{bmatrix}$$

In general, operators on R^2 and R^3 that map each vector into its symmetric image about some line or plane are called ***reflection operators***. Such operators are linear. Tables 2 and 3 list some of the common reflection operators.

Projection Operators

Consider the operator $T: R^2 \to R^2$ that maps each vector into its orthogonal projection on the x-axis (Figure 4.2.3). The equations relating the components of $\mathbf{x}$ and $\mathbf{w} = T(\mathbf{x})$ are

$$\begin{aligned} w_1 &= x = x + 0y \\ w_2 &= 0 = 0x + 0y \end{aligned} \tag{12}$$

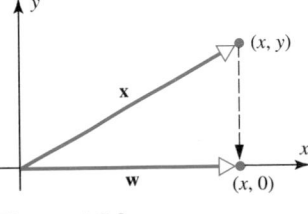

or, in matrix form,

$$\begin{bmatrix} w_1 \\ w_2 \end{bmatrix} = \begin{bmatrix} 1 & 0 \\ 0 & 0 \end{bmatrix} \begin{bmatrix} x \\ y \end{bmatrix} \tag{13}$$

Figure 4.2.3

Table 2

Operator	Illustration	Equations	Standard Matrix
Reflection about the y-axis	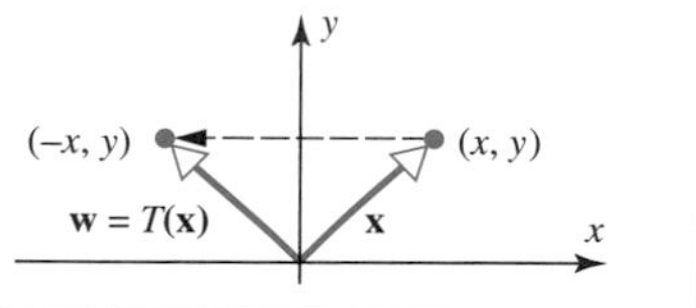	$w_1 = -x$ $w_2 = y$	$\begin{bmatrix} -1 & 0 \\ 0 & 1 \end{bmatrix}$
Reflection about the x-axis	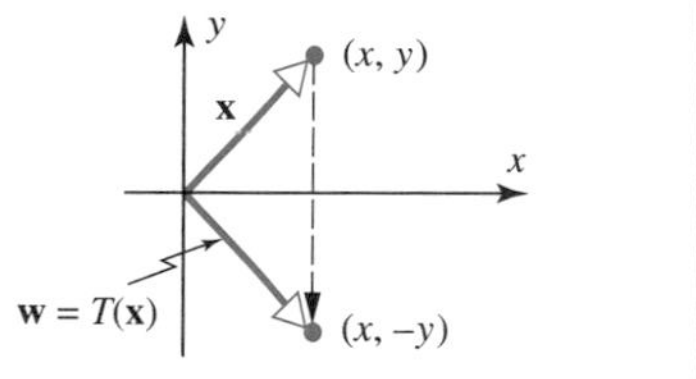	$w_1 = x$ $w_2 = -y$	$\begin{bmatrix} 1 & 0 \\ 0 & -1 \end{bmatrix}$
Reflection about the line $y = x$	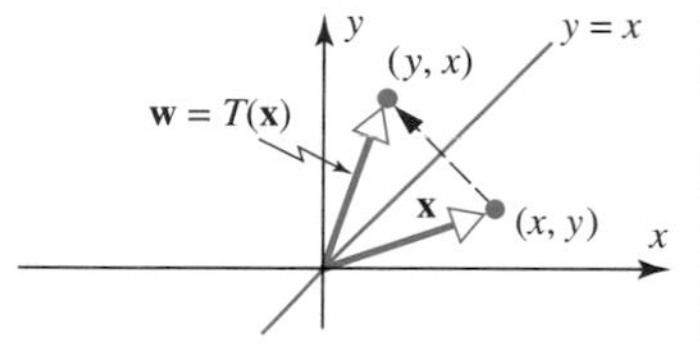	$w_1 = y$ $w_2 = x$	$\begin{bmatrix} 0 & 1 \\ 1 & 0 \end{bmatrix}$

Table 3

Operator	Illustration	Equations	Standard Matrix
Reflection about the xy-plane	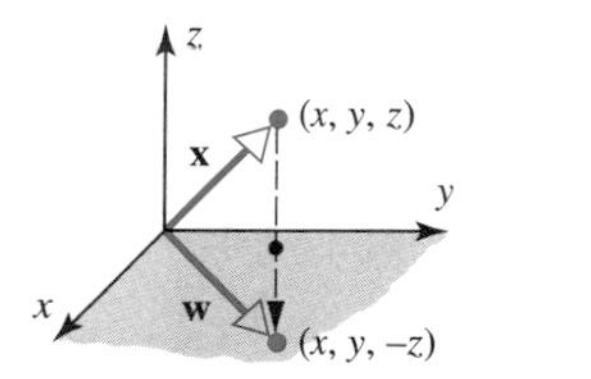	$w_1 = x$ $w_2 = y$ $w_3 = -z$	$\begin{bmatrix} 1 & 0 & 0 \\ 0 & 1 & 0 \\ 0 & 0 & -1 \end{bmatrix}$
Reflection about the xz-plane	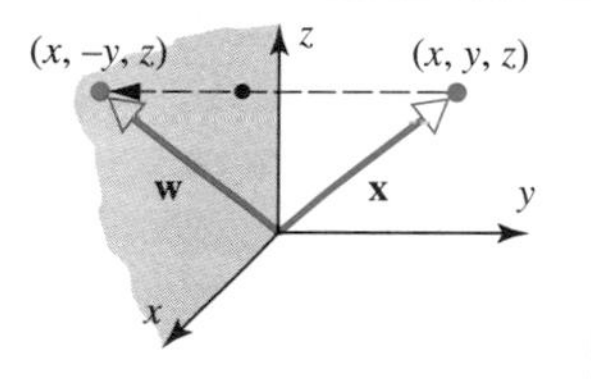	$w_1 = x$ $w_2 = -y$ $w_3 = z$	$\begin{bmatrix} 1 & 0 & 0 \\ 0 & -1 & 0 \\ 0 & 0 & 1 \end{bmatrix}$
Reflection about the yz-plane	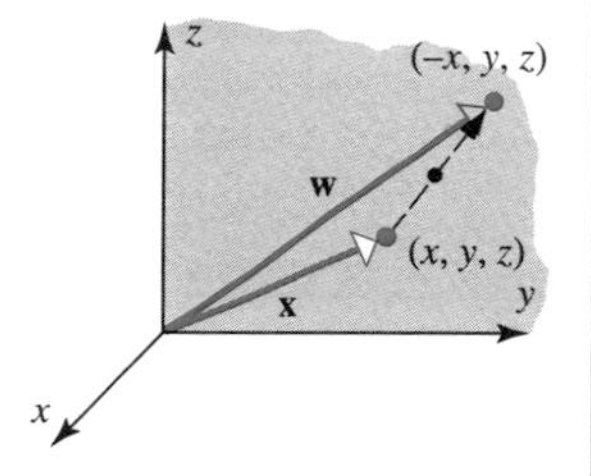	$w_1 = -x$ $w_2 = y$ $w_3 = z$	$\begin{bmatrix} -1 & 0 & 0 \\ 0 & 1 & 0 \\ 0 & 0 & 1 \end{bmatrix}$

The equations in (12) are linear, so T is a linear operator, and from (13) the standard matrix for T is

$$[T] = \begin{bmatrix} 1 & 0 \\ 0 & 0 \end{bmatrix}$$

In general, a ***projection operator*** (more precisely, an ***orthogonal projection operator***) on R^2 or R^3 is any operator that maps each vector into its orthogonal projection on a line or plane through the origin. It can be shown that such operators are linear. Some of the basic projection operators on R^2 and R^3 are listed in Tables 4 and 5.

Table 4

Operator	Illustration	Equations	Standard Matrix
Orthogonal projection on the x-axis		$w_1 = x$ $w_2 = 0$	$\begin{bmatrix} 1 & 0 \\ 0 & 0 \end{bmatrix}$
Orthogonal projection on the y-axis		$w_1 = 0$ $w_2 = y$	$\begin{bmatrix} 0 & 0 \\ 0 & 1 \end{bmatrix}$

Table 5

Operator	Illustration	Equations	Standard Matrix
Orthogonal projection on the xy-plane		$w_1 = x$ $w_2 = y$ $w_3 = 0$	$\begin{bmatrix} 1 & 0 & 0 \\ 0 & 1 & 0 \\ 0 & 0 & 0 \end{bmatrix}$
Orthogonal projection on the xz-plane		$w_1 = x$ $w_2 = 0$ $w_3 = z$	$\begin{bmatrix} 1 & 0 & 0 \\ 0 & 0 & 0 \\ 0 & 0 & 1 \end{bmatrix}$
Orthogonal projection on the yz-plane		$w_1 = 0$ $w_2 = y$ $w_3 = z$	$\begin{bmatrix} 0 & 0 & 0 \\ 0 & 1 & 0 \\ 0 & 0 & 1 \end{bmatrix}$

Rotation Operators

An operator that rotates each vector in R^2 through a fixed angle θ is called a ***rotation operator*** on R^2. Table 6 gives the formula for the rotation operators on R^2. To show how this is derived, consider the rotation operator that rotates each vector counterclockwise through a fixed positive angle θ. To find equations relating $\mathbf{x}$ and $\mathbf{w} = T(\mathbf{x})$, let ϕ be

Table 6

Operator	Illustration	Equations	Standard Matrix
Rotation through an angle θ	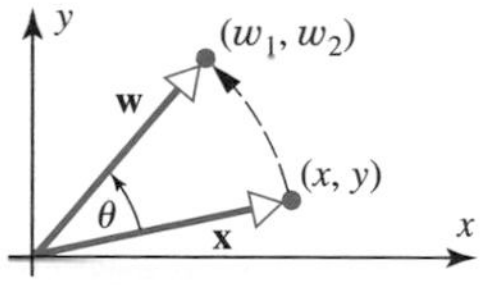	$w_1 = x \cos\theta - y \sin\theta$ $w_2 = x \sin\theta + y \cos\theta$	$\begin{bmatrix} \cos\theta & -\sin\theta \\ \sin\theta & \cos\theta \end{bmatrix}$

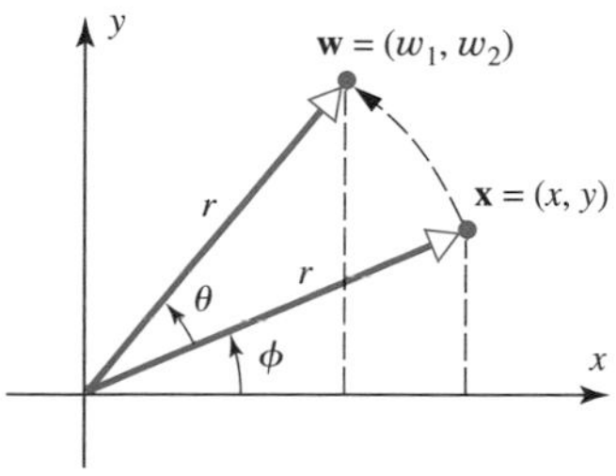

Figure 4.2.4

the angle from the positive x-axis to $\mathbf{x}$, and let r be the common length of $\mathbf{x}$ and $\mathbf{w}$ (Figure 4.2.4).

Then, from basic trigonometry,

$$x = r \cos\phi, \qquad y = r \sin\phi \tag{14}$$

and

$$w_1 = r \cos(\theta + \phi), \qquad w_2 = r \sin(\theta + \phi) \tag{15}$$

Using trigonometric identities on (15) yields

$$w_1 = r \cos\theta \cos\phi - r \sin\theta \sin\phi$$
$$w_2 = r \sin\theta \cos\phi + r \cos\theta \sin\phi$$

and substituting (14) yields

$$w_1 = x \cos\theta - y \sin\theta$$
$$w_2 = x \sin\theta + y \cos\theta \tag{16}$$

The equations in (16) are linear, so T is a linear operator; moreover, it follows from these equations that the standard matrix for T is

$$[T] = \begin{bmatrix} \cos\theta & -\sin\theta \\ \sin\theta & \cos\theta \end{bmatrix}$$

EXAMPLE 5 Rotation

If each vector in R^2 is rotated through an angle of $\pi/6\,(= 30°)$, then the image $\mathbf{w}$ of a vector

$$\mathbf{x} = \begin{bmatrix} x \\ y \end{bmatrix}$$

is

$$\mathbf{w} = \begin{bmatrix} \cos \pi/6 & -\sin \pi/6 \\ \sin \pi/6 & \cos \pi/6 \end{bmatrix} \begin{bmatrix} x \\ y \end{bmatrix} = \begin{bmatrix} \sqrt{3}/2 & -1/2 \\ 1/2 & \sqrt{3}/2 \end{bmatrix} \begin{bmatrix} x \\ y \end{bmatrix} = \begin{bmatrix} \dfrac{\sqrt{3}}{2}x - \dfrac{1}{2}y \\ \dfrac{1}{2}x + \dfrac{\sqrt{3}}{2}y \end{bmatrix}$$

For example, the image of the vector

$$\mathbf{x} = \begin{bmatrix} 1 \\ 1 \end{bmatrix} \quad \text{is} \quad \mathbf{w} = \begin{bmatrix} \dfrac{\sqrt{3}-1}{2} \\ \dfrac{1+\sqrt{3}}{2} \end{bmatrix} \quad \blacklozenge$$

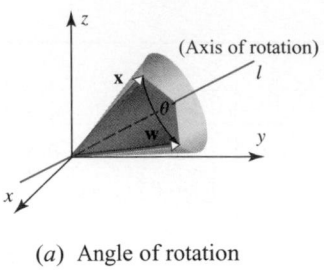

(*a*) Angle of rotation

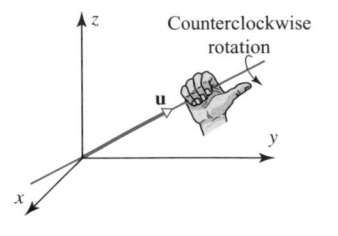

(*b*) Right-hand rule

Figure 4.2.5

A rotation of vectors in R^3 is usually described in relation to a ray emanating from the origin, called the **axis of rotation**. As a vector revolves around the axis of rotation, it sweeps out some portion of a cone (Figure 4.2.5*a*). The **angle of rotation**, which is measured in the base of the cone, is described as "clockwise" or "counterclockwise" in relation to a viewpoint that is along the axis of rotation *looking toward the origin*. For example, in Figure 4.2.5*a* the vector **w** results from rotating the vector **x** counterclockwise around the axis *l* through an angle θ. As in R^2, angles are *positive* if they are generated by counterclockwise rotations and *negative* if they are generated by clockwise rotations.

The most common way of describing a general axis of rotation is to specify a nonzero vector **u** that runs along the axis of rotation and has its initial point at the origin. The counterclockwise direction for a rotation about the axis can then be determined by a "right-hand rule" (Figure 4.2.5*b*): If the thumb of the right hand points in the direction of **u**, then the cupped fingers point in a counterclockwise direction.

A **rotation operator** on R^3 is a linear operator that rotates each vector in R^3 about some rotation axis through a fixed angle θ. In Table 7 we have described the rotation operators on R^3 whose axes of rotation are the positive coordinate axes. For each of these rotations one of the components is unchanged by the rotation, and the relationships between the other components can be derived by the same procedure used to derive (16). For example, in the rotation about the z-axis, the z-components of **x** and $\mathbf{w} = T(\mathbf{x})$ are the same, and the x- and y-components are related as in (16). This yields the rotation equation shown in the last row of Table 7.

Table 7

Operator	Illustration	Equations	Standard Matrix
Counterclockwise rotation about the positive x-axis through an angle θ		$w_1 = x$ $w_2 = y\cos\theta - z\sin\theta$ $w_3 = y\sin\theta + z\cos\theta$	$\begin{bmatrix} 1 & 0 & 0 \\ 0 & \cos\theta & -\sin\theta \\ 0 & \sin\theta & \cos\theta \end{bmatrix}$
Counterclockwise rotation about the positive y-axis through an angle θ		$w_1 = x\cos\theta + z\sin\theta$ $w_2 = y$ $w_3 = -x\sin\theta + z\cos\theta$	$\begin{bmatrix} \cos\theta & 0 & \sin\theta \\ 0 & 1 & 0 \\ -\sin\theta & 0 & \cos\theta \end{bmatrix}$
Counterclockwise rotation about the positive z-axis through an angle θ		$w_1 = x\cos\theta - y\sin\theta$ $w_2 = x\sin\theta + y\cos\theta$ $w_3 = z$	$\begin{bmatrix} \cos\theta & -\sin\theta & 0 \\ \sin\theta & \cos\theta & 0 \\ 0 & 0 & 1 \end{bmatrix}$

Yaw, Pitch, and Roll

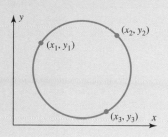

In aeronautics and astronautics, the orientation of an aircraft or space shuttle relative to an xyz-coordinate system is often described in terms of angles called **yaw**, **pitch**, and **roll**. If, for example, an aircraft is flying along the y-axis and the xy-plane defines the horizontal, then the aircraft's angle of rotation about the z-axis is called the **yaw**, its angle of rotation about the x-axis is called the **pitch**, and its angle of rotation about the y-axis is called the **roll**. A combination of yaw, pitch, and roll can be achieved by a single rotation about some axis through the origin. This is, in fact, how a space shuttle makes attitude adjustments—it doesn't perform each rotation separately; it calculates one axis, and rotates about that axis to get the correct orientation. Such rotation maneuvers are used to align an antenna, point the nose toward a celestial object, or position a payload bay for docking.

For completeness, we note that the standard matrix for a counterclockwise rotation through an angle θ about an axis in R^3, which is determined by an arbitrary *unit vector* $\mathbf{u} = (a, b, c)$ that has its initial point at the origin, is

$$\begin{bmatrix} a^2(1 - \cos\theta) + \cos\theta & ab(1 - \cos\theta) - c\sin\theta & ac(1 - \cos\theta) + b\sin\theta \\ ab(1 - \cos\theta) + c\sin\theta & b^2(1 - \cos\theta) + \cos\theta & bc(1 - \cos\theta) - a\sin\theta \\ ac(1 - \cos\theta) - b\sin\theta & bc(1 - \cos\theta) + a\sin\theta & c^2(1 - \cos\theta) + \cos\theta \end{bmatrix} \quad (17)$$

The derivation can be found in the book *Principles of Interactive Computer Graphics*, by W. M. Newman and R. F. Sproull (New York: McGraw-Hill, 1979). The reader may find it instructive to derive the results in Table 7 as special cases of this more general result.

Dilation and Contraction Operators

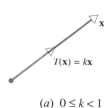

$T(\mathbf{x}) = k\mathbf{x}$

(a) $0 \le k < 1$

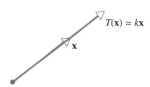

$T(\mathbf{x}) = k\mathbf{x}$

(b) $k > 1$

Figure 4.2.6

If k is a nonnegative scalar, then the operator $T(\mathbf{x}) = k\mathbf{x}$ on R^2 or R^3 is called a **contraction with factor k** if $0 \le k \le 1$ and a **dilation with factor k** if $k \ge 1$. The geometric effect of a contraction is to compress each vector by a factor of k (Figure 4.2.6a), and the effect of a dilation is to stretch each vector by a factor of k (Figure 4.2.6b). A contraction compresses R^2 or R^3 uniformly toward the origin from all directions, and a dilation stretches R^2 or R^3 uniformly away from the origin in all directions.

The most extreme contraction occurs when $k = 0$, in which case $T(\mathbf{x}) = k\mathbf{x}$ reduces to the zero operator $T(\mathbf{x}) = \mathbf{0}$, which compresses every vector into a single point (the origin). If $k = 1$, then $T(\mathbf{x}) = k\mathbf{x}$ reduces to the identity operator $T(\mathbf{x}) = \mathbf{x}$, which leaves each vector unchanged; this can be regarded as either a contraction or a dilation. Tables 8 and 9 list the dilation and contraction operators on R^2 and R^3.

Table 8

Operator	Illustration	Equations	Standard Matrix
Contraction with factor k on R^2 ($0 \le k \le 1$)		$w_1 = kx$ $w_2 = ky$	$\begin{bmatrix} k & 0 \\ 0 & k \end{bmatrix}$
Dilation with factor k on R^2 ($k \ge 1$)		$w_1 = kx$ $w_2 = ky$	

Rotations in R^3

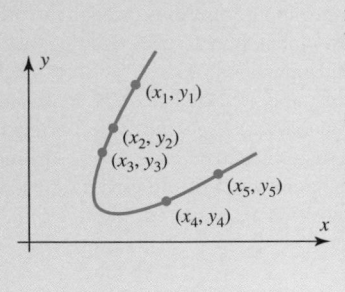

A familiar example of a rotation in R^3 is the rotation of the Earth about its axis through the North and South Poles. For simplicity, we will assume that the Earth is a sphere. Since the Sun rises in the east and sets in the west, we know that the Earth rotates from west to east. However, to an observer above the North Pole the rotation will appear counterclockwise, and to an observer below the South Pole it will appear clockwise. Thus, when a rotation in R^3 is described as clockwise or counterclockwise, a direction of view along the axis of rotation must also be stated.

There are some other facts about the Earth's rotation that are useful for understanding gen-

eral rotations in R^3. For example, as the Earth rotates about its axis, the North and South Poles remain fixed, as do all other points that lie on the axis of rotation. Thus, the axis of rotation can be thought of as the line of fixed points in the Earth's rotation. Moreover, all points on the Earth that are not on the axis of rotation move in circular paths that are centered on the axis and lie in planes that are perpendicular to the axis. For example, the points in the Equatorial Plane move within the Equatorial Plane in circles about the Earth's center.

Table 9

Operator	Illustration	Equations	Standard Matrix
Contraction with factor k on R^3 $(0 \leq k \leq 1)$		$w_1 = kx$ $w_2 = ky$ $w_3 = kz$	
Dilation with factor k on R^3 $(k \geq 1)$		$w_1 = kx$ $w_2 = ky$ $w_3 = kz$	$\begin{bmatrix} k & 0 & 0 \\ 0 & k & 0 \\ 0 & 0 & k \end{bmatrix}$

Compositions of Linear Transformations

If $T_A : R^n \rightarrow R^k$ and $T_B : R^k \rightarrow R^m$ are linear transformations, then for each $\mathbf{x}$ in R^n one can first compute $T_A(\mathbf{x})$, which is a vector in R^k, and then one can compute $T_B(T_A(\mathbf{x}))$, which is a vector in R^m. Thus, the application of T_A followed by T_B produces a transformation from R^n to R^m. This transformation is called the ***composition of T_B with T_A*** and is denoted by $T_B \circ T_A$ (read "T_B circle T_A"). Thus

$$(T_B \circ T_A)(\mathbf{x}) = T_B(T_A(\mathbf{x})) \tag{18}$$

The composition $T_B \circ T_A$ is linear since

$$(T_B \circ T_A)(\mathbf{x}) = T_B(T_A(\mathbf{x})) = B(A\mathbf{x}) = (BA)\mathbf{x} \tag{19}$$

so $T_B \circ T_A$ is multiplication by BA, which is a linear transformation. Formula (19) also tells us that the standard matrix for $T_B \circ T_A$ is BA. This is expressed by the formula

$$T_B \circ T_A = T_{BA} \tag{20}$$

REMARK Formula (20) captures an important idea: *Multiplying matrices is equivalent to composing the corresponding linear transformations in the right-to-left order of the factors.*

There is an alternative form of Formula (20): If $T_1: R^n \rightarrow R^k$ and $T_2: R^k \rightarrow R^m$ are linear transformations, then because the standard matrix for the composition $T_2 \circ T_1$ is the product of the standard matrices of T_2 and T_1, we have

$$[T_2 \circ T_1] = [T_2][T_1] \tag{21}$$

EXAMPLE 6 Composition of Two Rotations

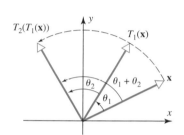

Figure 4.2.7

Let $T_1: R^2 \rightarrow R^2$ and $T_2: R^2 \rightarrow R^2$ be the linear operators that rotate vectors through the angles θ_1 and θ_2, respectively. Thus the operation

$$(T_2 \circ T_1)(\mathbf{x}) = T_2(T_1(\mathbf{x}))$$

first rotates $\mathbf{x}$ through the angle θ_1, then rotates $T_1(\mathbf{x})$ through the angle θ_2. It follows that the net effect of $T_2 \circ T_1$ is to rotate each vector in R^2 through the angle $\theta_1 + \theta_2$ (Figure 4.2.7).

Thus the standard matrices for these linear operators are

$$[T_1] = \begin{bmatrix} \cos\theta_1 & -\sin\theta_1 \\ \sin\theta_1 & \cos\theta_1 \end{bmatrix}, \qquad [T_2] = \begin{bmatrix} \cos\theta_2 & -\sin\theta_2 \\ \sin\theta_2 & \cos\theta_2 \end{bmatrix},$$

$$[T_2 \circ T_1] = \begin{bmatrix} \cos(\theta_1 + \theta_2) & -\sin(\theta_1 + \theta_2) \\ \sin(\theta_1 + \theta_2) & \cos(\theta_1 + \theta_2) \end{bmatrix}$$

These matrices should satisfy (21). With the help of some basic trigonometric identities, we can show that this is so as follows:

$$
\begin{aligned}
[T_2][T_1] &= \begin{bmatrix} \cos\theta_2 & -\sin\theta_2 \\ \sin\theta_2 & \cos\theta_2 \end{bmatrix} \begin{bmatrix} \cos\theta_1 & -\sin\theta_1 \\ \sin\theta_1 & \cos\theta_1 \end{bmatrix} \\[2mm]
&= \begin{bmatrix} \cos\theta_2\cos\theta_1 - \sin\theta_2\sin\theta_1 & -(\cos\theta_2\sin\theta_1 + \sin\theta_2\cos\theta_1) \\ \sin\theta_2\cos\theta_1 + \cos\theta_2\sin\theta_1 & -\sin\theta_2\sin\theta_1 + \cos\theta_2\cos\theta_1 \end{bmatrix} \\[2mm]
&= \begin{bmatrix} \cos(\theta_1 + \theta_2) & -\sin(\theta_1 + \theta_2) \\ \sin(\theta_1 + \theta_2) & \cos(\theta_1 + \theta_2) \end{bmatrix} \\[2mm]
&= [T_2 \circ T_1] \quad \blacklozenge
\end{aligned}
$$

REMARK In general, the order in which linear transformations are composed matters. This is to be expected, since the composition of two linear transformations corresponds to the multiplication of their standard matrices, and we know that the order in which matrices are multiplied makes a difference.

EXAMPLE 7 Composition Is Not Commutative

Let $T_1: R^2 \rightarrow R^2$ be the reflection operator about the line $y = x$, and let $T_2: R^2 \rightarrow R^2$ be the orthogonal projection on the y-axis. Figure 4.2.8 illustrates graphically that $T_1 \circ T_2$

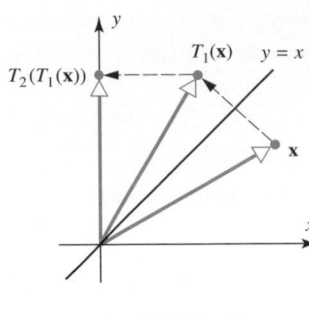

(a) $T_2 \circ T_1$

(b) $T_1 \circ T_2$

Figure 4.2.8

and $T_2 \circ T_1$ have different effects on a vector $\mathbf{x}$. This same conclusion can be reached by showing that the standard matrices for T_1 and T_2 do not commute:

$$[T_1 \circ T_2] = [T_1][T_2] = \begin{bmatrix} 0 & 1 \\ 1 & 0 \end{bmatrix} \begin{bmatrix} 0 & 0 \\ 0 & 1 \end{bmatrix} = \begin{bmatrix} 0 & 1 \\ 0 & 0 \end{bmatrix}$$

$$[T_2 \circ T_1] = [T_2][T_1] = \begin{bmatrix} 0 & 0 \\ 0 & 1 \end{bmatrix} \begin{bmatrix} 0 & 1 \\ 1 & 0 \end{bmatrix} = \begin{bmatrix} 0 & 0 \\ 1 & 0 \end{bmatrix}$$

so $[T_2 \circ T_1] \neq [T_1 \circ T_2]$. ◆

EXAMPLE 8 Composition of Two Reflections

Let $T_1: R^2 \to R^2$ be the reflection about the y-axis, and let $T_2: R^2 \to R^2$ be the reflection about the x-axis. In this case $T_1 \circ T_2$ and $T_2 \circ T_1$ are the same; both map each vector $\mathbf{x} = (x, y)$ into its negative $-\mathbf{x} = (-x, -y)$ (Figure 4.2.9):

$$(T_1 \circ T_2)(x, y) = T_1(x, -y) = (-x, -y)$$
$$(T_2 \circ T_1)(x, y) = T_2(-x, y) = (-x, -y)$$

The equality of $T_1 \circ T_2$ and $T_2 \circ T_1$ can also be deduced by showing that the standard matrices for T_1 and T_2 commute:

$$[T_1 \circ T_2] = [T_1][T_2] = \begin{bmatrix} -1 & 0 \\ 0 & 1 \end{bmatrix} \begin{bmatrix} 1 & 0 \\ 0 & -1 \end{bmatrix} = \begin{bmatrix} -1 & 0 \\ 0 & -1 \end{bmatrix}$$

$$[T_2 \circ T_1] = [T_2][T_1] = \begin{bmatrix} 1 & 0 \\ 0 & -1 \end{bmatrix} \begin{bmatrix} -1 & 0 \\ 0 & 1 \end{bmatrix} = \begin{bmatrix} -1 & 0 \\ 0 & -1 \end{bmatrix}$$

The operator $T(\mathbf{x}) = -\mathbf{x}$ on R^2 or R^3 is called the ***reflection about the origin***. As the computations above show, the standard matrix for this operator on R^2 is

$$[T] = \begin{bmatrix} -1 & 0 \\ 0 & -1 \end{bmatrix} ◆$$

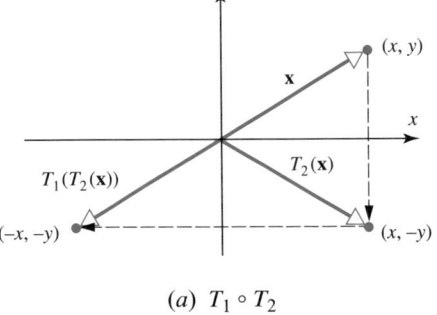
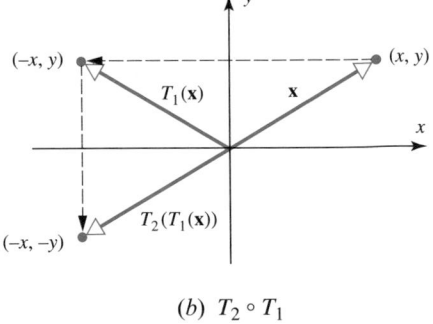

(a) $T_1 \circ T_2$

(b) $T_2 \circ T_1$

Figure 4.2.9

Compositions of Three or More Linear Transformations

Compositions can be defined for three or more linear transformations. For example, consider the linear transformations

$$T_1: R^n \to R^k, \qquad T_2: R^k \to R^l, \qquad T_3: R^l \to R^m$$

We define the composition $(T_3 \circ T_2 \circ T_1): R^n \rightarrow R^m$ by

$$(T_3 \circ T_2 \circ T_1)(\mathbf{x}) = T_3(T_2(T_1(\mathbf{x})))$$

It can be shown that this composition is a linear transformation and that the standard matrix for $T_3 \circ T_2 \circ T_1$ is related to the standard matrices for T_1, T_2, and T_3 by

$$[T_3 \circ T_2 \circ T_1] = [T_3][T_2][T_1] \qquad (22)$$

which is a generalization of (21). If the standard matrices for T_1, T_2, and T_3 are denoted by A, B, and C, respectively, then we also have the following generalization of (20):

$$T_C \circ T_B \circ T_A = T_{CBA} \qquad (23)$$

EXAMPLE 9 Composition of Three Transformations

Find the standard matrix for the linear operator $T: R^3 \rightarrow R^3$ that first rotates a vector counterclockwise about the z-axis through an angle θ, then reflects the resulting vector about the yz-plane, and then projects that vector orthogonally onto the xy-plane.

Solution

The linear transformation T can be expressed as the composition

$$T = T_3 \circ T_2 \circ T_1$$

where T_1 is the rotation about the z-axis, T_2 is the reflection about the yz-plane, and T_3 is the orthogonal projection on the xy-plane. From Tables 3, 5, and 7, the standard matrices for these linear transformations are

$$[T_1] = \begin{bmatrix} \cos\theta & -\sin\theta & 0 \\ \sin\theta & \cos\theta & 0 \\ 0 & 0 & 1 \end{bmatrix}, \qquad [T_2] = \begin{bmatrix} -1 & 0 & 0 \\ 0 & 1 & 0 \\ 0 & 0 & 1 \end{bmatrix}, \qquad [T_3] = \begin{bmatrix} 1 & 0 & 0 \\ 0 & 1 & 0 \\ 0 & 0 & 0 \end{bmatrix}$$

Thus, from (22) the standard matrix for T is $[T] = [T_3][T_2][T_1]$; that is,

$$[T] = \begin{bmatrix} 1 & 0 & 0 \\ 0 & 1 & 0 \\ 0 & 0 & 0 \end{bmatrix} \begin{bmatrix} -1 & 0 & 0 \\ 0 & 1 & 0 \\ 0 & 0 & 1 \end{bmatrix} \begin{bmatrix} \cos\theta & -\sin\theta & 0 \\ \sin\theta & \cos\theta & 0 \\ 0 & 0 & 1 \end{bmatrix}$$

$$= \begin{bmatrix} -\cos\theta & \sin\theta & 0 \\ \sin\theta & \cos\theta & 0 \\ 0 & 0 & 0 \end{bmatrix} \blacklozenge$$

EXERCISE SET 4.2

1. Find the domain and codomain of the transformation defined by the equations, and determine whether the transformation is linear.

(a) $w_1 = 3x_1 - 2x_2 + 4x_3$
$w_2 = 5x_1 - 8x_2 + x_3$

(b) $w_1 = 2x_1 x_2 - x_2$
$w_2 = x_1 + 3x_1 x_2$
$w_3 = x_1 + x_2$

(c) $w_1 = 5x_1 - x_2 + x_3$
$w_2 = -x_1 + x_2 + 7x_3$
$w_3 = 2x_1 - 4x_2 - x_3$

(d) $w_1 = x_1^2 - 3x_2 + x_3 - 2x_4$
$w_2 = 3x_1 - 4x_2 - x_3^2 + x_4$

2. Find the standard matrix for the linear transformation defined by the equations.

(a) $w_1 = 2x_1 - 3x_2 + x_4$
$w_2 = 3x_1 + 5x_2 - x_4$

(b) $w_1 = 7x_1 + 2x_2 - 8x_3$
$w_2 = \quad - x_2 + 5x_3$
$w_3 = 4x_1 + 7x_2 - \quad x_3$

(c) $w_1 = -x_1 + \quad x_2$
$w_2 = 3x_1 - 2x_2$
$w_3 = 5x_1 - 7x_2$

(d) $w_1 = x_1$
$w_2 = x_1 + x_2$
$w_3 = x_1 + x_2 + x_3$
$w_4 = x_1 + x_2 + x_3 + x_4$

3. Find the standard matrix for the linear operator $T: R^3 \to R^3$ given by

$$w_1 = 3x_1 + 5x_2 - x_3$$
$$w_2 = 4x_1 - \quad x_2 + x_3$$
$$w_3 = 3x_1 + 2x_2 - x_3$$

and then calculate $T(-1, 2, 4)$ by directly substituting in the equations and also by matrix multiplication.

4. Find the standard matrix for the linear operator T defined by the formula.

(a) $T(x_1, x_2) = (2x_1 - x_2, x_1 + x_2)$

(b) $T(x_1, x_2) = (x_1, x_2)$

(c) $T(x_1, x_2, x_3) = (x_1 + 2x_2 + x_3, x_1 + 5x_2, x_3)$

(d) $T(x_1, x_2, x_3) = (4x_1, 7x_2, -8x_3)$

5. Find the standard matrix for the linear transformation T defined by the formula.

(a) $T(x_1, x_2) = (x_2, -x_1, x_1 + 3x_2, x_1 - x_2)$

(b) $T(x_1, x_2, x_3, x_4) = (7x_1 + 2x_2 - x_3 + x_4, x_2 + x_3, -x_1)$

(c) $T(x_1, x_2, x_3) = (0, 0, 0, 0, 0)$

(d) $T(x_1, x_2, x_3, x_4) = (x_4, x_1, x_3, x_2, x_1 - x_3)$

6. In each part, the standard matrix $[T]$ of a linear transformation T is given. Use it to find $T(x)$. [Express the answers in matrix form.]

(a) $[T] = \begin{bmatrix} 1 & 2 \\ 3 & 4 \end{bmatrix}$; $\mathbf{x} = \begin{bmatrix} 3 \\ -2 \end{bmatrix}$

(b) $[T] = \begin{bmatrix} -1 & 2 & 0 \\ 3 & 1 & 5 \end{bmatrix}$; $\mathbf{x} = \begin{bmatrix} -1 \\ 1 \\ 3 \end{bmatrix}$

(c) $[T] = \begin{bmatrix} -2 & 1 & 4 \\ 3 & 5 & 7 \\ 6 & 0 & -1 \end{bmatrix}$; $\mathbf{x} = \begin{bmatrix} x_1 \\ x_2 \\ x_3 \end{bmatrix}$

(d) $[T] = \begin{bmatrix} -1 & 1 \\ 2 & 4 \\ 7 & 8 \end{bmatrix}$; $\mathbf{x} = \begin{bmatrix} x_1 \\ x_2 \end{bmatrix}$

7. In each part, use the standard matrix for T to find $T(\mathbf{x})$; then check the result by calculating $T(\mathbf{x})$ directly.

(a) $T(x_1, x_2) = (-x_1 + x_2, x_2)$; $\mathbf{x} = (-1, 4)$

(b) $T(x_1, x_2, x_3) = (2x_1 - x_2 + x_3, x_2 + x_3, 0)$; $\mathbf{x} = (2, 1, -3)$

8. Use matrix multiplication to find the reflection of $(-1, 2)$ about

(a) the x-axis   (b) the y-axis   (c) the line $y = x$

9. Use matrix multiplication to find the reflection of $(2, -5, 3)$ about

(a) the xy-plane   (b) the xz-plane   (c) the yz-plane

10. Use matrix multiplication to find the orthogonal projection of $(2, -5)$ on

(a) the x-axis   (b) the y-axis

11. Use matrix multiplication to find the orthogonal projection of $(-2, 1, 3)$ on

(a) the xy-plane   (b) the xz-plane   (c) the yz-plane

12. Use matrix multiplication to find the image of the vector $(3, -4)$ when it is rotated through an angle of

   (a) $\theta = 30°$ (b) $\theta = -60°$ (c) $\theta = 45°$ (d) $\theta = 90°$

13. Use matrix multiplication to find the image of the vector $(-2, 1, 2)$ if it is rotated

   (a) $30°$ about the x-axis   (b) $45°$ about the y-axis   (c) $90°$ about the z-axis

14. Find the standard matrix for the linear operator that rotates a vector in R^3 through an angle of $-60°$ about

   (a) the x-axis   (b) the y-axis   (c) the z-axis

15. Use matrix multiplication to find the image of the vector $(-2, 1, 2)$ if it is rotated

   (a) $-30°$ about the x-axis   (b) $-45°$ about the y-axis   (c) $-90°$ about the z-axis

16. Find the standard matrix for the stated composition of linear operators on R^2.

   (a) A rotation of $90°$, followed by a reflection about the line $y = x$.

   (b) An orthogonal projection on the y-axis, followed by a contraction with factor $k = \frac{1}{2}$.

   (c) A reflection about the x-axis, followed by a dilation with factor $k = 3$.

17. Find the standard matrix for the stated composition of linear operators on R^2.

   (a) A rotation of $60°$, followed by an orthogonal projection on the x-axis, followed by a reflection about the line $y = x$.

   (b) A dilation with factor $k = 2$, followed by a rotation of $45°$, followed by a reflection about the y-axis.

   (c) A rotation of $15°$, followed by a rotation of $105°$, followed by a rotation of $60°$.

18. Find the standard matrix for the stated composition of linear operators on R^3.

   (a) A reflection about the yz-plane, followed by an orthogonal projection on the xz-plane.

   (b) A rotation of $45°$ about the y-axis, followed by a dilation with factor $k = \sqrt{2}$.

   (c) An orthogonal projection on the xy-plane, followed by a reflection about the yz-plane.

19. Find the standard matrix for the stated composition of linear operators on R^3.

   (a) A rotation of $30°$ about the x-axis, followed by a rotation of $30°$ about the z-axis, followed by a contraction with factor $k = \frac{1}{4}$.

   (b) A reflection about the xy-plane, followed by a reflection about the xz-plane, followed by an orthogonal projection on the yz-plane.

   (c) A rotation of $270°$ about the x-axis, followed by a rotation of $90°$ about the y-axis, followed by a rotation of $180°$ about the z-axis.

20. Determine whether $T_1 \circ T_2 = T_2 \circ T_1$.

   (a) $T_1: R^2 \to R^2$ is the orthogonal projection on the x-axis, and $T_2: R^2 \to R^2$ is the orthogonal projection on the y-axis.

   (b) $T_1: R^2 \to R^2$ is the rotation through an angle θ_1, and $T_2: R^2 \to R^2$ is the rotation through an angle θ_2.

   (c) $T_1: R^2 \to R^2$ is the orthogonal projection on the x-axis, and $T_2: R^2 \to R^2$ is the rotation through an angle θ.

21. Determine whether $T_1 \circ T_2 = T_2 \circ T_1$.

   (a) $T_1: R^3 \to R^3$ is a dilation by a factor k, and $T_2: R^3 \to R^3$ is the rotation about the z-axis through an angle θ.

   (b) $T_1: R^3 \to R^3$ is the rotation about the x-axis through an angle θ_1, and $T_2: R^3 \to R^3$ is the rotation about the z-axis through an angle θ_2.

22. In R^3 the **orthogonal projections** on the x-axis, y-axis, and z-axis are defined by

$$T_1(x, y, z) = (x, 0, 0), \qquad T_2(x, y, z) = (0, y, 0), \qquad T_3(x, y, z) = (0, 0, z)$$

respectively.

(a) Show that the orthogonal projections on the coordinate axes are linear operators, and find their standard matrices.

(b) Show that if $T: R^3 \rightarrow R^3$ is an orthogonal projection on one of the coordinate axes, then for every vector $\mathbf{x}$ in R^3, the vectors $T(\mathbf{x})$ and $\mathbf{x} - T(\mathbf{x})$ are orthogonal vectors.

(c) Make a sketch showing $\mathbf{x}$ and $\mathbf{x} - T(\mathbf{x})$ in the case where T is the orthogonal projection on the x-axis.

23. Derive the standard matrices for the rotations about the x-axis, y-axis, and z-axis in R^3 from Formula (17).

24. Use Formula (17) to find the standard matrix for a rotation of $\pi/2$ radians about the axis determined by the vector $\mathbf{v} = (1, 1, 1)$.

Note Formula (17) requires that the vector defining the axis of rotation have length 1.

25. Verify Formula (21) for the given linear transformations.

(a) $T_1(x_1, x_2) = (x_1 + x_2, x_1 - x_2)$ and $T_2(x_1, x_2) = (3x_1, 2x_1 + 4x_2)$

(b) $T_1(x_1, x_2) = (4x_1, -2x_1 + x_2, -x_1 - 3x_2)$ and
$T_2(x_1, x_2, x_3) = (x_1 + 2x_2 - x_3, 4x_1 - x_3)$

(c) $T_1(x_1, x_2, x_3) = (-x_1 + x_2, -x_2 + x_3, -x_3 + x_1)$ and
$T_2(x_1, x_2, x_3) = (-2x_1, 3x_3, -4x_2)$

26. It can be proved that if A is a 2×2 matrix with $\det(A) = 1$ and such that the column vectors of A are orthogonal and have length 1, then multiplication by A is a rotation through some angle θ. Verify that

$$A = \begin{bmatrix} -1/\sqrt{2} & -1/\sqrt{2} \\ 1/\sqrt{2} & -1/\sqrt{2} \end{bmatrix}$$

satisfies the stated conditions and find the angle of rotation.

27. The result stated in Exercise 26 is also true in R^3: It can be proved that if A is a 3×3 matrix with $\det(A) = 1$ and such that the column vectors of A are pairwise orthogonal and have length 1, then multiplication by A is a rotation about some axis of rotation through some angle θ. Use Formula (17) to show that if A satisfies the stated conditions, then the angle of rotation satisfies the equation

$$\cos \theta = \frac{\text{tr}(A) - 1}{2}$$

28. Let A be a 3×3 matrix (other than the identity matrix) satisfying the conditions stated in Exercise 27. It can be shown that if $\mathbf{x}$ is any nonzero vector in R^3, then the vector $\mathbf{u} = A\mathbf{x} + A^T\mathbf{x} + [1 - \text{tr}(A)]\mathbf{x}$ determines an axis of rotation when $\mathbf{u}$ is positioned with its initial point at the origin. [See "The Axis of Rotation: Analysis, Algebra, Geometry," by Dan Kalman, *Mathematics Magazine*, Vol. 62, No. 4, October 1989.]

(a) Show that multiplication by

$$A = \begin{bmatrix} \frac{1}{9} & -\frac{4}{9} & \frac{8}{9} \\ \frac{8}{9} & \frac{4}{9} & \frac{1}{9} \\ -\frac{4}{9} & \frac{7}{9} & \frac{4}{9} \end{bmatrix}$$

is a rotation.

(b) Find a vector of length 1 that defines an axis for the rotation.

(c) Use the result in Exercise 27 to find the angle of rotation about the axis obtained in part (b).

Discussion & Discovery

29. In words, describe the geometric effect of multiplying a vector $\mathbf{x}$ by the matrix A.

(a) $A = \begin{bmatrix} 2 & 0 \\ 0 & 0 \end{bmatrix}$ (b) $A = \begin{bmatrix} 2 & 0 \\ 0 & -2 \end{bmatrix}$

30. In words, describe the geometric effect of multiplying a vector **x** by the matrix A.

(a) $A = \begin{bmatrix} 2 & 0 \\ 0 & 3 \end{bmatrix}$ (b) $A = \begin{bmatrix} \sqrt{3}/2 & -1/2 \\ 1/2 & \sqrt{3}/2 \end{bmatrix}$

31. In words, describe the geometric effect of multiplying a vector **x** by the matrix

$$A = \begin{bmatrix} \cos^2\theta - \sin^2\theta & -2\sin\theta\cos\theta \\ 2\sin\theta\cos\theta & \cos^2\theta - \sin^2\theta \end{bmatrix}$$

32. If multiplication by A rotates a vector **x** in the xy-plane through an angle θ, what is the effect of multiplying **x** by A^T? Explain your reasoning.

33. Let $\mathbf{x}_0$ be a nonzero column vector in R^2, and suppose that $T: R^2 \to R^2$ is the transformation defined by $T(\mathbf{x}) = \mathbf{x}_0 + R_\theta\mathbf{x}$, where R_θ is the standard matrix of the rotation of R^2 about the origin through the angle θ. Give a geometric description of this transformation. Is it a linear transformation? Explain.

34. A function of the form $f(x) = mx + b$ is commonly called a "linear function" because the graph of $y = mx + b$ is a line. Is f a linear transformation on R?

35. Let $\mathbf{x} = \mathbf{x}_0 + t\mathbf{v}$ be a line in R^n, and let $T: R^n \to R^n$ be a linear operator on R^n. What kind of geometric object is the image of this line under the operator T? Explain your reasoning.

4.3

PROPERTIES OF LINEAR TRANSFORMATIONS FROM R^n TO R^m

In this section we shall investigate the relationship between the invertibility of a matrix and properties of the corresponding matrix transformation. We shall also obtain a characterization of linear transformations from R^n to R^m that will form the basis for more general linear transformations to be discussed in subsequent sections, and we shall discuss some geometric properties of eigenvectors.

One-to-One Linear Transformations

Linear transformations that map distinct vectors (or points) into distinct vectors (or points) are of special importance. One example of such a transformation is the linear operator $T: R^2 \to R^2$ that rotates each vector through an angle θ. It is obvious geometrically that if **u** and **v** are distinct vectors in R^2, then so are the rotated vectors $T(\mathbf{u})$ and $T(\mathbf{v})$ (Figure 4.3.1).

In contrast, if $T: R^3 \to R^3$ is the orthogonal projection of R^3 on the xy-plane, then distinct points on the same vertical line are mapped into the same point in the xy-plane (Figure 4.3.2).

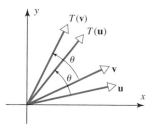

Figure 4.3.1 Distinct vectors **u** and **v** are rotated into distinct vectors $T(\mathbf{u})$ and $T(\mathbf{v})$.

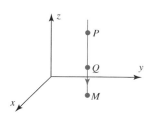

Figure 4.3.2 The distinct points P and Q are mapped into the same point M.

> ### DEFINITION
>
> A linear transformation $T: R^n \to R^m$ is said to be **one-to-one** if T maps distinct vectors (points) in R^n into distinct vectors (points) in R^m.

REMARK It follows from this definition that for each vector $\mathbf{w}$ in the range of a one-to-one linear transformation T, there is exactly one vector $\mathbf{x}$ such that $T(\mathbf{x}) = \mathbf{w}$.

EXAMPLE 1 One-to-One Linear Transformations

In the terminology of the preceding definition, the rotation operator of Figure 4.3.1 is one-to-one, but the orthogonal projection operator of Figure 4.3.2 is not. ◆

Let A be an $n \times n$ matrix, and let $T_A: R^n \to R^n$ be multiplication by A. We shall now investigate relationships between the invertibility of A and properties of T_A.

Recall from Theorem 2.3.6 (with $\mathbf{w}$ in place of $\mathbf{b}$) that the following are equivalent:

- A is invertible.
- $A\mathbf{x} = \mathbf{w}$ is consistent for every $n \times 1$ matrix $\mathbf{w}$.
- $A\mathbf{x} = \mathbf{w}$ has exactly one solution for every $n \times 1$ matrix $\mathbf{w}$.

However, the last of these statements is actually stronger than necessary. One can show that the following are equivalent (Exercise 24):

- A is invertible.
- $A\mathbf{x} = \mathbf{w}$ is consistent for every $n \times 1$ matrix $\mathbf{w}$.
- $A\mathbf{x} = \mathbf{w}$ has exactly one solution when the system is consistent.

Translating these into the corresponding statements about the linear operator T_A, we deduce that the following are equivalent:

- A is invertible.
- For every vector $\mathbf{w}$ in R^n, there is some vector $\mathbf{x}$ in R^n such that $T_A(\mathbf{x}) = \mathbf{w}$. Stated another way, the range of T_A is all of R^n.
- For every vector $\mathbf{w}$ in the range of T_A, there is exactly one vector $\mathbf{x}$ in R^n such that $T_A(\mathbf{x}) = \mathbf{w}$. Stated another way, T_A is one-to-one.

In summary, we have established the following theorem about linear operators on R^n.

THEOREM 4.3.1

> ### Equivalent Statements
>
> *If A is an $n \times n$ matrix and $T_A: R^n \to R^n$ is multiplication by A, then the following statements are equivalent.*
>
> *(a) A is invertible.*   *(b) The range of T_A is R^n.*   *(c) T_A is one-to-one.*

EXAMPLE 2 Applying Theorem 4.3.1

In Example 1 we observed that the rotation operator $T: R^2 \to R^2$ illustrated in Figure 4.3.1 is one-to-one. It follows from Theorem 4.3.1 that the range of T must be all of

R^2 and that the standard matrix for T must be invertible. To show that the range of T is all of R^2, we must show that every vector $\mathbf{w}$ in R^2 is the image of some vector $\mathbf{x}$ under T. But this is clearly so, since the vector $\mathbf{x}$ obtained by rotating $\mathbf{w}$ through the angle $-\theta$ maps into $\mathbf{w}$ when rotated through the angle θ. Moreover, from Table 6 of Section 4.2, the standard matrix for T is

$$[T] = \begin{bmatrix} \cos\theta & -\sin\theta \\ \sin\theta & \cos\theta \end{bmatrix}$$

which is invertible, since

$$\det[T] = \begin{vmatrix} \cos\theta & -\sin\theta \\ \sin\theta & \cos\theta \end{vmatrix} = \cos^2\theta + \sin^2\theta = 1 \neq 0 \quad \blacklozenge$$

EXAMPLE 3 Applying Theorem 4.3.1

In Example 1 we observed that the projection operator $T: R^3 \to R^3$ illustrated in Figure 4.3.2 is not one-to-one. It follows from Theorem 4.3.1 that the range of T is *not* all of R^3 and that the standard matrix for T is not invertible. To show directly that the range of T is not all of R^3, we must find a vector $\mathbf{w}$ in R^3 that is not the image of any vector $\mathbf{x}$ under T. But any vector $\mathbf{w}$ outside of the xy-plane has this property, since all images under T lie in the xy-plane. Moreover, from Table 5 of Section 4.2, the standard matrix for T is

$$[T] = \begin{bmatrix} 1 & 0 & 0 \\ 0 & 1 & 0 \\ 0 & 0 & 0 \end{bmatrix}$$

which is not invertible, since $\det[T] = 0$. $\quad \blacklozenge$

Inverse of a One-to-One Linear Operator

If $T_A: R^n \to R^n$ is a one-to-one linear operator, then from Theorem 4.3.1 the matrix A is invertible. Thus, $T_{A^{-1}}: R^n \to R^n$ is itself a linear operator; it is called the *inverse of* T_A. The linear operators T_A and $T_{A^{-1}}$ cancel the effect of one another in the sense that for all $\mathbf{x}$ in R^n,

$$T_A(T_{A^{-1}}(\mathbf{x})) = AA^{-1}\mathbf{x} = I\mathbf{x} = \mathbf{x}$$
$$T_{A^{-1}}(T_A(\mathbf{x})) = A^{-1}A\mathbf{x} = I\mathbf{x} = \mathbf{x}$$

or, equivalently,

$$T_A \circ T_{A^{-1}} = T_{AA^{-1}} = T_I$$
$$T_{A^{-1}} \circ T_A = T_{A^{-1}A} = T_I$$

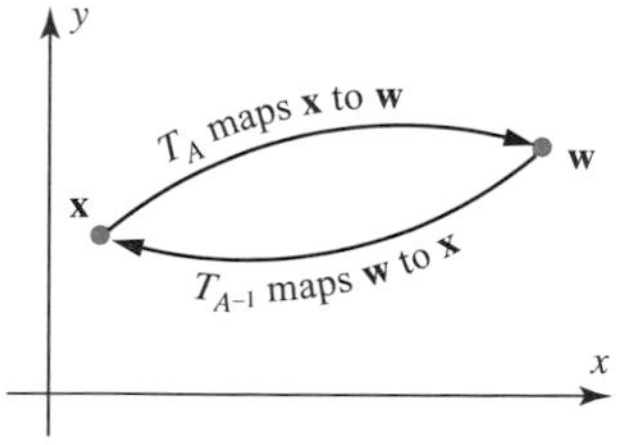

From a more geometric viewpoint, if $\mathbf{w}$ is the image of $\mathbf{x}$ under T_A, then $T_{A^{-1}}$ maps $\mathbf{w}$ back into $\mathbf{x}$, since

$$T_{A^{-1}}(\mathbf{w}) = T_{A^{-1}}(T_A(\mathbf{x})) = \mathbf{x}$$

Figure 4.3.3

(Figure 4.3.3).

Before turning to an example, it will be helpful to touch on a notational matter. When a one-to-one linear operator on R^n is written as $T: R^n \to R^n$ (rather than $T_A: R^n \to R^n$), then the inverse of the operator T is denoted by T^{-1} (rather than $T_{A^{-1}}$). Since the standard matrix for T^{-1} is the inverse of the standard matrix for T, we have

$$[T^{-1}] = [T]^{-1} \tag{1}$$

EXAMPLE 4 Standard Matrix for T^{-1}

Let $T: R^2 \to R^2$ be the operator that rotates each vector in R^2 through the angle θ, so from Table 6 of Section 4.2,

$$[T] = \begin{bmatrix} \cos\theta & -\sin\theta \\ \sin\theta & \cos\theta \end{bmatrix} \tag{2}$$

It is evident geometrically that to undo the effect of T, one must rotate each vector in R^2 through the angle $-\theta$. But this is exactly what the operator T^{-1} does, since the standard matrix for T^{-1} is

$$[T^{-1}] = [T]^{-1} = \begin{bmatrix} \cos\theta & \sin\theta \\ -\sin\theta & \cos\theta \end{bmatrix} = \begin{bmatrix} \cos(-\theta) & -\sin(-\theta) \\ \sin(-\theta) & \cos(-\theta) \end{bmatrix}$$

(verify), which is identical to (2) except that θ is replaced by $-\theta$. ◆

EXAMPLE 5 Finding T^{-1}

Show that the linear operator $T: R^2 \to R^2$ defined by the equations

$$w_1 = 2x_1 + x_2$$
$$w_2 = 3x_1 + 4x_2$$

is one-to-one, and find $T^{-1}(w_1, w_2)$.

Solution

The matrix form of these equations is

$$\begin{bmatrix} w_1 \\ w_2 \end{bmatrix} = \begin{bmatrix} 2 & 1 \\ 3 & 4 \end{bmatrix} \begin{bmatrix} x_1 \\ x_2 \end{bmatrix}$$

so the standard matrix for T is

$$[T] = \begin{bmatrix} 2 & 1 \\ 3 & 4 \end{bmatrix}$$

This matrix is invertible (so T is one-to-one) and the standard matrix for T^{-1} is

$$[T^{-1}] = [T]^{-1} = \begin{bmatrix} \frac{4}{5} & -\frac{1}{5} \\ -\frac{3}{5} & \frac{2}{5} \end{bmatrix}$$

Thus

$$[T^{-1}]\begin{bmatrix} w_1 \\ w_2 \end{bmatrix} = \begin{bmatrix} \frac{4}{5} & -\frac{1}{5} \\ -\frac{3}{5} & \frac{2}{5} \end{bmatrix}\begin{bmatrix} w_1 \\ w_2 \end{bmatrix} = \begin{bmatrix} \frac{4}{5}w_1 - \frac{1}{5}w_2 \\ -\frac{3}{5}w_1 + \frac{2}{5}w_2 \end{bmatrix}$$

from which we conclude that

$$T^{-1}(w_1, w_2) = (\tfrac{4}{5}w_1 - \tfrac{1}{5}w_2, -\tfrac{3}{5}w_1 + \tfrac{2}{5}w_2) \quad ◆$$

Linearity Properties In the preceding section we defined a transformation $T: R^n \to R^m$ to be linear if the equations relating $\mathbf{x}$ and $\mathbf{w} = T(\mathbf{x})$ are linear equations. The following theorem provides an alternative characterization of linearity. This theorem is fundamental and will be the basis for extending the concept of a linear transformation to more general settings later in this text.

THEOREM 4.3.2

> ## Properties of Linear Transformations
>
> *A transformation $T: R^n \to R^m$ is linear if and only if the following relationships hold for all vectors $\mathbf{u}$ and $\mathbf{v}$ in R^n and for every scalar c.*
>
> (a) $T(\mathbf{u} + \mathbf{v}) = T(\mathbf{u}) + T(\mathbf{v})$ (b) $T(c\mathbf{u}) = cT(\mathbf{u})$

Proof Assume first that T is a linear transformation, and let A be the standard matrix for T. It follows from the basic arithmetic properties of matrices that

$$T(\mathbf{u} + \mathbf{v}) = A(\mathbf{u} + \mathbf{v}) = A\mathbf{u} + A\mathbf{v} = T(\mathbf{u}) + T(\mathbf{v})$$

and

$$T(c\mathbf{u}) = A(c\mathbf{u}) = c(A\mathbf{u}) = cT(\mathbf{u})$$

Conversely, assume that properties (a) and (b) hold for the transformation T. We can prove that T is linear by finding a matrix A with the property that

$$T(\mathbf{x}) = A\mathbf{x} \tag{3}$$

for all vectors $\mathbf{x}$ in R^n. This will show that T is multiplication by A and therefore linear. But before we can produce this matrix, we need to observe that property (a) can be extended to three or more terms; for example, if $\mathbf{u}$, $\mathbf{v}$, and $\mathbf{w}$ are any vectors in R^n, then by first grouping $\mathbf{v}$ and $\mathbf{w}$ and applying property (a), we obtain

$$T(\mathbf{u} + \mathbf{v} + \mathbf{w}) = T(\mathbf{u} + (\mathbf{v} + \mathbf{w})) = T(\mathbf{u}) + T(\mathbf{v} + \mathbf{w}) = T(\mathbf{u}) + T(\mathbf{v}) + T(\mathbf{w})$$

More generally, for any vectors $\mathbf{v}_1, \mathbf{v}_2, \ldots, \mathbf{v}_k$ in R^n, we have

$$T(\mathbf{v}_1 + \mathbf{v}_2 + \cdots + \mathbf{v}_k) = T(\mathbf{v}_1) + T(\mathbf{v}_2) + \cdots + T(\mathbf{v}_k)$$

Now, to find the matrix A, let $\mathbf{e}_1, \mathbf{e}_2, \ldots, \mathbf{e}_n$ be the vectors

$$\mathbf{e}_1 = \begin{bmatrix} 1 \\ 0 \\ 0 \\ \vdots \\ 0 \end{bmatrix}, \quad \mathbf{e}_2 = \begin{bmatrix} 0 \\ 1 \\ 0 \\ \vdots \\ 0 \end{bmatrix}, \ldots, \quad \mathbf{e}_n = \begin{bmatrix} 0 \\ 0 \\ 0 \\ \vdots \\ 1 \end{bmatrix} \tag{4}$$

and let A be the matrix whose successive column vectors are $T(\mathbf{e}_1), T(\mathbf{e}_2), \ldots, T(\mathbf{e}_n)$; that is,

$$A = [T(\mathbf{e}_1) \mid T(\mathbf{e}_2) \mid \cdots \mid T(\mathbf{e}_n)] \tag{5}$$

If

$$\mathbf{x} = \begin{bmatrix} x_1 \\ x_2 \\ \vdots \\ x_n \end{bmatrix}$$

is any vector in R^n, then as discussed in Section 1.3, the product $A\mathbf{x}$ is a linear combination of the column vectors of A with coefficients from $\mathbf{x}$, so

$$\begin{aligned} A\mathbf{x} &= x_1 T(\mathbf{e}_1) + x_2 T(\mathbf{e}_2) + \cdots + x_n T(\mathbf{e}_n) \\ &= T(x_1\mathbf{e}_1) + T(x_2\mathbf{e}_2) + \cdots + T(x_n\mathbf{e}_n) \quad \longleftarrow \text{Property } (b) \\ &= T(x_1\mathbf{e}_1 + x_2\mathbf{e}_2 + \cdots + x_n\mathbf{e}_n) \quad \longleftarrow \text{Property } (a) \text{ for } n \text{ terms} \\ &= T(\mathbf{x}) \end{aligned}$$

which completes the proof. ∎

Expression (5) is important in its own right, since it provides an explicit formula for the standard matrix of a linear operator $T: R^n \to R^m$ in terms of the images of the vectors $\mathbf{e}_1, \mathbf{e}_2, \ldots, \mathbf{e}_n$ under T. For reasons that will be discussed later, the vectors $\mathbf{e}_1, \mathbf{e}_2, \ldots, \mathbf{e}_n$ in (4) are called the **standard basis** vectors for R^n. In R^2 and R^3 these are the vectors of length 1 along the coordinate axes (Figure 4.3.4).

Because of its importance, we shall state (5) as a theorem for future reference.

THEOREM 4.3.3

If $T: R^n \to R^m$ is a linear transformation, and $\mathbf{e}_1, \mathbf{e}_2, \ldots, \mathbf{e}_n$ are the standard basis vectors for R^n, then the standard matrix for T is

$$[T] = [T(\mathbf{e}_1) \mid T(\mathbf{e}_2) \mid \cdots \mid T(\mathbf{e}_n)] \tag{6}$$

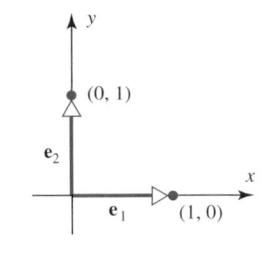

(a) Standard basis for R^2

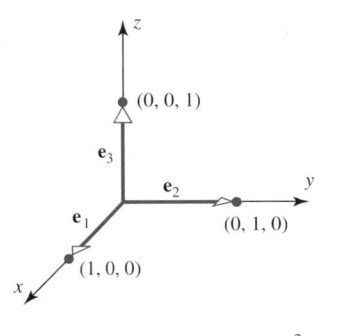

(b) Standard basis for R^3

Figure 4.3.4

Formula (6) is a powerful tool for finding standard matrices and analyzing the geometric effect of a linear transformation. For example, suppose that $T: R^3 \to R^3$ is the orthogonal projection on the xy-plane. Referring to Figure 4.3.4, it is evident geometrically that

$$T(\mathbf{e}_1) = \mathbf{e}_1 = \begin{bmatrix} 1 \\ 0 \\ 0 \end{bmatrix}, \qquad T(\mathbf{e}_2) = \mathbf{e}_2 = \begin{bmatrix} 0 \\ 1 \\ 0 \end{bmatrix}, \qquad T(\mathbf{e}_3) = \mathbf{0} = \begin{bmatrix} 0 \\ 0 \\ 0 \end{bmatrix}$$

so by (6),

$$[T] = \begin{bmatrix} 1 & 0 & 0 \\ 0 & 1 & 0 \\ 0 & 0 & 0 \end{bmatrix}$$

which agrees with the result in Table 5 of Section 4.2.

Using (6) another way, suppose that $T_A: R^3 \to R^2$ is multiplication by

$$A = \begin{bmatrix} -1 & 2 & 1 \\ 3 & 0 & 6 \end{bmatrix}$$

The images of the standard basis vectors can be read directly from the columns of the matrix A:

$$T_A \left(\begin{bmatrix} 1 \\ 0 \\ 0 \end{bmatrix} \right) = \begin{bmatrix} -1 \\ 3 \end{bmatrix}, \qquad T_A \left(\begin{bmatrix} 0 \\ 1 \\ 0 \end{bmatrix} \right) = \begin{bmatrix} 2 \\ 0 \end{bmatrix}, \qquad T_A \left(\begin{bmatrix} 0 \\ 0 \\ 1 \end{bmatrix} \right) = \begin{bmatrix} 1 \\ 6 \end{bmatrix}$$

EXAMPLE 6 Standard Matrix for a Projection Operator

Let l be the line in the xy-plane that passes through the origin and makes an angle θ with the positive x-axis, where $0 \le \theta < \pi$. As illustrated in Figure 4.3.5a, let $T: R^2 \to R^2$ be a linear operator that maps each vector into its orthogonal projection on l.

(a) Find the standard matrix for T.

(b) Find the orthogonal projection of the vector $\mathbf{x} = (1, 5)$ onto the line through the origin that makes an angle of $\theta = \pi/6$ with the positive x-axis.

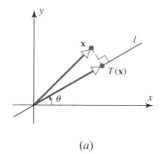

(a)

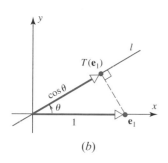

(b)

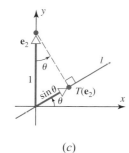

(c)

Figure 4.3.5

Solution (a)

From (6),

$$[T] = [T(\mathbf{e}_1) \mid T(\mathbf{e}_2)]$$

where $\mathbf{e}_1$ and $\mathbf{e}_2$ are the standard basis vectors for R^2. We consider the case where $0 \le \theta \le \pi/2$; the case where $\pi/2 < \theta < \pi$ is similar. Referring to Figure 4.3.5b, we have $\|T(\mathbf{e}_1)\| = \cos\theta$, so

$$T(\mathbf{e}_1) = \begin{bmatrix} \|T(\mathbf{e}_1)\|\cos\theta \\ \|T(\mathbf{e}_1)\|\sin\theta \end{bmatrix} = \begin{bmatrix} \cos^2\theta \\ \sin\theta\cos\theta \end{bmatrix}$$

and referring to Figure 4.3.5c, we have $\|T(\mathbf{e}_2)\| = \sin\theta$, so

$$T(\mathbf{e}_2) = \begin{bmatrix} \|T(\mathbf{e}_2)\|\cos\theta \\ \|T(\mathbf{e}_2)\|\sin\theta \end{bmatrix} = \begin{bmatrix} \sin\theta\cos\theta \\ \sin^2\theta \end{bmatrix}$$

Thus the standard matrix for T is

$$[T] = \begin{bmatrix} \cos^2\theta & \sin\theta\cos\theta \\ \sin\theta\cos\theta & \sin^2\theta \end{bmatrix}$$

Solution (b)

Since $\sin\pi/6 = 1/2$ and $\cos\pi/6 = \sqrt{3}/2$, it follows from part (a) that the standard matrix for this projection operator is

$$[T] = \begin{bmatrix} 3/4 & \sqrt{3}/4 \\ \sqrt{3}/4 & 1/4 \end{bmatrix}$$

Thus

$$T\left(\begin{bmatrix} 1 \\ 5 \end{bmatrix}\right) = \begin{bmatrix} 3/4 & \sqrt{3}/4 \\ \sqrt{3}/4 & 1/4 \end{bmatrix}\begin{bmatrix} 1 \\ 5 \end{bmatrix} = \begin{bmatrix} \dfrac{3 + 5\sqrt{3}}{4} \\ \dfrac{\sqrt{3} + 5}{4} \end{bmatrix}$$

or, in point notation,

$$T(1, 5) = \left(\frac{3 + 5\sqrt{3}}{4}, \frac{\sqrt{3} + 5}{4}\right) \quad \blacklozenge$$

Geometric Interpretation of Eigenvectors

Recall from Section 2.3 that if A is an $n \times n$ matrix, then λ is called an *eigenvalue* of A if there is a nonzero vector $\mathbf{x}$ such that

$$A\mathbf{x} = \lambda\mathbf{x} \quad \text{or, equivalently,} \quad (\lambda I - A)\mathbf{x} = \mathbf{0}$$

The nonzero vectors $\mathbf{x}$ satisfying this equation are called the *eigenvectors* of A corresponding to λ.

Eigenvalues and eigenvectors can also be defined for linear operators on R^n; the definitions parallel those for matrices.

DEFINITION

If $T: R^n \to R^n$ is a linear operator, then a scalar λ is called an ***eigenvalue of T*** if there is a nonzero $\mathbf{x}$ in R^n such that

$$T(\mathbf{x}) = \lambda\mathbf{x} \tag{7}$$

Those nonzero vectors $\mathbf{x}$ that satisfy this equation are called the ***eigenvectors of T corresponding to*** λ.

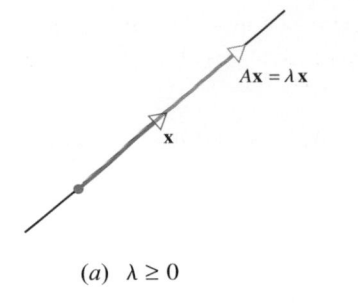

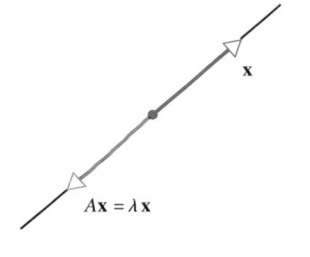

Figure 4.3.6

Observe that if A is the standard matrix for T, then (7) can be written as

$$A\mathbf{x} = \lambda\mathbf{x}$$

from which it follows that

- The eigenvalues of T are precisely the eigenvalues of its standard matrix A.
- $\mathbf{x}$ is an eigenvector of T corresponding to λ if and only if $\mathbf{x}$ is an eigenvector of A corresponding to λ.

If λ is an eigenvalue of A and $\mathbf{x}$ is a corresponding eigenvector, then $A\mathbf{x} = \lambda\mathbf{x}$, so multiplication by A maps $\mathbf{x}$ into a scalar multiple of itself. In R^2 and R^3, this means that *multiplication by A maps each eigenvector $\mathbf{x}$ into a vector that lies on the same line as $\mathbf{x}$* (Figure 4.3.6).

Recall from Section 4.2 that if $\lambda \geq 0$, then the linear operator $A\mathbf{x} = \lambda\mathbf{x}$ compresses $\mathbf{x}$ by a factor of λ if $0 \leq \lambda \leq 1$ or stretches $\mathbf{x}$ by a factor of λ if $\lambda \geq 1$. If $\lambda < 0$, then $A\mathbf{x} = \lambda\mathbf{x}$ reverses the direction of $\mathbf{x}$ and compresses the reversed vector by a factor of $|\lambda|$ if $0 \leq |\lambda| \leq 1$ or stretches the reversed vector by a factor of $|\lambda|$ if $|\lambda| \geq 1$ (Figure 4.3.7).

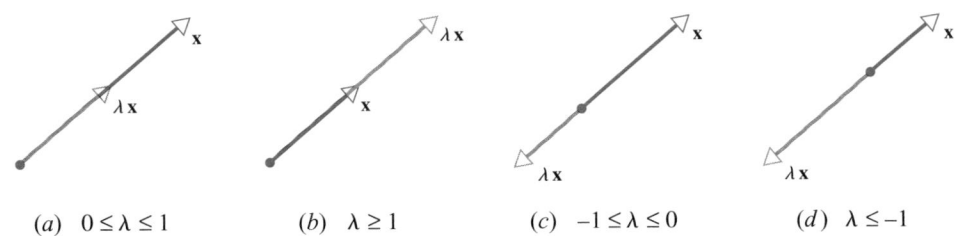

(a) $0 \leq \lambda \leq 1$ (b) $\lambda \geq 1$ (c) $-1 \leq \lambda \leq 0$ (d) $\lambda \leq -1$

Figure 4.3.7

EXAMPLE 7 Eigenvalues of a Linear Operator

Let $T: R^2 \to R^2$ be the linear operator that rotates each vector through an angle θ. It is evident geometrically that unless θ is a multiple of π, T does not map any nonzero vector $\mathbf{x}$ onto the same line as $\mathbf{x}$; consequently, T has no real eigenvalues. But if θ *is* a multiple of π, then every nonzero vector $\mathbf{x}$ is mapped onto the same line as $\mathbf{x}$, so *every* nonzero vector is an eigenvector of T. Let us verify these geometric observations algebraically. The standard matrix for T is

$$A = \begin{bmatrix} \cos\theta & -\sin\theta \\ \sin\theta & \cos\theta \end{bmatrix}$$

As discussed in Section 2.3, the eigenvalues of this matrix are the solutions of the characteristic equation

$$\det(\lambda I - A) = \begin{vmatrix} \lambda - \cos\theta & \sin\theta \\ -\sin\theta & \lambda - \cos\theta \end{vmatrix} = 0$$

that is,

$$(\lambda - \cos\theta)^2 + \sin^2\theta = 0 \tag{8}$$

But if θ is not a multiple of π, then $\sin^2\theta > 0$, so this equation has no real solution for λ, and consequently A has no real eigenvalues.[†] If θ *is* a multiple of π, then $\sin\theta = 0$ and

---

[†]There are applications that require complex scalars and vectors with complex components. In such cases, complex eigenvalues and eigenvectors with complex components are allowed. However, these have no direct geometric significance here. In later chapters we will discuss such eigenvalues and eigenvectors, but until explicitly stated otherwise, it will be assumed that only real eigenvalues and eigenvectors with real components are to be considered.

either $\cos\theta = 1$ or $\cos\theta = -1$, depending on the particular multiple of π. In the case where $\sin\theta = 0$ and $\cos\theta = 1$, the characteristic equation (8) becomes $(\lambda - 1)^2 = 0$, so $\lambda = 1$ is the only eigenvalue of A. In this case the matrix A is

$$A = \begin{bmatrix} 1 & 0 \\ 0 & 1 \end{bmatrix} = I$$

Thus, for all $\mathbf{x}$ in R^2,

$$T(\mathbf{x}) = A\mathbf{x} = I\mathbf{x} = \mathbf{x}$$

so T maps every vector to itself, and hence to the same line. In the case where $\sin\theta = 0$ and $\cos\theta = -1$, the characteristic equation (8) becomes $(\lambda + 1)^2 = 0$, so $\lambda = -1$ is the only eigenvalue of A. In this case the matrix A is

$$A = \begin{bmatrix} -1 & 0 \\ 0 & -1 \end{bmatrix} = -I$$

Thus, for all $\mathbf{x}$ in R^2,

$$T(\mathbf{x}) = A\mathbf{x} = -I\mathbf{x} = -\mathbf{x}$$

so T maps every vector to its negative, and hence to the same line as $\mathbf{x}$. ◆

EXAMPLE 8 Eigenvalues of a Linear Operator

Let $T: R^3 \to R^3$ be the orthogonal projection on the xy-plane. Vectors in the xy-plane are mapped into themselves under T, so each nonzero vector in the xy-plane is an eigenvector corresponding to the eigenvalue $\lambda = 1$. Every vector $\mathbf{x}$ along the z-axis is mapped into $\mathbf{0}$ under T, which is on the same line as $\mathbf{x}$, so every nonzero vector on the z-axis is an eigenvector corresponding to the eigenvalue $\lambda = 0$. Vectors that are not in the xy-plane or along the z-axis are not mapped into scalar multiples of themselves, so there are no other eigenvectors or eigenvalues.

To verify these geometric observations algebraically, recall from Table 5 of Section 4.2 that the standard matrix for T is

$$A = \begin{bmatrix} 1 & 0 & 0 \\ 0 & 1 & 0 \\ 0 & 0 & 0 \end{bmatrix}$$

The characteristic equation of A is

$$\det(\lambda I - A) = \begin{vmatrix} \lambda - 1 & 0 & 0 \\ 0 & \lambda - 1 & 0 \\ 0 & 0 & \lambda \end{vmatrix} = 0 \quad \text{or} \quad (\lambda - 1)^2 \lambda = 0$$

which has the solutions $\lambda = 0$ and $\lambda = 1$ anticipated above.

As discussed in Section 2.3, the eigenvectors of the matrix A corresponding to an eigenvalue λ are the nonzero solutions of

$$\begin{bmatrix} \lambda - 1 & 0 & 0 \\ 0 & \lambda - 1 & 0 \\ 0 & 0 & \lambda \end{bmatrix} \begin{bmatrix} x_1 \\ x_2 \\ x_3 \end{bmatrix} = \begin{bmatrix} 0 \\ 0 \\ 0 \end{bmatrix} \tag{9}$$

If $\lambda = 0$, this system is

$$\begin{bmatrix} -1 & 0 & 0 \\ 0 & -1 & 0 \\ 0 & 0 & 0 \end{bmatrix} \begin{bmatrix} x_1 \\ x_2 \\ x_3 \end{bmatrix} = \begin{bmatrix} 0 \\ 0 \\ 0 \end{bmatrix}$$

which has the solutions $x_1 = 0$, $x_2 = 0$, $x_3 = t$ (verify), or, in matrix form,

$$\begin{bmatrix} x_1 \\ x_2 \\ x_3 \end{bmatrix} = \begin{bmatrix} 0 \\ 0 \\ t \end{bmatrix}$$

As anticipated, these are the vectors along the z-axis. If $\lambda = 1$, then system (9) is

$$\begin{bmatrix} 0 & 0 & 0 \\ 0 & 0 & 0 \\ 0 & 0 & 1 \end{bmatrix} \begin{bmatrix} x_1 \\ x_2 \\ x_3 \end{bmatrix} = \begin{bmatrix} 0 \\ 0 \\ 0 \end{bmatrix}$$

which has the solutions $x_1 = s$, $x_2 = t$, $x_3 = 0$ (verify), or, in matrix form,

$$\begin{bmatrix} x_1 \\ x_2 \\ x_3 \end{bmatrix} = \begin{bmatrix} s \\ t \\ 0 \end{bmatrix}$$

As anticipated, these are the vectors in the xy-plane. ◆

Summary

In Theorem 2.3.6 we listed six results that are equivalent to the invertibility of a matrix A. We conclude this section by merging Theorem 4.3.1 with that list to produce the following theorem that relates all of the major topics we have studied thus far.

THEOREM 4.3.4

Equivalent Statements

If A is an $n \times n$ matrix, and if $T_A: R^n \to R^n$ is multiplication by A, then the following are equivalent.

(a) *A is invertible.*
(b) *$A\mathbf{x} = \mathbf{0}$ has only the trivial solution.*
(c) *The reduced row-echelon form of A is I_n.*
(d) *A is expressible as a product of elementary matrices.*
(e) *$A\mathbf{x} = \mathbf{b}$ is consistent for every $n \times 1$ matrix $\mathbf{b}$.*
(f) *$A\mathbf{x} = \mathbf{b}$ has exactly one solution for every $n \times 1$ matrix $\mathbf{b}$.*
(g) *$\det(A) \neq 0$.*
(h) *The range of T_A is R^n.*
(i) *T_A is one-to-one.*

EXERCISE SET 4.3

1. By inspection, determine whether the linear operator is one-to-one.

   (a) the orthogonal projection on the x-axis in R^2

   (b) the reflection about the y-axis in R^2

   (c) the reflection about the line $y = x$ in R^2

   (d) a contraction with factor $k > 0$ in R^2

   (e) a rotation about the z-axis in R^3

   (f) a reflection about the xy-plane in R^3

   (g) a dilation with factor $k > 0$ in R^3

2. Find the standard matrix for the linear operator defined by the equations, and use Theorem 4.3.4 to determine whether the operator is one-to-one.

   (a) $w_1 = 8x_1 + 4x_2$
   $w_2 = 2x_1 + x_2$

   (b) $w_1 = 2x_1 - 3x_2$
   $w_2 = 5x_1 + x_2$

   (c) $w_1 = -x_1 + 3x_2 + 2x_3$
   $w_2 = 2x_1 \quad\quad + 4x_3$
   $w_3 = x_1 + 3x_2 + 6x_3$

   (d) $w_1 = x_1 + 2x_2 + 3x_3$
   $w_2 = 2x_1 + 5x_2 + 3x_3$
   $w_3 = x_1 \quad\quad + 8x_3$

3. Show that the range of the linear operator defined by the equations

$$w_1 = 4x_1 - 2x_2$$
$$w_2 = 2x_1 - x_2$$

   is not all of R^2, and find a vector that is not in the range.

4. Show that the range of the linear operator defined by the equations

$$w_1 = x_1 - 2x_2 + x_3$$
$$w_2 = 5x_1 - x_2 + 3x_3$$
$$w_3 = 4x_1 + x_2 + 2x_3$$

   is not all of R^3, and find a vector that is not in the range.

5. Determine whether the linear operator $T: R^2 \to R^2$ defined by the equations is one-to-one; if so, find the standard matrix for the inverse operator, and find $T^{-1}(w_1, w_2)$.

   (a) $w_1 = x_1 + 2x_2$
   $w_2 = -x_1 + x_2$

   (b) $w_1 = 4x_1 - 6x_2$
   $w_2 = -2x_1 + 3x_2$

   (c) $w_1 = -x_2$
   $w_2 = -x_1$

   (d) $w_1 = 3x_1$
   $w_2 - -5x_1$

6. Determine whether the linear operator $T: R^3 \to R^3$ defined by the equations is one-to-one; if so, find the standard matrix for the inverse operator, and find $T^{-1}(w_1, w_2, w_3)$.

   (a) $w_1 = x_1 - 2x_2 + 2x_3$
   $w_2 = 2x_1 + x_2 + x_3$
   $w_3 = x_1 + x_2$

   (b) $w_1 = x_1 - 3x_2 + 4x_3$
   $w_2 = -x_1 + x_2 + x_3$
   $w_3 = \quad - 2x_2 + 5x_3$

   (c) $w_1 = x_1 + 4x_2 - x_3$
   $w_2 = 2x_1 + 7x_2 + x_3$
   $w_3 = x_1 + 3x_2$

   (d) $w_1 = x_1 + 2x_2 + x_3$
   $w_2 = -2x_1 + x_2 + 4x_3$
   $w_3 = 7x_1 + 4x_2 - 5x_3$

7. By inspection, determine the inverse of the given one-to-one linear operator.

   (a) the reflection about the x-axis in R^2

   (b) the rotation through an angle of $\pi/4$ in R^2

   (c) the dilation by a factor of 3 in R^2

   (d) the reflection about the yz-plane in R^3

   (e) the contraction by a factor of $\frac{1}{5}$ in R^3

In Exercises 8 and 9 use Theorem 4.3.2 to determine whether $T: R^2 \to R^2$ is a linear operator.

8. (a) $T(x, y) = (2x, y)$
   (b) $T(x, y) = (x^2, y)$
   (c) $T(x, y) = (-y, x)$
   (d) $T(x, y) = (x, 0)$

9. (a) $T(x, y) = (2x + y, x - y)$
   (b) $T(x, y) = (x + 1, y)$
   (c) $T(x, y) = (y, y)$
   (d) $T(x, y) = (\sqrt[3]{x}, \sqrt[3]{y})$

In Exercises 10 and 11 use Theorem 4.3.2 to determine whether $T: R^3 \to R^2$ is a linear transformation.

10. (a) $T(x, y, z) = (x, x + y + z)$
    (b) $T(x, y, z) = (1, 1)$

11. (a) $T(x, y, z) = (0, 0)$
    (b) $T(x, y, z) = (3x - 4y, 2x - 5z)$

12. In each part, use Theorem 4.3.3 to find the standard matrix for the linear operator from the images of the standard basis vectors.

    (a) the reflection operators on R^2 in Table 2 of Section 4.2

    (b) the reflection operators on R^3 in Table 3 of Section 4.2

    (c) the projection operators on R^2 in Table 4 of Section 4.2

    (d) the projection operators on R^3 in Table 5 of Section 4.2

    (e) the rotation operators on R^2 in Table 6 of Section 4.2

    (f) the dilation and contraction operators on R^3 in Table 9 of Section 4.2

13. Use Theorem 4.3.3 to find the standard matrix for $T: R^2 \rightarrow R^2$ from the images of the standard basis vectors.

    (a) $T: R^2 \rightarrow R^2$ projects a vector orthogonally onto the x-axis and then reflects that vector about the y-axis.

    (b) $T: R^2 \rightarrow R^2$ reflects a vector about the line $y = x$ and then reflects that vector about the x-axis.

    (c) $T: R^2 \rightarrow R^2$ dilates a vector by a factor of 3, then reflects that vector about the line $y = x$, and then projects that vector orthogonally onto the y-axis.

14. Use Theorem 4.3.3 to find the standard matrix for $T: R^3 \rightarrow R^3$ from the images of the standard basis vectors.

    (a) $T: R^3 \rightarrow R^3$ reflects a vector about the xz-plane and then contracts that vector by a factor of $\frac{1}{5}$.

    (b) $T: R^3 \rightarrow R^3$ projects a vector orthogonally onto the xz-plane and then projects that vector orthogonally onto the xy-plane.

    (c) $T: R^3 \rightarrow R^3$ reflects a vector about the xy-plane, then reflects that vector about the xz-plane, and then reflects that vector about the yz-plane.

15. Let $T_A: R^3 \rightarrow R^3$ be multiplication by

$$A = \begin{bmatrix} -1 & 3 & 0 \\ 2 & 1 & 2 \\ 4 & 5 & -3 \end{bmatrix}$$

and let $\mathbf{e}_1$, $\mathbf{e}_2$, and $\mathbf{e}_3$ be the standard basis vectors for R^3. Find the following vectors by inspection.

    (a) $T_A(\mathbf{e}_1)$, $T_A(\mathbf{e}_2)$, and $T_A(\mathbf{e}_3)$ (b) $T_A(\mathbf{e}_1 + \mathbf{e}_2 + \mathbf{e}_3)$ (c) $T_A(7\mathbf{e}_3)$

16. Determine whether multiplication by A is a one-to-one linear transformation.

    (a) $A = \begin{bmatrix} 1 & -1 \\ 2 & 0 \\ 3 & -4 \end{bmatrix}$ (b) $A = \begin{bmatrix} 1 & 2 & 3 \\ -1 & 0 & -4 \end{bmatrix}$ (c) $\begin{bmatrix} 1 & 2 & 1 \\ 0 & 1 & 1 \\ 1 & 1 & 0 \\ 1 & 0 & -1 \end{bmatrix}$

17. Use the result in Example 6 to find the orthogonal projection of $\mathbf{x}$ onto the line through the origin that makes an angle θ with the positive x-axis.

    (a) $\mathbf{x} = (-1, 2)$; $\theta = 45°$ (b) $\mathbf{x} = (1, 0)$; $\theta = 30°$ (c) $\mathbf{x} = (1, 5)$; $\theta = 120°$

18. Use the type of argument given in Example 8 to find the eigenvalues and corresponding eigenvectors of T. Check your conclusions by calculating the eigenvalues and corresponding eigenvectors from the standard matrix for T.

    (a) $T: R^2 \rightarrow R^2$ is the reflection about the x-axis.

    (b) $T: R^2 \rightarrow R^2$ is the reflection about the line $y = x$.

    (c) $T: R^2 \rightarrow R^2$ is the orthogonal projection on the x-axis.

    (d) $T: R^2 \rightarrow R^2$ is the contraction by a factor of $\frac{1}{2}$.

19. Follow the directions of Exercise 18.

    (a) $T: R^3 \rightarrow R^3$ is the reflection about the yz-plane.

    (b) $T: R^3 \rightarrow R^3$ is the orthogonal projection on the xz-plane.

    (c) $T: R^3 \rightarrow R^3$ is the dilation by a factor of 2.

    (d) $T: R^3 \rightarrow R^3$ is a rotation of $\pi/4$ about the z-axis.

20. (a) Is a composition of one-to-one linear transformations one-to-one? Justify your conclusion.

    (b) Can the composition of a one-to-one linear transformation and a linear transformation that is not one-to-one be one-to-one? Account for both possible orders of composition and justify your conclusion.

21. Show that $T(x, y) = (0, 0)$ defines a linear operator on R^2 but $T(x, y) = (1, 1)$ does not.

22. (a) Prove that if $T: R^n \rightarrow R^m$ is a linear transformation, then $T(\mathbf{0}) = \mathbf{0}$—that is, T maps the zero vector in R^n into the zero vector in R^m.

    (b) The converse of this is not true. Find an example of a function that satisfies $T(\mathbf{0}) = \mathbf{0}$ but is not a linear transformation.

23. Let l be the line in the xy-plane that passes through the origin and makes an angle θ with the positive x-axis, where $0 \le \theta < \pi$. Let $T: R^2 \rightarrow R^2$ be the linear operator that reflects each vector about l (see the accompanying figure).

    (a) Use the method of Example 6 to find the standard matrix for T.

    (b) Find the reflection of the vector $\mathbf{x} = (1, 5)$ about the line l through the origin that makes an angle of $\theta = 30°$ with the positive x-axis.

24. Prove: An $n \times n$ matrix A is invertible if and only if the linear system $A\mathbf{x} = \mathbf{w}$ has exactly one solution for every vector $\mathbf{w}$ in R^n for which the system is consistent.

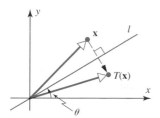

Figure Ex-23

Discussion & Discovery

25. Indicate whether each statement is always true or sometimes false. Justify your answer by giving a logical argument or a counterexample.

    (a) If T maps R^n into R^m, and $T(\mathbf{0}) = \mathbf{0}$, then T is linear.

    (b) If $T: R^n \rightarrow R^m$ is a one-to-one linear transformation, then there are no distinct vectors $\mathbf{u}$ and $\mathbf{v}$ in R^n such that $T(\mathbf{u} - \mathbf{v}) = \mathbf{0}$.

    (c) If $T: R^n \rightarrow R^n$ is a linear operator, and if $T(\mathbf{x}) = 2\mathbf{x}$ for some vector $\mathbf{x}$, then $\lambda = 2$ is an eigenvalue of T.

    (d) If T maps R^n into R^m, and if $T(c_1\mathbf{u} + c_2\mathbf{v}) = c_1 T(\mathbf{u}) + c_2 T(\mathbf{v})$ for all scalars c_1 and c_2 and for all vectors $\mathbf{u}$ and $\mathbf{v}$ in R^n, then T is linear.

26. Indicate whether each statement is always true, sometimes true, or always false.

    (a) If $T: R^n \rightarrow R^m$ is a linear transformation and $m > n$, then T is one-to-one.

    (b) If $T: R^n \rightarrow R^m$ is a linear transformation and $m < n$, then T is one-to-one.

    (c) If $T: R^n \rightarrow R^m$ is a linear transformation and $m = n$, then T is one-to-one.

27. Let A be an $n \times n$ matrix such that $\det(A) = 0$, and let $T: R^n \rightarrow R^n$ be multiplication by A.

    (a) What can you say about the range of the linear operator T? Give an example that illustrates your conclusion.

    (b) What can you say about the number of vectors that T maps into $\mathbf{0}$?

28. In each part, make a conjecture about the eigenvectors and eigenvalues of the matrix A corresponding to the given transformation by considering the geometric properties of multiplication by A. Confirm each of your conjectures with computations.

    (a) Reflection about the line $y = c$.   (b) Contraction by a factor of $\frac{1}{2}$.

4.4
LINEAR TRANS-FORMATIONS AND POLYNOMIALS

In this section we shall apply our new knowledge of linear transformations to polynomials. This is the beginning of a general strategy of using our ideas about R^n to solve problems that are in different, yet somehow analogous, settings.

Polynomials and Vectors

Suppose that we have a polynomial function, say

$$p(x) = ax^2 + bx + c$$

where x is a real-valued variable. To form the related function $2p(x)$, we multiply each of its coefficients by 2:

$$2p(x) = 2ax^2 + 2bx + 2c$$

That is, if the coefficients of the polynomial $p(x)$ are a, b, c in descending order of the power of x with which they are associated, then $2p(x)$ is also a polynomial, and its coefficients are $2a$, $2b$, $2c$ in the same order.

Similarly, if $q(x) = dx^2 + ex + f$ is another polynomial function, then $p(x) + q(x)$ is also a polynomial, and its coefficients are $a + d$, $b + e$, $c + f$. We add polynomials by adding corresponding coefficients.

This suggests that associating a polynomial with the vector consisting of its coefficients may be useful.

EXAMPLE 1 Correspondence between Polynomials and Vectors

Consider the quadratic function $p(x) = ax^2 + bx + c$. Define the vector

$$\mathbf{z} = \begin{bmatrix} a \\ b \\ c \end{bmatrix}$$

consisting of the coefficients of this polynomial in descending order of the corresponding power of x. Then multiplication of $p(x)$ by a scalar s gives $sp(x) = sax^2 + sbx + sc$, and this corresponds exactly to the scalar multiple

$$s\mathbf{z} = \begin{bmatrix} sa \\ sb \\ sc \end{bmatrix}$$

of $\mathbf{z}$. Similarly, $p(x) + p(x)$ is $2ax^2 + 2bx + 2c$, and this corresponds exactly to the vector sum $\mathbf{z} + \mathbf{z}$:

$$\mathbf{z} + \mathbf{z} = \begin{bmatrix} a \\ b \\ c \end{bmatrix} + \begin{bmatrix} a \\ b \\ c \end{bmatrix}$$

$$= \begin{bmatrix} 2a \\ 2b \\ 2c \end{bmatrix} \quad \blacklozenge$$

In general, given a polynomial $p(x) = a_n x^n + a_{n-1} x^{n-1} + \cdots + a_1 x + a_0$, we associate with it the vector

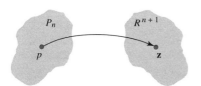

Figure 4.4.1 The vector z is associated with the polynomial p.

$$\mathbf{z} = \begin{bmatrix} a_n \\ a_{n-1} \\ \vdots \\ a_1 \\ a_0 \end{bmatrix}$$

in R^{n+1} (Figure 4.4.1). It is then possible to view operations like $p(x) \to 2p(x)$ as being equivalent to a linear transformation on R^{n+1}, namely $T(\mathbf{z}) = 2\mathbf{z}$. We can perform the desired operations in R^{n+1} rather than on the polynomials themselves.

EXAMPLE 2   Addition of Polynomials by Adding Vectors

Let $p(x) = 4x^3 - 2x + 1$ and $q(x) = 3x^3 - 3x^2 + x$. Then to compute $r(x) = 4p(x) - 2q(x)$, we could define

$$\mathbf{u} = \begin{bmatrix} 4 \\ 0 \\ -2 \\ 1 \end{bmatrix}, \qquad \mathbf{v} = \begin{bmatrix} 3 \\ -3 \\ 1 \\ 0 \end{bmatrix}$$

and perform the corresponding operation on these vectors:

$$4\mathbf{u} - 2\mathbf{v} = 4\begin{bmatrix} 4 \\ 0 \\ -2 \\ 1 \end{bmatrix} - 2\begin{bmatrix} 3 \\ -3 \\ 1 \\ 0 \end{bmatrix}$$

$$= \begin{bmatrix} 10 \\ 6 \\ -10 \\ 4 \end{bmatrix}$$

Hence $r(x) = 10x^3 + 6x^2 - 10x + 4$. ◆

This association between polynomials of degree n and vectors in R^{n+1} would be useful for someone writing a computer program to perform polynomial computations, as in a computer algebra system. The coefficients of polynomial functions could be stored as vectors, and computations could be performed on these vectors.

For convenience, we define P_n to be the set of all polynomials of degree at most n (including the zero polynomial, all the coefficients of which are zero). This is also called the *space* of polynomials of degree at most n. The use of the word *space* indicates that this set has some sort of structure to it. The structure of P_n will be explored in Chapter 8.

Calculus Required   ### EXAMPLE 3   Differentiation of Polynomials

---

Differentiation takes polynomials of degree n to polynomials of degree $n - 1$, so the corresponding transformation on vectors must take vectors in R^{n+1} to vectors in R^n. Hence, if differentiation corresponds to a linear transformation, it must be represented by a $n \times (n + 1)$ matrix. For example, if p is an element of P_2—that is,

$$p(x) = ax^2 + bx + c$$

for some real numbers a, b, and c—then

$$\frac{d}{dx} p(x) = 2ax + b$$

Evidently, if $p(x)$ in P_2 corresponds to the vector (a, b, c) in R^3, then its derivative is in P_1 and corresponds to the vector $(2a, b)$ in R^2. Note that

$$\begin{bmatrix} 2a \\ b \end{bmatrix} = \begin{bmatrix} 2 & 0 & 0 \\ 0 & 1 & 0 \end{bmatrix} \begin{bmatrix} a \\ b \\ c \end{bmatrix}$$

The operation differentiation, $D: P_2 \rightarrow P_1$, corresponds to a linear transformation $T_A: R^3 \rightarrow R^2$, where

$$A = \begin{bmatrix} 2 & 0 & 0 \\ 0 & 1 & 0 \end{bmatrix} \blacklozenge$$

Some transformations from P_n to P_m do not correspond to linear transformations from R^{n+1} to R^{m+1}. For example, if we consider the transformation of $ax^2 + bx + c$ in P_2 to $|a|$ in P_0, the space of all constants (viewed as polynomials of degree zero, plus the zero polynomial), then we find that there is no matrix that maps (a, b, c) in R^3 to $|a|$ in R. Other transformations may correspond to transformations that are not *quite* linear, in the following sense.

> **DEFINITION**
>
> An ***affine transformation*** from R^n to R^m is a mapping of the form $S(\mathbf{u}) = T(\mathbf{u}) + \mathbf{f}$, where T is a linear transformation from R^n to R^m and $\mathbf{f}$ is a (constant) vector in R^m.

The affine transformation S is a linear transformation if $\mathbf{f}$ is the zero vector. Otherwise, it isn't linear, because it doesn't satisfy Theorem 4.3.2. This may seem surprising because the form of S looks like a natural generalization of an equation describing a line, but linear transformations satisfy the *Principle of Superposition*

$$T(c_1 \mathbf{u} + c_2 \mathbf{v}) = c_1 T(\mathbf{u}) + c_2 T(\mathbf{v})$$

for any scalars c_1, c_2 and any vectors $\mathbf{u}$, $\mathbf{v}$ in their domain. (This is just a restatement of Theorem 4.3.2.) Affine transformations with $\mathbf{f}$ nonzero don't have this property.

EXAMPLE 4   Affine Transformations

The mapping

$$S(\mathbf{u}) = \begin{bmatrix} 0 & 1 \\ -1 & 0 \end{bmatrix} \mathbf{u} + \begin{bmatrix} 1 \\ 1 \end{bmatrix}$$

is an affine transformation on R^2. If $\mathbf{u} = (a, b)$, then

$$S(\mathbf{u}) = \begin{bmatrix} 0 & 1 \\ -1 & 0 \end{bmatrix} \begin{bmatrix} a \\ b \end{bmatrix} + \begin{bmatrix} 1 \\ 1 \end{bmatrix}$$

$$= \begin{bmatrix} b + 1 \\ -a + 1 \end{bmatrix}$$

The corresponding operation from P_1 to P_1 takes $ax + b$ to $(b + 1)x - a + 1$. $\blacklozenge$

The relationship between an action on P_n and its corresponding action on the vector of coefficients in R^{n+1}, and the similarities between P_n and R^{n+1}, will be explored in more detail later in this text.

Interpolating Polynomials

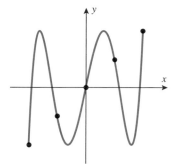

Figure 4.4.2   Interpolation

Consider the problem of interpolating a polynomial to a set of $n + 1$ points $(x_0, y_0), \ldots, (x_n, y_n)$. That is, we seek to find a curve $p(x) = a_m x^m + a_{m-1} x^{m-1} + \cdots + a_1 x + a_0$ of minimum degree that goes through each of these data points (Figure 4.4.2). Such a curve must satisfy

$$
\begin{aligned}
y_0 &= a_m x_0^m + a_{m-1} x_0^{m-1} + \cdots + a_1 x_0 + a_0 \\
y_1 &= a_m x_1^m + a_{m-1} x_1^{m-1} + \cdots + a_1 x_1 + a_0 \\
&\ \ \vdots \qquad \vdots \qquad \vdots \qquad\qquad \vdots \quad \vdots \\
y_n &= a_m x_n^m + a_{m-1} x_n^{m-1} + \cdots + a_1 x_n + a_0
\end{aligned}
$$

Because the x_i are known, this leads to the following matrix system:

$$
\begin{bmatrix}
1 & x_0 & x_0^2 & \cdots & x_0^m \\
1 & x_1 & x_1^2 & \cdots & x_1^m \\
\vdots & \vdots & \vdots & \cdots & \vdots \\
1 & x_{n-1} & x_{n-1}^2 & \cdots & x_{n-1}^m \\
1 & x_n & x_n^2 & \cdots & x_n^m
\end{bmatrix}
\begin{bmatrix}
a_0 \\ a_1 \\ \vdots \\ a_{m-1} \\ a_m
\end{bmatrix}
=
\begin{bmatrix}
y_0 \\ y_1 \\ \vdots \\ y_{n-1} \\ y_n
\end{bmatrix}
$$

Note that this is a square system when $n = m$. Taking $n = m$ gives the following system for the coefficients of the interpolating polynomial $p(x)$:

$$
\begin{bmatrix}
1 & x_0 & x_0^2 & \cdots & x_0^n \\
1 & x_1 & x_1^2 & \cdots & x_1^n \\
\vdots & \vdots & \vdots & \cdots & \vdots \\
1 & x_{n-1} & x_{n-1}^2 & \cdots & x_{n-1}^n \\
1 & x_n & x_n^2 & \cdots & x_n^n
\end{bmatrix}
\begin{bmatrix}
a_0 \\ a_1 \\ \vdots \\ a_{n-1} \\ a_n
\end{bmatrix}
=
\begin{bmatrix}
y_0 \\ y_1 \\ \vdots \\ y_{n-1} \\ y_n
\end{bmatrix}
\tag{1}
$$

The matrix in (1) is known as a ***Vandermonde matrix***; column j is the second column raised elementwise to the $j - 1$ power. The linear system in (1) is said to be a ***Vandermonde system***.

EXAMPLE 5   Interpolating a Cubic

To interpolate a polynomial to the data $(-2, 11), (-1, 2), (1, 2), (2, -1)$, we form the Vandermonde system (1):

$$
\begin{bmatrix}
1 & x_0 & x_0^2 & x_0^3 \\
1 & x_1 & x_1^2 & x_1^3 \\
1 & x_2 & x_2^2 & x_2^3 \\
1 & x_3 & x_3^2 & x_3^3
\end{bmatrix}
\begin{bmatrix}
a_0 \\ a_1 \\ a_2 \\ a_3
\end{bmatrix}
=
\begin{bmatrix}
y_0 \\ y_1 \\ y_2 \\ y_3
\end{bmatrix}
$$

For this data, we have

$$
\begin{bmatrix}
1 & -2 & 4 & -8 \\
1 & -1 & 1 & -1 \\
1 & 1 & 1 & 1 \\
1 & 2 & 4 & 8
\end{bmatrix}
\begin{bmatrix}
a_0 \\ a_1 \\ a_2 \\ a_3
\end{bmatrix}
=
\begin{bmatrix}
11 \\ 2 \\ 2 \\ -1
\end{bmatrix}
$$

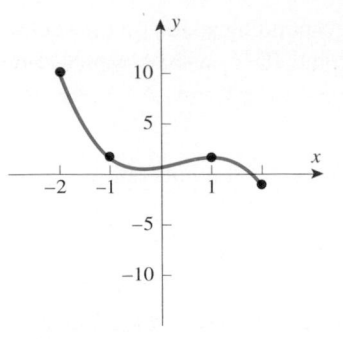

Figure 4.4.3 The interpolant of Example 5

The solution, found by Gaussian elimination, is

$$\begin{bmatrix} a_0 \\ a_1 \\ a_2 \\ a_3 \end{bmatrix} = \begin{bmatrix} 1 \\ 1 \\ 1 \\ -1 \end{bmatrix}$$

and so the interpolant is $p(x) = -x^3 + x^2 + x + 1$. This is plotted in Figure 4.4.3, together with the data points, and we see that $p(x)$ does indeed interpolate the data, as required. ◆

Newton Form

The interpolating polynomial $p(x) = a_n x^n + a_{n-1} x^{n-1} + \cdots + a_1 x + a_0$ is said to be written in its natural, or standard, form. But there is convenience in using other forms. For example, suppose we seek a cubic interpolant to the data (x_0, y_0), (x_1, y_1), (x_2, y_2), (x_3, y_3). If we write

$$p(x) = a_3 x^3 + a_2 x^2 + a_1 x + a_0 \tag{2}$$

in the equivalent form

$$p(x) = a_3(x - x_0)^3 + a_2(x - x_0)^2 + a_1(x - x_0) + a_0$$

then the interpolation condition $p(x_0) = y_0$ immediately gives $a_0 = y_0$. This reduces the size of the system that must be solved from $(n + 1) \times (n + 1)$ to $n \times n$. That is not much of a savings, but if we take this idea further, we may write (2) in the equivalent form

$$p(x) = b_3(x - x_0)(x - x_1)(x - x_2) + b_2(x - x_0)(x - x_1) + b_1(x - x_0) + b_0 \tag{3}$$

which is called the **Newton form** of the interpolant. Set $h_i = x_i - x_{i-1}$ for $i = 1, 2, 3$. The interpolation conditions give

$$p(x_0) = b_0$$
$$p(x_1) = b_1 h_1 + b_0$$
$$p(x_2) = b_2(h_1 + h_2)h_2 + b_1(h_1 + h_2) + b_0$$
$$p(x_3) = b_3(h_1 + h_2 + h_3)(h_2 + h_3)h_3 + b_2(h_1 + h_2 + h_3)(h_2 + h_3) + b_1(h_1 + h_2 + h_3) + b_0$$

that is,

$$\begin{bmatrix} 1 & 0 & 0 & 0 \\ 1 & h_1 & 0 & 0 \\ 1 & h_1 + h_2 & (h_1 + h_2)h_2 & 0 \\ 1 & h_1 + h_2 + h_3 & (h_1 + h_2 + h_3)(h_2 + h_3) & (h_1 + h_2 + h_3)(h_2 + h_3)h_3 \end{bmatrix} \begin{bmatrix} b_0 \\ b_1 \\ b_2 \\ b_3 \end{bmatrix} = \begin{bmatrix} y_0 \\ y_1 \\ y_2 \\ y_3 \end{bmatrix} \tag{4}$$

Unlike the Vandermonde system (1), this system has a lower triangular coefficient matrix. This is a much simpler system. We may solve for the coefficients very easily and efficiently by **forward-substitution**, in analogy with back-substitution. In the case of equally spaced points arranged in increasing order, we have $h_i = h > 0$, so (4) becomes

$$\begin{bmatrix} 1 & 0 & 0 & 0 \\ 1 & h & 0 & 0 \\ 1 & 2h & 2h^2 & 0 \\ 1 & 3h & 6h^2 & 6h^3 \end{bmatrix} \begin{bmatrix} b_0 \\ b_1 \\ b_2 \\ b_3 \end{bmatrix} = \begin{bmatrix} y_0 \\ y_1 \\ y_2 \\ y_3 \end{bmatrix}$$

Note that the determinant of (4) is nonzero exactly when h_i is nonzero for each i, so there exists a unique interpolant whenever the x_i are distinct. Because the Vandermonde system computes a different form of the same interpolant, it too must have a unique solution exactly when the x_i are distinct.

EXAMPLE 6   Interpolating a Cubic in Newton Form

To interpolate a polynomial in Newton form to the data $(-2, 11), (-1, 2), (1, 2), (2, -1)$ of Example 5, we form the system (4):

$$\begin{bmatrix} 1 & 0 & 0 & 0 \\ 1 & 1 & 0 & 0 \\ 1 & 3 & 6 & 0 \\ 1 & 4 & 12 & 12 \end{bmatrix} \begin{bmatrix} b_0 \\ b_1 \\ b_2 \\ b_3 \end{bmatrix} = \begin{bmatrix} 11 \\ 2 \\ 2 \\ -1 \end{bmatrix}$$

The solution, found by forward-substitution, is

$$\begin{aligned} b_0 &= 11 \\ b_0 + b_1 &= 2 & b_1 &= -9 \\ b_0 + 3b_1 + 6b_2 &= 2 & b_2 &= 3 \\ b_0 + 4b_1 + 12b_2 + 12b_3 &= -1 & b_3 &= -1 \end{aligned}$$

and so, from (3), the interpolant is

$$\begin{aligned} p(x) &= -1 \cdot (x+2)(x+1)(x-1) + 3 \cdot (x+2)(x+1) + (-9) \cdot (x+2) + 11 \\ &= -(x+2)(x+1)(x-1) + 3(x+2)(x+1) - 9(x+2) + 11 \; \blacklozenge \end{aligned}$$

Converting between Forms

The Newton form offers other advantages, but now we turn to the following question: If we have the coefficients of the interpolating polynomial in Newton form, what are the coefficients in the standard form? For example, if we know the coefficients in

$$p(x) = b_3(x - x_0)(x - x_1)(x - x_2) + b_2(x - x_0)(x - x_1) + b_1(x - x_0) + b_0$$

because we have solved (4) in order to avoid having to solve the more complicated Vandermonde system (1), how can we get the coefficients in (2),

$$p(x) = a_3 x^3 + a_2 x^2 + a_1 x + a_0$$

from b_0, b_1, b_2, b_3? Expanding the products in (3) gives

$$\begin{aligned} p(x) &= b_3(x - x_0)(x - x_1)(x - x_2) + b_2(x - x_0)(x - x_1) + b_1(x - x_0) + b_0 \\ &= b_3 x^3 + (b_2 - b_3(x_0 + x_1 + x_2))x^2 \\ &\quad + (b_1 - b_2(x_0 + x_1) + b_3(x_0 x_1 + x_0 x_2 + x_1 x_2))x \\ &\quad + b_0 - x_0 b_1 + x_0 x_1 b_2 - x_0 x_1 x_2 b_3 \end{aligned}$$

so

$$\begin{aligned} a_0 &= b_0 - x_0 b_1 + x_0 x_1 b_2 - x_0 x_1 x_2 b_3 \\ a_1 &= b_1 - b_2(x_0 + x_1) + b_3(x_0 x_1 + x_0 x_2 + x_1 x_2) \\ a_2 &= b_2 - b_3(x_0 + x_1 + x_2) \\ a_3 &= b_3 \end{aligned}$$

This can be expressed as

$$
\begin{bmatrix} a_0 \\ a_1 \\ a_2 \\ a_3 \end{bmatrix} = \begin{bmatrix} 1 & -x_0 & x_0x_1 & -x_0x_1x_2 \\ 0 & 1 & -(x_0+x_1) & x_0x_1 + x_0x_2 + x_1x_2 \\ 0 & 0 & 1 & -(x_0+x_1+x_2) \\ 0 & 0 & 0 & 1 \end{bmatrix} \begin{bmatrix} b_0 \\ b_1 \\ b_2 \\ b_3 \end{bmatrix} \tag{5}
$$

This is an important result! Solving the Vandermonde system (1) by Gaussian elimination would require us to form an $n \times n$ matrix that might have no nonzero entries and then to solve it using a number of arithmetic operations that grows in proportion to n^3 for large n. But solving the lower triangular system (4) requires an amount of work that grows in proportion to n^2 for large n, and using (5) to compute the coefficients a_0, a_1, a_2, a_3 also requires an amount of work that grows in proportion to n^2 for large n. Hence, for large n, the latter approach is an order of magnitude more efficient. The two-step procedure of solving (4) and then using the linear transformation (5) is a superior approach to solving (1) when n is large (Figure 4.4.4).

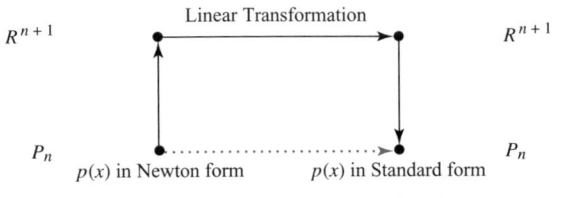

Figure 4.4.4   Indirect route to conversion from Newton form to standard form

EXAMPLE 7   Changing Forms

In Example 5 we found that $a_0 = 1$, $a_1 = 1$, $a_2 = 1$, $a_3 = -1$, whereas in Example 6 we found that $b_0 = 11$, $b_1 = -9$, $b_2 = 3$, $b_3 = -1$ for the same data. From (5), with $x_0 = -2$, $x_1 = -1$, $x_1 = 1$, we expect that

$$
\begin{bmatrix} 1 \\ 1 \\ 1 \\ -1 \end{bmatrix} = \begin{bmatrix} 1 & 2 & 2 & -2 \\ 0 & 1 & 3 & -1 \\ 0 & 0 & 1 & 2 \\ 0 & 0 & 0 & 1 \end{bmatrix} \begin{bmatrix} 11 \\ -9 \\ 3 \\ -1 \end{bmatrix}
$$

which checks. ◆

There is another approach to solving (1), based on the Fast Fourier Transform, that also requires an amount of work proportional to n^2. The point for now is to see that the use of linear transformations on R^{n+1} can help us perform computations involving polynomials. The original problem—to fit a polynomial of minimum degree to a set of data points—was not couched in the language of linear algebra at all. But rephrasing it in those terms and using matrices and the notation of linear transformations on R^{n+1} has allowed us to see when a unique solution must exist, how to compute it efficiently, and how to transform it among various forms.

**EXERCISE SET
4.4**

1. Identify the operations on polynomials that correspond to the following operations on vectors. Give the resulting polynomial.

   (a) $\begin{bmatrix} 1 \\ 2 \\ -1 \end{bmatrix} - 2\begin{bmatrix} 3 \\ 0 \\ 2 \end{bmatrix}$ (b) $5\begin{bmatrix} 4 \\ 3 \\ 0 \end{bmatrix} + 6\begin{bmatrix} 1 \\ 2 \\ 1 \end{bmatrix}$

   (c) $\begin{bmatrix} 1 \\ 2 \\ 1 \\ -2 \\ 1 \end{bmatrix} - \begin{bmatrix} 0 \\ 2 \\ 0 \\ -2 \\ 0 \end{bmatrix}$ (d) $\pi\begin{bmatrix} 4 \\ -3 \\ 7 \\ 1 \end{bmatrix}$

2. (a) Consider the operation on P_2 that takes $ax^2 + bx + c$ to $cx^2 + bx + a$. Does it correspond to a linear transformation from R^3 to R^3? If so, what is its matrix?

   (b) Consider the operation on P_3 that takes $ax^3 + bx^2 + cx + d$ to $cx^3 - bx^2 - ax + d$. Does it correspond to a linear transformation from R^3 to R^3? If so, what is its matrix?

3. (a) Consider the transformation of $ax^2 + bx + c$ in P_2 to $|a|$ in P_0. Show that it does not correspond to a linear transformation by showing that there is no matrix that maps (a, b, c) in R^3 to $|a|$ in R.

   (b) Does the transformation of $ax^2 + bx + c$ in P_2 to a in P_0 correspond to a linear transformation from R^3 to R?

4. (a) Consider the operation $M: P_2 \to P_3$ that takes $p(x)$ in P_2 to $xp(x)$ in P_3. Does this correspond to a linear transformation from R^3 to R^4? If so, what is its matrix?

   (b) Consider the operation $N: P_2 \to P_3$ that takes $p(x)$ in P_n to $(x-1)p(x)$ in P_{n+1}. Does this correspond to a linear transformation from R^3 to R^4? If so, what is its matrix?

   (c) Consider the operation $W: P_2 \to P_3$ that takes $p(x)$ in P_n to $xp(x) + 1$ in P_{n+1}. Does this correspond to a linear transformation from R^3 to R^4? If so, what is its matrix?

5. **(For Readers Who Have Studied Calculus)** What matrix corresponds to differentiation in each case?

   (a) $D: P_3 \to P_2$ (b) $D: P_4 \to P_3$ (c) $D: P_5 \to P_4$

6. **(For Readers Who Have Studied Calculus)** What matrix corresponds to differentiation in each case, assuming we represent $p(x) = a_n x^n + a_{n-1}x^{n-1} + \cdots + a_1 x + a_0$ as the vector $(a_0, a_1, \ldots, a_{n-1}, a_n)$?

   Note This is the opposite of the ordering of coefficients we have been using.

   (a) $D: P_3 \to P_2$ (b) $D: P_4 \to P_3$ (c) $D: P_5 \to P_4$

7. Consider the following matrices. What is the corresponding transformation on polynomials? Indicate the domain P_i and the codomain P_j.

   (a) $\begin{bmatrix} 1 & 1 \\ 1 & -1 \end{bmatrix}$ (b) $\begin{bmatrix} 1 & 0 \\ 1 & 1 \\ 2 & -1 \end{bmatrix}$ (c) $\begin{bmatrix} 1 & 0 & 2 & -1 \\ 2 & 1 & 1 & 3 \end{bmatrix}$

   (d) $\begin{bmatrix} 0 & 0 & 0 \\ 0 & 1 & 0 \\ 0 & 0 & 0 \end{bmatrix}$ (e) $\begin{bmatrix} 0 & 1 & 0 \end{bmatrix}$

8. Consider the space of all functions of the form $a + b\cos(x) + c\sin(x)$ where a, b, c are scalars.

   (a) What matrix, if any, corresponds to the change of variables $x \to x - \pi/2$, assuming that we represent a function in this space as the vector (a, b, c)?

   (b) What matrix corresponds to differentiation of functions on this space?

9. Consider the space of all functions of the form $a + bt + ce^t + de^{-t}$, where a, b, c, d are scalars.

   (a) What function in the space corresponds to the sum of $(1, 2, 3, 4)$ and $(-1, -2, 0, -1)$, assuming that we represent a function in this space as the vector (a, b, c, d)?

   (b) Is $\cosh(t)$ in this space? That is, does $\cosh(t)$ correspond to some choice of a, b, c, d?

   (c) What matrix corresponds to differentiation of functions on this space?

10. Show that the Principle of Superposition is equivalent to Theorem 4.3.2.

11. Show that an affine transformation with **f** nonzero is not a linear transformation.

12. Find a quadratic interpolant to the data $(-1, 2), (0, 0), (1, 2)$ using the Vandermonde system approach.

13. (a) Find a quadratic interpolant to the data $(-2, 1), (0, 1), (1, 4)$ using the Vandermonde system approach from (1).

    (b) Repeat using the Newton approach from (4).

14. (a) Find a polynomial interpolant to the data $(-1, 0), (0, 0), (1, 0), (2, 6)$ using the Vandermonde system approach from (1).

    (b) Repeat using the Newton approach from (4).

    (c) Use (5) to get your answer in part (a) from your answer in part (b).

    (d) Use (5) to get your answer in part (b) from your answer in part (a) by finding the inverse of the matrix.

    (e) What happens if you change the data to $(-1, 0), (0, 0), (1, 0), (2, 0)$?

15. (a) Find a polynomial interpolant to the data $(-2, -10), (-1, 2), (1, 2), (2, 14)$ using the Vandermonde system approach from (1).

    (b) Repeat using the Newton approach from (4).

    (c) Use (5) to get your answer in part (a) from your answer in part (b).

    (d) Use (5) to get your answer in part (b) from your answer in part (a) by finding the inverse of the matrix.

16. Show that the determinant of the 2×2 Vandermonde matrix

$$\begin{bmatrix} 1 & a \\ 1 & b \end{bmatrix}$$

can be written as $(b - a)$ and that the determinant of the 3×3 Vandermonde matrix

$$\det \begin{bmatrix} 1 & a & a^2 \\ 1 & b & b^2 \\ 1 & c & c^2 \end{bmatrix}$$

can be written as $(b - a)(c - a)(c - b)$. Conclude that a unique straight line can be fit through any two points $(x_0, y_0), (x_1, y_1)$ with x_0 and x_1 distinct, and that a unique parabola (which may be degenerate, such as a line) can be fit through any three points $(x_0, y_0), (x_1, y_1), (x_2, y_2)$ with $x_0, x_1,$ and x_2 distinct.

17. (a) What form does (5) take for lines?

    (b) What form does (5) take for quadratics?

    (c) What form does (5) take for quartics?

Discussion & Discovery

18. **(For Readers Who Have Studied Calculus)**

    (a) Does indefinite integration of functions in P_n correspond to some linear transformation from R^{n+1} to R^{n+2}?

(b) Does definite integration (from $x = 0$ to $x = 1$) of functions in P_n correspond to some linear transformation from R^{n+1} to R?

19. **(For Readers Who Have Studied Calculus)**

(a) What matrix corresponds to second differentiation of functions from P_2 (giving functions in P_0)?

(b) What matrix corresponds to second differentiation of functions from P_3 (giving functions in P_1)?

(c) Is the matrix for second differentiation the square of the matrix for (first) differentiation?

20. Consider the transformation from P_2 to P_2 associated with the matrix

$$\begin{bmatrix} 0 & 0 & 0 \\ 0 & 1 & 0 \\ 0 & 0 & 0 \end{bmatrix}$$

and the transformation from P_2 to P_0 associated with the matrix

$$\begin{bmatrix} 0 & 1 & 0 \end{bmatrix}$$

These differ only in their codomains. Comment on this difference. In what ways (if any) is it important?

21. The third major technique for polynomial interpolation is interpolation using **Lagrange interpolating polynomials**. Given a set of distinct x-values $x_0, x_1, \ldots, x_n$, define the $n + 1$ Lagrange interpolating polynomials for these values by (for $i = 0, 1, \ldots, n$)

$$L_i(x) = \frac{(x - x_0)(x - x_1) \cdots (x - x_{i-1})(x - x_{i+1}) \cdots (x - x_n)}{(x_i - x_0)(x_i - x_1) \cdots (x_i - x_{i-1})(x_i - x_{i+1}) \cdots (x_i - x_n)}$$

Note that $L_i(x)$ is a polynomial of exact degree n and that $L_i(x_j) = 0$ if $i \neq j$, and $L_i(x_i) = 1$. It follows that we can write the polynomial interpolant to $(x_0, y_0), \ldots, (x_n, y_n)$ in the form

$$p(x) = c_0 L_0(x) + c_1 L_1(x) + \cdots + c_n L_n(x)$$

where $c_i = y_i$, $i = 0, 1, \ldots, n$.

(a) Verify that $p(x) = y_0 L_0(x) + y_1 L_1(x) + \cdots + y_n L_n(x)$ is the unique interpolating polynomial for this data.

(b) What is the linear system for the coefficients $c_0, c_1, \ldots, c_n$, corresponding to (1) for the Vandermonde approach and to (4) for the Newton approach?

(c) Compare the three approaches to polynomial interpolation that we have seen. Which is most efficient with respect to finding the coefficients? Which is most efficient with respect to evaluating the interpolant somewhere between data points?

22. Generalize the result in Problem 16 by finding a formula for the determinant of an $n \times n$ Vandermonde matrix for arbitrary n.

23. The **norm** of a linear transformation $T_A : R^n \to R^n$ can be defined by

$$\|T\|_E = \max \frac{\|T(\mathbf{x})\|}{\|\mathbf{x}\|}$$

where the maximum is taken over all nonzero $\mathbf{x}$ in R^n. (The subscript indicates that the norm of the linear transformation on the left is found using the Euclidean vector norm on the right.) It is a fact that the largest value is always achieved—that is, there is always some $\mathbf{x}_0$ in R^n such that $\|T\|_E = (\|T(\mathbf{x}_0)\|/\|\mathbf{x}_0\|)$. What are the norms of the linear transformations T_A with the following matrices?

(a) $\begin{bmatrix} 2 & 0 \\ 0 & 1 \end{bmatrix}$ (b) $\begin{bmatrix} 1 & 0 \\ 0 & -1 \end{bmatrix}$ (c) $\begin{bmatrix} 2 & 0 \\ 0 & -3 \end{bmatrix}$ (d) $\begin{bmatrix} 1/\sqrt{2} & 1/\sqrt{2} \\ 1/\sqrt{2} & -1/\sqrt{2} \end{bmatrix}$

CHAPTER 4

Technology Exercises

The following exercises are designed to be solved using a technology utility. Typically, this will be MATLAB, *Mathematica*, Maple, Derive, or Mathcad, but it may also be some other type of linear algebra software or a scientific calculator with some linear algebra capabilities. For each exercise you will need to read the relevant documentation for the particular utility you are using. The goal of these exercises is to provide you with a basic proficiency with your technology utility. Once you have mastered the techniques in these exercises, you will be able to use your technology utility to solve many of the problems in the regular exercise sets.

Section 4.1　T1.　**(Vector Operations in R^n)** With most technology utilities, the commands for operating on vectors in R^n are the same as those for operating on vectors in R^2 and R^3, and the command for computing a dot product produces the Euclidean inner product in R^n. Use your utility to perform computations in Exercises 1, 3, and 9 of Section 4.1.

Section 4.2　T1.　**(Rotations)** Find the standard matrix for the linear operator on R^3 that performs a counterclockwise rotation of $45°$ about the x-axis, followed by a counterclockwise rotation of $60°$ about the y-axis, followed by a counterclockwise rotation of $30°$ about the z-axis. Then find the image of the point $(1, 1, 1)$ under this operator.

Section 4.3　T1.　**(Projections)** Use your utility to perform the computations for $\theta = \pi/6$ in Example 6. Then project the vectors $(1, 1)$ and $(1, -5)$. Repeat for $\theta = \pi/4, \pi/3, \pi/2, \pi$.

Section 4.4　T1.　**(Interpolation)** Most technology utilities have a command that performs polynomial interpolation. Read your documentation, and find the command or commands for fitting a polynomial interpolant to given data. Then use it (or them) to confirm the result of Example 5.

General Vector Spaces

CHAPTER CONTENTS

INTRODUCTION: In the last chapter we generalized vectors from 2- and 3-space to vectors in *n*-space. In this chapter we shall generalize the concept of vector still further. We shall state a set of axioms that, if satisfied by a class of objects, will entitle those objects to be called "vectors." These generalized vectors will include, among other things, various kinds of matrices and functions. Our work in this chapter is not an idle exercise in theoretical mathematics; it will provide a powerful tool for extending our geometric visualization to a wide variety of important mathematical problems where geometric intuition would not otherwise be available. We can visualize vectors in R^2 and R^3 as arrows, which enables us to draw or form mental pictures to help solve problems. Because the axioms we give to define our new kinds of vectors will be based on properties of vectors in R^2 and R^3, the new vectors will have many familiar properties. Consequently, when we want to solve a problem involving our new kinds of vectors, say matrices or functions, we may be able to get a foothold on the problem by visualizing what the corresponding problem would be like in R^2 and R^3.

5.1
REAL VECTOR SPACES

In this section we shall extend the concept of a vector by extracting the most important properties of familiar vectors and turning them into axioms. Thus, when a set of objects satisfies these axioms, they will automatically have the most important properties of familiar vectors, thereby making it reasonable to regard these objects as new kinds of vectors.

Vector Space Axioms

The following definition consists of ten axioms. As you read each axiom, keep in mind that you have already seen each of them as parts of various definitions and theorems in the preceding two chapters (for instance, see Theorem 4.1.1). Remember, too, that you do not prove axioms; they are simply the "rules of the game."

DEFINITION

Let V be an arbitrary nonempty set of objects on which two operations are defined: addition, and multiplication by scalars (numbers). By ***addition*** we mean a rule for associating with each pair of objects $\mathbf{u}$ and $\mathbf{v}$ in V an object $\mathbf{u} + \mathbf{v}$, called the ***sum*** of $\mathbf{u}$ and $\mathbf{v}$; by ***scalar multiplication*** we mean a rule for associating with each scalar k and each object $\mathbf{u}$ in V an object $k\mathbf{u}$, called the ***scalar multiple*** of $\mathbf{u}$ by k. If the following axioms are satisfied by all objects $\mathbf{u}$, $\mathbf{v}$, $\mathbf{w}$ in V and all scalars k and m, then we call V a ***vector space*** and we call the objects in V ***vectors***.

1. If $\mathbf{u}$ and $\mathbf{v}$ are objects in V, then $\mathbf{u} + \mathbf{v}$ is in V.
2. $\mathbf{u} + \mathbf{v} = \mathbf{v} + \mathbf{u}$
3. $\mathbf{u} + (\mathbf{v} + \mathbf{w}) = (\mathbf{u} + \mathbf{v}) + \mathbf{w}$
4. There is an object $\mathbf{0}$ in V, called a ***zero vector*** for V, such that $\mathbf{0} + \mathbf{u} = \mathbf{u} + \mathbf{0} = \mathbf{u}$ for all $\mathbf{u}$ in V.
5. For each $\mathbf{u}$ in V, there is an object $-\mathbf{u}$ in V, called a ***negative*** of $\mathbf{u}$, such that $\mathbf{u} + (-\mathbf{u}) = (-\mathbf{u}) + \mathbf{u} = \mathbf{0}$.
6. If k is any scalar and $\mathbf{u}$ is any object in V, then $k\mathbf{u}$ is in V.
7. $k(\mathbf{u} + \mathbf{v}) = k\mathbf{u} + k\mathbf{v}$
8. $(k + m)\mathbf{u} = k\mathbf{u} + m\mathbf{u}$
9. $k(m\mathbf{u}) = (km)(\mathbf{u})$
10. $1\mathbf{u} = \mathbf{u}$

REMARK Depending on the application, scalars may be real numbers or complex numbers. Vector spaces in which the scalars are complex numbers are called ***complex vector spaces***, and those in which the scalars must be real are called ***real vector spaces***. In Chapter 10 we shall discuss complex vector spaces; until then, *all of our scalars will be real numbers.*

The reader should keep in mind that the definition of a vector space specifies neither the nature of the vectors nor the operations. Any kind of object can be a vector, and the operations of addition and scalar multiplication may not have any relationship or similarity to the standard vector operations on R^n. The only requirement is that the ten vector space axioms be satisfied. Some authors use the notations $\oplus$ and $\odot$ for vector addition and scalar multiplication to distinguish these operations from addition and multiplication of real numbers; we will not use this convention, however.

Examples of Vector Spaces

The following examples will illustrate the variety of possible vector spaces. In each example we will specify a nonempty set V and two operations, addition and scalar multiplication; then we shall verify that the ten vector space axioms are satisfied, thereby entitling V, with the specified operations, to be called a vector space.

EXAMPLE 1 R^n Is a Vector Space

The set $V = R^n$ with the standard operations of addition and scalar multiplication defined in Section 4.1 is a vector space. Axioms 1 and 6 follow from the definitions of the standard operations on R^n; the remaining axioms follow from Theorem 4.1.1. ◆

The three most important special cases of R^n are R (the real numbers), R^2 (the vectors in the plane), and R^3 (the vectors in 3-space).

EXAMPLE 2 A Vector Space of 2 × 2 Matrices

Show that the set V of all 2×2 matrices with real entries is a vector space if addition is defined to be matrix addition and scalar multiplication is defined to be matrix scalar multiplication.

Solution

In this example we will find it convenient to verify the axioms in the following order: 1, 6, 2, 3, 7, 8, 9, 4, 5, and 10. Let

$$\mathbf{u} = \begin{bmatrix} u_{11} & u_{12} \\ u_{21} & u_{22} \end{bmatrix} \quad \text{and} \quad \mathbf{v} = \begin{bmatrix} v_{11} & v_{12} \\ v_{21} & v_{22} \end{bmatrix}$$

To prove Axiom 1, we must show that $\mathbf{u} + \mathbf{v}$ is an object in V; that is, we must show that $\mathbf{u} + \mathbf{v}$ is a 2×2 matrix. But this follows from the definition of matrix addition, since

$$\mathbf{u} + \mathbf{v} = \begin{bmatrix} u_{11} & u_{12} \\ u_{21} & u_{22} \end{bmatrix} + \begin{bmatrix} v_{11} & v_{12} \\ v_{21} & v_{22} \end{bmatrix} = \begin{bmatrix} u_{11} + v_{11} & u_{12} + v_{12} \\ u_{21} + v_{21} & u_{22} + v_{22} \end{bmatrix}$$

Similarly, Axiom 6 holds because for any real number k, we have

$$k\mathbf{u} = k\begin{bmatrix} u_{11} & u_{12} \\ u_{21} & u_{22} \end{bmatrix} = \begin{bmatrix} ku_{11} & ku_{12} \\ ku_{21} & ku_{22} \end{bmatrix}$$

so $k\mathbf{u}$ is a 2×2 matrix and consequently is an object in V.

Axiom 2 follows from Theorem 1.4.1a since

$$\mathbf{u} + \mathbf{v} = \begin{bmatrix} u_{11} & u_{12} \\ u_{21} & u_{22} \end{bmatrix} + \begin{bmatrix} v_{11} & v_{12} \\ v_{21} & v_{22} \end{bmatrix} = \begin{bmatrix} v_{11} & v_{12} \\ v_{21} & v_{22} \end{bmatrix} + \begin{bmatrix} u_{11} & u_{12} \\ u_{21} & u_{22} \end{bmatrix} = \mathbf{v} + \mathbf{u}$$

Similarly, Axiom 3 follows from part (b) of that theorem; and Axioms 7, 8, and 9 follow from parts (h), (j), and (l), respectively.

To prove Axiom 4, we must find an object $\mathbf{0}$ in V such that $\mathbf{0} + \mathbf{u} = \mathbf{u} + \mathbf{0} = \mathbf{u}$ for all $\mathbf{u}$ in V. This can be done by defining $\mathbf{0}$ to be

$$\mathbf{0} = \begin{bmatrix} 0 & 0 \\ 0 & 0 \end{bmatrix}$$

With this definition,

$$\mathbf{0} + \mathbf{u} = \begin{bmatrix} 0 & 0 \\ 0 & 0 \end{bmatrix} + \begin{bmatrix} u_{11} & u_{12} \\ u_{21} & u_{22} \end{bmatrix} = \begin{bmatrix} u_{11} & u_{12} \\ u_{21} & u_{22} \end{bmatrix} = \mathbf{u}$$

and similarly $\mathbf{u} + \mathbf{0} = \mathbf{u}$. To prove Axiom 5, we must show that each object $\mathbf{u}$ in V has a negative $-\mathbf{u}$ such that $\mathbf{u} + (-\mathbf{u}) = \mathbf{0}$ and $(-\mathbf{u}) + \mathbf{u} = \mathbf{0}$. This can be done by defining the negative of $\mathbf{u}$ to be

$$-\mathbf{u} = \begin{bmatrix} -u_{11} & -u_{12} \\ -u_{21} & -u_{22} \end{bmatrix}$$

With this definition,

$$\mathbf{u} + (-\mathbf{u}) = \begin{bmatrix} u_{11} & u_{12} \\ u_{21} & u_{22} \end{bmatrix} + \begin{bmatrix} -u_{11} & -u_{12} \\ -u_{21} & -u_{22} \end{bmatrix} = \begin{bmatrix} 0 & 0 \\ 0 & 0 \end{bmatrix} = \mathbf{0}$$

and similarly $(-\mathbf{u}) + \mathbf{u} = \mathbf{0}$. Finally, Axiom 10 is a simple computation:

$$1\mathbf{u} = 1 \begin{bmatrix} u_{11} & u_{12} \\ u_{21} & u_{22} \end{bmatrix} = \begin{bmatrix} u_{11} & u_{12} \\ u_{21} & u_{22} \end{bmatrix} = \mathbf{u} \quad \blacklozenge$$

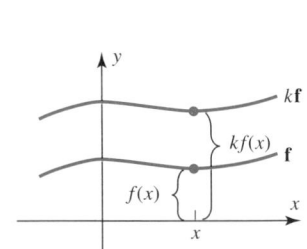

(a)

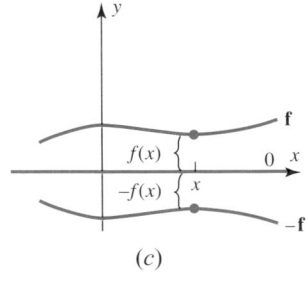

(b)

(c)

Figure 5.1.1

EXAMPLE 3   A Vector Space of $m \times n$ Matrices

Example 2 is a special case of a more general class of vector spaces. The arguments in that example can be adapted to show that the set V of all $m \times n$ matrices with real entries, together with the operations of matrix addition and scalar multiplication, is a vector space. The $m \times n$ zero matrix is the zero vector $\mathbf{0}$, and if $\mathbf{u}$ is the $m \times n$ matrix U, then the matrix $-U$ is the negative $-\mathbf{u}$ of the vector $\mathbf{u}$. We shall denote this vector space by the symbol M_{mn}. $\blacklozenge$

EXAMPLE 4   A Vector Space of Real-Valued Functions

Let V be the set of real-valued functions defined on the entire real line $(-\infty, \infty)$. If $\mathbf{f} = f(x)$ and $\mathbf{g} = g(x)$ are two such functions and k is any real number, define the sum function $\mathbf{f} + \mathbf{g}$ and the scalar multiple $k\mathbf{f}$, respectively, by

$$(\mathbf{f} + \mathbf{g})(x) = f(x) + g(x) \quad \text{and} \quad (k\mathbf{f})(x) = kf(x)$$

In other words, the value of the function $\mathbf{f} + \mathbf{g}$ at x is obtained by adding together the values of $\mathbf{f}$ and $\mathbf{g}$ at x (Figure 5.1.1a). Similarly, the value of $k\mathbf{f}$ at x is k times the value of $\mathbf{f}$ at x (Figure 5.1.1b). In the exercises we shall ask you to show that V is a vector space with respect to these operations. This vector space is denoted by $F(-\infty, \infty)$. If $\mathbf{f}$ and $\mathbf{g}$ are vectors in this space, then to say that $\mathbf{f} = \mathbf{g}$ is equivalent to saying that $f(x) = g(x)$ for all x in the interval $(-\infty, \infty)$.

The vector $\mathbf{0}$ in $F(-\infty, \infty)$ is the constant function that is identically zero for all values of x. The graph of this function is the line that coincides with the x-axis. The negative of a vector $\mathbf{f}$ is the function $-\mathbf{f} = -f(x)$. Geometrically, the graph of $-\mathbf{f}$ is the reflection of the graph of $\mathbf{f}$ across the x-axis (Figure 5.1.1c). $\blacklozenge$

REMARK   In the preceding example we focused on the interval $(-\infty, \infty)$. Had we restricted our attention to some closed interval $[a, b]$ or some open interval (a, b), the

Real Vector Spaces • • • **225**

functions defined on those intervals with the operations stated in the example would also have produced vector spaces. Those vector spaces are denoted by $F[a, b]$ and $F(a, b)$, respectively.

EXAMPLE 5   A Set That Is Not a Vector Space

Let $V = R^2$ and define addition and scalar multiplication operations as follows: If $\mathbf{u} = (u_1, u_2)$ and $\mathbf{v} = (v_1, v_2)$, then define

$$\mathbf{u} + \mathbf{v} = (u_1 + v_1, u_2 + v_2)$$

and if k is any real number, then define

$$k\mathbf{u} = (ku_1, 0)$$

For example, if $\mathbf{u} = (2, 4)$, $\mathbf{v} = (-3, 5)$, and $k = 7$, then

$$\mathbf{u} + \mathbf{v} = (2 + (-3), 4 + 5) = (-1, 9)$$
$$k\mathbf{u} = 7\mathbf{u} = (7 \cdot 2, 0) = (14, 0)$$

The addition operation is the standard addition operation on R^2, but the scalar multiplication operation is not the standard scalar multiplication. In the exercises we will ask you to show that the first nine vector space axioms are satisfied; however, there are values of $\mathbf{u}$ for which Axiom 10 fails to hold. For example, if $\mathbf{u} = (u_1, u_2)$ is such that $u_2 \neq 0$, then

$$1\mathbf{u} = 1(u_1, u_2) = (1 \cdot u_1, 0) = (u_1, 0) \neq \mathbf{u}$$

Thus V is not a vector space with the stated operations. ◆

EXAMPLE 6   Every Plane through the Origin Is a Vector Space

Let V be any plane through the origin in R^3. We shall show that the points in V form a vector space under the standard addition and scalar multiplication operations for vectors in R^3. From Example 1, we know that R^3 itself is a vector space under these operations. Thus Axioms 2, 3, 7, 8, 9, and 10 hold for all points in R^3 and consequently for all points in the plane V. We therefore need only show that Axioms 1, 4, 5, and 6 are satisfied.

Since the plane V passes through the origin, it has an equation of the form

$$ax + by + cz = 0 \tag{1}$$

(Theorem 3.5.1). Thus, if $\mathbf{u} = (u_1, u_2, u_3)$ and $\mathbf{v} = (v_1, v_2, v_3)$ are points in V, then $au_1 + bu_2 + cu_3 = 0$ and $av_1 + bv_2 + cv_3 = 0$. Adding these equations gives

$$a(u_1 + v_1) + b(u_2 + v_2) + c(u_3 + v_3) = 0$$

This equality tells us that the coordinates of the point

$$\mathbf{u} + \mathbf{v} = (u_1 + v_1, u_2 + v_2, u_3 + v_3)$$

satisfy (1); thus $\mathbf{u} + \mathbf{v}$ lies in the plane V. This proves that Axiom 1 is satisfied. The verifications of Axioms 4 and 6 are left as exercises; however, we shall prove that Axiom 5 is satisfied. Multiplying $au_1 + bu_2 + cu_3 = 0$ through by -1 gives

$$a(-u_1) + b(-u_2) + c(-u_3) = 0$$

Thus $-\mathbf{u} = (-u_1, -u_2, -u_3)$ lies in V. This establishes Axiom 5. ◆

EXAMPLE 7 The Zero Vector Space

Let V consist of a single object, which we denote by $\mathbf{0}$, and define

$$\mathbf{0} + \mathbf{0} = \mathbf{0} \quad \text{and} \quad k\mathbf{0} = \mathbf{0}$$

for all scalars k. It is easy to check that all the vector space axioms are satisfied. We call this the *zero vector space*. ◆

Some Properties of Vectors

As we progress, we shall add more examples of vector spaces to our list. We conclude this section with a theorem that gives a useful list of vector properties.

THEOREM 5.1.1

> *Let V be a vector space, $\mathbf{u}$ a vector in V, and k a scalar; then:*
>
> (a) $0\mathbf{u} = \mathbf{0}$
> (b) $k\mathbf{0} = \mathbf{0}$
> (c) $(-1)\mathbf{u} = -\mathbf{u}$
> (d) *If $k\mathbf{u} = \mathbf{0}$, then $k = 0$ or $\mathbf{u} = \mathbf{0}$.*

We shall prove parts (a) and (c) and leave proofs of the remaining parts as exercises.

Proof (a) We can write

$$0\mathbf{u} + 0\mathbf{u} = (0 + 0)\mathbf{u} \quad \text{[Axiom 8]}$$
$$= 0\mathbf{u} \qquad \text{[Property of the number 0]}$$

By Axiom 5 the vector $0\mathbf{u}$ has a negative, $-0\mathbf{u}$. Adding this negative to both sides above yields

$$[0\mathbf{u} + 0\mathbf{u}] + (-0\mathbf{u}) = 0\mathbf{u} + (-0\mathbf{u})$$

or

$$0\mathbf{u} + [0\mathbf{u} + (-0\mathbf{u})] = 0\mathbf{u} + (-0\mathbf{u}) \quad \text{[Axiom 3]}$$
$$0\mathbf{u} + \mathbf{0} = \mathbf{0} \qquad \text{[Axiom 5]}$$
$$0\mathbf{u} = \mathbf{0} \qquad \text{[Axiom 4]}$$

Proof (c) To show that $(-1)\mathbf{u} = -\mathbf{u}$, we must demonstrate that $\mathbf{u} + (-1)\mathbf{u} = \mathbf{0}$. To see this, observe that

$$\mathbf{u} + (-1)\mathbf{u} = 1\mathbf{u} + (-1)\mathbf{u} \quad \text{[Axiom 10]}$$
$$= (1 + (-1))\mathbf{u} \quad \text{[Axiom 8]}$$
$$= 0\mathbf{u} \qquad \text{[Property of numbers]}$$
$$= \mathbf{0} \qquad \text{[Part (a) above]} \quad ■$$

EXERCISE SET 5.1

In Exercises 1–16 a set of objects is given, together with operations of addition and scalar multiplication. Determine which sets are vector spaces under the given operations. For those that are not vector spaces, list all axioms that fail to hold.

1. The set of all triples of real numbers (x, y, z) with the operations

$$(x, y, z) + (x', y', z') = (x + x', y + y', z + z') \quad \text{and} \quad k(x, y, z) = (kx, y, z)$$

2. The set of all triples of real numbers (x, y, z) with the operations

$$(x, y, z) + (x', y', z') = (x + x', y + y', z + z') \quad \text{and} \quad k(x, y, z) = (0, 0, 0)$$

3. The set of all pairs of real numbers (x, y) with the operations

$$(x, y) + (x', y') = (x + x', y + y') \quad \text{and} \quad k(x, y) = (2kx, 2ky)$$

4. The set of all real numbers x with the standard operations of addition and multiplication.

5. The set of all pairs of real numbers of the form $(x, 0)$ with the standard operations on R^2.

6. The set of all pairs of real numbers of the form (x, y), where $x \geq 0$, with the standard operations on R^2.

7. The set of all n-tuples of real numbers of the form $(x, x, \ldots, x)$ with the standard operations on R^n.

8. The set of all pairs of real numbers (x, y) with the operations

$$(x, y) + (x', y') = (x + x' + 1, y + y' + 1) \quad \text{and} \quad k(x, y) = (kx, ky)$$

9. The set of all 2×2 matrices of the form

$$\begin{bmatrix} a & 1 \\ 1 & b \end{bmatrix}$$

with the standard matrix addition and scalar multiplication.

10. The set of all 2×2 matrices of the form

$$\begin{bmatrix} a & 0 \\ 0 & b \end{bmatrix}$$

with the standard matrix addition and scalar multiplication.

11. The set of all real-valued functions f defined everywhere on the real line and such that $f(1) = 0$, with the operations defined in Example 4.

12. The set of all 2×2 matrices of the form

$$\begin{bmatrix} a & a + b \\ a + b & b \end{bmatrix}$$

with matrix addition and scalar multiplication.

13. The set of all pairs of real numbers of the form $(1, x)$ with the operations

$$(1, y) + (1, y') = (1, y + y') \quad \text{and} \quad k(1, y) = (1, ky)$$

14. The set of polynomials of the form $a + bx$ with the operations

$$(a_0 + a_1 x) + (b_0 + b_1 x) = (a_0 + b_0) + (a_1 + b_1)x \quad \text{and} \quad k(a_0 + a_1 x) = (ka_0) + (ka_1)x$$

15. The set of all positive real numbers with the operations

$$x + y = xy \quad \text{and} \quad kx = x^k$$

16. The set of all pairs of real numbers (x, y) with the operations

$$(x, y) + (x', y') = (xx', yy') \quad \text{and} \quad k(x, y) = (kx, ky)$$

17. Show that the following sets with the given operations fail to be vector spaces by identifying all axioms that fail to hold.

   (a) The set of all triples of real numbers with the standard vector addition but with scalar multiplication defined by $k(x, y, z) = (k^2 x, k^2 y, k^2 z)$.

   (b) The set of all triples of real numbers with addition defined by $(x, y, z) + (u, v, w) = (z + w, y + v, x + u)$ and standard scalar multiplication.

   (c) The set of all 2×2 invertible matrices with the standard matrix addition and scalar multiplication.

18. Show that the set of all 2×2 matrices of the form $\begin{bmatrix} a & 1 \\ 1 & b \end{bmatrix}$ with addition defined by

$$\begin{bmatrix} a & 1 \\ 1 & b \end{bmatrix} + \begin{bmatrix} c & 1 \\ 1 & d \end{bmatrix} = \begin{bmatrix} a+c & 1 \\ 1 & b+d \end{bmatrix}$$ and scalar multiplication defined by $k \begin{bmatrix} a & 1 \\ 1 & b \end{bmatrix} =$

$\begin{bmatrix} ka & 1 \\ 1 & kb \end{bmatrix}$ is a vector space. What is the zero vector in this space?

19. (a) Show that the set of all points in R^2 lying on a line is a vector space, with respect to the standard operations of vector addition and scalar multiplication, exactly when the line passes through the origin.

(b) Show that the set of all points in R^3 lying on a plane is a vector space, with respect to the standard operations of vector addition and scalar multiplication, exactly when the plane passes through the origin.

20. Consider the set of all 2×2 invertible matrices with vector addition defined to be matrix *multiplication* and the standard scalar multiplication. Is this a vector space?

21. Show that the first nine vector space axioms are satisfied if $V = R^2$ has the addition and scalar multiplication operations defined in Example 5.

22. Prove that a line passing through the origin in R^3 is a vector space under the standard operations on R^3.

23. Complete the unfinished details of Example 4.

24. Complete the unfinished details of Example 6.

Discussion & Discovery

25. We showed in Example 6 that every plane in R^3 that passes through the origin is a vector space under the standard operations on R^3. Is the same true for planes that do not pass through the origin? Explain your reasoning.

26. It was shown in Exercise 14 above that the set of polynomials of degree 1 or less is a vector space under the operations stated in that exercise. Is the set of polynomials whose degree is exactly 1 a vector space under those operations? Explain your reasoning.

27. Consider the set whose only element is the moon. Is this set a vector space under the operations moon + moon = moon and k(moon) = moon for every real number k? Explain your reasoning.

28. Do you think that it is possible to have a vector space with exactly two distinct vectors in it? Explain your reasoning.

29. The following is a proof of part (*b*) of Theorem 5.1.1. Justify each step by filling in the blank line with the word *hypothesis* or by specifying the number of one of the vector space axioms given in this section.

Hypothesis: Let **u** be any vector in a vector space V, **0** the zero vector in V, and k a scalar.

Conclusion: Then $k\mathbf{0} = \mathbf{0}$.

Proof: (1) First, $k\mathbf{0} + k\mathbf{u} = k(\mathbf{0} + \mathbf{u})$. _____

    (2) $\qquad\qquad = k\mathbf{u}$ _____

    (3) Since $k\mathbf{u}$ is in V, $-k\mathbf{u}$ is in V. _____

    (4) Therefore, $(k\mathbf{0} + k\mathbf{u}) + (-k\mathbf{u}) = k\mathbf{u} + (-k\mathbf{u})$. _____

    (5) $\qquad k\mathbf{0} + (k\mathbf{u} + (-k\mathbf{u})) = k\mathbf{u} + (-k\mathbf{u})$ _____

    (6) $\qquad\qquad\qquad k\mathbf{0} + \mathbf{0} = \mathbf{0}$ _____

    (7) Finally, $\qquad\qquad\qquad k\mathbf{0} = \mathbf{0}$. _____

30. Prove part (*d*) of Theorem 5.1.1.

31. The following is a proof that the cancellation law for addition holds in a vector space. Justify each step by filling in the blank line with the word *hypothesis* or by specifying the number of one of the vector space axioms given in this section.

Hypothesis: Let **u**, **v**, and **w** be vectors in a vector space V and suppose that $\mathbf{u} + \mathbf{w} = \mathbf{v} + \mathbf{w}$.

Conclusion: Then $\mathbf{u} = \mathbf{v}$.

Proof: (1) First, $(\mathbf{u} + \mathbf{w}) + (-\mathbf{w})$ and $(\mathbf{v} + \mathbf{w}) + (-\mathbf{w})$ are vectors in V. _____

(2) Then $(\mathbf{u} + \mathbf{w}) + (-\mathbf{w}) = (\mathbf{v} + \mathbf{w}) + (-\mathbf{w})$. _____

(3) The left side of the equality in step (2) is

$$(\mathbf{u} + \mathbf{w}) + (-\mathbf{w}) = \mathbf{u} + (\mathbf{w} + (-\mathbf{w})) \text{ _____}$$
$$= \mathbf{u} \text{ _____}$$

(4) The right side of the equality in step (2) is

$$(\mathbf{v} + \mathbf{w}) + (-\mathbf{w}) = \mathbf{v} + (\mathbf{w} + (-\mathbf{w})) \text{ _____}$$
$$= \mathbf{v} \text{ _____}$$

From the equality in step (2), it follows from steps (3) and (4) that $\mathbf{u} = \mathbf{v}$.

32. Do you think it is possible for a vector space to have two different zero vectors? That is, is it possible to have two *different* vectors $\mathbf{0}_1$ and $\mathbf{0}_2$ such that these vectors both satisfy Axiom 4? Explain your reasoning.

33. Do you think that it is possible for a vector **u** in a vector space to have two different negatives? That is, is it possible to have two *different* vectors $(-\mathbf{u})_1$ and $(-\mathbf{u})_2$, both of which satisfy Axiom 5? Explain your reasoning.

34. The set of ten axioms of a vector space is not an independent set because Axiom 2 can be deduced from other axioms in the set. Using the expression

$$(\mathbf{u} + \mathbf{v}) + (\mathbf{v} + \mathbf{u})$$

and Axiom 7 as a starting point, prove that $\mathbf{u} + \mathbf{v} = \mathbf{v} + \mathbf{u}$.

Hint You can use Theorem 5.1.1 since the proof of each part of that theorem does not use Axiom 2.

5.2
SUBSPACES

It is possible for one vector space to be contained within another vector space. For example, we showed in the preceding section that planes through the origin are vector spaces that are contained in the vector space R^3. In this section we shall study this important concept in detail.

A subset of a vector space V that is itself a vector space with respect to the operations of vector addition and scalar multiplication defined on V is given a special name.

> **DEFINITION**
>
> A subset W of a vector space V is called a **subspace** of V if W is itself a vector space under the addition and scalar multiplication defined on V.

In general, one must verify the ten vector space axioms to show that a set W with addition and scalar multiplication forms a vector space. However, if W is part of a larger set V that is already known to be a vector space, then certain axioms need not be verified for W because they are "inherited" from V. For example, there is no need to check that $\mathbf{u} + \mathbf{v} = \mathbf{v} + \mathbf{u}$ (Axiom 2) for W because this holds for all vectors in V and consequently for all vectors in W. Other axioms inherited by W from V are 3, 7, 8, 9, and 10. Thus, to show that a set W is a subspace of a vector space V, we need only verify Axioms 1, 4, 5, and 6. The following theorem shows that even Axioms 4 and 5 can be omitted.

THEOREM 5.2.1

If W is a set of one or more vectors from a vector space V, then W is a subspace of V if and only if the following conditions hold.

(a) If $\mathbf{u}$ and $\mathbf{v}$ are vectors in W, then $\mathbf{u} + \mathbf{v}$ is in W.

(b) If k is any scalar and $\mathbf{u}$ is any vector in W, then $k\mathbf{u}$ is in W.

Proof If W is a subspace of V, then all the vector space axioms are satisfied; in particular, Axioms 1 and 6 hold. But these are precisely conditions (a) and (b).

Conversely, assume conditions (a) and (b) hold. Since these conditions are vector space Axioms 1 and 6, we need only show that W satisfies the remaining eight axioms. Axioms 2, 3, 7, 8, 9, and 10 are automatically satisfied by the vectors in W since they are satisfied by all vectors in V. Therefore, to complete the proof, we need only verify that Axioms 4 and 5 are satisfied by vectors in W.

Let $\mathbf{u}$ be any vector in W. By condition (b), $k\mathbf{u}$ is in W for every scalar k. Setting $k = 0$, it follows from Theorem 5.1.1 that $0\mathbf{u} = \mathbf{0}$ is in W, and setting $k = -1$, it follows that $(-1)\mathbf{u} = -\mathbf{u}$ is in W. ∎

REMARK A set W of one or more vectors from a vector space V is said to be **closed under addition** if condition (a) in Theorem 5.2.1 holds and **closed under scalar multiplication** if condition (b) holds. Thus Theorem 5.2.1 states that W is a subspace of V if and only if W is closed under addition and closed under scalar multiplication.

EXAMPLE 1 Testing for a Subspace

In Example 6 of Section 5.1 we verified the ten vector space axioms to show that the points in a plane through the origin of R^3 form a subspace of R^3. In light of Theorem 5.2.1 we can see that much of that work was unnecessary; it would have been sufficient to verify that the plane is closed under addition and scalar multiplication (Axioms 1 and 6). In Section 5.1 we verified those two axioms algebraically; however, they can also be proved geometrically as follows: Let W be any plane through the origin, and let $\mathbf{u}$ and $\mathbf{v}$ be any vectors in W. Then $\mathbf{u} + \mathbf{v}$ must lie in W because it is the diagonal of the parallelogram determined by $\mathbf{u}$ and $\mathbf{v}$ (Figure 5.2.1), and $k\mathbf{u}$ must lie in W for any scalar k because $k\mathbf{u}$ lies on a line through $\mathbf{u}$. Thus W is closed under addition and scalar multiplication, so it is a subspace of R^3. ◆

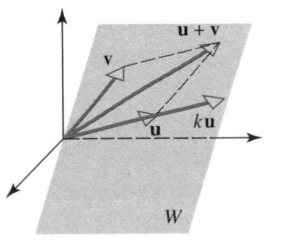

Figure 5.2.1 The vectors $\mathbf{u} + \mathbf{v}$ and $k\mathbf{u}$ both lie in the same plane as $\mathbf{u}$ and $\mathbf{v}$.

EXAMPLE 2 Lines through the Origin Are Subspaces

Show that a line through the origin of R^3 is a subspace of R^3.

Solution

Let W be a line through the origin of R^3. It is evident geometrically that the sum of two vectors on this line also lies on the line and that a scalar multiple of a vector on the

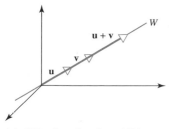

(a) W is closed under addition.

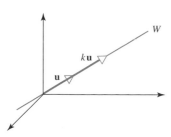

(b) W is closed under scalar multiplication.

Figure 5.2.2

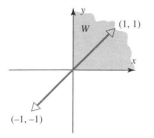

Figure 5.2.3 *W is not closed under scalar multiplication.*

line is on the line as well (Figure 5.2.2). Thus W is closed under addition and scalar multiplication, so it is a subspace of R^3. In the exercises we will ask you to prove this result algebraically using parametric equations for the line. ◆

EXAMPLE 3   A Subset of R^2 That Is Not a Subspace

Let W be the set of all points (x, y) in R^2 such that $x \geq 0$ and $y \geq 0$. These are the points in the first quadrant. The set W is *not* a subspace of R^2 since it is not closed under scalar multiplication. For example, $\mathbf{v} = (1, 1)$ lies in W, but its negative $(-1)\mathbf{v} = -\mathbf{v} = (-1, -1)$ does not (Figure 5.2.3). ◆

Every nonzero vector space V has at least two subspaces: V itself is a subspace, and the set $\{\mathbf{0}\}$ consisting of just the zero vector in V is a subspace called the ***zero subspace***. Combining this with Examples 1 and 2, we obtain the following list of subspaces of R^2 and R^3:

Subspaces of R^2	**Subspaces of R^3**
• $\{\mathbf{0}\}$	• $\{\mathbf{0}\}$
• Lines through the origin	• Lines through the origin
• R^2	• Planes through the origin
	• R^3

Later, we will show that these are the only subspaces of R^2 and R^3.

EXAMPLE 4   Subspaces of M_{nn}

From Theorem 1.7.2, the sum of two symmetric matrices is symmetric, and a scalar multiple of a symmetric matrix is symmetric. Thus the set of $n \times n$ symmetric matrices is a subspace of the vector space M_{nn} of all $n \times n$ matrices. Similarly, the set of $n \times n$ upper triangular matrices, the set of $n \times n$ lower triangular matrices, and the set of $n \times n$ diagonal matrices all form subspaces of M_{nn}, since each of these sets is closed under addition and scalar multiplication. ◆

EXAMPLE 5   A Subspace of Polynomials of Degree $\leq n$

Let n be a nonnegative integer, and let W consist of all functions expressible in the form

$$p(x) = a_0 + a_1 x + \cdots + a_n x^n \tag{1}$$

where $a_0, \ldots, a_n$ are real numbers. Thus W consists of all real polynomials of degree n or less. The set W is a subspace of the vector space of all real-valued functions discussed in Example 4 of the preceding section. To see this, let $\mathbf{p}$ and $\mathbf{q}$ be the polynomials

$$p(x) = a_0 + a_1 x + \cdots + a_n x^n \quad \text{and} \quad q(x) = b_0 + b_1 x + \cdots + b_n x^n$$

Then

$$(\mathbf{p} + \mathbf{q})(x) = p(x) + q(x) = (a_0 + b_0) + (a_1 + b_1)x + \cdots + (a_n + b_n)x^n$$

The CMYK Color Model

Color magazines and books are printed using what is called a **CMYK** *color model*. Colors in this model are created using four colored inks: cyan (C), magenta (M), yellow (Y), and black (K). The colors can be created either by mixing inks of the four types and printing with the mixed inks (the *spot color method*) or by printing dot patterns (called *rosettes*) with the four colors and allowing the reader's eye and perception process to create the desired color combination (the *process color method*). There is a numbering system for commercial inks, called the *Pantone Matching System*, that assigns every commercial ink color a number in accordance with its percentages of cyan, magenta, yellows, and black. One way to represent a Pantone color is by associating the four base colors with the vectors

$$\mathbf{c} = (1, 0, 0, 0) \quad \text{(pure cyan)}$$
$$\mathbf{m} = (0, 1, 0, 0) \quad \text{(pure magenta)}$$
$$\mathbf{y} = (0, 0, 1, 0) \quad \text{(pure yellow)}$$
$$\mathbf{k} = (0, 0, 0, 1) \quad \text{(pure black)}$$

in R^4 and describing the ink color as a linear combination of these using coefficients between 0 and 1, inclusive. Thus, an ink color $\mathbf{p}$ is represented as a linear combination of the form

$$\mathbf{p} = c_1\mathbf{c} + c_2\mathbf{m} + c_3\mathbf{y} + c_4\mathbf{k} = (c_1, c_2, c_3, c_4)$$

where $0 \le c_i \le 1$. The set of all such linear combinations is called **CMYK** *space*, although it is not a subspace of R^4. (Why?) For example, Pantone color 876CVC is a mixture of 38% cyan, 59% magenta, 73% yellow, and 7% black; Pantone color 216CVC is a mixture of 0% cyan, 83% magenta, 34% yellow, and 47% black; and Pantone color 328CVC is a mixture of 100% cyan, 0% magenta, 47% yellow, and 30% black. We can denote these colors by $\mathbf{p}_{876} = (0.38, 0.59, 0.73, 0.07)$, $\mathbf{p}_{216} = (0, 0.83, 0.34, 0.47)$, and $\mathbf{p}_{328} = (1, 0, 0.47, 0.30)$, respectively.

and

$$(k\mathbf{p})(x) = kp(x) = (ka_0) + (ka_1)x + \cdots + (ka_n)x^n$$

These functions have the form given in (1), so $\mathbf{p} + \mathbf{q}$ and $k\mathbf{p}$ lie in W. As in Section 4.4, we shall denote the vector space W in this example by the symbol P_n. ◆

Calculus Required   **EXAMPLE 6   Subspaces of Functions Continuous on $(-\infty, \infty)$**

Recall from calculus that if $\mathbf{f}$ and $\mathbf{g}$ are continuous functions on the interval $(-\infty, \infty)$ and k is a constant, then $\mathbf{f} + \mathbf{g}$ and $k\mathbf{f}$ are also continuous. Thus the continuous functions on the interval $(-\infty, \infty)$ form a subspace of $F(-\infty, \infty)$, since they are closed under addition and scalar multiplication. We denote this subspace by $C(-\infty, \infty)$. Similarly, if $\mathbf{f}$ and $\mathbf{g}$ have continuous first derivatives on $(-\infty, \infty)$, then so do $\mathbf{f} + \mathbf{g}$ and $k\mathbf{f}$. Thus the functions with continuous first derivatives on $(-\infty, \infty)$ form a subspace of $F(-\infty, \infty)$. We denote this subspace by $C^1(-\infty, \infty)$, where the superscript 1 is used to emphasize the *first* derivative. However, it is a theorem of calculus that every differentiable function is continuous, so $C^1(-\infty, \infty)$ is actually a subspace of $C(-\infty, \infty)$.

To take this a step further, for each positive integer m, the functions with continuous mth derivatives on $(-\infty, \infty)$ form a subspace of $C^1(-\infty, \infty)$ as do the functions that have continuous derivatives of all orders. We denote the subspace of functions with continuous mth derivatives on $(-\infty, \infty)$ by $C^m(-\infty, \infty)$, and we denote the subspace of functions that have continuous derivatives of all orders on $(-\infty, \infty)$ by $C^\infty(-\infty, \infty)$. Finally, it is a theorem of calculus that polynomials have continuous derivatives of all orders, so P_n is a subspace of $C^\infty(-\infty, \infty)$. The hierarchy of subspaces discussed in this example is illustrated in Figure 5.2.4. ◆

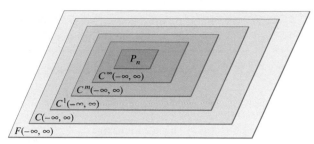

Figure 5.2.4

REMARK In the preceding example we focused on the interval $(-\infty, \infty)$. Had we focused on a closed interval $[a, b]$, then the subspaces corresponding to those defined in the example would be denoted by $C[a, b]$, $C^m[a, b]$, and $C^\infty[a, b]$. Similarly, on an open interval (a, b) they would be denoted by $C(a, b)$, $C^m(a, b)$, and $C^\infty(a, b)$.

Solution Spaces of Homogeneous Systems

If $A\mathbf{x} = \mathbf{b}$ is a system of linear equations, then each vector $\mathbf{x}$ that satisfies this equation is called a ***solution vector*** of the system. The following theorem shows that the solution vectors of a *homogeneous* linear system form a vector space, which we shall call the ***solution space*** of the system.

THEOREM 5.2.2

> *If $A\mathbf{x} = \mathbf{0}$ is a homogeneous linear system of m equations in n unknowns, then the set of solution vectors is a subspace of R^n.*

Proof Let W be the set of solution vectors. There is at least one vector in W, namely **0**. To show that W is closed under addition and scalar multiplication, we must show that if $\mathbf{x}$ and $\mathbf{x}'$ are any solution vectors and k is any scalar, then $\mathbf{x} + \mathbf{x}'$ and $k\mathbf{x}$ are also solution vectors. But if $\mathbf{x}$ and $\mathbf{x}'$ are solution vectors, then

$$A\mathbf{x} = \mathbf{0} \quad \text{and} \quad A\mathbf{x}' = \mathbf{0}$$

from which it follows that

$$A(\mathbf{x} + \mathbf{x}') = A\mathbf{x} + A\mathbf{x}' = \mathbf{0} + \mathbf{0} = \mathbf{0}$$

and

$$A(k\mathbf{x}) = kA\mathbf{x} = k\mathbf{0} = \mathbf{0}$$

which proves that $\mathbf{x} + \mathbf{x}'$ and $k\mathbf{x}$ are solution vectors. ∎

EXAMPLE 7 Solution Spaces That Are Subspaces of R^3

Consider the linear systems

(a) $\begin{bmatrix} 1 & -2 & 3 \\ 2 & -4 & 6 \\ 3 & -6 & 9 \end{bmatrix} \begin{bmatrix} x \\ y \\ z \end{bmatrix} = \begin{bmatrix} 0 \\ 0 \\ 0 \end{bmatrix}$ (b) $\begin{bmatrix} 1 & -2 & 3 \\ -3 & 7 & -8 \\ -2 & 4 & -6 \end{bmatrix} \begin{bmatrix} x \\ y \\ z \end{bmatrix} = \begin{bmatrix} 0 \\ 0 \\ 0 \end{bmatrix}$

(c) $\begin{bmatrix} 1 & -2 & 3 \\ -3 & 7 & -8 \\ 4 & 1 & 2 \end{bmatrix} \begin{bmatrix} x \\ y \\ z \end{bmatrix} = \begin{bmatrix} 0 \\ 0 \\ 0 \end{bmatrix}$ (d) $\begin{bmatrix} 0 & 0 & 0 \\ 0 & 0 & 0 \\ 0 & 0 & 0 \end{bmatrix} \begin{bmatrix} x \\ y \\ z \end{bmatrix} = \begin{bmatrix} 0 \\ 0 \\ 0 \end{bmatrix}$

Each of these systems has three unknowns, so the solutions form subspaces of R^3. Geometrically, this means that each solution space must be the origin only, a line through the origin, a plane through the origin, or all of R^3. We shall now verify that this is so (leaving it to the reader to solve the systems).

Solution

(a) The solutions are
$$x = 2s - 3t, \qquad y = s, \qquad z = t$$
from which it follows that
$$x = 2y - 3z \quad \text{or} \quad x - 2y + 3z = 0$$
This is the equation of the plane through the origin with $\mathbf{n} = (1, -2, 3)$ as a normal vector.

(b) The solutions are
$$x = -5t, \qquad y = -t, \qquad z = t$$
which are parametric equations for the line through the origin parallel to the vector $\mathbf{v} = (-5, -1, 1)$.

(c) The solution is $x = 0, y = 0, z = 0$, so the solution space is the origin only—that is, $\{\mathbf{0}\}$.

(d) The solutions are
$$x = r, \qquad y = s, \qquad z = t$$
where r, s, and t have arbitrary values, so the solution space is all of R^3. ◆

In Section 1.3 we introduced the concept of a linear combination of column vectors. The following definition extends this idea to more general vectors.

DEFINITION

A vector $\mathbf{w}$ is called a *linear combination* of the vectors $\mathbf{v}_1, \mathbf{v}_2, \ldots, \mathbf{v}_r$ if it can be expressed in the form
$$\mathbf{w} = k_1\mathbf{v}_1 + k_2\mathbf{v}_2 + \cdots + k_r\mathbf{v}_r$$
where $k_1, k_2, \ldots, k_r$ are scalars.

REMARK   If $r = 1$, then the equation in the preceding definition reduces to $\mathbf{w} = k_1\mathbf{v}_1$; that is, $\mathbf{w}$ is a linear combination of a single vector $\mathbf{v}_1$ if it is a scalar multiple of $\mathbf{v}_1$.

EXAMPLE 8   Vectors in R^3 Are Linear Combinations of i, j, and k

---

Every vector $\mathbf{v} = (a, b, c)$ in R^3 is expressible as a linear combination of the standard basis vectors
$$\mathbf{i} = (1, 0, 0), \qquad \mathbf{j} = (0, 1, 0), \qquad \mathbf{k} = (0, 0, 1)$$
since
$$\mathbf{v} = (a, b, c) = a(1, 0, 0) + b(0, 1, 0) + c(0, 0, 1) = a\mathbf{i} + b\mathbf{j} + c\mathbf{k} \quad ◆$$

EXAMPLE 9 Checking a Linear Combination

Consider the vectors $\mathbf{u} = (1, 2, -1)$ and $\mathbf{v} = (6, 4, 2)$ in R^3. Show that $\mathbf{w} = (9, 2, 7)$ is a linear combination of $\mathbf{u}$ and $\mathbf{v}$ and that $\mathbf{w}' = (4, -1, 8)$ is *not* a linear combination of $\mathbf{u}$ and $\mathbf{v}$.

Solution

In order for $\mathbf{w}$ to be a linear combination of $\mathbf{u}$ and $\mathbf{v}$, there must be scalars k_1 and k_2 such that $\mathbf{w} = k_1\mathbf{u} + k_2\mathbf{v}$; that is,

$$(9, 2, 7) = k_1(1, 2, -1) + k_2(6, 4, 2)$$

or

$$(9, 2, 7) = (k_1 + 6k_2, 2k_1 + 4k_2, -k_1 + 2k_2)$$

Equating corresponding components gives

$$k_1 + 6k_2 = 9$$
$$2k_1 + 4k_2 = 2$$
$$-k_1 + 2k_2 = 7$$

Solving this system using Gaussian elimination yields $k_1 = -3$, $k_2 = 2$, so

$$\mathbf{w} = -3\mathbf{u} + 2\mathbf{v}$$

Similarly, for $\mathbf{w}'$ to be a linear combination of $\mathbf{u}$ and $\mathbf{v}$, there must be scalars k_1 and k_2 such that $\mathbf{w}' = k_1\mathbf{u} + k_2\mathbf{v}$; that is,

$$(4, -1, 8) = k_1(1, 2, -1) + k_2(6, 4, 2)$$

or

$$(4, -1, 8) = (k_1 + 6k_2, 2k_1 + 4k_2, -k_1 + 2k_2)$$

Equating corresponding components gives

$$k_1 + 6k_2 = 4$$
$$2k_1 + 4k_2 = -1$$
$$-k_1 + 2k_2 = 8$$

This system of equations is inconsistent (verify), so no such scalars k_1 and k_2 exist. Consequently, $\mathbf{w}'$ is not a linear combination of $\mathbf{u}$ and $\mathbf{v}$. ◆

Spanning

If $\mathbf{v}_1, \mathbf{v}_2, \ldots, \mathbf{v}_r$ are vectors in a vector space V, then generally some vectors in V may be linear combinations of $\mathbf{v}_1, \mathbf{v}_2, \ldots, \mathbf{v}_r$ and others may not. The following theorem shows that if we construct a set W consisting of all those vectors that are expressible as linear combinations of $\mathbf{v}_1, \mathbf{v}_2, \ldots, \mathbf{v}_r$, then W forms a subspace of V.

THEOREM 5.2.3

If $\mathbf{v}_1, \mathbf{v}_2, \ldots, \mathbf{v}_r$ are vectors in a vector space V, then

(a) The set W of all linear combinations of $\mathbf{v}_1, \mathbf{v}_2, \ldots, \mathbf{v}_r$ is a subspace of V.

(b) W is the smallest subspace of V that contains $\mathbf{v}_1, \mathbf{v}_2, \ldots, \mathbf{v}_r$ in the sense that every other subspace of V that contains $\mathbf{v}_1, \mathbf{v}_2, \ldots, \mathbf{v}_r$ must contain W.

Proof (*a*)   To show that W is a subspace of V, we must prove that it is closed under addition and scalar multiplication. There is at least one vector in W—namely $\mathbf{0}$, since $\mathbf{0} = 0\mathbf{v}_1 + 0\mathbf{v}_2 + \cdots + 0\mathbf{v}_r$. If $\mathbf{u}$ and $\mathbf{v}$ are vectors in W, then

$$\mathbf{u} = c_1\mathbf{v}_1 + c_2\mathbf{v}_2 + \cdots + c_r\mathbf{v}_r$$

and

$$\mathbf{v} = k_1\mathbf{v}_1 + k_2\mathbf{v}_2 + \cdots + k_r\mathbf{v}_r$$

where $c_1, c_2, \ldots, c_r, k_1, k_2, \ldots, k_r$ are scalars. Therefore,

$$\mathbf{u} + \mathbf{v} = (c_1 + k_1)\mathbf{v}_1 + (c_2 + k_2)\mathbf{v}_2 + \cdots + (c_r + k_r)\mathbf{v}_r$$

and, for any scalar k,

$$k\mathbf{u} = (kc_1)\mathbf{v}_1 + (kc_2)\mathbf{v}_2 + \cdots + (kc_r)\mathbf{v}_r$$

Thus $\mathbf{u} + \mathbf{v}$ and $k\mathbf{u}$ are linear combinations of $\mathbf{v}_1, \mathbf{v}_2, \ldots, \mathbf{v}_r$ and consequently lie in W. Therefore, W is closed under addition and scalar multiplication.

Proof (*b*)   Each vector $\mathbf{v}_i$ is a linear combination of $\mathbf{v}_1, \mathbf{v}_2, \ldots, \mathbf{v}_r$ since we can write

$$\mathbf{v}_i = 0\mathbf{v}_1 + 0\mathbf{v}_2 + \cdots + 1\mathbf{v}_i + \cdots + 0\mathbf{v}_r$$

Therefore, the subspace W contains each of the vectors $\mathbf{v}_1, \mathbf{v}_2, \ldots, \mathbf{v}_r$. Let W' be any other subspace that contains $\mathbf{v}_1, \mathbf{v}_2, \ldots, \mathbf{v}_r$. Since W' is closed under addition and scalar multiplication, it must contain all linear combinations of $\mathbf{v}_1, \mathbf{v}_2, \ldots, \mathbf{v}_r$. Thus, W' contains each vector of W. $\blacksquare$

We make the following definition.

DEFINITION

If $S = \{\mathbf{v}_1, \mathbf{v}_2, \ldots, \mathbf{v}_r\}$ is a set of vectors in a vector space V, then the subspace W of V consisting of all linear combinations of the vectors in S is called the ***space spanned*** by $\mathbf{v}_1, \mathbf{v}_2, \ldots, \mathbf{v}_r$, and we say that the vectors $\mathbf{v}_1, \mathbf{v}_2, \ldots, \mathbf{v}_r$ ***span*** W. To indicate that W is the space spanned by the vectors in the set $S = \{\mathbf{v}_1, \mathbf{v}_2, \ldots, \mathbf{v}_r\}$, we write

$$W = \text{span}(S) \quad \text{or} \quad W = \text{span}\{\mathbf{v}_1, \mathbf{v}_2, \ldots, \mathbf{v}_r\}$$

EXAMPLE 10   Spaces Spanned by One or Two Vectors

If $\mathbf{v}_1$ and $\mathbf{v}_2$ are noncollinear vectors in R^3 with their initial points at the origin, then $\text{span}\{\mathbf{v}_1, \mathbf{v}_2\}$, which consists of all linear combinations $k_1\mathbf{v}_1 + k_2\mathbf{v}_2$, is the plane determined by $\mathbf{v}_1$ and $\mathbf{v}_2$ (see Figure 5.2.5*a*). Similarly, if $\mathbf{v}$ is a nonzero vector in R^2 or R^3, then $\text{span}\{\mathbf{v}\}$, which is the set of all scalar multiples $k\mathbf{v}$, is the line determined by $\mathbf{v}$ (see Figure 5.2.5*b*). $\blacklozenge$

EXAMPLE 11   Spanning Set for P_n

The polynomials $1, x, x^2, \ldots, x^n$ span the vector space P_n defined in Example 5 since each polynomial $\mathbf{p}$ in P_n can be written as

$$\mathbf{p} = a_0 + a_1 x + \cdots + a_n x^n$$

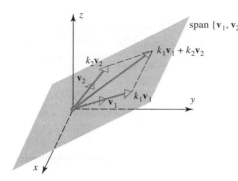

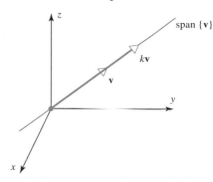

(*a*) Span $\{\mathbf{v}_1, \mathbf{v}_2\}$ is the plane through the origin determined by $\mathbf{v}_1$ and $\mathbf{v}_2$.

(*b*) Span $\{\mathbf{v}\}$ is the line through the origin determined by $\mathbf{v}$.

Figure 5.2.5

which is a linear combination of $1, x, x^2, \ldots, x^n$. We can denote this by writing

$$P_n = \text{span}\{1, x, x^2, \ldots, x^n\} \quad \blacklozenge$$

EXAMPLE 12 Three Vectors That Do Not Span R^3

Determine whether $\mathbf{v}_1 = (1, 1, 2)$, $\mathbf{v}_2 = (1, 0, 1)$, and $\mathbf{v}_3 = (2, 1, 3)$ span the vector space R^3.

Solution

We must determine whether an arbitrary vector $\mathbf{b} = (b_1, b_2, b_3)$ in R^3 can be expressed as a linear combination

$$\mathbf{b} = k_1\mathbf{v}_1 + k_2\mathbf{v}_2 + k_3\mathbf{v}_3$$

of the vectors $\mathbf{v}_1$, $\mathbf{v}_2$, and $\mathbf{v}_3$. Expressing this equation in terms of components gives

$$(b_1, b_2, b_3) = k_1(1, 1, 2) + k_2(1, 0, 1) + k_3(2, 1, 3)$$

or

$$(b_1, b_2, b_3) = (k_1 + k_2 + 2k_3, k_1 + k_3, 2k_1 + k_2 + 3k_3)$$

or

$$
\begin{aligned}
k_1 + k_2 + 2k_3 &= b_1 \\
k_1 \quad\;\; + k_3 &= b_2 \\
2k_1 + k_2 + 3k_3 &= b_3
\end{aligned}
$$

The problem thus reduces to determining whether this system is consistent for all values of b_1, b_2, and b_3. By parts (*e*) and (*g*) of Theorem 4.3.4, this system is consistent for all b_1, b_2, and b_3 if and only if the coefficient matrix

$$A = \begin{bmatrix} 1 & 1 & 2 \\ 1 & 0 & 1 \\ 2 & 1 & 3 \end{bmatrix}$$

has a nonzero determinant. However, $\det(A) = 0$ (verify), so $\mathbf{v}_1$, $\mathbf{v}_2$, and $\mathbf{v}_3$ do not span R^3. $\blacklozenge$

Spanning sets are not unique. For example, any two noncollinear vectors that lie in the plane shown in Figure 5.2.5 will span that same plane, and any nonzero vector on the line in that figure will span the same line. We leave the proof of the following useful theorem as an exercise.

THEOREM 5.2.4

If $S = \{\mathbf{v}_1, \mathbf{v}_2, \ldots, \mathbf{v}_r\}$ and $S' = \{\mathbf{w}_1, \mathbf{w}_2, \ldots, \mathbf{w}_k\}$ are two sets of vectors in a vector space V, then

$$span\{\mathbf{v}_1, \mathbf{v}_2, \ldots, \mathbf{v}_r\} = span\{\mathbf{w}_1, \mathbf{w}_2, \ldots, \mathbf{w}_k\}$$

if and only if each vector in S is a linear combination of those in S' and each vector in S' is a linear combination of those in S.

EXERCISE SET 5.2

1. Use Theorem 5.2.1 to determine which of the following are subspaces of R^3.

   (a) all vectors of the form $(a, 0, 0)$

   (b) all vectors of the form $(a, 1, 1)$

   (c) all vectors of the form (a, b, c), where $b = a + c$

   (d) all vectors of the form (a, b, c), where $b = a + c + 1$

   (e) all vectors of the form $(a, b, 0)$

2. Use Theorem 5.2.1 to determine which of the following are subspaces of M_{22}.

   (a) all 2×2 matrices with integer entries

   (b) all matrices

   $$\begin{bmatrix} a & b \\ c & d \end{bmatrix}$$

   where $a + b + c + d = 0$

   (c) all 2×2 matrices A such that $\det(A) = 0$

   (d) all matrices of the form

   $$\begin{bmatrix} a & b \\ 0 & c \end{bmatrix}$$

   (e) all matrices of the form

   $$\begin{bmatrix} a & a \\ -a & -a \end{bmatrix}$$

3. Use Theorem 5.2.1 to determine which of the following are subspaces of P_3.

   (a) all polynomials $a_0 + a_1 x + a_2 x^2 + a_3 x^3$ for which $a_0 = 0$

   (b) all polynomials $a_0 + a_1 x + a_2 x^2 + a_3 x^3$ for which $a_0 + a_1 + a_2 + a_3 = 0$

   (c) all polynomials $a_0 + a_1 x + a_2 x^2 + a_3 x^3$ for which $a_0, a_1, a_2,$ and a_3 are integers

   (d) all polynomials of the form $a_0 + a_1 x$, where a_0 and a_1 are real numbers

4. Use Theorem 5.2.1 to determine which of the following are subspaces of the space $F(-\infty, \infty)$.

   (a) all f such that $f(x) \leq 0$ for all x (b) all f such that $f(0) = 0$

   (c) all f such that $f(0) = 2$ (d) all constant functions

   (e) all f of the form $k_1 + k_2 \sin x$, where k_1 and k_2 are real numbers

5. Use Theorem 5.2.1 to determine which of the following are subspaces of M_{nn}.

   (a) all $n \times n$ matrices A such that $\operatorname{tr}(A) = 0$

   (b) all $n \times n$ matrices A such that $A^T = -A$

(c) all $n \times n$ matrices A such that the linear system $A\mathbf{x} = \mathbf{0}$ has only the trivial solution

(d) all $n \times n$ matrices A such that $AB = BA$ for a fixed $n \times n$ matrix B

6. Determine whether the solution space of the system $A\mathbf{x} = \mathbf{0}$ is a line through the origin, a plane through the origin, or the origin only. If it is a plane, find an equation for it; if it is a line, find parametric equations for it.

(a) $A = \begin{bmatrix} -1 & 1 & 1 \\ 3 & -1 & 0 \\ 2 & -4 & -5 \end{bmatrix}$　(b) $A = \begin{bmatrix} 1 & -2 & 3 \\ -3 & 6 & 9 \\ -2 & 4 & -6 \end{bmatrix}$

(c) $A = \begin{bmatrix} 1 & 2 & 3 \\ 2 & 5 & 3 \\ 1 & 0 & 8 \end{bmatrix}$　(d) $A = \begin{bmatrix} 1 & 2 & -6 \\ 1 & 4 & 4 \\ 3 & 10 & 6 \end{bmatrix}$

(e) $A = \begin{bmatrix} 1 & -1 & 1 \\ 2 & -1 & 4 \\ 3 & 1 & 11 \end{bmatrix}$　(f) $A = \begin{bmatrix} 1 & -3 & 1 \\ 2 & -6 & 2 \\ 3 & -9 & 3 \end{bmatrix}$

7. Which of the following are linear combinations of $\mathbf{u} = (0, -2, 2)$ and $\mathbf{v} = (1, 3, -1)$?

(a) $(2, 2, 2)$　　(b) $(3, 1, 5)$　　(c) $(0, 4, 5)$　　(d) $(0, 0, 0)$

8. Express the following as linear combinations of $\mathbf{u} = (2, 1, 4)$, $\mathbf{v} = (1, -1, 3)$, and $\mathbf{w} = (3, 2, 5)$.

(a) $(-9, -7, -15)$　　(b) $(6, 11, 6)$　　(c) $(0, 0, 0)$　　(d) $(7, 8, 9)$

9. Express the following as linear combinations of $\mathbf{p}_1 = 2 + x + 4x^2$, $\mathbf{p}_2 = 1 - x + 3x^2$, and $\mathbf{p}_3 = 3 + 2x + 5x^2$.

(a) $-9 - 7x - 15x^2$　　(b) $6 + 11x + 6x^2$　　(c) 0　　(d) $7 + 8x + 9x^2$

10. Which of the following are linear combinations of

$$A = \begin{bmatrix} 4 & 0 \\ -2 & -2 \end{bmatrix}, \quad B = \begin{bmatrix} 1 & -1 \\ 2 & 3 \end{bmatrix}, \quad C = \begin{bmatrix} 0 & 2 \\ 1 & 4 \end{bmatrix}?$$

(a) $\begin{bmatrix} 6 & -8 \\ -1 & -8 \end{bmatrix}$　(b) $\begin{bmatrix} 0 & 0 \\ 0 & 0 \end{bmatrix}$　(c) $\begin{bmatrix} 6 & 0 \\ 3 & 8 \end{bmatrix}$　(d) $\begin{bmatrix} -1 & 5 \\ 7 & 1 \end{bmatrix}$

11. In each part, determine whether the given vectors span R^3.

(a) $\mathbf{v}_1 = (2, 2, 2)$, $\mathbf{v}_2 = (0, 0, 3)$, $\mathbf{v}_3 = (0, 1, 1)$

(b) $\mathbf{v}_1 = (2, -1, 3)$, $\mathbf{v}_2 = (4, 1, 2)$, $\mathbf{v}_3 = (8, -1, 8)$

(c) $\mathbf{v}_1 = (3, 1, 4)$, $\mathbf{v}_2 = (2, -3, 5)$, $\mathbf{v}_3 = (5, -2, 9)$, $\mathbf{v}_4 = (1, 4, -1)$

(d) $\mathbf{v}_1 = (1, 2, 6)$, $\mathbf{v}_2 = (3, 4, 1)$, $\mathbf{v}_3 = (4, 3, 1)$, $\mathbf{v}_4 = (3, 3, 1)$

12. Let $\mathbf{f} = \cos^2 x$ and $\mathbf{g} = \sin^2 x$. Which of the following lie in the space spanned by $\mathbf{f}$ and $\mathbf{g}$?

(a) $\cos 2x$　　(b) $3 + x^2$　　(c) 1　　(d) $\sin x$　　(e) 0

13. Determine whether the following polynomials span P_2.

$$\mathbf{p}_1 = 1 - x + 2x^2, \qquad \mathbf{p}_2 = 3 + x, \qquad \mathbf{p}_3 = 5 - x + 4x^2, \qquad \mathbf{p}_4 = -2 - 2x + 2x^2$$

14. Let $\mathbf{v}_1 = (2, 1, 0, 3)$, $\mathbf{v}_2 = (3, -1, 5, 2)$, and $\mathbf{v}_3 = (-1, 0, 2, 1)$. Which of the following vectors are in span$\{\mathbf{v}_1, \mathbf{v}_2, \mathbf{v}_3\}$?

(a) $(2, 3, -7, 3)$　　(b) $(0, 0, 0, 0)$　　(c) $(1, 1, 1, 1)$　　(d) $(-4, 6, -13, 4)$

15. Find an equation for the plane spanned by the vectors $\mathbf{u} = (-1, 1, 1)$ and $\mathbf{v} = (3, 4, 4)$.

16. Find parametric equations for the line spanned by the vector $\mathbf{u} = (3, -2, 5)$.

17. Show that the solution vectors of a consistent nonhomogeneous system of m linear equations in n unknowns do not form a subspace of R^n.

18. Prove Theorem 5.2.4.

19. Use Theorem 5.2.4 to show that $\mathbf{v}_1 = (1, 6, 4)$, $\mathbf{v}_2 = (2, 4, -1)$, $\mathbf{v}_3 = (-1, 2, 5)$, and $\mathbf{w}_1 = (1, -2, -5)$, $\mathbf{w}_2 = (0, 8, 9)$ span the same subspace of R^3.

20. A line L through the origin in R^3 can be represented by parametric equations of the form $x = at$, $y = bt$, and $z = ct$. Use these equations to show that L is a subspace of R^3; that is, show that if $\mathbf{v}_1 = (x_1, y_1, z_1)$ and $\mathbf{v}_2 = (x_2, y_2, z_2)$ are points on L and k is any real number, then $k\mathbf{v}_1$ and $\mathbf{v}_1 + \mathbf{v}_2$ are also points on L.

21. **(For Readers Who Have Studied Calculus)** Show that the following sets of functions are subspaces of $F(-\infty, \infty)$.

    (a) all everywhere continuous functions

    (b) all everywhere differentiable functions

    (c) all everywhere differentiable functions that satisfy $\mathbf{f}' + 2\mathbf{f} = \mathbf{0}$

22. **(For Readers Who Have Studied Calculus)** Show that the set of continuous functions $\mathbf{f} = f(x)$ on $[a, b]$ such that

$$\int_a^b f(x)\, dx = 0$$

is a subspace of $C[a, b]$.

Discussion
& Discovery

23. Indicate whether each statement is always true or sometimes false. Justify your answer by giving a logical argument or a counterexample.

    (a) If $A\mathbf{x} = \mathbf{b}$ is any consistent linear system of m equations in n unknowns, then the solution set is a subspace of R^n.

    (b) If W is a set of one or more vectors from a vector space V, and if $k\mathbf{u} + \mathbf{v}$ is a vector in W for all vectors $\mathbf{u}$ and $\mathbf{v}$ in W and for all scalars k, then W is a subspace of V.

    (c) If S is a finite set of vectors in a vector space V, then span(S) must be closed under addition and scalar multiplication.

    (d) The intersection of two subspaces of a vector space V is also a subspace of V.

    (e) If span(S_1) = span(S_2), then $S_1 = S_2$.

24. (a) Under what conditions will two vectors in R^3 span a plane? A line?

    (b) Under what conditions will it be true that span$\{\mathbf{u}\}$ = span$\{\mathbf{v}\}$? Explain.

    (c) If $A\mathbf{x} = \mathbf{b}$ is a consistent system of m equations in n unknowns, under what conditions will it be true that the solution set is a subspace of R^n? Explain.

25. Recall that lines through the origin are subspaces of R^2. If W_1 is the line $y = x$ and W_2 is the line $y = -x$, is the union $W_1 \cup W_2$ a subspace of R^2? Explain your reasoning.

26. (a) Let M_{22} be the vector space of 2×2 matrices. Find four matrices that span M_{22}.

    (b) In words, describe a set of matrices that spans M_{nn}.

27. We showed in Example 8 that the vectors $\mathbf{i}, \mathbf{j}, \mathbf{k}$ span R^3. However, spanning sets are not unique. What geometric property must a set of three vectors in R^3 have if they are to span R^3?

5.3

LINEAR INDEPENDENCE

In the preceding section we learned that a set of vectors $S = \{\mathbf{v}_1, \mathbf{v}_2, \ldots, \mathbf{v}_r\}$ spans a given vector space V if every vector in V is expressible as a linear combination of the vectors in S. In general, there may be more than one way to express a vector in V as a linear combination of vectors in a spanning set. In this section we shall study conditions under which each vector in V is expressible as a linear combination of the spanning vectors in exactly one way. Spanning sets with this property play a fundamental role in the study of vector spaces.

> **DEFINITION**
>
> If $S = \{\mathbf{v}_1, \mathbf{v}_2, \ldots, \mathbf{v}_r\}$ is a nonempty set of vectors, then the vector equation
>
> $$k_1\mathbf{v}_1 + k_2\mathbf{v}_2 + \cdots + k_r\mathbf{v}_r = \mathbf{0}$$
>
> has at least one solution, namely
>
> $$k_1 = 0, \quad k_2 = 0, \ldots, \quad k_r = 0$$
>
> If this is the only solution, then S is called a ***linearly independent*** set. If there are other solutions, then S is called a ***linearly dependent*** set.

EXAMPLE 1   A Linearly Dependent Set

If $\mathbf{v}_1 = (2, -1, 0, 3)$, $\mathbf{v}_2 = (1, 2, 5, -1)$, and $\mathbf{v}_3 = (7, -1, 5, 8)$, then the set of vectors $S = \{\mathbf{v}_1, \mathbf{v}_2, \mathbf{v}_3\}$ is linearly dependent, since $3\mathbf{v}_1 + \mathbf{v}_2 - \mathbf{v}_3 = \mathbf{0}$. $\blacklozenge$

EXAMPLE 2   A Linearly Dependent Set

The polynomials

$$\mathbf{p}_1 = 1 - x, \quad \mathbf{p}_2 = 5 + 3x - 2x^2, \quad \text{and} \quad \mathbf{p}_3 = 1 + 3x - x^2$$

form a linearly dependent set in P_2 since $3\mathbf{p}_1 - \mathbf{p}_2 + 2\mathbf{p}_3 = \mathbf{0}$. $\blacklozenge$

EXAMPLE 3   Linearly Independent Sets

Consider the vectors $\mathbf{i} = (1, 0, 0)$, $\mathbf{j} = (0, 1, 0)$, and $\mathbf{k} = (0, 0, 1)$ in R^3. In terms of components, the vector equation

$$k_1\mathbf{i} + k_2\mathbf{j} + k_3\mathbf{k} = \mathbf{0}$$

becomes

$$k_1(1, 0, 0) + k_2(0, 1, 0) + k_3(0, 0, 1) = (0, 0, 0)$$

or, equivalently,

$$(k_1, k_2, k_3) = (0, 0, 0)$$

This implies that $k_1 = 0$, $k_2 = 0$, and $k_3 = 0$, so the set $S = \{\mathbf{i}, \mathbf{j}, \mathbf{k}\}$ is linearly independent. A similar argument can be used to show that the vectors

$$\mathbf{e}_1 = (1, 0, 0, \ldots, 0), \quad \mathbf{e}_2 = (0, 1, 0, \ldots, 0), \ldots, \quad \mathbf{e}_n = (0, 0, 0, \ldots, 1)$$

form a linearly independent set in R^n. $\blacklozenge$

EXAMPLE 4   Determining Linear Independence/Dependence

Determine whether the vectors

$$\mathbf{v}_1 = (1, -2, 3), \qquad \mathbf{v}_2 = (5, 6, -1), \qquad \mathbf{v}_3 = (3, 2, 1)$$

form a linearly dependent set or a linearly independent set.

Solution

In terms of components, the vector equation

$$k_1\mathbf{v}_1 + k_2\mathbf{v}_2 + k_3\mathbf{v}_3 = \mathbf{0}$$

becomes

$$k_1(1, -2, 3) + k_2(5, 6, -1) + k_3(3, 2, 1) = (0, 0, 0)$$

or, equivalently,

$$(k_1 + 5k_2 + 3k_3, -2k_1 + 6k_2 + 2k_3, 3k_1 - k_2 + k_3) = (0, 0, 0)$$

Equating corresponding components gives

$$k_1 + 5k_2 + 3k_3 = 0$$
$$-2k_1 + 6k_2 + 2k_3 = 0$$
$$3k_1 - k_2 + k_3 = 0$$

Thus $\mathbf{v}_1$, $\mathbf{v}_2$, and $\mathbf{v}_3$ form a linearly dependent set if this system has a nontrivial solution, or a linearly independent set if it has only the trivial solution. Solving this system using Gaussian elimination yields

$$k_1 = -\tfrac{1}{2}t, \qquad k_2 = -\tfrac{1}{2}t, \qquad k_3 = t$$

Thus the system has nontrivial solutions and $\mathbf{v}_1$, $\mathbf{v}_2$, and $\mathbf{v}_3$ form a linearly dependent set. Alternatively, we could show the existence of nontrivial solutions without solving the system by showing that the coefficient matrix has determinant zero and consequently is not invertible (verify). ◆

EXAMPLE 5 Linearly Independent Set in P_n

Show that the polynomials

$$1, x, x^2, \ldots, x^n$$

form a linearly independent set of vectors in P_n.

Solution

Let $\mathbf{p}_0 = 1$, $\mathbf{p}_1 = x$, $\mathbf{p}_2 = x^2, \ldots, \mathbf{p}_n = x^n$ and assume that some linear combination of these polynomials is zero, say

$$a_0\mathbf{p}_0 + a_1\mathbf{p}_1 + a_2\mathbf{p}_2 + \cdots + a_n\mathbf{p}_n = \mathbf{0}$$

or, equivalently,

$$a_0 + a_1x + a_2x^2 + \cdots + a_nx^n = 0 \quad \text{for all } x \text{ in } (-\infty, \infty) \tag{1}$$

We must show that

$$a_0 = a_1 = a_2 = \cdots = a_n = 0$$

To see that this is so, recall from algebra that a *nonzero* polynomial of degree n has at most n distinct roots. But this implies that $a_0 = a_1 = a_2 = \cdots = a_n = 0$; otherwise, it would follow from (1) that $a_0 + a_1x + a_2x^2 + \cdots + a_nx^n$ is a nonzero polynomial with infinitely many roots. ◆

The term *linearly dependent* suggests that the vectors "depend" on each other in some way. The following theorem shows that this is in fact the case.

THEOREM 5.3.1

> *A set S with two or more vectors is*
>
> (a) *Linearly dependent if and only if at least one of the vectors in S is expressible as a linear combination of the other vectors in S.*
> (b) *Linearly independent if and only if no vector in S is expressible as a linear combination of the other vectors in S.*

We shall prove part (a) and leave the proof of part (b) as an exercise.

Proof **(a)**   Let $S = \{\mathbf{v}_1, \mathbf{v}_2, \ldots, \mathbf{v}_r\}$ be a set with two or more vectors. If we assume that S is linearly dependent, then there are scalars $k_1, k_2, \ldots, k_r$, not all zero, such that

$$k_1\mathbf{v}_1 + k_2\mathbf{v}_2 + \cdots + k_r\mathbf{v}_r = \mathbf{0} \tag{2}$$

To be specific, suppose that $k_1 \neq 0$. Then (2) can be rewritten as

$$\mathbf{v}_1 = \left(-\frac{k_2}{k_1}\right)\mathbf{v}_2 + \cdots + \left(-\frac{k_r}{k_1}\right)\mathbf{v}_r$$

which expresses $\mathbf{v}_1$ as a linear combination of the other vectors in S. Similarly, if $k_j \neq 0$ in (2) for some $j = 2, 3, \ldots, r$, then $\mathbf{v}_j$ is expressible as a linear combination of the other vectors in S.

Conversely, let us assume that at least one of the vectors in S is expressible as a linear combination of the other vectors. To be specific, suppose that

$$\mathbf{v}_1 = c_2\mathbf{v}_2 + c_3\mathbf{v}_3 + \cdots + c_r\mathbf{v}_r$$

so

$$\mathbf{v}_1 - c_2\mathbf{v}_2 - c_3\mathbf{v}_3 - \cdots - c_r\mathbf{v}_r = \mathbf{0}$$

It follows that S is linearly dependent since the equation

$$k_1\mathbf{v}_1 + k_2\mathbf{v}_2 + \cdots + k_r\mathbf{v}_r = \mathbf{0}$$

is satisfied by

$$k_1 = 1, \quad k_2 = -c_2, \ldots, \quad k_r = -c_r$$

which are not all zero. The proof in the case where some vector other than $\mathbf{v}_1$ is expressible as a linear combination of the other vectors in S is similar. ∎

EXAMPLE 6   Example 1 Revisited

In Example 1 we saw that the vectors

$$\mathbf{v}_1 = (2, -1, 0, 3), \quad \mathbf{v}_2 = (1, 2, 5, -1), \quad \text{and} \quad \mathbf{v}_3 = (7, -1, 5, 8)$$

form a linearly dependent set. It follows from Theorem 5.3.1 that at least one of these vectors is expressible as a linear combination of the other two. In this example each vector is expressible as a linear combination of the other two since it follows from the equation $3\mathbf{v}_1 + \mathbf{v}_2 - \mathbf{v}_3 = \mathbf{0}$ (see Example 1) that

$$\mathbf{v}_1 = -\tfrac{1}{3}\mathbf{v}_2 + \tfrac{1}{3}\mathbf{v}_3, \quad \mathbf{v}_2 = -3\mathbf{v}_1 + \mathbf{v}_3, \quad \text{and} \quad \mathbf{v}_3 = 3\mathbf{v}_1 + \mathbf{v}_2 \quad ◆$$

EXAMPLE 7   Example 3 Revisited

In Example 3 we saw that the vectors $\mathbf{i} = (1, 0, 0)$, $\mathbf{j} = (0, 1, 0)$, and $\mathbf{k} = (0, 0, 1)$ form a linearly independent set. Thus it follows from Theorem 5.3.1 that none of these vectors is expressible as a linear combination of the other two. To see directly that this is so, suppose that $\mathbf{k}$ is expressible as

$$\mathbf{k} = k_1\mathbf{i} + k_2\mathbf{j}$$

Then, in terms of components,

$$(0, 0, 1) = k_1(1, 0, 0) + k_2(0, 1, 0) \quad \text{or} \quad (0, 0, 1) = (k_1, k_2, 0)$$

But the last equation is not satisfied by any values of k_1 and k_2, so $\mathbf{k}$ cannot be expressed as a linear combination of $\mathbf{i}$ and $\mathbf{j}$. Similarly, $\mathbf{i}$ is not expressible as a linear combination of $\mathbf{j}$ and $\mathbf{k}$, and $\mathbf{j}$ is not expressible as a linear combination of $\mathbf{i}$ and $\mathbf{k}$. ◆

The following theorem gives two simple facts about linear independence that are important to know.

THEOREM 5.3.2

> (a) *A finite set of vectors that contains the zero vector is linearly dependent.*
> (b) *A set with exactly two vectors is linearly independent if and only if neither vector is a scalar multiple of the other.*

We shall prove part (a) and leave the proof of part (b) as an exercise.

***Proof** (a)*   For any vectors $\mathbf{v}_1, \mathbf{v}_2, \ldots, \mathbf{v}_r$, the set $S = \{\mathbf{v}_1, \mathbf{v}_2, \ldots, \mathbf{v}_r, \mathbf{0}\}$ is linearly dependent since the equation

$$0\mathbf{v}_1 + 0\mathbf{v}_2 + \cdots + 0\mathbf{v}_r + 1(\mathbf{0}) = \mathbf{0}$$

expresses $\mathbf{0}$ as a linear combination of the vectors in S with coefficients that are not all zero. ∎

EXAMPLE 8   Using Theorem 5.3.2*b*

The functions $\mathbf{f}_1 = x$ and $\mathbf{f}_2 = \sin x$ form a linearly independent set of vectors in $F(-\infty, \infty)$, since neither function is a constant multiple of the other. ◆

Geometric Interpretation of Linear Independence

Linear independence has some useful geometric interpretations in R^2 and R^3:

- In R^2 or R^3, a set of two vectors is linearly independent if and only if the vectors do not lie on the same line when they are placed with their initial points at the origin (Figure 5.3.1).

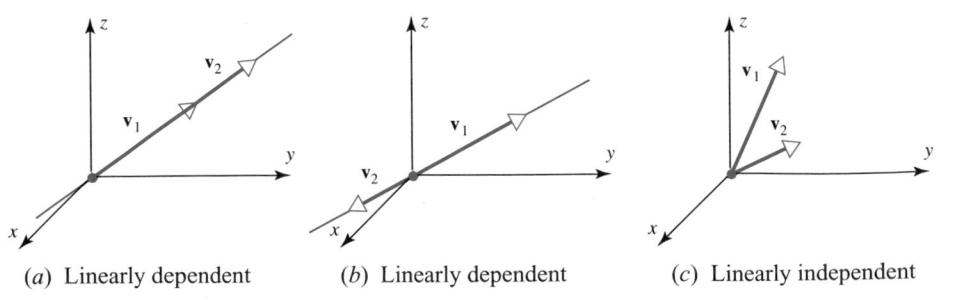

(a) Linearly dependent   (b) Linearly dependent   (c) Linearly independent

Figure 5.3.1

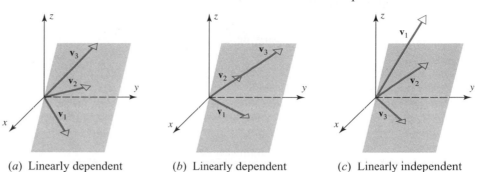

(*a*) Linearly dependent   (*b*) Linearly dependent   (*c*) Linearly independent

Figure 5.3.2

- In R^3, a set of three vectors is linearly independent if and only if the vectors do not lie in the same plane when they are placed with their initial points at the origin (Figure 5.3.2).

The first result follows from the fact that two vectors are linearly independent if and only if neither vector is a scalar multiple of the other. Geometrically, this is equivalent to stating that the vectors do not lie on the same line when they are positioned with their initial points at the origin.

The second result follows from the fact that three vectors are linearly independent if and only if none of the vectors is a linear combination of the other two. Geometrically, this is equivalent to stating that none of the vectors lies in the same plane as the other two, or, alternatively, that the three vectors do not lie in a common plane when they are positioned with their initial points at the origin (why?).

The next theorem shows that a linearly independent set in R^n can contain at most n vectors.

THEOREM 5.3.3

Let $S = \{\mathbf{v}_1, \mathbf{v}_2, \ldots, \mathbf{v}_r\}$ be a set of vectors in R^n. If $r > n$, then S is linearly dependent.

Proof Suppose that

$$\mathbf{v}_1 = (v_{11}, v_{12}, \ldots, v_{1n})$$
$$\mathbf{v}_2 = (v_{21}, v_{22}, \ldots, v_{2n})$$
$$\vdots \qquad\qquad \vdots$$
$$\mathbf{v}_r = (v_{r1}, v_{r2}, \ldots, v_{rn})$$

Consider the equation

$$k_1\mathbf{v}_1 + k_2\mathbf{v}_2 + \cdots + k_r\mathbf{v}_r = \mathbf{0}$$

If, as illustrated in Example 4, we express both sides of this equation in terms of components and then equate corresponding components, we obtain the system

$$v_{11}k_1 + v_{21}k_2 + \cdots + v_{r1}k_r = 0$$
$$v_{12}k_1 + v_{22}k_2 + \cdots + v_{r2}k_r = 0$$
$$\vdots \qquad \vdots \qquad\quad \vdots \qquad \vdots$$
$$v_{1n}k_1 + v_{2n}k_2 + \cdots + v_{rn}k_r = 0$$

This is a homogeneous system of n equations in the r unknowns $k_1, \ldots, k_r$. Since $r > n$, it follows from Theorem 1.2.1 that the system has nontrivial solutions. Therefore, $S = \{\mathbf{v}_1, \mathbf{v}_2, \ldots, \mathbf{v}_r\}$ is a linearly dependent set. ∎

REMARK The preceding theorem tells us that a set in R^2 with more than two vectors is linearly dependent and a set in R^3 with more than three vectors is linearly dependent.

Linear Independence of Functions

Calculus Required

Sometimes linear dependence of functions can be deduced from known identities. For example, the functions

$$\mathbf{f}_1 = \sin^2 x, \quad \mathbf{f}_2 = \cos^2 x, \quad \text{and} \quad \mathbf{f}_3 = 5$$

form a linearly dependent set in $F(-\infty, \infty)$, since the equation

$$5\mathbf{f}_1 + 5\mathbf{f}_2 - \mathbf{f}_3 = 5\sin^2 x + 5\cos^2 x - 5 = 5(\sin^2 x + \cos^2 x) - 5 = \mathbf{0}$$

expresses $\mathbf{0}$ as a linear combination of $\mathbf{f}_1, \mathbf{f}_2$, and $\mathbf{f}_3$ with coefficients that are not all zero. However, it is only in special situations that such identities can be applied. Although there is no general method that can be used to establish linear independence or linear dependence of functions in $F(-\infty, \infty)$, we shall now develop a theorem that can sometimes be used to show that a given set of functions is linearly independent.

If $\mathbf{f}_1 = f_1(x), \mathbf{f}_2 = f_2(x), \ldots, \mathbf{f}_n = f_n(x)$ are $n - 1$ times differentiable functions on the interval $(-\infty, \infty)$, then the determinant

$$W(x) = \begin{vmatrix} f_1(x) & f_2(x) & \cdots & f_n(x) \\ f_1'(x) & f_2'(x) & \cdots & f_n'(x) \\ \vdots & \vdots & & \vdots \\ f_1^{(n-1)}(x) & f_2^{(n-1)}(x) & \cdots & f_n^{(n-1)}(x) \end{vmatrix}$$

is called the **_Wronskian_** of $f_1, f_2, \ldots, f_n$. As we shall now show, this determinant is useful for ascertaining whether the functions $\mathbf{f}_1, \mathbf{f}_2, \ldots, \mathbf{f}_n$ form a linearly independent set of vectors in the vector space $C^{(n-1)}(-\infty, \infty)$.

Suppose, for the moment, that $\mathbf{f}_1, \mathbf{f}_2, \ldots, \mathbf{f}_n$ are linearly dependent vectors in $C^{(n-1)}(-\infty, \infty)$. Then there exist scalars $k_1, k_2, \ldots, k_n$, *not all zero*, such that

$$k_1 f_1(x) + k_2 f_2(x) + \cdots + k_n f_n(x) = 0$$

for all x in the interval $(-\infty, \infty)$. Combining this equation with the equations obtained by $n - 1$ successive differentiations yields

$$
\begin{aligned}
k_1 f_1(x) \quad &+ k_2 f_2(x) \quad + \cdots + k_n f_n(x) \quad &= 0 \\
k_1 f_1'(x) \quad &+ k_2 f_2'(x) \quad + \cdots + k_n f_n'(x) \quad &= 0 \\
\vdots \qquad & \qquad \vdots \qquad\qquad \vdots \qquad\qquad \vdots \\
k_1 f_1^{(n-1)}(x) &+ k_2 f_2^{(n-1)}(x) + \cdots + k_n f_n^{(n-1)}(x) &= 0
\end{aligned}
$$

Thus, the linear dependence of $\mathbf{f}_1, \mathbf{f}_2, \ldots, \mathbf{f}_n$ implies that the linear system

$$\begin{bmatrix} f_1(x) & f_2(x) & \cdots & f_n(x) \\ f_1'(x) & f_2'(x) & \cdots & f_n'(x) \\ \vdots & \vdots & & \vdots \\ f_1^{(n-1)}(x) & f_2^{(n-1)}(x) & \cdots & f_n^{(n-1)}(x) \end{bmatrix} \begin{bmatrix} k_1 \\ k_2 \\ \vdots \\ k_n \end{bmatrix} = \begin{bmatrix} 0 \\ 0 \\ \vdots \\ 0 \end{bmatrix}$$

has a nontrivial solution for *every* x in the interval $(-\infty, \infty)$. This implies in turn that for every x in $(-\infty, \infty)$ the coefficient matrix is not invertible, or, equivalently, that its determinant (the Wronskian) is zero for every x in $(-\infty, \infty)$. Thus, if the Wronskian is *not* identically zero on $(-\infty, \infty)$, then the functions $\mathbf{f}_1, \mathbf{f}_2, \ldots, \mathbf{f}_n$ must be linearly independent vectors in $C^{(n-1)}(-\infty, \infty)$. This is the content of the following theorem.

Józef Maria Hoëne-Wroński

Józef Maria Hoëne-Wroński *(1776–1853)* was a Polish-French mathematician and philosopher. Wroński received his early education in Poznán and Warsaw. He served as an artillery officer in the Prussian army in a national uprising in 1794, was taken prisoner by the Russian army, and on his release studied philosophy at various German universities. He became a French citizen in 1800 and eventually settled in Paris, where he did research in analysis leading to some controversial mathematical papers and relatedly to a famous court trial over financial matters. Several years thereafter, his proposed research on the determination of longitude at sea was rebuffed by the British Board of Longitude, and Wrónski turned to studies in Messianic philosophy. In the 1830s he investigated the feasibility of caterpillar vehicles to compete with trains, with no luck, and spent his last years in poverty. Much of his mathematical work was fraught with errors and imprecision, but it often contained valuable isolated results and ideas. Some writers attribute this lifelong pattern of argumentation to psychopathic tendencies and to an exaggeration of the importance of his own work.

THEOREM 5.3.4

> *If the functions* $\mathbf{f}_1, \mathbf{f}_2, \ldots, \mathbf{f}_n$ *have* $n-1$ *continuous derivatives on the interval* $(-\infty, \infty)$, *and if the Wronskian of these functions is not identically zero on* $(-\infty, \infty)$, *then these functions form a linearly independent set of vectors in* $C^{(n-1)}(-\infty, \infty)$.

EXAMPLE 9 Linearly Independent Set in $C^1(-\infty, \infty)$

Show that the functions $\mathbf{f}_1 = x$ and $\mathbf{f}_2 = \sin x$ form a linearly independent set of vectors in $C^1(-\infty, \infty)$.

Solution

In Example 8 we showed that these vectors form a linearly independent set by noting that neither vector is a scalar multiple of the other. However, for illustrative purposes, we shall obtain this same result using Theorem 5.3.4. The Wronskian is

$$W(x) = \begin{vmatrix} x & \sin x \\ 1 & \cos x \end{vmatrix} = x \cos x - \sin x$$

This function does not have value zero for all x in the interval $(-\infty, \infty)$, as can be seen by evaluating it at $x = \pi/2$, so $\mathbf{f}_1$ and $\mathbf{f}_2$ form a linearly independent set. ◆

EXAMPLE 10 Linearly Independent Set in $C^2(-\infty, \infty)$

Show that $\mathbf{f}_1 = 1$, $\mathbf{f}_2 = e^x$, and $\mathbf{f}_3 = e^{2x}$ form a linearly independent set of vectors in $C^2(-\infty, \infty)$.

Solution

The Wronskian is

$$W(x) = \begin{vmatrix} 1 & e^x & e^{2x} \\ 0 & e^x & 2e^{2x} \\ 0 & e^x & 4e^{2x} \end{vmatrix} = 2e^{3x}$$

This function does not have value zero for all x (in fact, for any x) in the interval $(-\infty, \infty)$, so $\mathbf{f}_1$, $\mathbf{f}_2$, and $\mathbf{f}_3$ form a linearly independent set. ◆

REMARK The converse of Theorem 5.3.4 is false. If the Wronskian of $\mathbf{f}_1, \mathbf{f}_2, \ldots, \mathbf{f}_n$ is identically zero on $(-\infty, \infty)$, then no conclusion can be reached about the linear independence of $\{\mathbf{f}_1, \mathbf{f}_2, \ldots, \mathbf{f}_n\}$; this set of vectors may be linearly independent or linearly dependent.

EXERCISE SET
5.3

1. Explain why the following are linearly dependent sets of vectors. (Solve this problem by inspection.)

   (a) $\mathbf{u}_1 = (-1, 2, 4)$ and $\mathbf{u}_2 = (5, -10, -20)$ in R^3

   (b) $\mathbf{u}_1 = (3, -1)$, $\mathbf{u}_2 = (4, 5)$, $\mathbf{u}_3 = (-4, 7)$ in R^2

   (c) $\mathbf{p}_1 = 3 - 2x + x^2$ and $\mathbf{p}_2 = 6 - 4x + 2x^2$ in P_2

   (d) $A = \begin{bmatrix} -3 & 4 \\ 2 & 0 \end{bmatrix}$ and $B = \begin{bmatrix} 3 & -4 \\ -2 & 0 \end{bmatrix}$ in M_{22}

2. Which of the following sets of vectors in R^3 are linearly dependent?

   (a) $(4, -1, 2)$, $(-4, 10, 2)$ (b) $(-3, 0, 4)$, $(5, -1, 2)$, $(1, 1, 3)$

   (c) $(8, -1, 3)$, $(4, 0, 1)$ (d) $(-2, 0, 1)$, $(3, 2, 5)$, $(6, -1, 1)$, $(7, 0, -2)$

3. Which of the following sets of vectors in R^4 are linearly dependent?

   (a) $(3, 8, 7, -3)$, $(1, 5, 3, -1)$, $(2, -1, 2, 6)$, $(1, 4, 0, 3)$

   (b) $(0, 0, 2, 2)$, $(3, 3, 0, 0)$, $(1, 1, 0, -1)$

   (c) $(0, 3, -3, -6)$, $(-2, 0, 0, -6)$, $(0, -4, -2, -2)$, $(0, -8, 4, -4)$

   (d) $(3, 0, -3, 6)$, $(0, 2, 3, 1)$, $(0, -2, -2, 0)$, $(-2, 1, 2, 1)$

4. Which of the following sets of vectors in P_2 are linearly dependent?

   (a) $2 - x + 4x^2$, $3 + 6x + 2x^2$, $2 + 10x - 4x^2$

   (b) $3 + x + x^2$, $2 - x + 5x^2$, $4 - 3x^2$

   (c) $6 - x^2$, $1 + x + 4x^2$

   (d) $1 + 3x + 3x^2$, $x + 4x^2$, $5 + 6x + 3x^2$, $7 + 2x - x^2$

5. Assume that $\mathbf{v}_1$, $\mathbf{v}_2$, and $\mathbf{v}_3$ are vectors in R^3 that have their initial points at the origin. In each part, determine whether the three vectors lie in a plane.

   (a) $\mathbf{v}_1 = (2, -2, 0)$, $\mathbf{v}_2 = (6, 1, 4)$, $\mathbf{v}_3 = (2, 0, -4)$

   (b) $\mathbf{v}_1 = (-6, 7, 2)$, $\mathbf{v}_2 = (3, 2, 4)$, $\mathbf{v}_3 = (4, -1, 2)$

6. Assume that $\mathbf{v}_1$, $\mathbf{v}_2$, and $\mathbf{v}_3$ are vectors in R^3 that have their initial points at the origin. In each part, determine whether the three vectors lie on the same line.

   (a) $\mathbf{v}_1 = (-1, 2, 3)$, $\mathbf{v}_2 = (2, -4, -6)$, $\mathbf{v}_3 = (-3, 6, 0)$

   (b) $\mathbf{v}_1 = (2, -1, 4)$, $\mathbf{v}_2 = (4, 2, 3)$, $\mathbf{v}_3 = (2, 7, -6)$

   (c) $\mathbf{v}_1 = (4, 6, 8)$, $\mathbf{v}_2 = (2, 3, 4)$, $\mathbf{v}_3 = (-2, -3, -4)$

7. (a) Show that the vectors $\mathbf{v}_1 = (0, 3, 1, -1)$, $\mathbf{v}_2 = (6, 0, 5, 1)$, and $\mathbf{v}_3 = (4, -7, 1, 3)$ form a linearly dependent set in R^4.

   (b) Express each vector as a linear combination of the other two.

8. (a) Show that the vectors $\mathbf{v}_1 = (1, 2, 3, 4)$, $\mathbf{v}_2 = (0, 1, 0, -1)$, and $\mathbf{v}_3 = (1, 3, 3, 3)$, form a linearly dependent set in R^4.

   (b) Express each vector as a linear combination of the other two.

9. For which real values of λ do the following vectors form a linearly dependent set in R^3?

$$\mathbf{v}_1 = \left(\lambda, -\tfrac{1}{2}, -\tfrac{1}{2}\right), \qquad \mathbf{v}_2 = \left(-\tfrac{1}{2}, \lambda, -\tfrac{1}{2}\right), \qquad \mathbf{v}_3 = \left(-\tfrac{1}{2}, -\tfrac{1}{2}, \lambda\right)$$

10. Show that if $\{\mathbf{v}_1, \mathbf{v}_2, \mathbf{v}_3\}$ is a linearly independent set of vectors, then so are $\{\mathbf{v}_1, \mathbf{v}_2\}$, $\{\mathbf{v}_1, \mathbf{v}_3\}$, $\{\mathbf{v}_2, \mathbf{v}_3\}$, $\{\mathbf{v}_1\}$, $\{\mathbf{v}_2\}$, and $\{\mathbf{v}_3\}$.

11. Show that if $S = \{\mathbf{v}_1, \mathbf{v}_2, \ldots, \mathbf{v}_r\}$ is a linearly independent set of vectors, then so is every nonempty subset of S.

12. Show that if $\{\mathbf{v}_1, \mathbf{v}_2, \mathbf{v}_3\}$ is a linearly dependent set of vectors in a vector space V, and $\mathbf{v}_4$ is any vector in V, then $\{\mathbf{v}_1, \mathbf{v}_2, \mathbf{v}_3, \mathbf{v}_4\}$ is also linearly dependent.

13. Show that if $\{\mathbf{v}_1, \mathbf{v}_2, \ldots, \mathbf{v}_r\}$ is a linearly dependent set of vectors in a vector space V, and if $\mathbf{v}_{r+1}, \ldots, \mathbf{v}_n$ are any vectors in V, then $\{\mathbf{v}_1, \mathbf{v}_2, \ldots, \mathbf{v}_r, \mathbf{v}_{r+1}, \ldots, \mathbf{v}_n\}$ is also linearly dependent.

14. Show that every set with more than three vectors from P_2 is linearly dependent.

15. Show that if $\{\mathbf{v}_1, \mathbf{v}_2\}$ is linearly independent and $\mathbf{v}_3$ does not lie in span $\{\mathbf{v}_1, \mathbf{v}_2\}$, then $\{\mathbf{v}_1, \mathbf{v}_2, \mathbf{v}_3\}$ is linearly independent.

16. Prove: For any vectors $\mathbf{u}$, $\mathbf{v}$, and $\mathbf{w}$, the vectors $\mathbf{u} - \mathbf{v}$, $\mathbf{v} - \mathbf{w}$, and $\mathbf{w} - \mathbf{u}$ form a linearly dependent set.

17. Prove: The space spanned by two vectors in R^3 is a line through the origin, a plane through the origin, or the origin itself.

18. Under what conditions is a set with one vector linearly independent?

19. Are the vectors $\mathbf{v}_1$, $\mathbf{v}_2$, and $\mathbf{v}_3$ in part (a) of the accompanying figure linearly independent? What about those in part (b)? Explain.

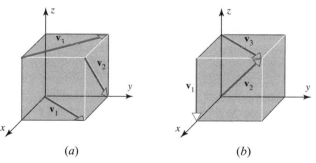

(a) (b)

Figure Ex-19

20. Use appropriate identities, where required, to determine which of the following sets of vectors in $F(-\infty, \infty)$ are linearly dependent.

   (a) 6, $3\sin^2 x$, $2\cos^2 x$ (b) x, $\cos x$ (c) 1, $\sin x$, $\sin 2x$

   (d) $\cos 2x$, $\sin^2 x$, $\cos^2 x$ (e) $(3 - x)^2$, $x^2 - 6x$, 5 (f) 0, $\cos^3 \pi x$, $\sin^5 3\pi x$

21. **(For Readers Who Have Studied Calculus)** Use the Wronskian to show that the following sets of vectors are linearly independent.

   (a) 1, x, e^x (b) $\sin x$, $\cos x$, $x \sin x$ (c) e^x, xe^x, $x^2 e^x$ (d) 1, x, x^2

22. Use part (*a*) of Theorem 5.3.1 to prove part (*b*).

23. Prove part (*b*) of Theorem 5.3.2.

Discussion
&
Discovery

24. Indicate whether each statement is always true or sometimes false. Justify your answer by giving a logical argument or a counterexample.

(a) The set of 2×2 matrices that contain exactly two 1's and two 0's is a linearly independent set in M_{22}.

(b) If $\{\mathbf{v}_1, \mathbf{v}_2\}$ is a linearly dependent set, then each vector is a scalar multiple of the other.

(c) If $\{\mathbf{v}_1, \mathbf{v}_2, \mathbf{v}_3\}$ is a linearly independent set, then so is the set $\{k\mathbf{v}_1, k\mathbf{v}_2, k\mathbf{v}_3\}$ for every nonzero scalar k.

(d) The converse of Theorem 5.3.2*a* is also true.

25. Show that if $\{\mathbf{v}_1, \mathbf{v}_2, \mathbf{v}_3\}$ is a linearly dependent set with nonzero vectors, then each vector in the set is expressible as a linear combination of the other two.

26. Theorem 5.3.3 implies that four nonzero vectors in R^3 must be linearly dependent. Give an informal geometric argument to explain this result.

27. (a) In Example 3 we showed that the mutually orthogonal vectors $\mathbf{i}, \mathbf{j}$, and $\mathbf{k}$ form a linearly independent set of vectors in R^3. Do you think that every set of three nonzero mutually orthogonal vectors in R^3 is linearly independent? Justify your conclusion with a geometric argument.

(b) Justify your conclusion with an algebraic argument.

Hint Use dot products.

5.4
BASIS AND DIMENSION

We usually think of a line as being one-dimensional, a plane as two-dimensional, and the space around us as three-dimensional. It is the primary purpose of this section to make this intuitive notion of "dimension" more precise.

Nonrectangular Coordinate Systems

In plane analytic geometry we learned to associate a point P in the plane with a pair of coordinates (a, b) by projecting P onto a pair of perpendicular coordinate axes (Figure 5.4.1*a*). By this process, each point in the plane is assigned a unique set of coordinates, and conversely, each pair of coordinates is associated with a unique point in the plane. We describe this by saying that the coordinate system establishes a ***one-to-one correspondence*** between points in the plane and ordered pairs of real numbers. Although perpendicular coordinate axes are the most common, any two nonparallel lines can be used to define a coordinate system in the plane. For example, in Figure 5.4.1*b*, we have attached a pair of coordinates (a, b) to the point P by projecting P parallel to the nonperpendicular coordinate axes. Similarly, in 3-space any three noncoplanar coordinate axes can be used to define a coordinate system (Figure 5.4.1*c*).

Our first objective in this section is to extend the concept of a coordinate system to general vector spaces. As a start, it will be helpful to reformulate the notion of a coordinate system in 2-space or 3-space using vectors rather than coordinate axes to specify the coordinate system. This can be done by replacing each coordinate axis with a vector of length 1 that points in the positive direction of the axis. In Figure 5.4.2*a*, for example, $\mathbf{v}_1$ and $\mathbf{v}_2$ are such vectors. As illustrated in that figure, if P is any point in the plane, the vector $\overrightarrow{OP}$ can be written as a linear combination of $\mathbf{v}_1$ and $\mathbf{v}_2$ by projecting P

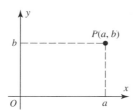

(a) Coordinates of P in a rectangular coordinate system in 2-space

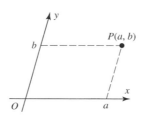

(b) Coordinates of P in a nonrectangular coordinate system in 2-space

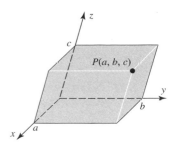

(c) Coordinates of P in a nonrectangular coordinate system in 3-space

Figure 5.4.1

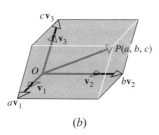

(a)

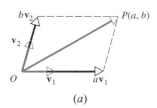

(b)

Figure 5.4.2

parallel to $\mathbf{v}_1$ and $\mathbf{v}_2$ to make $\overrightarrow{OP}$ the diagonal of a parallelogram determined by vectors $a\mathbf{v}_1$ and $b\mathbf{v}_2$:

$$\overrightarrow{OP} = a\mathbf{v}_1 + b\mathbf{v}_2$$

It is evident that the numbers a and b in this vector formula are precisely the coordinates of P in the coordinate system of Figure 5.4.1b. Similarly, the coordinates (a, b, c) of the point P in Figure 5.4.1c can be obtained by expressing $\overrightarrow{OP}$ as a linear combination of the vectors shown in Figure 5.4.2b.

Informally stated, vectors that specify a coordinate system are called "basis vectors" for that system. Although we used basis vectors of length 1 in the preceding discussion, we shall see in a moment that this is not essential—nonzero vectors of any length will suffice.

The scales of measurement along the coordinate axes are essential ingredients of any coordinate system. Usually, one tries to use the same scale on each axis and to have the integer points on the axes spaced 1 unit of distance apart. However, this is not always practical or appropriate: Unequal scales or scales in which the integral points are more or less than 1 unit apart may be required to fit a particular graph on a printed page or to represent physical quantities with diverse units in the same coordinate system (time in seconds on one axis and temperature in hundreds of degrees on another, for example). When a coordinate system is specified by a set of basis vectors, then the lengths of those vectors correspond to the distances between successive integer points on the coordinate axes (Figure 5.4.3). Thus it is the directions of the basis vectors that define the positive directions of the coordinate axes and the lengths of the basis vectors that establish the scales of measurement.

The following key definition will make the preceding ideas more precise and enable us to extend the concept of a coordinate system to general vector spaces.

DEFINITION

If V is any vector space and $S = \{\mathbf{v}_1, \mathbf{v}_2, \ldots, \mathbf{v}_n\}$ is a set of vectors in V, then S is called a ***basis*** for V if the following two conditions hold:

(a) S is linearly independent.

(b) S spans V.

A basis is the vector space generalization of a coordinate system in 2-space and 3-space. The following theorem will help us to see why this is so.

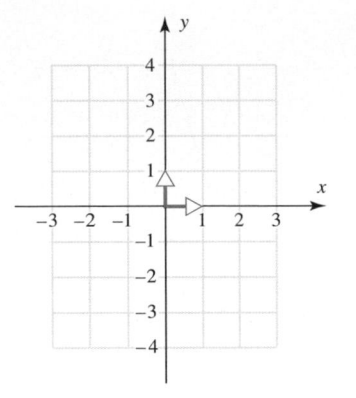

(a) Equal scales. Perpendicular axes.

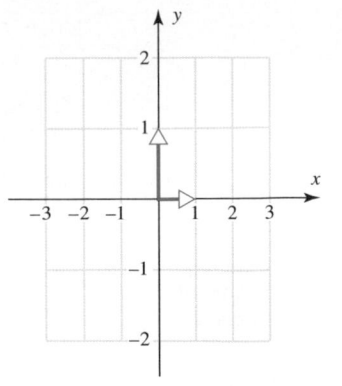

(b) Unequal scales. Perpendicular axes.

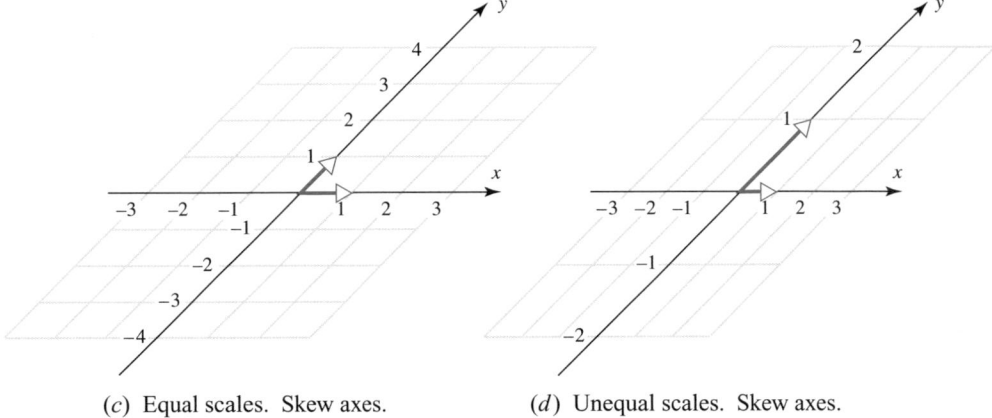

(c) Equal scales. Skew axes.

(d) Unequal scales. Skew axes.

Figure 5.4.3

THEOREM 5.4.1

Uniqueness of Basis Representation

If $S = \{v_1, v_2, \ldots, v_n\}$ is a basis for a vector space V, then every vector v in V can be expressed in the form $v = c_1v_1 + c_2v_2 + \cdots + c_nv_n$ in exactly one way.

Proof Since S spans V, it follows from the definition of a spanning set that every vector in V is expressible as a linear combination of the vectors in S. To see that there is only *one* way to express a vector as a linear combination of the vectors in S, suppose that some vector v can be written as

$$v = c_1v_1 + c_2v_2 + \cdots + c_nv_n$$

and also as

$$v = k_1v_1 + k_2v_2 + \cdots + k_nv_n$$

Subtracting the second equation from the first gives

$$0 = (c_1 - k_1)v_1 + (c_2 - k_2)v_2 + \cdots + (c_n - k_n)v_n$$

Since the right side of this equation is a linear combination of vectors in S, the linear independence of S implies that

$$c_1 - k_1 = 0, \quad c_2 - k_2 = 0, \ldots, \quad c_n - k_n = 0$$

that is,

$$c_1 = k_1, \quad c_2 = k_2, \ldots, \quad c_n = k_n$$

Thus, the two expressions for **v** are the same. ∎

Coordinates Relative to a Basis

If $S = \{\mathbf{v}_1, \mathbf{v}_2, \ldots, \mathbf{v}_n\}$ is a basis for a vector space V, and

$$\mathbf{v} = c_1\mathbf{v}_1 + c_2\mathbf{v}_2 + \cdots + c_n\mathbf{v}_n$$

is the expression for a vector **v** in terms of the basis S, then the scalars $c_1, c_2, \ldots, c_n$ are called the ***coordinates*** of **v** relative to the basis S. The vector $(c_1, c_2, \ldots, c_n)$ in R^n constructed from these coordinates is called the ***coordinate vector of* v *relative to* S**; it is denoted by

$$(\mathbf{v})_S = (c_1, c_2, \ldots, c_n)$$

REMARK   It should be noted that coordinate vectors depend not only on the basis S but also on the order in which the basis vectors are written; a change in the order of the basis vectors results in a corresponding change of order for the entries in the coordinate vectors.

EXAMPLE 1   Standard Basis for R^3

In Example 3 of the preceding section, we showed that if

$$\mathbf{i} = (1, 0, 0), \quad \mathbf{j} = (0, 1, 0), \quad \text{and} \quad \mathbf{k} = (0, 0, 1)$$

then $S = \{\mathbf{i}, \mathbf{j}, \mathbf{k}\}$ is a linearly independent set in R^3. This set also spans R^3 since any vector $\mathbf{v} = (a, b, c)$ in R^3 can be written as

$$\mathbf{v} = (a, b, c) = a(1, 0, 0) + b(0, 1, 0) + c(0, 0, 1) = a\mathbf{i} + b\mathbf{j} + c\mathbf{k} \qquad (1)$$

Thus S is a basis for R^3; it is called the ***standard basis*** for R^3. Looking at the coefficients of **i**, **j**, and **k** in (1), it follows that the coordinates of **v** relative to the standard basis are a, b, and c, so

$$(\mathbf{v})_S = (a, b, c)$$

Comparing this result to (1), we see that

$$\mathbf{v} = (\mathbf{v})_S$$

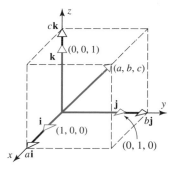

Figure 5.4.4

This equation states that the components of a vector **v** relative to a rectangular xyz-coordinate system and the coordinates of **v** relative to the standard basis are the same; thus, the coordinate system and the basis produce precisely the same one-to-one correspondence between points in 3-space and ordered triples of real numbers (Figure 5.4.4). ◆

The results in the preceding example are a special case of those in the next example.

EXAMPLE 2   Standard Basis for R^n

In Example 3 of the preceding section, we showed that if

$$\mathbf{e}_1 = (1, 0, 0, \ldots, 0), \quad \mathbf{e}_2 = (0, 1, 0, \ldots, 0), \ldots, \quad \mathbf{e}_n = (0, 0, 0, \ldots, 1)$$

then

$$S = \{\mathbf{e}_1, \mathbf{e}_2, \ldots, \mathbf{e}_n\}$$

is a linearly independent set in R^n. Moreover, this set also spans R^n since any vector $\mathbf{v} = (v_1, v_2, \ldots, v_n)$ in R^n can be written as

$$\mathbf{v} = v_1\mathbf{e}_1 + v_2\mathbf{e}_2 + \cdots + v_n\mathbf{e}_n \tag{2}$$

Thus S is a basis for R^n; it is called the ***standard basis for R^n***. It follows from (2) that the coordinates of $\mathbf{v} = (v_1, v_2, \ldots, v_n)$ relative to the standard basis are $v_1, v_2, \ldots, v_n$, so

$$(\mathbf{v})_S = (v_1, v_2, \ldots, v_n)$$

As in Example 1, we have $\mathbf{v} = (\mathbf{v})_S$, so a vector $\mathbf{v}$ and its coordinate vector relative to the standard basis for R^n are the same. ◆

REMARK We will see in a subsequent example that a vector and its coordinate vector need not be the same; the equality that we observed in the two preceding examples is a special situation that occurs only with the standard basis for R^n.

REMARK In R^2 and R^3, the standard basis vectors are commonly denoted by $\mathbf{i}$, $\mathbf{j}$, and $\mathbf{k}$, rather than by $\mathbf{e}_1$, $\mathbf{e}_2$, and $\mathbf{e}_3$. We shall use both notations, depending on the particular situation.

EXAMPLE 3 Demonstrating That a Set of Vectors Is a Basis

Let $\mathbf{v}_1 = (1, 2, 1)$, $\mathbf{v}_2 = (2, 9, 0)$, and $\mathbf{v}_3 = (3, 3, 4)$. Show that the set $S = \{\mathbf{v}_1, \mathbf{v}_2, \mathbf{v}_3\}$ is a basis for R^3.

Solution

To show that the set S spans R^3, we must show that an arbitrary vector $\mathbf{b} = (b_1, b_2, b_3)$ can be expressed as a linear combination

$$\mathbf{b} = c_1\mathbf{v}_1 + c_2\mathbf{v}_2 + c_3\mathbf{v}_3$$

of the vectors in S. Expressing this equation in terms of components gives

$$(b_1, b_2, b_3) = c_1(1, 2, 1) + c_2(2, 9, 0) + c_3(3, 3, 4)$$

or

$$(b_1, b_2, b_3) = (c_1 + 2c_2 + 3c_3, 2c_1 + 9c_2 + 3c_3, c_1 + 4c_3)$$

or, on equating corresponding components,

$$\begin{aligned} c_1 + 2c_2 + 3c_3 &= b_1 \\ 2c_1 + 9c_2 + 3c_3 &= b_2 \\ c_1 \qquad\;\; + 4c_3 &= b_3 \end{aligned} \tag{3}$$

Thus, to show that S spans R^3, we must demonstrate that system (3) has a solution for all choices of $\mathbf{b} = (b_1, b_2, b_3)$.

To prove that S is linearly independent, we must show that the only solution of

$$c_1\mathbf{v}_1 + c_2\mathbf{v}_2 + c_3\mathbf{v}_3 = \mathbf{0} \tag{4}$$

is $c_1 = c_2 = c_3 = 0$. As above, if (4) is expressed in terms of components, the verification of independence reduces to showing that the homogeneous system

$$c_1 + 2c_2 + 3c_3 = 0$$
$$2c_1 + 9c_2 + 3c_3 = 0 \tag{5}$$
$$c_1 \quad\quad + 4c_3 = 0$$

has only the trivial solution. Observe that systems (3) and (5) have the same coefficient matrix. Thus, by parts (b), (e), and (g) of Theorem 4.3.4, we can simultaneously prove that S is linearly independent and spans R^3 by demonstrating that in systems (3) and (5), the matrix of coefficients has a nonzero determinant. From

$$A = \begin{bmatrix} 1 & 2 & 3 \\ 2 & 9 & 3 \\ 1 & 0 & 4 \end{bmatrix} \quad \text{we find} \quad \det(A) = \begin{vmatrix} 1 & 2 & 3 \\ 2 & 9 & 3 \\ 1 & 0 & 4 \end{vmatrix} = -1$$

and so S is a basis for R^3. ◆

EXAMPLE 4   Representing a Vector Using Two Bases

Let $S = \{\mathbf{v}_1, \mathbf{v}_2, \mathbf{v}_3\}$ be the basis for R^3 in the preceding example.

(a)   Find the coordinate vector of $\mathbf{v} = (5, -1, 9)$ with respect to S.

(b)   Find the vector $\mathbf{v}$ in R^3 whose coordinate vector with respect to the basis S is $(\mathbf{v})_S = (-1, 3, 2)$.

Solution (a)

We must find scalars c_1, c_2, c_3 such that

$$\mathbf{v} = c_1 \mathbf{v}_1 + c_2 \mathbf{v}_2 + c_3 \mathbf{v}_3$$

or, in terms of components,

$$(5, -1, 9) = c_1(1, 2, 1) + c_2(2, 9, 0) + c_3(3, 3, 4)$$

Equating corresponding components gives

$$c_1 + 2c_2 + 3c_3 = \quad 5$$
$$2c_1 + 9c_2 + 3c_3 = -1$$
$$c_1 \quad\quad + 4c_3 = \quad 9$$

Solving this system, we obtain $c_1 = 1$, $c_2 = -1$, $c_3 = 2$ (verify). Therefore,

$$(\mathbf{v})_S = (1, -1, 2)$$

Solution (b)

Using the definition of the coordinate vector $(\mathbf{v})_S$, we obtain

$$\mathbf{v} = (-1)\mathbf{v}_1 + 3\mathbf{v}_2 + 2\mathbf{v}_3$$
$$= (-1)(1, 2, 1) + 3(2, 9, 0) + 2(3, 3, 4) = (11, 31, 7) \quad ◆$$

EXAMPLE 5   Standard Basis for P_n

(a)   Show that $S = \{1, x, x^2, \ldots, x^n\}$ is a basis for the vector space P_n of polynomials of the form $a_0 + a_1 x + \cdots + a_n x^n$.

(b) Find the coordinate vector of the polynomial $\mathbf{p} = a_0 + a_1 x + a_2 x^2$ relative to the basis $S = \{1, x, x^2\}$ for P_2.

Solution (a)

We showed that S spans P_n in Example 11 of Section 5.2, and we showed that S is a linearly independent set in Example 5 of Section 5.3. Thus S is a basis for P_n; it is called the **standard basis for P_n**.

Solution (b)

The coordinates of $\mathbf{p} = a_0 + a_1 x + a_2 x^2$ are the scalar coefficients of the basis vectors 1, x, and x^2, so $(\mathbf{p})_S = (a_0, a_1, a_2)$. ◆

EXAMPLE 6 Standard Basis for M_{mn}

Let

$$M_1 = \begin{bmatrix} 1 & 0 \\ 0 & 0 \end{bmatrix}, \qquad M_2 = \begin{bmatrix} 0 & 1 \\ 0 & 0 \end{bmatrix}, \qquad M_3 = \begin{bmatrix} 0 & 0 \\ 1 & 0 \end{bmatrix}, \qquad M_4 = \begin{bmatrix} 0 & 0 \\ 0 & 1 \end{bmatrix}$$

The set $S = \{M_1, M_2, M_3, M_4\}$ is a basis for the vector space M_{22} of 2×2 matrices. To see that S spans M_{22}, note that an arbitrary vector (matrix)

$$\begin{bmatrix} a & b \\ c & d \end{bmatrix}$$

can be written as

$$\begin{bmatrix} a & b \\ c & d \end{bmatrix} = a \begin{bmatrix} 1 & 0 \\ 0 & 0 \end{bmatrix} + b \begin{bmatrix} 0 & 1 \\ 0 & 0 \end{bmatrix} + c \begin{bmatrix} 0 & 0 \\ 1 & 0 \end{bmatrix} + d \begin{bmatrix} 0 & 0 \\ 0 & 1 \end{bmatrix}$$
$$= aM_1 + bM_2 + cM_3 + dM_4$$

To see that S is linearly independent, assume that

$$aM_1 + bM_2 + cM_3 + dM_4 = 0$$

That is,

$$a \begin{bmatrix} 1 & 0 \\ 0 & 0 \end{bmatrix} + b \begin{bmatrix} 0 & 1 \\ 0 & 0 \end{bmatrix} + c \begin{bmatrix} 0 & 0 \\ 1 & 0 \end{bmatrix} + d \begin{bmatrix} 0 & 0 \\ 0 & 1 \end{bmatrix} = \begin{bmatrix} 0 & 0 \\ 0 & 0 \end{bmatrix}$$

It follows that

$$\begin{bmatrix} a & b \\ c & d \end{bmatrix} = \begin{bmatrix} 0 & 0 \\ 0 & 0 \end{bmatrix}$$

Thus $a = b = c = d = 0$, so S is linearly independent. The basis S in this example is called the **standard basis for M_{22}**. More generally, the **standard basis for M_{mn}** consists of the mn different matrices with a single 1 and zeros for the remaining entries. ◆

EXAMPLE 7 Basis for the Subspace span(S)

If $S = \{\mathbf{v}_1, \mathbf{v}_2, \ldots, \mathbf{v}_r\}$ is a *linearly independent* set in a vector space V, then S is a basis for the subspace span(S) since the set S spans span(S) by definition of span(S). ◆

> **DEFINITION**
>
> A nonzero vector space V is called *finite-dimensional* if it contains a finite set of vectors $\{v_1, v_2, \ldots, v_n\}$ that forms a basis. If no such set exists, V is called *infinite-dimensional*. In addition, we shall regard the zero vector space to be finite-dimensional.

EXAMPLE 8 Some Finite- and Infinite-Dimensional Spaces

By Examples 2, 5, and 6, the vector spaces R^n, P_n, and M_{mn} are finite-dimensional. The vector spaces $F(-\infty, \infty)$, $C(-\infty, \infty)$, $C^m(-\infty, \infty)$, and $C^\infty(-\infty, \infty)$ are infinite-dimensional (Exercise 24). ◆

The next theorem will provide the key to the concept of dimension.

THEOREM 5.4.2

> *Let V be a finite-dimensional vector space, and let $\{v_1, v_2, \ldots, v_n\}$ be any basis.*
>
> (a) *If a set has more than n vectors, then it is linearly dependent.*
> (b) *If a set has fewer than n vectors, then it does not span V.*

Proof (a) Let $S' = \{w_1, w_2, \ldots, w_m\}$ be any set of m vectors in V, where $m > n$. We want to show that S' is linearly dependent. Since $S = \{v_1, v_2, \ldots, v_n\}$ is a basis, each w_i can be expressed as a linear combination of the vectors in S, say

$$
\begin{aligned}
w_1 &= a_{11}v_1 + a_{21}v_2 + \cdots + a_{n1}v_n \\
w_2 &= a_{12}v_1 + a_{22}v_2 + \cdots + a_{n2}v_n \\
&\ \ \vdots \qquad \vdots \qquad \vdots \qquad \quad \vdots \\
w_m &= a_{1m}v_1 + a_{2m}v_2 + \cdots + a_{nm}v_n
\end{aligned}
\tag{6}
$$

To show that S' is linearly dependent, we must find scalars $k_1, k_2, \ldots, k_m$, not all zero, such that

$$
k_1 w_1 + k_2 w_2 + \cdots + k_m w_m = 0 \tag{7}
$$

Using the equations in (6), we can rewrite (7) as

$$
\begin{aligned}
(k_1 a_{11} + k_2 a_{12} + \cdots + k_m a_{1m})v_1 & \\
+ (k_1 a_{21} + k_2 a_{22} + \cdots + k_m a_{2m})v_2 & \\
\ddots \qquad\qquad\qquad & \\
+ (k_1 a_{n1} + k_2 a_{n2} + \cdots + k_m a_{nm})v_n &= 0
\end{aligned}
$$

Thus, from the linear independence of S, the problem of proving that S' is a linearly dependent set reduces to showing there are scalars $k_1, k_2, \ldots, k_m$, not all zero, that satisfy

$$
\begin{aligned}
a_{11}k_1 + a_{12}k_2 + \cdots + a_{1m}k_m &= 0 \\
a_{21}k_1 + a_{22}k_2 + \cdots + a_{2m}k_m &= 0 \\
\vdots \qquad \vdots \qquad\quad \vdots \qquad \vdots& \\
a_{n1}k_1 + a_{n2}k_2 + \cdots + a_{nm}k_m &= 0
\end{aligned}
\tag{8}
$$

But (8) has more unknowns than equations, so the proof is complete since Theorem 1.2.1 guarantees the existence of nontrivial solutions.

Proof (b) Let $S' = \{\mathbf{w}_1, \mathbf{w}_2, \ldots, \mathbf{w}_m\}$ be any set of m vectors in V, where $m < n$. We want to show that S' does not span V. The proof will be by contradiction: We will show that assuming S' spans V leads to a contradiction of the linear independence of $\{\mathbf{v}_1, \mathbf{v}_2, \ldots, \mathbf{v}_n\}$.

If S' spans V, then every vector in V is a linear combination of the vectors in S'. In particular, each basis vector $\mathbf{v}_i$ is a linear combination of the vectors in S', say

$$
\begin{aligned}
\mathbf{v}_1 &= a_{11}\mathbf{w}_1 + a_{21}\mathbf{w}_2 + \cdots + a_{m1}\mathbf{w}_m \\
\mathbf{v}_2 &= a_{12}\mathbf{w}_1 + a_{22}\mathbf{w}_2 + \cdots + a_{m2}\mathbf{w}_m \\
&\;\;\vdots \qquad\quad \vdots \qquad\quad \vdots \qquad\qquad \vdots \\
\mathbf{v}_n &= a_{1n}\mathbf{w}_1 + a_{2n}\mathbf{w}_2 + \cdots + a_{mn}\mathbf{w}_m
\end{aligned}
\tag{9}
$$

To obtain our contradiction, we will show that there are scalars $k_1, k_2, \ldots, k_n$, not all zero, such that

$$
k_1\mathbf{v}_1 + k_2\mathbf{v}_2 + \cdots + k_n\mathbf{v}_n = \mathbf{0}
\tag{10}
$$

But observe that (9) and (10) have the same form as (6) and (7) except that m and n are interchanged and the $\mathbf{w}$'s and $\mathbf{v}$'s are interchanged. Thus the computations that led to (8) now yield

$$
\begin{aligned}
a_{11}k_1 + a_{12}k_2 + \cdots + a_{1n}k_n &= 0 \\
a_{21}k_1 + a_{22}k_2 + \cdots + a_{2n}k_n &= 0 \\
\vdots \qquad \vdots \qquad\qquad \vdots \qquad \vdots \\
a_{m1}k_1 + a_{m2}k_2 + \cdots + a_{mn}k_n &= 0
\end{aligned}
$$

This linear system has more unknowns than equations and hence has nontrivial solutions by Theorem 1.2.1. ∎

It follows from the preceding theorem that if $S = \{\mathbf{v}_1, \mathbf{v}_2, \ldots, \mathbf{v}_n\}$ is any basis for a vector space V, then all sets in V that simultaneously span V and are linearly independent must have precisely n vectors. Thus, all bases for V must have the same number of vectors as the arbitrary basis S. This yields the following result, which is one of the most important in linear algebra.

THEOREM 5.4.3

> *All bases for a finite-dimensional vector space have the same number of vectors.*

To see how this theorem is related to the concept of "dimension," recall that the standard basis for R^n has n vectors (Example 2). Thus Theorem 5.4.3 implies that all bases for R^n have n vectors. In particular, every basis for R^3 has three vectors, every basis for R^2 has two vectors, and every basis for R^1 ($=R$) has one vector. Intuitively, R^3 is three-dimensional, R^2 (a plane) is two-dimensional, and R (a line) is one-dimensional. Thus, for familiar vector spaces, the number of vectors in a basis is the same as the dimension. This suggests the following definition.

> **DEFINITION**
>
> The ***dimension*** of a finite-dimensional vector space V, denoted by $\dim(V)$, is defined to be the number of vectors in a basis for V. In addition, we define the zero vector space to have dimension zero.

REMARK From here on we shall follow a common convention of regarding the empty set to be a basis for the zero vector space. This is consistent with the preceding definition, since the empty set has no vectors and the zero vector space has dimension zero.

EXAMPLE 9   Dimensions of Some Vector Spaces

$$\dim(R^n) = n \qquad \text{[The standard basis has } n \text{ vectors (Example 2).]}$$

$$\dim(P_n) = n + 1 \quad \text{[The standard basis has } n + 1 \text{ vectors (Example 5).]}$$

$$\dim(M_{mn}) = mn \quad \text{[The standard basis has } mn \text{ vectors (Example 6).]} \quad \blacklozenge$$

EXAMPLE 10   Dimension of a Solution Space

Determine a basis for and the dimension of the solution space of the homogeneous system

$$
\begin{aligned}
2x_1 + 2x_2 - \ x_3 \qquad\quad + x_5 &= 0 \\
-x_1 - \ x_2 + 2x_3 - 3x_4 + x_5 &= 0 \\
x_1 + \ x_2 - 2x_3 \qquad\quad - x_5 &= 0 \\
x_3 + \ x_4 + x_5 &= 0
\end{aligned}
$$

Solution

In Example 7 of Section 1.2 it was shown that the general solution of the given system is

$$x_1 = -s - t, \qquad x_2 = s, \qquad x_3 = -t, \qquad x_4 = 0, \qquad x_5 = t$$

Therefore, the solution vectors can be written as

$$
\begin{bmatrix} x_1 \\ x_2 \\ x_3 \\ x_4 \\ x_5 \end{bmatrix}
=
\begin{bmatrix} -s - t \\ s \\ -t \\ 0 \\ t \end{bmatrix}
=
\begin{bmatrix} -s \\ s \\ 0 \\ 0 \\ 0 \end{bmatrix}
+
\begin{bmatrix} -t \\ 0 \\ -t \\ 0 \\ t \end{bmatrix}
= s
\begin{bmatrix} -1 \\ 1 \\ 0 \\ 0 \\ 0 \end{bmatrix}
+ t
\begin{bmatrix} 1 \\ 0 \\ -1 \\ 0 \\ 1 \end{bmatrix}
$$

which shows that the vectors

$$
\mathbf{v}_1 =
\begin{bmatrix} -1 \\ 1 \\ 0 \\ 0 \\ 0 \end{bmatrix}
\quad \text{and} \quad
\mathbf{v}_2 =
\begin{bmatrix} -1 \\ 0 \\ -1 \\ 0 \\ 1 \end{bmatrix}
$$

span the solution space. Since they are also linearly independent (verify), $\{\mathbf{v}_1, \mathbf{v}_2\}$ is a basis, and the solution space is two-dimensional. $\blacklozenge$

Some Fundamental Theorems

We shall devote the remainder of this section to a series of theorems that reveal the subtle interrelationships among the concepts of spanning, linear independence, basis, and dimension. These theorems are not idle exercises in mathematical theory—they are essential to the understanding of vector spaces, and many practical applications of linear algebra build on them.

The following theorem, which we call the *Plus/Minus Theorem* (our own name), establishes two basic principles on which most of the theorems to follow will rely.

THEOREM 5.4.4

> **Plus/Minus Theorem**
>
> *Let S be a nonempty set of vectors in a vector space V.*
>
> (a) *If S is a linearly independent set, and if* **v** *is a vector in V that is outside of span(S), then the set S ∪ {***v***} that results by inserting* **v** *into S is still linearly independent.*
>
> (b) *If* **v** *is a vector in S that is expressible as a linear combination of other vectors in S, and if S − {***v***} denotes the set obtained by removing* **v** *from S, then S and S − {***v***} span the same space; that is,*
>
> $$span(S) = span(S - \{\mathbf{v}\})$$

We shall defer the proof to the end of the section, so that we may move more immediately to the consequences of the theorem. However, the theorem can be visualized in R^3 as follows:

(a)   A set S of two linearly independent vectors in R^3 spans a plane through the origin. If we enlarge S by inserting any vector **v** outside of this plane (Figure 5.4.5*a*), then the resulting set of three vectors is still linearly independent since none of the three vectors lies in the same plane as the other two.

(b)   If S is a set of three noncollinear vectors in R^3 that lie in a common plane through the origin (Figure 5.4.5*b*, *c*), then the three vectors span the plane. However, if we remove from S any vector **v** that is a linear combination of the other two, then the remaining set of two vectors still spans the plane.

In general, to show that a set of vectors $\{\mathbf{v}_1, \mathbf{v}_2, \ldots, \mathbf{v}_n\}$ is a basis for a vector space V, we must show that the vectors are linearly independent and span V. However, if we happen to know that V has dimension n (so that $\{\mathbf{v}_1, \mathbf{v}_2, \ldots, \mathbf{v}_n\}$ contains the right number of vectors for a basis), then it suffices to check *either* linear independence *or* spanning—the remaining condition will hold automatically. This is the content of the following theorem.

THEOREM 5.4.5

> *If V is an n-dimensional vector space, and if S is a set in V with exactly n vectors, then S is a basis for V if either S spans V or S is linearly independent.*

Proof   Assume that S has exactly n vectors and spans V. To prove that S is a basis, we must show that S is a linearly independent set. But if this is not so, then some vector **v** in S is a linear combination of the remaining vectors. If we remove this vector from S, then it follows from the Plus/Minus Theorem (5.4.4*b*) that the remaining set of $n - 1$ vectors still spans V. But this is impossible, since it follows from Theorem 5.4.2*b* that no

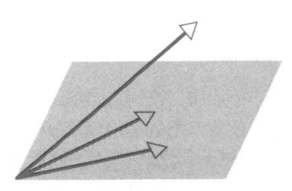

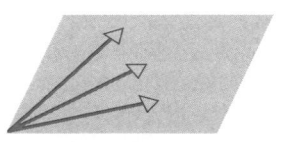

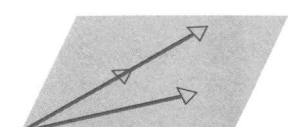

(*a*) None of the three vectors lies in the same plane as the other two.

(*b*) Any of the vectors can be removed, and the remaining two will still span the plane.

(*c*) Either of the collinear vectors can be removed, and the remaining two will still span the plane.

Figure 5.4.5

set with fewer than n vectors can span an n-dimensional vector space. Thus S is linearly independent.

Assume that S has exactly n vectors and is a linearly independent set. To prove that S is a basis, we must show that S spans V. But if this is not so, then there is some vector **v** in V that is not in span(S). If we insert this vector into S, then it follows from the Plus/Minus Theorem (5.4.4a) that this set of $n + 1$ vectors is still linearly independent. But this is impossible, since it follows from Theorem 5.4.2a that no set with more than n vectors in an n-dimensional vector space can be linearly independent. Thus S spans V. ■

EXAMPLE 11   Checking for a Basis

(a) Show that $\mathbf{v}_1 = (-3, 7)$ and $\mathbf{v}_2 = (5, 5)$ form a basis for R^2 by inspection.
(b) Show that $\mathbf{v}_1 = (2, 0, -1)$, $\mathbf{v}_2 = (4, 0, 7)$, and $\mathbf{v}_3 = (-1, 1, 4)$ form a basis for R^3 by inspection.

Solution (a)

Since neither vector is a scalar multiple of the other, the two vectors form a linearly independent set in the two-dimensional space R^2, and hence they form a basis by Theorem 5.4.5.

Solution (b)

The vectors $\mathbf{v}_1$ and $\mathbf{v}_2$ form a linearly independent set in the xz-plane (why?). The vector $\mathbf{v}_3$ is outside of the xz-plane, so the set $\{\mathbf{v}_1, \mathbf{v}_2, \mathbf{v}_3\}$ is also linearly independent. Since R^3 is three-dimensional, Theorem 5.4.5 implies that $\{\mathbf{v}_1, \mathbf{v}_2, \mathbf{v}_3\}$ is a basis for R^3. ◆

The following theorem shows that for a finite-dimensional vector space V, every set that spans V contains a basis for V within it, and every linearly independent set in V is part of some basis for V.

THEOREM 5.4.6

Let S be a finite set of vectors in a finite-dimensional vector space V.

(a) If S spans V but is not a basis for V, then S can be reduced to a basis for V by removing appropriate vectors from S.
(b) If S is a linearly independent set that is not already a basis for V, then S can be enlarged to a basis for V by inserting appropriate vectors into S.

Proof (a)   If S is a set of vectors that spans V but is not a basis for V, then S is a linearly dependent set. Thus some vector **v** in S is expressible as a linear combination of the other vectors in S. By the Plus/Minus Theorem (5.4.4b), we can remove **v** from S, and the resulting set S' will still span V. If S' is linearly independent, then S' is a basis for V, and we are done. If S' is linearly dependent, then we can remove some appropriate vector from S' to produce a set S'' that still spans V. We can continue removing vectors in this way until we finally arrive at a set of vectors in S that is linearly independent and spans V. This subset of S is a basis for V.

Proof (b)   Suppose that $\dim(V) = n$. If S is a linearly independent set that is not already a basis for V, then S fails to span V, and there is some vector **v** in V that is not in span(S). By the Plus/Minus Theorem (5.4.4a), we can insert **v** into S, and the resulting set S' will still be linearly independent. If S' spans V, then S' is a basis for V, and we are finished. If S' does not span V, then we can insert an appropriate vector into S' to produce a set S'' that is still linearly independent. We can continue inserting vectors in this way until we reach a set with n linearly independent vectors in V. This set will be a basis for V by Theorem 5.4.5. ∎

It can be proved (Exercise 30) that any subspace of a finite-dimensional vector space is finite-dimensional. We conclude this section with a theorem showing that the dimension of a subspace of a finite-dimensional vector space V cannot exceed the dimension of V itself and that the only way a subspace can have the same dimension as V is if the subspace is the entire vector space V. Figure 5.4.6 illustrates this idea in R^3. In that figure, observe that successively larger subspaces increase in dimension.

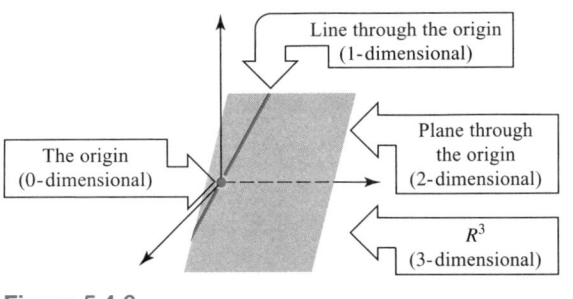

Figure 5.4.6

THEOREM 5.4.7

If W is a subspace of a finite-dimensional vector space V, then $\dim(W) \leq \dim(V)$; moreover, if $\dim(W) = \dim(V)$, then $W = V$.

Proof Since V is finite-dimensional, so is W by Exercise 30. Accordingly, suppose that $S = \{\mathbf{w}_1, \mathbf{w}_2, \ldots, \mathbf{w}_m\}$ is a basis for W. Either S is also a basis for V or it is not. If it is, then $\dim(W) = \dim(V) = m$. If it is not, then by Theorem 5.4.6b, vectors can be added to the linearly independent set S to make it into a basis for V, so $\dim(W) < \dim(V)$. Thus $\dim(W) \leq \dim(V)$ in all cases. If $\dim(W) = \dim(V)$, then S is a set of m linearly independent vectors in the m-dimensional vector space V; hence S is a basis for V by Theorem 5.4.5. This implies that $W = V$ (why?). ∎

Additional Proofs

Proof of Theorem 5.4.4a Assume that $S = \{\mathbf{v}_1, \mathbf{v}_2, \ldots, \mathbf{v}_r\}$ is a linearly independent set of vectors in V, and $\mathbf{v}$ is a vector in V outside of span(S). To show that $S' = \{\mathbf{v}_1, \mathbf{v}_2, \ldots, \mathbf{v}_r, \mathbf{v}\}$ is a linearly independent set, we must show that the only scalars that satisfy

$$k_1\mathbf{v}_1 + k_2\mathbf{v}_2 + \cdots + k_r\mathbf{v}_r + k_{r+1}\mathbf{v} = \mathbf{0} \tag{11}$$

are $k_1 = k_2 = \cdots = k_r = k_{r+1} = 0$. But we must have $k_{r+1} = 0$; otherwise, we could solve (11) for $\mathbf{v}$ as a linear combination of $\mathbf{v}_1, \mathbf{v}_2, \ldots, \mathbf{v}_r$, contradicting the assumption that $\mathbf{v}$ is outside of span(S). Thus (11) simplifies to

$$k_1\mathbf{v}_1 + k_2\mathbf{v}_2 + \cdots + k_r\mathbf{v}_r = \mathbf{0} \tag{12}$$

which, by the linear independence of $\{\mathbf{v}_1, \mathbf{v}_2, \ldots, \mathbf{v}_r\}$, implies that

$$k_1 = k_2 = \cdots = k_r = 0$$

Proof of Theorem 5.4.4b Assume that $S = \{\mathbf{v}_1, \mathbf{v}_2, \ldots, \mathbf{v}_r\}$ is a set of vectors in V, and to be specific, suppose that $\mathbf{v}_r$ is a linear combination of $\mathbf{v}_1, \mathbf{v}_2, \ldots, \mathbf{v}_{r-1}$, say

$$\mathbf{v}_r = c_1\mathbf{v}_1 + c_2\mathbf{v}_2 + \cdots + c_{r-1}\mathbf{v}_{r-1} \tag{13}$$

We want to show that if $\mathbf{v}_r$ is removed from S, then the remaining set of vectors $\{\mathbf{v}_1, \mathbf{v}_2, \ldots, \mathbf{v}_{r-1}\}$ still spans span(S); that is, we must show that every vector $\mathbf{w}$ in span(S) is expressible as a linear combination of $\{\mathbf{v}_1, \mathbf{v}_2, \ldots, \mathbf{v}_{r-1}\}$. But if $\mathbf{w}$ is in span(S), then $\mathbf{w}$ is expressible in the form

$$\mathbf{w} - k_1\mathbf{v}_1 + k_2\mathbf{v}_2 + \cdots + k_{r-1}\mathbf{v}_{r-1} + k_r\mathbf{v}_r$$

or, on substituting (13),

$$\mathbf{w} = k_1\mathbf{v}_1 + k_2\mathbf{v}_2 + \cdots + k_{r-1}\mathbf{v}_{r-1} + k_r(c_1\mathbf{v}_1 + c_2\mathbf{v}_2 + \cdots + c_{r-1}\mathbf{v}_{r-1})$$

which expresses $\mathbf{w}$ as a linear combination of $\mathbf{v}_1, \mathbf{v}_2, \ldots, \mathbf{v}_{r-1}$. ∎

EXERCISE SET 5.4

1. Explain why the following sets of vectors are *not* bases for the indicated vector spaces. (Solve this problem by inspection.)

   (a) $\mathbf{u}_1 = (1, 2)$, $\mathbf{u}_2 = (0, 3)$, $\mathbf{u}_3 = (2, 7)$ for R^2

   (b) $\mathbf{u}_1 = (-1, 3, 2)$, $\mathbf{u}_2 = (6, 1, 1)$ for R^3

   (c) $\mathbf{p}_1 = 1 + x + x^2$, $\mathbf{p}_2 = x - 1$ for P_2

   (d) $A = \begin{bmatrix} 1 & 1 \\ 2 & 3 \end{bmatrix}$, $B = \begin{bmatrix} 6 & 0 \\ -1 & 4 \end{bmatrix}$, $C = \begin{bmatrix} 3 & 0 \\ 1 & 7 \end{bmatrix}$, $D - \begin{bmatrix} 5 & 1 \\ 4 & 2 \end{bmatrix}$, $E = \begin{bmatrix} 7 & 1 \\ 2 & 9 \end{bmatrix}$

   for M_{22}

2. Which of the following sets of vectors are bases for R^2?

   (a) $(2, 1)$, $(3, 0)$ (b) $(4, 1)$, $(-7, -8)$

   (c) $(0, 0)$, $(1, 3)$ (d) $(3, 9)$, $(-4, -12)$

3. Which of the following sets of vectors are bases for R^3?

   (a) $(1, 0, 0)$, $(2, 2, 0)$, $(3, 3, 3)$ (b) $(3, 1, -4)$, $(2, 5, 6)$, $(1, 4, 8)$

   (c) $(2, -3, 1)$, $(4, 1, 1)$, $(0, -7, 1)$ (d) $(1, 6, 4)$, $(2, 4, -1)$, $(-1, 2, 5)$

4. Which of the following sets of vectors are bases for P_2?

   (a) $1 - 3x + 2x^2$, $1 + x + 4x^2$, $1 - 7x$

   (b) $4 + 6x + x^2$, $-1 + 4x + 2x^2$, $5 + 2x - x^2$

   (c) $1 + x + x^2$, $x + x^2$, x^2

   (d) $-4 + x + 3x^2$, $6 + 5x + 2x^2$, $8 + 4x + x^2$

5. Show that the following set of vectors is a basis for M_{22}.

   $$\begin{bmatrix} 3 & 6 \\ 3 & -6 \end{bmatrix}, \quad \begin{bmatrix} 0 & -1 \\ -1 & 0 \end{bmatrix}, \quad \begin{bmatrix} 0 & -8 \\ -12 & -4 \end{bmatrix}, \quad \begin{bmatrix} 1 & 0 \\ -1 & 2 \end{bmatrix}$$

6. Let V be the space spanned by $\mathbf{v}_1 = \cos^2 x$, $\mathbf{v}_2 = \sin^2 x$, $\mathbf{v}_3 = \cos 2x$.

   (a) Show that $S = \{\mathbf{v}_1, \mathbf{v}_2, \mathbf{v}_3\}$ is not a basis for V.   (b) Find a basis for V.

7. Find the coordinate vector of $\mathbf{w}$ relative to the basis $S = \{\mathbf{u}_1, \mathbf{u}_2\}$ for R^2.

   (a) $\mathbf{u}_1 = (1, 0)$, $\mathbf{u}_2 = (0, 1)$; $\mathbf{w} = (3, -7)$

   (b) $\mathbf{u}_1 = (2, -4)$, $\mathbf{u}_2 = (3, 8)$; $\mathbf{w} = (1, 1)$

   (c) $\mathbf{u}_1 = (1, 1)$, $\mathbf{u}_2 = (0, 2)$; $\mathbf{w} = (a, b)$

8. Find the coordinate vector of **w** relative to the basis $S = \{\mathbf{u}_1, \mathbf{u}_2\}$ of R^2.

   (a) $\mathbf{u}_1 = (1, -1)$, $\mathbf{u}_2 = (1, 1)$; $\mathbf{w} = (1, 0)$

   (b) $\mathbf{u}_1 = (1, -1)$, $\mathbf{u}_2 = (1, 1)$; $\mathbf{w} = (0, 1)$

   (c) $\mathbf{u}_1 = (1, -1)$, $\mathbf{u}_2 = (1, 1)$; $\mathbf{w} = (1, 1)$

9. Find the coordinate vector of **v** relative to the basis $S = \{\mathbf{v}_1, \mathbf{v}_2, \mathbf{v}_3\}$.

   (a) $\mathbf{v} = (2, -1, 3)$; $\mathbf{v}_1 = (1, 0, 0)$, $\mathbf{v}_2 = (2, 2, 0)$, $\mathbf{v}_3 = (3, 3, 3)$

   (b) $\mathbf{v} = (5, -12, 3)$; $\mathbf{v}_1 = (1, 2, 3)$, $\mathbf{v}_2 = (-4, 5, 6)$, $\mathbf{v}_3 = (7, -8, 9)$

10. Find the coordinate vector of **p** relative to the basis $S = \{\mathbf{p}_1, \mathbf{p}_2, \mathbf{p}_3\}$.

   (a) $\mathbf{p} = 4 - 3x + x^2$; $\mathbf{p}_1 = 1$, $\mathbf{p}_2 = x$, $\mathbf{p}_3 = x^2$

   (b) $\mathbf{p} = 2 - x + x^2$; $\mathbf{p}_1 = 1 + x$, $\mathbf{p}_2 = 1 + x^2$, $\mathbf{p}_3 = x + x^2$

11. Find the coordinate vector of A relative to the basis $S = \{A_1, A_2, A_3, A_4\}$.

$$A = \begin{bmatrix} 2 & 0 \\ -1 & 3 \end{bmatrix}; \quad A_1 = \begin{bmatrix} -1 & 1 \\ 0 & 0 \end{bmatrix}, \quad A_2 = \begin{bmatrix} 1 & 1 \\ 0 & 0 \end{bmatrix}$$

$$A_3 = \begin{bmatrix} 0 & 0 \\ 1 & 0 \end{bmatrix}, \quad A_4 = \begin{bmatrix} 0 & 0 \\ 0 & 1 \end{bmatrix}$$

In Exercises 12–17 determine the dimension of and a basis for the solution space of the system.

12. $\begin{aligned} x_1 + x_2 - x_3 &= 0 \\ -2x_1 - x_2 + 2x_3 &= 0 \\ -x_1 \phantom{{}- x_2} + x_3 &= 0 \end{aligned}$
      13. $\begin{aligned} 3x_1 + x_2 + x_3 + x_4 &= 0 \\ 5x_1 - x_2 + x_3 - x_4 &= 0 \end{aligned}$

14. $\begin{aligned} x_1 - 4x_2 + 3x_3 - x_4 &= 0 \\ 2x_1 - 8x_2 + 6x_3 - 2x_4 &= 0 \end{aligned}$
      15. $\begin{aligned} x_1 - 3x_2 + x_3 &= 0 \\ 2x_1 - 6x_2 + 2x_3 &= 0 \\ 3x_1 - 9x_2 + 3x_3 &= 0 \end{aligned}$

16. $\begin{aligned} 2x_1 + x_2 + 3x_3 &= 0 \\ x_1 \phantom{{}+ x_2} + 5x_3 &= 0 \\ x_2 + x_3 &= 0 \end{aligned}$
      17. $\begin{aligned} x + y + z &= 0 \\ 3x + 2y - 2z &= 0 \\ 4x + 3y - z &= 0 \\ 6x + 5y + z &= 0 \end{aligned}$

18. Determine bases for the following subspaces of R^3.

   (a) the plane $3x - 2y + 5z = 0$

   (b) the plane $x - y = 0$

   (c) the line $x = 2t$, $y = -t$, $z = 4t$

   (d) all vectors of the form (a, b, c), where $b = a + c$

19. Determine the dimensions of the following subspaces of R^4.

   (a) all vectors of the form $(a, b, c, 0)$

   (b) all vectors of the form (a, b, c, d), where $d = a + b$ and $c = a - b$

   (c) all vectors of the form (a, b, c, d), where $a = b = c = d$

20. Determine the dimension of the subspace of P_3 consisting of all polynomials $a_0 + a_1x + a_2x^2 + a_3x^3$ for which $a_0 = 0$.

21. Find a standard basis vector that can be added to the set $\{\mathbf{v}_1, \mathbf{v}_2\}$ to produce a basis for R^3.

   (a) $\mathbf{v}_1 = (-1, 2, 3)$, $\mathbf{v}_2 = (1, -2, -2)$ (b) $\mathbf{v}_1 = (1, -1, 0)$, $\mathbf{v}_2 = (3, 1, -2)$

22. Find standard basis vectors that can be added to the set $\{\mathbf{v}_1, \mathbf{v}_2\}$ to produce a basis for R^4.

$$\mathbf{v}_1 = (1, -4, 2, -3), \qquad \mathbf{v}_2 = (-3, 8, -4, 6)$$

23. Let $\{\mathbf{v}_1, \mathbf{v}_2, \mathbf{v}_3\}$ be a basis for a vector space V. Show that $\{\mathbf{u}_1, \mathbf{u}_2, \mathbf{u}_3\}$ is also a basis, where $\mathbf{u}_1 = \mathbf{v}_1$, $\mathbf{u}_2 = \mathbf{v}_1 + \mathbf{v}_2$, and $\mathbf{u}_3 = \mathbf{v}_1 + \mathbf{v}_2 + \mathbf{v}_3$.

24. (a) Show that for every positive integer n, one can find $n + 1$ linearly independent vectors in $F(-\infty, \infty)$.

    Hint Look for polynomials.

    (b) Use the result in part (a) to prove that $F(-\infty, \infty)$ is infinite-dimensional.

    (c) Prove that $C(-\infty, \infty)$, $C^m(-\infty, \infty)$, and $C^\infty(-\infty, \infty)$ are infinite-dimensional vector spaces.

25. Let S be a basis for an n-dimensional vector space V. Show that if $\mathbf{v}_1, \mathbf{v}_2, \ldots, \mathbf{v}_r$ form a linearly independent set of vectors in V, then the coordinate vectors $(\mathbf{v}_1)_S, (\mathbf{v}_2)_S, \ldots, (\mathbf{v}_r)_S$ form a linearly independent set in R^n, and conversely.

26. Using the notation from Exercise 25, show that if $\mathbf{v}_1, \mathbf{v}_2, \ldots, \mathbf{v}_r$ span V, then the coordinate vectors $(\mathbf{v}_1)_S, (\mathbf{v}_2)_S, \ldots, (\mathbf{v}_r)_S$ span R^n, and conversely.

27. Find a basis for the subspace of P_2 spanned by the given vectors.

    (a) $-1 + x - 2x^2$, $3 + 3x + 6x^2$, 9

    (b) $1 + x$, x^2, $-2 + 2x^2$, $-3x$

    (c) $1 + x - 3x^2$, $2 + 2x - 6x^2$, $3 + 3x - 9x^2$

    Hint Let S be the standard basis for P_2 and work with the coordinate vectors relative to S; note Exercises 25 and 26.

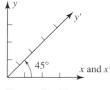

Figure Ex-28

28. The accompanying figure shows a rectangular xy-coordinate system and an $x'y'$-coordinate system with skewed axes. Assuming that 1-unit scales are used on all the axes, find the $x'y'$-coordinates of the points whose xy-coordinates are given.

    (a) $(1, 1)$ (b) $(1, 0)$ (c) $(0, 1)$ (d) (a, b)

29. The accompanying figure shows a rectangular xy-coordinate system determined by the unit basis vectors $\mathbf{i}$ and $\mathbf{j}$ and an $x'y'$-coordinate system determined by unit basis vectors $\mathbf{u}_1$ and $\mathbf{u}_2$. Find the $x'y'$-coordinates of the points whose xy-coordinates are given.

    (a) $(\sqrt{3}, 1)$ (b) $(1, 0)$ (c) $(0, 1)$ (d) (a, b)

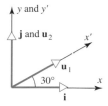

Figure Ex-29

30. Prove: Any subspace of a finite-dimensional vector space is finite-dimensional.

Discussion & Discovery

31. The basis that we gave for M_{22} in Example 6 consisted of noninvertible matrices. Do you think that there is a basis for M_{22} consisting of invertible matrices? Justify your answer.

32. (a) The vector space of all diagonal $n \times n$ matrices has dimension _____.

    (b) The vector space of all symmetric $n \times n$ matrices has dimension _____.

    (c) The vector space of all upper triangular $n \times n$ matrices has dimension _____.

33. (a) For a 3×3 matrix A, explain in words why the set $I_3, A, A^2, \ldots, A^9$ must be linearly dependent if the ten matrices are distinct.

    (b) State a corresponding result for an $n \times n$ matrix A.

34. State the two parts of Theorem 5.4.2 in contrapositive form. [See Exercise 34 of Section 1.4.]

35. (a) The equation $x_1 + x_2 + \cdots + x_n = 0$ can be viewed as a linear system of one equation in n unknowns. Make a conjecture about the dimension of its solution space.

    (b) Confirm your conjecture by finding a basis.

36. (a) Show that the set W of polynomials in P_2 such that $p(1) = 0$ is a subspace of P_2.

    (b) Make a conjecture about the dimension of W.

    (c) Confirm your conjecture by finding a basis for W.

5.5

ROW SPACE, COLUMN SPACE, AND NULLSPACE

In this section we shall study three important vector spaces that are associated with matrices. Our work here will provide us with a deeper understanding of the relationships between the solutions of a linear system of equations and properties of its coefficient matrix.

We begin with some definitions.

DEFINITION

For an $m \times n$ matrix

$$A = \begin{bmatrix} a_{11} & a_{12} & \cdots & a_{1n} \\ a_{21} & a_{22} & \cdots & a_{2n} \\ \vdots & \vdots & & \vdots \\ a_{m1} & a_{m2} & \cdots & a_{mn} \end{bmatrix}$$

the vectors

$$\mathbf{r}_1 = [a_{11} \quad a_{12} \quad \cdots \quad a_{1n}]$$
$$\mathbf{r}_2 = [a_{21} \quad a_{22} \quad \cdots \quad a_{2n}]$$
$$\vdots$$
$$\mathbf{r}_m = [a_{m1} \quad a_{m2} \quad \cdots \quad a_{mn}]$$

in R^n formed from the rows of A are called the **row vectors** of A, and the vectors

$$\mathbf{c}_1 = \begin{bmatrix} a_{11} \\ a_{21} \\ \vdots \\ a_{m1} \end{bmatrix}, \quad \mathbf{c}_2 = \begin{bmatrix} a_{12} \\ a_{22} \\ \vdots \\ a_{m2} \end{bmatrix}, \dots, \quad \mathbf{c}_n = \begin{bmatrix} a_{1n} \\ a_{2n} \\ \vdots \\ a_{mn} \end{bmatrix}$$

in R^m formed from the columns of A are called the **column vectors** of A.

EXAMPLE 1 Row and Column Vectors in a 2 × 3 Matrix

Let

$$A = \begin{bmatrix} 2 & 1 & 0 \\ 3 & -1 & 4 \end{bmatrix}$$

The row vectors of A are

$$\mathbf{r}_1 = [2 \quad 1 \quad 0] \quad \text{and} \quad \mathbf{r}_2 = [3 \quad -1 \quad 4]$$

and the column vectors of A are

$$\mathbf{c}_1 = \begin{bmatrix} 2 \\ 3 \end{bmatrix}, \quad \mathbf{c}_2 = \begin{bmatrix} 1 \\ -1 \end{bmatrix}, \quad \text{and} \quad \mathbf{c}_3 = \begin{bmatrix} 0 \\ 4 \end{bmatrix} \quad \blacklozenge$$

The following definition defines three important vector spaces associated with a matrix.

DEFINITION

If A is an $m \times n$ matrix, then the subspace of R^n spanned by the row vectors of A is called the **row space** of A, and the subspace of R^m spanned by the column vectors of A is called the **column space** of A. The solution space of the homogeneous system of equations $A\mathbf{x} = \mathbf{0}$, which is a subspace of R^n, is called the **nullspace** of A.

In this section and the next we shall be concerned with the following two general questions:

- What relationships exist between the solutions of a linear system $A\mathbf{x} = \mathbf{b}$ and the row space, column space, and nullspace of the coefficient matrix A?
- What relationships exist among the row space, column space, and nullspace of a matrix?

To investigate the first of these questions, suppose that

$$A = \begin{bmatrix} a_{11} & a_{12} & \cdots & a_{1n} \\ a_{21} & a_{22} & \cdots & a_{2n} \\ \vdots & \vdots & & \vdots \\ a_{m1} & a_{m2} & \cdots & a_{mn} \end{bmatrix} \quad \text{and} \quad \mathbf{x} = \begin{bmatrix} x_1 \\ x_2 \\ \vdots \\ x_n \end{bmatrix}$$

It follows from Formula (10) of Section 1.3 that if $\mathbf{c}_1, \mathbf{c}_2, \ldots, \mathbf{c}_n$ denote the column vectors of A, then the product $A\mathbf{x}$ can be expressed as a linear combination of these column vectors with coefficients from $\mathbf{x}$; that is,

$$A\mathbf{x} = x_1 \mathbf{c}_1 + x_2 \mathbf{c}_2 + \cdots + x_n \mathbf{c}_n \tag{1}$$

Thus a linear system, $A\mathbf{x} = \mathbf{b}$, of m equations in n unknowns can be written as

$$x_1 \mathbf{c}_1 + x_2 \mathbf{c}_2 + \cdots + x_n \mathbf{c}_n = \mathbf{b} \tag{2}$$

from which we conclude that $A\mathbf{x} = \mathbf{b}$ is consistent if and only if $\mathbf{b}$ is expressible as a linear combination of the column vectors of A or, equivalently, if and only if $\mathbf{b}$ is in the column space of A. This yields the following theorem.

THEOREM 5.5.1

> *A system of linear equations $A\mathbf{x} = \mathbf{b}$ is consistent if and only if $\mathbf{b}$ is in the column space of A.*

EXAMPLE 2 A Vector b in the Column Space of A

Let $A\mathbf{x} = \mathbf{b}$ be the linear system

$$\begin{bmatrix} -1 & 3 & 2 \\ 1 & 2 & -3 \\ 2 & 1 & -2 \end{bmatrix} \begin{bmatrix} x_1 \\ x_2 \\ x_3 \end{bmatrix} = \begin{bmatrix} 1 \\ -9 \\ -3 \end{bmatrix}$$

Show that $\mathbf{b}$ is in the column space of A, and express $\mathbf{b}$ as a linear combination of the column vectors of A.

Solution

Solving the system by Gaussian elimination yields (verify)

$$x_1 = 2, \qquad x_2 = -1, \qquad x_3 = 3$$

Since the system is consistent, $\mathbf{b}$ is in the column space of A. Moreover, from (2) and the solution obtained, it follows that

$$2 \begin{bmatrix} -1 \\ 1 \\ 2 \end{bmatrix} - \begin{bmatrix} 3 \\ 2 \\ 1 \end{bmatrix} + 3 \begin{bmatrix} 2 \\ -3 \\ -2 \end{bmatrix} = \begin{bmatrix} 1 \\ -9 \\ -3 \end{bmatrix} \quad \blacklozenge$$

The next theorem establishes a fundamental relationship between the solutions of a nonhomogeneous linear system $A\mathbf{x} = \mathbf{b}$ and those of the corresponding homogeneous linear system $A\mathbf{x} = \mathbf{0}$ with the same coefficient matrix.

THEOREM 5.5.2

> *If $\mathbf{x}_0$ denotes any single solution of a consistent linear system $A\mathbf{x} = \mathbf{b}$, and if $\mathbf{v}_1, \mathbf{v}_2, \ldots, \mathbf{v}_k$ form a basis for the nullspace of A—that is, the solution space of the homogeneous system $A\mathbf{x} = \mathbf{0}$—then every solution of $A\mathbf{x} = \mathbf{b}$ can be expressed in the form*
>
> $$\mathbf{x} = \mathbf{x}_0 + c_1\mathbf{v}_1 + c_2\mathbf{v}_2 + \cdots + c_k\mathbf{v}_k \tag{3}$$
>
> *and, conversely, for all choices of scalars $c_1, c_2, \ldots, c_k$, the vector $\mathbf{x}$ in this formula is a solution of $A\mathbf{x} = \mathbf{b}$.*

Proof Assume that $\mathbf{x}_0$ is any fixed solution of $A\mathbf{x} = \mathbf{b}$ and that $\mathbf{x}$ is an arbitrary solution. Then

$$A\mathbf{x}_0 = \mathbf{b} \quad \text{and} \quad A\mathbf{x} = \mathbf{b}$$

Subtracting these equations yields

$$A\mathbf{x} - A\mathbf{x}_0 = \mathbf{0} \quad \text{or} \quad A(\mathbf{x} - \mathbf{x}_0) = \mathbf{0}$$

which shows that $\mathbf{x} - \mathbf{x}_0$ is a solution of the homogeneous system $A\mathbf{x} = \mathbf{0}$. Since $\mathbf{v}_1, \mathbf{v}_2, \ldots, \mathbf{v}_k$ is a basis for the solution space of this system, we can express $\mathbf{x} - \mathbf{x}_0$ as a linear combination of these vectors, say

$$\mathbf{x} - \mathbf{x}_0 = c_1\mathbf{v}_1 + c_2\mathbf{v}_2 + \cdots + c_k\mathbf{v}_k$$

Thus,

$$\mathbf{x} = \mathbf{x}_0 + c_1\mathbf{v}_1 + c_2\mathbf{v}_2 + \cdots + c_k\mathbf{v}_k$$

which proves the first part of the theorem. Conversely, for all choices of the scalars $c_1, c_2, \ldots, c_k$ in (3), we have

$$A\mathbf{x} = A(\mathbf{x}_0 + c_1\mathbf{v}_1 + c_2\mathbf{v}_2 + \cdots + c_k\mathbf{v}_k)$$

or

$$A\mathbf{x} = A\mathbf{x}_0 + c_1(A\mathbf{v}_1) + c_2(A\mathbf{v}_2) + \cdots + c_k(A\mathbf{v}_k)$$

But $\mathbf{x}_0$ is a solution of the nonhomogeneous system, and $\mathbf{v}_1, \mathbf{v}_2, \ldots, \mathbf{v}_k$ are solutions of the homogeneous system, so the last equation implies that

$$A\mathbf{x} = \mathbf{b} + \mathbf{0} + \mathbf{0} + \cdots + \mathbf{0} = \mathbf{b}$$

which shows that $\mathbf{x}$ is a solution of $A\mathbf{x} = \mathbf{b}$. ■

General and Particular Solutions

There is some terminology associated with Formula (3). The vector $\mathbf{x}_0$ is called a ***particular solution*** of $A\mathbf{x} = \mathbf{b}$. The expression $\mathbf{x}_0 + c_1\mathbf{v}_1 + c_2\mathbf{v}_2 + \cdots + c_k\mathbf{v}_k$ is called the ***general solution*** of $A\mathbf{x} = \mathbf{b}$, and the expression $c_1\mathbf{v}_1 + c_2\mathbf{v}_2 + \cdots + c_k\mathbf{v}_k$ is called the ***general solution*** of $A\mathbf{x} = \mathbf{0}$. With this terminology, Formula (3) states that *the general solution of $A\mathbf{x} = \mathbf{b}$ is the sum of any particular solution of $A\mathbf{x} = \mathbf{b}$ and the general solution of $A\mathbf{x} = \mathbf{0}$.*

For linear systems with two or three unknowns, Theorem 5.5.2 has a nice geometric interpretation in R^2 and R^3. For example, consider the case where $A\mathbf{x} = \mathbf{0}$ and $A\mathbf{x} = \mathbf{b}$ are linear systems with two unknowns. The solutions of $A\mathbf{x} = \mathbf{0}$ form a subspace of R^2 and hence constitute a line through the origin, the origin only, or all of R^2. From Theorem 5.5.2, the solutions of $A\mathbf{x} = \mathbf{b}$ can be obtained by adding any particular solution of $A\mathbf{x} = \mathbf{b}$, say $\mathbf{x}_0$, to the solutions of $A\mathbf{x} = \mathbf{0}$. Assuming that $\mathbf{x}_0$ is positioned with its initial point at the origin, this has the geometric effect of translating the solution space of $A\mathbf{x} = \mathbf{0}$ so that the point at the origin is moved to the tip of $\mathbf{x}_0$ (Figure 5.5.1). This means that the solution vectors of $A\mathbf{x} = \mathbf{b}$ form a line through the tip of $\mathbf{x}_0$, the point at the tip of $\mathbf{x}_0$, or all of R^2. (Can you visualize the last case?) Similarly, for linear systems with three unknowns, the solutions of $A\mathbf{x} = \mathbf{b}$ constitute a plane through the tip of any particular solution $\mathbf{x}_0$, a line through the tip of $\mathbf{x}_0$, the point at the tip of $\mathbf{x}_0$, or all of R^3.

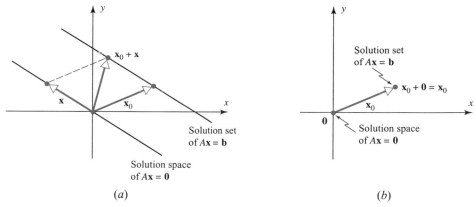

(a) (b)

Figure 5.5.1 Adding $\mathbf{x}_0$ to each vector $\mathbf{x}$ in the solution space of $A\mathbf{x} = \mathbf{0}$ translates the solution space.

EXAMPLE 3 General Solution of a Linear System $A\mathbf{x} = \mathbf{b}$

In Example 4 of Section 1.2 we solved the nonhomogeneous linear system

$$
\begin{aligned}
x_1 + 3x_2 - 2x_3 \quad\quad + 2x_5 \quad\quad\quad &= 0 \\
2x_1 + 6x_2 - 5x_3 - 2x_4 + 4x_5 - 3x_6 &= -1 \\
5x_3 + 10x_4 \quad\quad + 15x_6 &= 5 \\
2x_1 + 6x_2 \quad\quad + 8x_4 + 4x_5 + 18x_6 &= 6
\end{aligned}
\tag{4}
$$

and obtained

$$x_1 = -3r - 4s - 2t, \quad x_2 = r, \quad x_3 = -2s, \quad x_4 = s, \quad x_5 = t, \quad x_6 = \tfrac{1}{3}$$

This result can be written in vector form as

$$
\begin{bmatrix} x_1 \\ x_2 \\ x_3 \\ x_4 \\ x_5 \\ x_6 \end{bmatrix}
=
\begin{bmatrix} -3r - 4s - 2t \\ r \\ -2s \\ s \\ t \\ \tfrac{1}{3} \end{bmatrix}
=
\underbrace{\begin{bmatrix} 0 \\ 0 \\ 0 \\ 0 \\ 0 \\ \tfrac{1}{3} \end{bmatrix}}_{\mathbf{x}_0}
+
\underbrace{r\begin{bmatrix} -3 \\ 1 \\ 0 \\ 0 \\ 0 \\ 0 \end{bmatrix}
+ s\begin{bmatrix} -4 \\ 0 \\ -2 \\ 1 \\ 0 \\ 0 \end{bmatrix}
+ t\begin{bmatrix} -2 \\ 0 \\ 0 \\ 0 \\ 1 \\ 0 \end{bmatrix}}_{\mathbf{x}}
\tag{5}
$$

which is the general solution of (4). The vector $\mathbf{x}_0$ in (5) is a particular solution of (4); the linear combination $\mathbf{x}$ in (5) is the general solution of the homogeneous system

$$
\begin{aligned}
x_1 + 3x_2 - 2x_3 \quad\quad + 2x_5 \quad\quad\quad &= 0 \\
2x_1 + 6x_2 - 5x_3 - 2x_4 + 4x_5 - 3x_6 &= 0 \\
5x_3 + 10x_4 \quad\quad + 15x_6 &= 0 \\
2x_1 + 6x_2 \quad\quad + 8x_4 + 4x_5 + 18x_6 &= 0
\end{aligned}
$$

(verify). ◆

Bases for Row Spaces, Column Spaces, and Nullspaces

We first developed elementary row operations for the purpose of solving linear systems, and we know from that work that performing an elementary row operation on an augmented matrix does not change the solution set of the corresponding linear system. It follows that applying an elementary row operation to a matrix A does not change the solution set of the corresponding linear system $A\mathbf{x} = \mathbf{0}$, or, stated another way, it does not change the nullspace of A. Thus we have the following theorem.

THEOREM 5.5.3

> *Elementary row operations do not change the nullspace of a matrix.*

EXAMPLE 4 Basis for Nullspace

Find a basis for the nullspace of

$$A = \begin{bmatrix} 2 & 2 & -1 & 0 & 1 \\ -1 & -1 & 2 & -3 & 1 \\ 1 & 1 & -2 & 0 & -1 \\ 0 & 0 & 1 & 1 & 1 \end{bmatrix}$$

Solution

The nullspace of A is the solution space of the homogeneous system

$$\begin{aligned} 2x_1 + 2x_2 - \ x_3 \qquad\quad + x_5 &= 0 \\ -x_1 - \ x_2 + 2x_3 - 3x_4 + x_5 &= 0 \\ x_1 + \ x_2 - 2x_3 \qquad\quad - x_5 &= 0 \\ x_3 + \ x_4 + x_5 &= 0 \end{aligned}$$

In Example 10 of Section 5.4 we showed that the vectors

$$\mathbf{v}_1 = \begin{bmatrix} -1 \\ 1 \\ 0 \\ 0 \\ 0 \end{bmatrix} \quad \text{and} \quad \mathbf{v}_2 = \begin{bmatrix} -1 \\ 0 \\ -1 \\ 0 \\ 1 \end{bmatrix}$$

form a basis for this space. ◆

The following theorem is a companion to Theorem 5.5.3.

THEOREM 5.5.4

> *Elementary row operations do not change the row space of a matrix.*

Proof Suppose that the row vectors of a matrix A are $\mathbf{r}_1, \mathbf{r}_2, \ldots, \mathbf{r}_m$, and let B be obtained from A by performing an elementary row operation. We shall show that every vector in the row space of B is also in the row space of A and that, conversely, every vector in the row space of A is in the row space of B. We can then conclude that A and B have the same row space.

Consider the possibilities: If the row operation is a row interchange, then B and A have the same row vectors and consequently have the same row space. If the row operation is multiplication of a row by a nonzero scalar or the addition of a multiple of one row to another, then the row vectors $\mathbf{r}'_1, \mathbf{r}'_2, \ldots, \mathbf{r}'_m$ of B are linear combinations of

$\mathbf{r}_1, \mathbf{r}_2, \ldots, \mathbf{r}_m$; thus they lie in the row space of A. Since a vector space is closed under addition and scalar multiplication, all linear combinations of $\mathbf{r}'_1, \mathbf{r}'_2, \ldots, \mathbf{r}'_m$ will also lie in the row space of A. Therefore, each vector in the row space of B is in the row space of A.

Since B is obtained from A by performing a row operation, A can be obtained from B by performing the inverse operation (Section 1.5). Thus the argument above shows that the row space of A is contained in the row space of B. ■

In light of Theorems 5.5.3 and 5.5.4, one might anticipate that elementary row operations should not change the column space of a matrix. However, this is *not* so—elementary row operations can change the column space. For example, consider the matrix

$$A = \begin{bmatrix} 1 & 3 \\ 2 & 6 \end{bmatrix}$$

The second column is a scalar multiple of the first, so the column space of A consists of all scalar multiples of the first column vector. However, if we add -2 times the first row of A to the second row, we obtain

$$B = \begin{bmatrix} 1 & 3 \\ 0 & 0 \end{bmatrix}$$

Here again the second column is a scalar multiple of the first, so the column space of B consists of all scalar multiples of the first column vector. This is not the same as the column space of A.

Although elementary row operations can change the column space of a matrix, we shall show that whatever relationships of linear independence or linear dependence exist among the column vectors prior to a row operation will also hold for the corresponding columns of the matrix that results from that operation. To make this more precise, suppose a matrix B results from performing an elementary row operation on an $m \times n$ matrix A. By Theorem 5.5.3, the two homogeneous linear systems

$$A\mathbf{x} = \mathbf{0} \quad \text{and} \quad B\mathbf{x} = \mathbf{0}$$

have the same solution set. Thus the first system has a nontrivial solution if and only if the same is true of the second. But if the column vectors of A and B, respectively, are

$$\mathbf{c}_1, \mathbf{c}_2, \ldots, \mathbf{c}_n \quad \text{and} \quad \mathbf{c}'_1, \mathbf{c}'_2, \ldots, \mathbf{c}'_n$$

then from (2) the two systems can be rewritten as

$$x_1\mathbf{c}_1 + x_2\mathbf{c}_2 + \cdots + x_n\mathbf{c}_n = \mathbf{0} \tag{6}$$

and

$$x_1\mathbf{c}'_1 + x_2\mathbf{c}'_2 + \cdots + x_n\mathbf{c}'_n = \mathbf{0} \tag{7}$$

Thus (6) has a nontrivial solution for $x_1, x_2, \ldots, x_n$ if and only if the same is true of (7). This implies that the column vectors of A are linearly independent if and only if the same is true of B. Although we shall omit the proof, this conclusion also applies to any subset of the column vectors. Thus we have the following result.

THEOREM 5.5.5

> *If A and B are row equivalent matrices, then*
>
> *(a) A given set of column vectors of A is linearly independent if and only if the corresponding column vectors of B are linearly independent.*
>
> *(b) A given set of column vectors of A forms a basis for the column space of A if and only if the corresponding column vectors of B form a basis for the column space of B.*

The following theorem makes it possible to find bases for the row and column spaces of a matrix in row-echelon form by inspection.

THEOREM 5.5.6

> *If a matrix R is in row-echelon form, then the row vectors with the leading 1's (the nonzero row vectors) form a basis for the row space of R, and the column vectors with the leading 1's of the row vectors form a basis for the column space of R.*

Since this result is virtually self-evident when one looks at numerical examples, we shall omit the proof; the proof involves little more than an analysis of the positions of the 0's and 1's of R.

EXAMPLE 5 Bases for Row and Column Spaces

The matrix

$$R = \begin{bmatrix} 1 & -2 & 5 & 0 & 3 \\ 0 & 1 & 3 & 0 & 0 \\ 0 & 0 & 0 & 1 & 0 \\ 0 & 0 & 0 & 0 & 0 \end{bmatrix}$$

is in row-echelon form. From Theorem 5.5.6, the vectors

$$\mathbf{r}_1 = [1 \quad -2 \quad 5 \quad 0 \quad 3]$$
$$\mathbf{r}_2 = [0 \quad 1 \quad 3 \quad 0 \quad 0]$$
$$\mathbf{r}_3 = [0 \quad 0 \quad 0 \quad 1 \quad 0]$$

form a basis for the row space of R, and the vectors

$$\mathbf{c}_1 = \begin{bmatrix} 1 \\ 0 \\ 0 \\ 0 \end{bmatrix}, \qquad \mathbf{c}_2 = \begin{bmatrix} -2 \\ 1 \\ 0 \\ 0 \end{bmatrix}, \qquad \mathbf{c}_4 = \begin{bmatrix} 0 \\ 0 \\ 1 \\ 0 \end{bmatrix}$$

form a basis for the column space of R. ◆

EXAMPLE 6 Bases for Row and Column Spaces

Find bases for the row and column spaces of

$$A = \begin{bmatrix} 1 & -3 & 4 & -2 & 5 & 4 \\ 2 & -6 & 9 & -1 & 8 & 2 \\ 2 & -6 & 9 & -1 & 9 & 7 \\ -1 & 3 & -4 & 2 & -5 & -4 \end{bmatrix}$$

Solution

Since elementary row operations do not change the row space of a matrix, we can find a basis for the row space of A by finding a basis for the row space of any row-echelon form of A. Reducing A to row-echelon form, we obtain (verify)

$$R = \begin{bmatrix} 1 & -3 & 4 & -2 & 5 & 4 \\ 0 & 0 & 1 & 3 & -2 & -6 \\ 0 & 0 & 0 & 0 & 1 & 5 \\ 0 & 0 & 0 & 0 & 0 & 0 \end{bmatrix}$$

By Theorem 5.5.6, the nonzero row vectors of R form a basis for the row space of R and hence form a basis for the row space of A. These basis vectors are

$$\mathbf{r}_1 = [1 \quad -3 \quad 4 \quad -2 \quad 5 \quad 4]$$
$$\mathbf{r}_2 = [0 \quad 0 \quad 1 \quad 3 \quad -2 \quad -6]$$
$$\mathbf{r}_3 = [0 \quad 0 \quad 0 \quad 0 \quad 1 \quad 5]$$

Keeping in mind that A and R may have different column spaces, we cannot find a basis for the column space of A *directly* from the column vectors of R. However, it follows from Theorem 5.5.5b that if we can find a set of column vectors of R that forms a basis for the column space of R, then the *corresponding* column vectors of A will form a basis for the column space of A.

The first, third, and fifth columns of R contain the leading 1's of the row vectors, so

$$\mathbf{c}_1' = \begin{bmatrix} 1 \\ 0 \\ 0 \\ 0 \end{bmatrix}, \qquad \mathbf{c}_3' = \begin{bmatrix} 4 \\ 1 \\ 0 \\ 0 \end{bmatrix}, \qquad \mathbf{c}_5' = \begin{bmatrix} 5 \\ -2 \\ 1 \\ 0 \end{bmatrix}$$

form a basis for the column space of R; thus the corresponding column vectors of A—namely,

$$\mathbf{c}_1 = \begin{bmatrix} 1 \\ 2 \\ 2 \\ -1 \end{bmatrix}, \qquad \mathbf{c}_3 = \begin{bmatrix} 4 \\ 9 \\ 9 \\ -4 \end{bmatrix}, \qquad \mathbf{c}_5 = \begin{bmatrix} 5 \\ 8 \\ 9 \\ -5 \end{bmatrix}$$

form a basis for the column space of A. ◆

EXAMPLE 7 Basis for a Vector Space Using Row Operations

Find a basis for the space spanned by the vectors

$$\mathbf{v}_1 = (1, -2, 0, 0, 3), \qquad \mathbf{v}_2 = (2, -5, -3, -2, 6),$$
$$\mathbf{v}_3 = (0, 5, 15, 10, 0), \qquad \mathbf{v}_4 = (2, 6, 18, 8, 6)$$

Solution

Except for a variation in notation, the space spanned by these vectors is the row space of the matrix

$$\begin{bmatrix} 1 & -2 & 0 & 0 & 3 \\ 2 & -5 & -3 & -2 & 6 \\ 0 & 5 & 15 & 10 & 0 \\ 2 & 6 & 18 & 8 & 6 \end{bmatrix}$$

Reducing this matrix to row-echelon form, we obtain

$$\begin{bmatrix} 1 & -2 & 0 & 0 & 3 \\ 0 & 1 & 3 & 2 & 0 \\ 0 & 0 & 1 & 1 & 0 \\ 0 & 0 & 0 & 0 & 0 \end{bmatrix}$$

The nonzero row vectors in this matrix are

$$\mathbf{w}_1 = (1, -2, 0, 0, 3), \qquad \mathbf{w}_2 = (0, 1, 3, 2, 0), \qquad \mathbf{w}_3 = (0, 0, 1, 1, 0)$$

These vectors form a basis for the row space and consequently form a basis for the subspace of R^5 spanned by $\mathbf{v}_1, \mathbf{v}_2, \mathbf{v}_3,$ and $\mathbf{v}_4$. ◆

Observe that in Example 6 the basis vectors obtained for the column space of A consisted of column vectors of A, but the basis vectors obtained for the row space of A were not all row vectors of A. The following example illustrates a procedure for finding a basis for the row space of a matrix A that consists entirely of row vectors of A.

EXAMPLE 8   Basis for the Row Space of a Matrix

Find a basis for the row space of

$$A = \begin{bmatrix} 1 & -2 & 0 & 0 & 3 \\ 2 & -5 & -3 & -2 & 6 \\ 0 & 5 & 15 & 10 & 0 \\ 2 & 6 & 18 & 8 & 6 \end{bmatrix}$$

consisting entirely of row vectors from A.

Solution

We will transpose A, thereby converting the row space of A into the column space of A^T; then we will use the method of Example 6 to find a basis for the column space of A^T; and then we will transpose again to convert column vectors back to row vectors. Transposing A yields

$$A^T = \begin{bmatrix} 1 & 2 & 0 & 2 \\ -2 & -5 & 5 & 6 \\ 0 & -3 & 15 & 18 \\ 0 & -2 & 10 & 8 \\ 3 & 6 & 0 & 6 \end{bmatrix}$$

Reducing this matrix to row-echelon form yields

$$\begin{bmatrix} 1 & 2 & 0 & 2 \\ 0 & 1 & -5 & -10 \\ 0 & 0 & 0 & 1 \\ 0 & 0 & 0 & 0 \\ 0 & 0 & 0 & 0 \end{bmatrix}$$

The first, second, and fourth columns contain the leading 1's, so the corresponding column vectors in A^T form a basis for the column space of A^T; these are

$$\mathbf{c}_1 = \begin{bmatrix} 1 \\ -2 \\ 0 \\ 0 \\ 3 \end{bmatrix}, \quad \mathbf{c}_2 = \begin{bmatrix} 2 \\ -5 \\ -3 \\ -2 \\ 6 \end{bmatrix}, \quad \text{and} \quad \mathbf{c}_4 = \begin{bmatrix} 2 \\ 6 \\ 18 \\ 8 \\ 6 \end{bmatrix}$$

Transposing again and adjusting the notation appropriately yields the basis vectors

$$\mathbf{r}_1 = \begin{bmatrix} 1 & -2 & 0 & 0 & 3 \end{bmatrix}, \qquad \mathbf{r}_2 = \begin{bmatrix} 2 & -5 & -3 & -2 & 6 \end{bmatrix},$$

and

$$\mathbf{r}_4 = [2 \quad 6 \quad 18 \quad 8 \quad 6]$$

for the row space of A. ◆

We know from Theorem 5.5.5 that elementary row operations do not alter relationships of linear independence and linear dependence among the column vectors; however, Formulas (6) and (7) imply an even deeper result. Because these formulas actually have *the same scalar coefficients* $x_1, x_2, \ldots, x_n$, it follows that elementary row operations do not alter the *formulas* (linear combinations) that relate linearly dependent column vectors. We omit the formal proof.

EXAMPLE 9 Basis and Linear Combinations

(a) Find a subset of the vectors

$$\mathbf{v}_1 = (1, -2, 0, 3), \qquad \mathbf{v}_2 = (2, -5, -3, 6),$$
$$\mathbf{v}_3 = (0, 1, 3, 0), \qquad \mathbf{v}_4 = (2, -1, 4, -7), \qquad \mathbf{v}_5 = (5, -8, 1, 2)$$

that forms a basis for the space spanned by these vectors.

(b) Express each vector not in the basis as a linear combination of the basis vectors.

Solution (a)

We begin by constructing a matrix that has $\mathbf{v}_1, \mathbf{v}_2, \ldots, \mathbf{v}_5$ as its column vectors:

$$\begin{bmatrix} 1 & 2 & 0 & 2 & 5 \\ -2 & -5 & 1 & -1 & -8 \\ 0 & -3 & 3 & 4 & 1 \\ 3 & 6 & 0 & -7 & 2 \end{bmatrix} \tag{8}$$

$$\begin{array}{ccccc} \uparrow & \uparrow & \uparrow & \uparrow & \uparrow \\ \mathbf{v}_1 & \mathbf{v}_2 & \mathbf{v}_3 & \mathbf{v}_4 & \mathbf{v}_5 \end{array}$$

The first part of our problem can be solved by finding a basis for the column space of this matrix. Reducing the matrix to *reduced* row-echelon form and denoting the column vectors of the resulting matrix by $\mathbf{w}_1, \mathbf{w}_2, \mathbf{w}_3, \mathbf{w}_4,$ and $\mathbf{w}_5$ yields

$$\begin{bmatrix} 1 & 0 & 2 & 0 & 1 \\ 0 & 1 & -1 & 0 & 1 \\ 0 & 0 & 0 & 1 & 1 \\ 0 & 0 & 0 & 0 & 0 \end{bmatrix} \tag{9}$$

$$\begin{array}{ccccc} \uparrow & \uparrow & \uparrow & \uparrow & \uparrow \\ \mathbf{w}_1 & \mathbf{w}_2 & \mathbf{w}_3 & \mathbf{w}_4 & \mathbf{w}_5 \end{array}$$

The leading 1's occur in columns 1, 2, and 4, so by Theorem 5.5.6,

$$\{\mathbf{w}_1, \ \mathbf{w}_2, \ \mathbf{w}_4\}$$

is a basis for the column space of (9), and consequently,

$$\{\mathbf{v}_1, \ \mathbf{v}_2, \ \mathbf{v}_4\}$$

is a basis for the column space of (9).

Solution (b)

We shall start by expressing $\mathbf{w}_3$ and $\mathbf{w}_5$ as linear combinations of the basis vectors $\mathbf{w}_1$, $\mathbf{w}_2$, $\mathbf{w}_4$. The simplest way of doing this is to express $\mathbf{w}_3$ and $\mathbf{w}_5$ in terms of basis vectors with smaller subscripts. Thus we shall express $\mathbf{w}_3$ as a linear combination of $\mathbf{w}_1$ and $\mathbf{w}_2$, and we shall express $\mathbf{w}_5$ as a linear combination of $\mathbf{w}_1$, $\mathbf{w}_2$, and $\mathbf{w}_4$. By inspection of (9), these linear combinations are

$$\mathbf{w}_3 = 2\mathbf{w}_1 - \mathbf{w}_2$$
$$\mathbf{w}_5 = \mathbf{w}_1 + \mathbf{w}_2 + \mathbf{w}_4$$

We call these the **dependency equations**. The corresponding relationships in (8) are

$$\mathbf{v}_3 = 2\mathbf{v}_1 - \mathbf{v}_2$$
$$\mathbf{v}_5 = \mathbf{v}_1 + \mathbf{v}_2 + \mathbf{v}_4 \quad \blacklozenge$$

The procedure illustrated in the preceding example is sufficiently important that we shall summarize the steps:

Given a set of vectors $S = \{\mathbf{v}_1, \mathbf{v}_2, \ldots, \mathbf{v}_k\}$ in R^n, the following procedure produces a subset of these vectors that forms a basis for span(S) and expresses those vectors of S that are not in the basis as linear combinations of the basis vectors.

Step 1. Form the matrix A having $\mathbf{v}_1, \mathbf{v}_2, \ldots, \mathbf{v}_k$ as its column vectors.

Step 2. Reduce the matrix A to its reduced row-echelon form R, and let $\mathbf{w}_1, \mathbf{w}_2, \ldots, \mathbf{w}_k$ be the column vectors of R.

Step 3. Identify the columns that contain the leading 1's in R. The corresponding column vectors of A are the basis vectors for span(S).

Step 4. Express each column vector of R that does *not* contain a leading 1 as a linear combination of preceding column vectors that do contain leading 1's. (You will be able to do this by inspection.) This yields a set of dependency equations involving the column vectors of R. The corresponding equations for the column vectors of A express the vectors that are not in the basis as linear combinations of the basis vectors.

EXERCISE SET 5.5

1. List the row vectors and column vectors of the matrix

$$\begin{bmatrix} 2 & -1 & 0 & 1 \\ 3 & 5 & 7 & -1 \\ 1 & 4 & 2 & 7 \end{bmatrix}$$

2. Express the product $A\mathbf{x}$ as a linear combination of the column vectors of A.

(a) $\begin{bmatrix} 2 & 3 \\ -1 & 4 \end{bmatrix} \begin{bmatrix} 1 \\ 2 \end{bmatrix}$

(b) $\begin{bmatrix} 4 & 0 & -1 \\ 3 & 6 & 2 \\ 0 & -1 & 4 \end{bmatrix} \begin{bmatrix} -2 \\ 3 \\ 5 \end{bmatrix}$

(c) $\begin{bmatrix} -3 & 6 & 2 \\ 5 & -4 & 0 \\ 2 & 3 & -1 \\ 1 & 8 & 3 \end{bmatrix} \begin{bmatrix} -1 \\ 2 \\ 5 \end{bmatrix}$

(d) $\begin{bmatrix} 2 & 1 & 5 \\ 6 & 3 & -8 \end{bmatrix} \begin{bmatrix} 3 \\ 0 \\ -5 \end{bmatrix}$

3. Determine whether $\mathbf{b}$ is in the column space of A, and if so, express $\mathbf{b}$ as a linear combination of the column vectors of A.

(a) $A = \begin{bmatrix} 1 & 3 \\ 4 & -6 \end{bmatrix}$; $\mathbf{b} = \begin{bmatrix} -2 \\ 10 \end{bmatrix}$ 　　(b) $A = \begin{bmatrix} 1 & 1 & 2 \\ 1 & 0 & 1 \\ 2 & 1 & 3 \end{bmatrix}$; $\mathbf{b} = \begin{bmatrix} -1 \\ 0 \\ 2 \end{bmatrix}$

(c) $A = \begin{bmatrix} 1 & -1 & 1 \\ 9 & 3 & 1 \\ 1 & 1 & 1 \end{bmatrix}$; $\mathbf{b} = \begin{bmatrix} 5 \\ 1 \\ -1 \end{bmatrix}$ 　　(d) $A = \begin{bmatrix} 1 & -1 & 1 \\ 1 & 1 & -1 \\ -1 & -1 & 1 \end{bmatrix}$; $\mathbf{b} = \begin{bmatrix} 2 \\ 0 \\ 0 \end{bmatrix}$

(e) $A = \begin{bmatrix} 1 & 2 & 0 & 1 \\ 0 & 1 & 2 & 1 \\ 1 & 2 & 1 & 3 \\ 0 & 1 & 2 & 2 \end{bmatrix}$; $\mathbf{b} = \begin{bmatrix} 4 \\ 3 \\ 5 \\ 7 \end{bmatrix}$

4. Suppose that $x_1 = -1$, $x_2 = 2$, $x_3 = 4$, $x_4 = -3$ is a solution of a nonhomogeneous linear system $A\mathbf{x} = \mathbf{b}$ and that the solution set of the homogeneous system $A\mathbf{x} = \mathbf{0}$ is given by the formulas

$$x_1 = -3r + 4s, \qquad x_2 = r - s, \qquad x_3 = r, \qquad x_4 = s$$

(a) Find the vector form of the general solution of $A\mathbf{x} = \mathbf{0}$.

(b) Find the vector form of the general solution of $A\mathbf{x} = \mathbf{b}$.

5. Find the vector form of the general solution of the given linear system $A\mathbf{x} = \mathbf{b}$; then use that result to find the vector form of the general solution of $A\mathbf{x} = \mathbf{0}$.

(a) $\begin{aligned} x_1 - 3x_2 &= 1 \\ 2x_1 - 6x_2 &= 2 \end{aligned}$ 　　(b) $\begin{aligned} x_1 + x_2 + 2x_3 &= 5 \\ x_1 \quad\ \ + x_3 &= -2 \\ 2x_1 + x_2 + 3x_3 &= 3 \end{aligned}$

(c) $\begin{aligned} x_1 - 2x_2 + x_3 + 2x_4 &= -1 \\ 2x_1 - 4x_2 + 2x_3 + 4x_4 &= -2 \\ -x_1 + 2x_2 - x_3 - 2x_4 &= 1 \\ 3x_1 - 6x_2 + 3x_3 + 6x_4 &= -3 \end{aligned}$ 　　(d) $\begin{aligned} x_1 + 2x_2 - 3x_3 + x_4 &= 4 \\ -2x_1 + x_2 + 2x_3 + x_4 &= -1 \\ -x_1 + 3x_2 - x_3 + 2x_4 &= 3 \\ 4x_1 - 7x_2 \quad\ \ - 5x_4 &= -5 \end{aligned}$

6. Find a basis for the nullspace of A.

(a) $A = \begin{bmatrix} 1 & -1 & 3 \\ 5 & -4 & -4 \\ 7 & -6 & 2 \end{bmatrix}$ 　　(b) $A = \begin{bmatrix} 2 & 0 & -1 \\ 4 & 0 & -2 \\ 0 & 0 & 0 \end{bmatrix}$

(c) $A = \begin{bmatrix} 1 & 4 & 5 & 2 \\ 2 & 1 & 3 & 0 \\ -1 & 3 & 2 & 2 \end{bmatrix}$ 　　(d) $A = \begin{bmatrix} 1 & 4 & 5 & 6 & 9 \\ 3 & -2 & 1 & 4 & -1 \\ -1 & 0 & -1 & -2 & -1 \\ 2 & 3 & 5 & 7 & 8 \end{bmatrix}$

(e) $A = \begin{bmatrix} 1 & -3 & 2 & 2 & 1 \\ 0 & 3 & 6 & 0 & -3 \\ 2 & -3 & -2 & 4 & 4 \\ 3 & -6 & 0 & 6 & 5 \\ -2 & 9 & 2 & -4 & -5 \end{bmatrix}$

7. In each part, a matrix in row-echelon form is given. By inspection, find bases for the row and column spaces of A.

(a) $\begin{bmatrix} 1 & 0 & 2 \\ 0 & 0 & 1 \\ 0 & 0 & 0 \end{bmatrix}$ 　　(b) $\begin{bmatrix} 1 & -3 & 0 & 0 \\ 0 & 1 & 0 & 0 \\ 0 & 0 & 0 & 0 \\ 0 & 0 & 0 & 0 \end{bmatrix}$

(c) $\begin{bmatrix} 1 & 2 & 4 & 5 \\ 0 & 1 & -3 & 0 \\ 0 & 0 & 1 & -3 \\ 0 & 0 & 0 & 1 \\ 0 & 0 & 0 & 0 \end{bmatrix}$ 　　(d) $\begin{bmatrix} 1 & 2 & -1 & 5 \\ 0 & 1 & 4 & 3 \\ 0 & 0 & 1 & -7 \\ 0 & 0 & 0 & 1 \end{bmatrix}$

8. For the matrices in Exercise 6, find a basis for the row space of A by reducing the matrix to row-echelon form.

9. For the matrices in Exercise 6, find a basis for the column space of A.

10. For the matrices in Exercise 6, find a basis for the row space of A consisting entirely of row vectors of A.

11. Find a basis for the subspace of R^4 spanned by the given vectors.

    (a) $(1, 1, -4, -3)$, $(2, 0, 2, -2)$, $(2, -1, 3, 2)$

    (b) $(-1, 1, -2, 0)$, $(3, 3, 6, 0)$, $(9, 0, 0, 3)$

    (c) $(1, 1, 0, 0)$, $(0, 0, 1, 1)$, $(-2, 0, 2, 2)$, $(0, -3, 0, 3)$

12. Find a subset of the vectors that forms a basis for the space spanned by the vectors; then express each vector that is not in the basis as a linear combination of the basis vectors.

    (a) $\mathbf{v}_1 = (1, 0, 1, 1)$, $\mathbf{v}_2 = (-3, 3, 7, 1)$, $\mathbf{v}_3 = (-1, 3, 9, 3)$, $\mathbf{v}_4 = (-5, 3, 5, -1)$

    (b) $\mathbf{v}_1 = (1, -2, 0, 3)$, $\mathbf{v}_2 = (2, -4, 0, 6)$, $\mathbf{v}_3 = (-1, 1, 2, 0)$, $\mathbf{v}_4 = (0, -1, 2, 3)$

    (c) $\mathbf{v}_1 = (1, -1, 5, 2)$, $\mathbf{v}_2 = (-2, 3, 1, 0)$, $\mathbf{v}_3 = (4, -5, 9, 4)$, $\mathbf{v}_4 = (0, 4, 2, -3)$,
        $\mathbf{v}_5 = (-7, 18, 2, -8)$

13. Prove that the row vectors of an $n \times n$ invertible matrix A form a basis for R^n.

14. (a) Let

$$A = \begin{bmatrix} 0 & 1 & 0 \\ 1 & 0 & 0 \\ 0 & 0 & 0 \end{bmatrix}$$

and consider a rectangular xyz-coordinate system in 3-space. Show that the nullspace of A consists of all points on the z-axis and that the column space consists of all points in the xy-plane (see the accompanying figure).

    (b) Find a 3×3 matrix whose nullspace is the x-axis and whose column space is the yz-plane.

15. Find a 3×3 matrix whose nullspace is

    (a) a point   (b) a line   (c) a plane

Figure Ex-14

Discussion & Discovery

16. Indicate whether each statement is always true or sometimes false. Justify your answer by giving a logical argument or a counterexample.

    (a) If E is an elementary matrix, then A and EA must have the same nullspace.

    (b) If E is an elementary matrix, then A and EA must have the same row space.

    (c) If E is an elementary matrix, then A and EA must have the same column space.

    (d) If $A\mathbf{x} = \mathbf{b}$ does not have any solutions, then $\mathbf{b}$ is not in the column space of A.

    (e) The row space and nullspace of an invertible matrix are the same.

17. (a) Find all 2×2 matrices whose nullspace is the line $3x - 5y = 0$.

    (b) Sketch the nullspaces of the following matrices:

$$A = \begin{bmatrix} 1 & 4 \\ 0 & 5 \end{bmatrix}, \quad B = \begin{bmatrix} 1 & 0 \\ 0 & 5 \end{bmatrix}, \quad C = \begin{bmatrix} 6 & 2 \\ 3 & 1 \end{bmatrix}, \quad D = \begin{bmatrix} 0 & 0 \\ 0 & 0 \end{bmatrix}$$

18. The equation $x_1 + x_2 + x_3 = 1$ can be viewed as a linear system of one equation in three unknowns. Express its general solution as a particular solution plus the general solution of the corresponding homogeneous system. [Write the vectors in column form.]

19. Suppose that A and B are $n \times n$ matrices and A is invertible. Invent and prove a theorem that describes how the row spaces of AB and B are related.

5.6
RANK AND NULLITY

In the preceding section we investigated the relationships between systems of linear equations and the row space, column space, and nullspace of the coefficient matrix. In this section we shall be concerned with relationships between the dimensions of the row space, column space, and nullspace of a matrix and its transpose. The results we will obtain are fundamental and will provide deeper insights into linear systems and linear transformations.

Four Fundamental Matrix Spaces

If we consider a matrix A and its transpose A^T together, then there are six vector spaces of interest:

row space of A	row space of A^T
column space of A	column space of A^T
nullspace of A	nullspace of A^T

However, transposing a matrix converts row vectors into column vectors and column vectors into row vectors, so except for a difference in notation, the row space of A^T is the same as the column space of A, and the column space of A^T is the same as the row space of A. This leaves four vector spaces of interest:

row space of A	column space of A
nullspace of A	nullspace of A^T

These are known as the ***fundamental matrix spaces*** associated with A. If A is an $m \times n$ matrix, then the row space of A and the nullspace of A are subspaces of R^n, and the column space of A and the nullspace of A^T are subspaces of R^m. Our primary goal in this section is to establish relationships between the dimensions of these four vector spaces.

Row and Column Spaces Have Equal Dimensions

In Example 6 of Section 5.5, we found that the row and column spaces of the matrix

$$A = \begin{bmatrix} 1 & -3 & 4 & -2 & 5 & 4 \\ 2 & -6 & 9 & -1 & 8 & 2 \\ 2 & -6 & 9 & -1 & 9 & 7 \\ -1 & 3 & -4 & 2 & -5 & -4 \end{bmatrix}$$

each have three basis vectors; that is, both are three-dimensional. It is not accidental that these dimensions are the same; it is a consequence of the following general result.

THEOREM 5.6.1

> *If A is any matrix, then the row space and column space of A have the same dimension.*

Proof Let R be any row-echelon form of A. It follows from Theorem 5.5.4 that

$$\dim(\text{row space of } A) = \dim(\text{row space of } R)$$

and it follows from Theorem 5.5.5b that

$$\dim(\text{column space of } A) = \dim(\text{column space of } R)$$

Thus the proof will be complete if we can show that the row space and column space of R have the same dimension. But the dimension of the row space of R is the number of nonzero rows, and the dimension of the column space of R is the number of columns that contain leading 1's (Theorem 5.5.6). However, the nonzero rows are precisely the rows in which the leading 1's occur, so the number of leading 1's and the number of nonzero rows are the same. This shows that the row space and column space of R have the same dimension. ∎

The dimensions of the row space, column space, and nullspace of a matrix are such important numbers that there is some notation and terminology associated with them.

DEFINITION

The common dimension of the row space and column space of a matrix A is called the *rank* of A and is denoted by rank(A); the dimension of the nullspace of A is called the *nullity* of A and is denoted by nullity(A).

EXAMPLE 1 Rank and Nullity of a 4 × 6 Matrix

Find the rank and nullity of the matrix

$$A = \begin{bmatrix} -1 & 2 & 0 & 4 & 5 & -3 \\ 3 & -7 & 2 & 0 & 1 & 4 \\ 2 & -5 & 2 & 4 & 6 & 1 \\ 4 & -9 & 2 & -4 & -4 & 7 \end{bmatrix}$$

Solution

The reduced row-echelon form of A is

$$\begin{bmatrix} 1 & 0 & -4 & -28 & -37 & 13 \\ 0 & 1 & -2 & -12 & -16 & 5 \\ 0 & 0 & 0 & 0 & 0 & 0 \\ 0 & 0 & 0 & 0 & 0 & 0 \end{bmatrix} \tag{1}$$

(verify). Since there are two nonzero rows (or, equivalently, two leading 1's), the row space and column space are both two-dimensional, so rank $(A) = 2$. To find the nullity of A, we must find the dimension of the solution space of the linear system $A\mathbf{x} = \mathbf{0}$. This system can be solved by reducing the augmented matrix to reduced row-echelon form. The resulting matrix will be identical to (1), except that it will have an additional last column of zeros, and the corresponding system of equations will be

$$x_1 - 4x_3 - 28x_4 - 37x_5 + 13x_6 = 0$$
$$x_2 - 2x_3 - 12x_4 - 16x_5 + 5x_6 = 0$$

or, on solving for the leading variables,

$$x_1 = 4x_3 + 28x_4 + 37x_5 - 13x_6$$
$$x_2 = 2x_3 + 12x_4 + 16x_5 - 5x_6 \tag{2}$$

It follows that the general solution of the system is

$$x_1 = 4r + 28s + 37t - 13u$$
$$x_2 = 2r + 12s + 16t - 5u$$
$$x_3 = r$$
$$x_4 = s$$
$$x_5 = t$$
$$x_6 = u$$

or, equivalently,

$$
\begin{bmatrix} x_1 \\ x_2 \\ x_3 \\ x_4 \\ x_5 \\ x_6 \end{bmatrix} = r \begin{bmatrix} 4 \\ 2 \\ 1 \\ 0 \\ 0 \\ 0 \end{bmatrix} + s \begin{bmatrix} 28 \\ 12 \\ 0 \\ 1 \\ 0 \\ 0 \end{bmatrix} + t \begin{bmatrix} 37 \\ 16 \\ 0 \\ 0 \\ 1 \\ 0 \end{bmatrix} + u \begin{bmatrix} -13 \\ -5 \\ 0 \\ 0 \\ 0 \\ 1 \end{bmatrix} \tag{3}
$$

Because the four vectors on the right side of (3) form a basis for the solution space, nullity(A) = 4. ◆

The following theorem states that a matrix and its transpose have the same rank.

THEOREM 5.6.2

If A is any matrix, then rank(A) = rank(A^T).

Proof

$$\text{rank}(A) = \dim(\text{row space of } A) = \dim(\text{column space of } A^T) = \text{rank}(A^T). \quad \blacksquare$$

The following theorem establishes an important relationship between the rank and nullity of a matrix.

THEOREM 5.6.3

Dimension Theorem for Matrices

If A is a matrix with n columns, then

$$rank(A) + nullity(A) = n \tag{4}$$

Proof Since A has n columns, the homogeneous linear system $A\mathbf{x} = \mathbf{0}$ has n unknowns (variables). These fall into two categories: the leading variables and the free variables. Thus

$$\begin{bmatrix} \text{number of leading} \\ \text{variables} \end{bmatrix} + \begin{bmatrix} \text{number of free} \\ \text{variables} \end{bmatrix} = n$$

But the number of leading variables is the same as the number of leading 1's in the reduced row-echelon form of A, and this is the rank of A. Thus

$$\text{rank}(A) + \begin{bmatrix} \text{number of free} \\ \text{variables} \end{bmatrix} = n$$

The number of free variables is equal to the nullity of A. This is so because the nullity of A is the dimension of the solution space of $A\mathbf{x} = \mathbf{0}$, which is the same as the number of parameters in the general solution [see (3), for example], which is the same as the number of free variables. Thus

$$\text{rank}(A) + \text{nullity}(A) = n \quad \blacksquare$$

The proof of the preceding theorem contains two results that are of importance in their own right.

THEOREM 5.6.4

> *If A is an m × n matrix, then*
>
> *(a) rank(A) = the number of leading variables in the solution of* $A\mathbf{x} = \mathbf{0}$.
> *(b) nullity(A) = the number of parameters in the general solution of* $A\mathbf{x} = \mathbf{0}$.

EXAMPLE 2 The Sum of Rank and Nullity

The matrix

$$A = \begin{bmatrix} -1 & 2 & 0 & 4 & 5 & -3 \\ 3 & -7 & 2 & 0 & 1 & 4 \\ 2 & -5 & 2 & 4 & 6 & 1 \\ 4 & -9 & 2 & -4 & -4 & 7 \end{bmatrix}$$

has 6 columns, so

$$\text{rank}(A) + \text{nullity}(A) = 6$$

This is consistent with Example 1, where we showed that

$$\text{rank}(A) = 2 \quad \text{and} \quad \text{nullity}(A) = 4 \ \blacklozenge$$

EXAMPLE 3 Number of Parameters in a General Solution

Find the number of parameters in the general solution of $A\mathbf{x} = \mathbf{0}$ if A is a 5×7 matrix of rank 3.

Solution

From (4),

$$\text{nullity}(A) = n - \text{rank}(A) = 7 - 3 = 4$$

Thus there are four parameters. $\blacklozenge$

Suppose now that A is an $m \times n$ matrix of rank r; it follows from Theorem 5.6.2 that A^T is an $n \times m$ matrix of rank r. Applying Theorem 5.6.3 to A and A^T yields

$$\text{nullity}(A) = n - r, \qquad \text{nullity}(A^T) = m - r$$

from which we deduce the following table relating the dimensions of the four fundamental spaces of an $m \times n$ matrix A of rank r.

Fundamental Space	Dimension
Row space of A	r
Column space of A	r
Nullspace of A	$n - r$
Nullspace of A^T	$m - r$

Applications of Rank

The advent of the Internet has stimulated research on finding efficient methods for transmitting large amounts of digital data over communications lines with limited bandwidth. Digital data is commonly stored in matrix form, and many techniques for improving transmission speed use the rank of a matrix in some way. Rank plays a role because it measures the "redundancy" in a matrix in the sense that if A is an $m \times n$ matrix of rank k, then $n - k$ of the column vectors and $m - k$ of the row vectors can be expressed in terms of k linearly independent column or row vectors. The essential idea in many data compression schemes is to approximate the original data set by a data set with smaller rank that conveys nearly the same information, then eliminate redundant vectors in the approximating set to speed up the transmission time.

Maximum Value for Rank

If A is an $m \times n$ matrix, then the row vectors lie in R^n and the column vectors lie in R^m. This implies that the row space of A is at most n-dimensional and that the column space is at most m-dimensional. Since the row and column spaces have the same dimension (the rank of A), we must conclude that if $m \neq n$, then the rank of A is at most the smaller of the values of m and n. We denote this by writing

$$\text{rank}(A) \leq \min(m, n) \tag{5}$$

where $\min(m, n)$ denotes the smaller of the numbers m and n if $m \neq n$ or denotes their common value if $m = n$.

EXAMPLE 4   Maximum Value of Rank for a 7 × 4 Matrix

If A is a 7×4 matrix, then the rank of A is at most 4, and consequently, the seven row vectors must be linearly dependent. If A is a 4×7 matrix, then again the rank of A is at most 4, and consequently, the seven column vectors must be linearly dependent. ◆

Linear Systems of m Equations in n Unknowns

In earlier sections we obtained a wide range of theorems concerning linear systems of n equations in n unknowns. (See Theorem 4.3.4.) We shall now turn our attention to linear systems of m equations in n unknowns in which m and n need not be the same.

The following theorem specifies conditions under which a linear system of m equations in n unknowns is guaranteed to be consistent.

THEOREM 5.6.5

> **The Consistency Theorem**
>
> *If $A\mathbf{x} = \mathbf{b}$ is a linear system of m equations in n unknowns, then the following are equivalent.*
>
> *(a)* $A\mathbf{x} = \mathbf{b}$ *is consistent.*
> *(b)* $\mathbf{b}$ *is in the column space of A.*
> *(c)* *The coefficient matrix A and the augmented matrix $[A \mid \mathbf{b}]$ have the same rank.*

Proof   It suffices to prove the two equivalences $(a) \Leftrightarrow (b)$ and $(b) \Leftrightarrow (c)$, since it will then follow as a matter of logic that $(a) \Leftrightarrow (c)$.

(a) ⇔ (b)   See Theorem 5.5.1.

(b) $\Rightarrow$ **(c)**   We will show that if **b** is in the column space of A, then the column spaces of A and $[A \mid \mathbf{b}]$ are actually the same, from which it will follow that these two matrices have the same rank.

By definition, the column space of a matrix is the space spanned by its column vectors, so the column spaces of A and $[A \mid \mathbf{b}]$ can be expressed as

$$\text{span}\{\mathbf{c}_1, \mathbf{c}_2, \ldots, \mathbf{c}_n\} \quad \text{and} \quad \text{span}\{\mathbf{c}_1, \mathbf{c}_2, \ldots, \mathbf{c}_n, \mathbf{b}\}$$

respectively. If **b** is in the column space of A, then each vector in the set $\{\mathbf{c}_1, \mathbf{c}_2, \ldots, \mathbf{c}_n, \mathbf{b}\}$ is a linear combination of the vectors in $\{\mathbf{c}_1, \mathbf{c}_2, \ldots, \mathbf{c}_n\}$ and conversely (why?). Thus, from Theorem 5.2.4, the column spaces of A and $[A \mid \mathbf{b}]$ are the same.

(c) $\Rightarrow$ **(b)**   Assume that A and $[A \mid \mathbf{b}]$ have the same rank r. By Theorem 5.4.6a, there is some subset of the column vectors of A that forms a basis for the column space of A. Suppose that those column vectors are

$$\mathbf{c}_1', \mathbf{c}_2', \ldots, \mathbf{c}_r'$$

These r basis vectors also belong to the r-dimensional column space of $[A \mid \mathbf{b}]$; hence they also form a basis for the column space of $[A \mid \mathbf{b}]$ by Theorem 5.4.6a. This means that **b** is expressible as a linear combination of $\mathbf{c}_1', \mathbf{c}_2', \ldots, \mathbf{c}_r'$, and consequently **b** lies in the column space of A. ∎

It is not hard to visualize why this theorem is true if one views the rank of a matrix as the number of nonzero rows in its reduced row-echelon form. For example, the augmented matrix for the system

$$
\begin{aligned}
x_1 - 2x_2 - 3x_3 + 2x_4 &= -4 \\
-3x_1 + 7x_2 - x_3 + x_4 &= -3 \\
2x_1 - 5x_2 + 4x_3 - 3x_4 &= 7 \\
-3x_1 + 6x_2 + 9x_3 - 6x_4 &= -1
\end{aligned}
\quad \text{is} \quad
\begin{bmatrix}
1 & -2 & -3 & 2 & -4 \\
-3 & 7 & -1 & 1 & -3 \\
2 & -5 & 4 & -3 & 7 \\
-3 & 6 & 9 & -6 & -1
\end{bmatrix}
$$

which has the following reduced row-echelon form (verify):

$$
\begin{bmatrix}
1 & 0 & -23 & 16 & 0 \\
0 & 1 & -10 & 7 & 0 \\
0 & 0 & 0 & 0 & 1 \\
0 & 0 & 0 & 0 & 0
\end{bmatrix}
$$

We see from the third row in this matrix that the system is inconsistent. However, it is also because of this row that the reduced row-echelon form of the augmented matrix has fewer zero rows than the reduced row-echelon form of the coefficient matrix. This forces the coefficient matrix and the augmented matrix for the system to have different ranks.

The Consistency Theorem is concerned with conditions under which a linear system $A\mathbf{x} = \mathbf{b}$ is consistent for a *specific* vector **b**. The following theorem is concerned with conditions under which a linear system is consistent for *all possible* choices of **b**.

THEOREM 5.6.6

> *If $A\mathbf{x} = \mathbf{b}$ is a linear system of m equations in n unknowns, then the following are equivalent.*
>
> (a) $A\mathbf{x} = \mathbf{b}$ *is consistent for every $m \times 1$ matrix* **b**.
> (b) *The column vectors of A span R^m.*
> (c) $\text{rank}(A) = m$.

Proof It suffices to prove the two equivalences $(a) \Leftrightarrow (b)$ and $(a) \Leftrightarrow (c)$, since it will then follow as a matter of logic that $(b) \Leftrightarrow (c)$.

(a) $\Leftrightarrow$ (b) From Formula (2) of Section 5.5, the system $A\mathbf{x} = \mathbf{b}$ can be expressed as

$$x_1\mathbf{c}_1 + x_2\mathbf{c}_2 + \cdots + x_n\mathbf{c}_n = \mathbf{b}$$

from which we can conclude that $A\mathbf{x} = \mathbf{b}$ is consistent for every $m \times 1$ matrix $\mathbf{b}$ if and only if every such $\mathbf{b}$ is expressible as a linear combination of the column vectors $\mathbf{c}_1, \mathbf{c}_2, \ldots, \mathbf{c}_n$, or, equivalently, if and only if these column vectors span R^m.

(a) $\Rightarrow$ (c) From the assumption that $A\mathbf{x} = \mathbf{b}$ is consistent for every $m \times 1$ matrix $\mathbf{b}$, and from parts (a) and (b) of the Consistency Theorem (5.6.5), it follows that every vector $\mathbf{b}$ in R^m lies in the column space of A; that is, the column space of A is all of R^m. Thus $\text{rank}(A) = \dim(R^m) = m$.

(c) $\Rightarrow$ (a) From the assumption that $\text{rank}(A) = m$, it follows that the column space of A is a subspace of R^m of dimension m and hence must be all of R^m by Theorem 5.4.7. It now follows from parts (a) and (b) of the Consistency Theorem (5.6.5) that $A\mathbf{x} = \mathbf{b}$ is consistent for every vector $\mathbf{b}$ in R^m, since every such $\mathbf{b}$ is in the column space of A. ■

A linear system with more equations than unknowns is called an ***overdetermined linear system***. If $A\mathbf{x} = \mathbf{b}$ is an overdetermined linear system of m equations in n unknowns (so that $m > n$), then the column vectors of A cannot span R^m; it follows from the last theorem that *for a fixed $m \times n$ matrix A with $m > n$, the overdetermined linear system $A\mathbf{x} = \mathbf{b}$ cannot be consistent for every possible $\mathbf{b}$.*

EXAMPLE 5 An Overdetermined System

The linear system

$$\begin{aligned}
x_1 - 2x_2 &= b_1 \\
x_1 - x_2 &= b_2 \\
x_1 + x_2 &= b_3 \\
x_1 + 2x_2 &= b_4 \\
x_1 + 3x_2 &= b_5
\end{aligned}$$

is overdetermined, so it cannot be consistent for all possible values of b_1, b_2, b_3, b_4, and b_5. Exact conditions under which the system is consistent can be obtained by solving the linear system by Gauss–Jordan elimination. We leave it for the reader to show that the augmented matrix is row equivalent to

$$\begin{bmatrix}
1 & 0 & 2b_2 - b_1 \\
0 & 1 & b_2 - b_1 \\
0 & 0 & b_3 - 3b_2 + 2b_1 \\
0 & 0 & b_4 - 4b_2 + 3b_1 \\
0 & 0 & b_5 - 5b_2 + 4b_1
\end{bmatrix}$$

Thus, the system is consistent if and only if b_1, b_2, b_3, b_4, and b_5 satisfy the conditions

$$\begin{aligned}
2b_1 - 3b_2 + b_3 \qquad\qquad &= 0 \\
3b_1 - 4b_2 \qquad + b_4 \quad &= 0 \\
4b_1 - 5b_2 \qquad\qquad + b_5 &= 0
\end{aligned}$$

or, on solving this homogeneous linear system,

$$b_1 = 5r - 4s, \qquad b_2 = 4r - 3s, \qquad b_3 = 2r - s, \qquad b_4 = r, \qquad b_5 = s$$

where r and s are arbitrary. ◆

In Formula (3) of Theorem 5.5.2, the scalars $c_1, c_2, \ldots, c_k$ are the arbitrary parameters in the general solutions of both $A\mathbf{x} = \mathbf{b}$ and $A\mathbf{x} = \mathbf{0}$. Thus these two systems have the same number of parameters in their general solutions. Moreover, it follows from part (b) of Theorem 5.6.4 that the number of such parameters is nullity(A). This fact and the Dimension Theorem for Matrices (5.6.3) yield the following theorem.

THEOREM 5.6.7

If $A\mathbf{x} = \mathbf{b}$ is a consistent linear system of m equations in n unknowns, and if A has rank r, then the general solution of the system contains $n - r$ parameters.

EXAMPLE 6 Number of Parameters in a General Solution

If A is a 5×7 matrix with rank 4, and if $A\mathbf{x} = \mathbf{b}$ is a consistent linear system, then the general solution of the system contains $7 - 4 = 3$ parameters. ◆

In earlier sections we obtained a wide range of conditions under which a homogeneous linear system $A\mathbf{x} = \mathbf{0}$ of n equations in n unknowns is guaranteed to have only the trivial solution. (See Theorem 4.3.4.) The following theorem obtains some corresponding results for systems of m equations in n unknowns, where m and n may differ.

THEOREM 5.6.8

If A is an $m \times n$ matrix, then the following are equivalent.

(a) $A\mathbf{x} = \mathbf{0}$ *has only the trivial solution.*
(b) *The column vectors of A are linearly independent.*
(c) $A\mathbf{x} = \mathbf{b}$ *has at most one solution (none or one) for every $m \times 1$ matrix* $\mathbf{b}$.

Proof It suffices to prove the two equivalences (a) ⇔ (b) and (a) ⇔ (c), since it will then follow as a matter of logic that (b) ⇔ (c).

(a) ⇔ (b) If $\mathbf{c}_1, \mathbf{c}_2, \ldots, \mathbf{c}_n$ are the column vectors of A, then the linear system $A\mathbf{x} = \mathbf{0}$ can be written as

$$x_1\mathbf{c}_1 + x_2\mathbf{c}_2 + \cdots + x_n\mathbf{c}_n = \mathbf{0} \tag{6}$$

If $\mathbf{c}_1, \mathbf{c}_2, \ldots, \mathbf{c}_n$ are linearly independent vectors, then this equation is satisfied only by $x_1 = x_2 = \cdots = x_n = 0$, which means that $A\mathbf{x} = \mathbf{0}$ has only the trivial solution. Conversely, if $A\mathbf{x} = \mathbf{0}$ has only the trivial solution, then Equation (6) is satisfied only by $x_1 = x_2 = \cdots = x_n = 0$, which means that $\mathbf{c}_1, \mathbf{c}_2, \ldots, \mathbf{c}_n$ are linearly independent.

(a) ⇒ (c) Assume that $A\mathbf{x} = \mathbf{0}$ has only the trivial solution. Either $A\mathbf{x} = \mathbf{b}$ is consistent or it is not. If it is not consistent, then there are no solutions of $A\mathbf{x} = \mathbf{b}$, and we are done. If $A\mathbf{x} = \mathbf{b}$ is consistent, let $\mathbf{x}_0$ be any solution. From the discussion following Theorem 5.5.2 and the fact that $A\mathbf{x} = \mathbf{0}$ has only the trivial solution, we conclude that the general solution of $A\mathbf{x} = \mathbf{b}$ is $\mathbf{x}_0 + \mathbf{0} = \mathbf{x}_0$. Thus the only solution of $A\mathbf{x} = \mathbf{b}$ is $\mathbf{x}_0$.

(c) ⇒ (a)   Assume that $A\mathbf{x} = \mathbf{b}$ has at most one solution for every $m \times 1$ matrix $\mathbf{b}$. Then, in particular, $A\mathbf{x} = \mathbf{0}$ has at most one solution. Thus $A\mathbf{x} = \mathbf{0}$ has only the trivial solution.   ∎

A linear system with more unknowns than equations is called an ***underdetermined linear system***. If $A\mathbf{x} = \mathbf{b}$ is a consistent underdetermined linear system of m equations in n unknowns (so that $m < n$), then it follows from Theorem 5.6.7 that the general solution has at least one parameter (why?); hence *a consistent underdetermined linear system must have infinitely many solutions*. In particular, an underdetermined homogeneous linear system has infinitely many solutions, though this was already proved in Chapter 1 (Theorem 1.2.1).

EXAMPLE 7   An Underdetermined System

If A is a 5×7 matrix, then for every 7×1 matrix $\mathbf{b}$, the linear system $A\mathbf{x} = \mathbf{b}$ is underdetermined. Thus $A\mathbf{x} = \mathbf{b}$ must be consistent for some $\mathbf{b}$, and for each such $\mathbf{b}$ the general solution must have $7 - r$ parameters, where r is the rank of A.   ◆

Summary

In Theorem 4.3.4 we listed eight results that are equivalent to the invertibility of a matrix A. We conclude this section by adding eight more results to that list to produce the following theorem, which relates all of the major topics we have studied thus far.

THEOREM 5.6.9

> ### Equivalent Statements
>
> *If A is an $n \times n$ matrix, and if $T_A : R^n \to R^n$ is multiplication by A, then the following are equivalent.*
>
> *(a)   A is invertible.*
> *(b) $A\mathbf{x} = \mathbf{0}$ has only the trivial solution.*
> *(c)   The reduced row-echelon form of A is I_n.*
> *(d)   A is expressible as a product of elementary matrices.*
> *(e) $A\mathbf{x} = \mathbf{b}$ is consistent for every $n \times 1$ matrix $\mathbf{b}$.*
> *(f) $A\mathbf{x} = \mathbf{b}$ has exactly one solution for every $n \times 1$ matrix $\mathbf{b}$.*
> *(g) $\det(A) \neq 0$.*
> *(h)   The range of T_A is R^n.*
> *(i) T_A is one-to-one.*
> *(j)   The column vectors of A are linearly independent.*
> *(k)   The row vectors of A are linearly independent.*
> *(l)   The column vectors of A span R^n.*
> *(m)   The row vectors of A span R^n.*
> *(n)   The column vectors of A form a basis for R^n.*
> *(o)   The row vectors of A form a basis for R^n.*
> *(p)   A has rank n.*
> *(q)   A has nullity 0.*

Proof   We already know from Theorem 4.3.4 that statements (a) through (i) are equivalent. To complete the proof, we will show that (j) through (q) are equivalent to (b) by proving the sequence of implications $(b) \Rightarrow (j) \Rightarrow (k) \Rightarrow (l) \Rightarrow (m) \Rightarrow (n) \Rightarrow (o) \Rightarrow (p) \Rightarrow (q) \Rightarrow (b)$.

(b) ⇒ (j)   If $A\mathbf{x} = \mathbf{0}$ has only the trivial solution, then by Theorem 5.6.8, the column vectors of A are linearly independent.

(j) ⇒ (k) ⇒ (l) ⇒ (m) ⇒ (n) ⇒ (o)   This follows from Theorem 5.4.5 and the fact that R^n is an n-dimensional vector space. (The details are omitted.)

(o) ⇒ (p)   If the n row vectors of A form a basis for R^n, then the row space of A is n-dimensional and A has rank n.

(p) ⇒ (q)   This follows from the Dimension Theorem (5.6.3).

(q) ⇒ (b)   If A has nullity 0, then the solution space of $A\mathbf{x} = \mathbf{0}$ has dimension 0, which means that it contains only the zero vector. Hence $A\mathbf{x} = \mathbf{0}$ has only the trivial solution.

∎

EXERCISE SET
5.6

1. Verify that $\text{rank}(A) = \text{rank}(A^T)$.

$$A = \begin{bmatrix} 1 & 2 & 4 & 0 \\ -3 & 1 & 5 & 2 \\ -2 & 3 & 9 & 2 \end{bmatrix}$$

2. Find the rank and nullity of the matrix; then verify that the values obtained satisfy Formula (4) of the Dimension Theorem.

   (a) $A = \begin{bmatrix} 1 & -1 & 3 \\ 5 & -4 & -4 \\ 7 & -6 & 2 \end{bmatrix}$ (b) $A = \begin{bmatrix} 2 & 0 & -1 \\ 4 & 0 & -2 \\ 0 & 0 & 0 \end{bmatrix}$

   (c) $A = \begin{bmatrix} 1 & 4 & 5 & 2 \\ 2 & 1 & 3 & 0 \\ -1 & 3 & 2 & 2 \end{bmatrix}$ (d) $A = \begin{bmatrix} 1 & 4 & 5 & 6 & 9 \\ 3 & -2 & 1 & 4 & -1 \\ -1 & 0 & -1 & -2 & -1 \\ 2 & 3 & 5 & 7 & 8 \end{bmatrix}$

   (e) $A = \begin{bmatrix} 1 & -3 & 2 & 2 & 1 \\ 0 & 3 & 6 & 0 & -3 \\ 2 & -3 & -2 & 4 & 4 \\ 3 & -6 & 0 & 6 & 5 \\ -2 & 9 & 2 & -4 & -5 \end{bmatrix}$

3. In each part of Exercise 2, use the results obtained to find the number of leading variables and the number of parameters in the solution of $A\mathbf{x} = \mathbf{0}$ without solving the system.

4. In each part, use the information in the table to find the dimension of the row space, column space, and nullspace of A, and of the nullspace of A^T.

	(a)	(b)	(c)	(d)	(e)	(f)	(g)
Size of A	3×3	3×3	3×3	5×9	9×5	4×4	6×2
Rank(A)	3	2	1	2	2	0	2

5. In each part, find the largest possible value for the rank of A and the smallest possible value for the nullity of A.

   (a) A is 4×4 (b) A is 3×5 (c) A is 5×3

6. If A is an $m \times n$ matrix, what are the largest possible value for its rank and the smallest possible value for its nullity?

   Hint   See Exercise 5.

7. In each part, use the information in the table to determine whether the linear system $A\mathbf{x} = \mathbf{b}$ is consistent. If so, state the number of parameters in its general solution.

	(a)	(b)	(c)	(d)	(e)	(f)	(g)
Size of A	3×3	3×3	3×3	5×9	5×9	4×4	6×2
Rank(A)	3	2	1	2	2	0	2
Rank[$A \mid \mathbf{b}$]	3	3	1	2	3	0	2

8. For each of the matrices in Exercise 7, find the nullity of A, and determine the number of parameters in the general solution of the homogeneous linear system $A\mathbf{x} = \mathbf{0}$.

9. What conditions must be satisfied by b_1, b_2, b_3, b_4, and b_5 for the overdetermined linear system

$$x_1 - 3x_2 = b_1$$
$$x_1 - 2x_2 = b_2$$
$$x_1 + x_2 = b_3$$
$$x_1 - 4x_2 = b_4$$
$$x_1 + 5x_2 = b_5$$

to be consistent?

10. Let

$$A = \begin{bmatrix} a_{11} & a_{12} & a_{13} \\ a_{21} & a_{22} & a_{23} \end{bmatrix}$$

Show that A has rank 2 if and only if one or more of the determinants

$$\begin{vmatrix} a_{11} & a_{12} \\ a_{21} & a_{22} \end{vmatrix}, \quad \begin{vmatrix} a_{11} & a_{13} \\ a_{21} & a_{23} \end{vmatrix}, \quad \begin{vmatrix} a_{12} & a_{13} \\ a_{22} & a_{23} \end{vmatrix}$$

are nonzero.

11. Suppose that A is a 3×3 matrix whose nullspace is a line through the origin in 3-space. Can the row or column space of A also be a line through the origin? Explain.

12. Discuss how the rank of A varies with t.

(a) $A = \begin{bmatrix} 1 & 1 & t \\ 1 & t & 1 \\ t & 1 & 1 \end{bmatrix}$ (b) $A = \begin{bmatrix} t & 3 & -1 \\ 3 & 6 & -2 \\ -1 & -3 & t \end{bmatrix}$

13. Are there values of r and s for which

$$\begin{bmatrix} 1 & 0 & 0 \\ 0 & r-2 & 2 \\ 0 & s-1 & r+2 \\ 0 & 0 & 3 \end{bmatrix}$$

has rank 1 or 2? If so, find those values.

14. Use the result in Exercise 10 to show that the set of points (x, y, z) in R^3 for which the matrix

$$\begin{bmatrix} x & y & z \\ 1 & x & y \end{bmatrix}$$

has rank 1 is the curve with parametric equations $x = t$, $y = t^2$, $z = t^3$.

15. Prove: If $k \neq 0$, then A and kA have the same rank.

Discussion & Discovery

16. (a) Give an example of a 3×3 matrix whose column space is a plane through the origin in 3-space.

(b) What kind of geometric object is the nullspace of your matrix?

(c) What kind of geometric object is the row space of your matrix?

(d) In general, if the column space of a 3×3 matrix is a plane through the origin in 3-space, what can you say about the geometric properties of the nullspace and row space? Explain your reasoning.

17. Indicate whether each statement is always true or sometimes false. Justify your answer by giving a logical argument or a counterexample.

   (a) If A is not square, then the row vectors of A must be linearly dependent.

   (b) If A is square, then either the row vectors or the column vectors of A must be linearly independent.

   (c) If the row vectors and the column vectors of A are linearly independent, then A must be square.

   (d) Adding one additional column to a matrix A increases its rank by one.

18. (a) If A is a 3×5 matrix, then the number of leading 1's in the reduced row-echelon form of A is at most _____. Why?

   (b) If A is a 3×5 matrix, then the number of parameters in the general solution of $A\mathbf{x} = \mathbf{0}$ is at most _____. Why?

   (c) If A is a 5×3 matrix, then the number of leading 1's in the reduced row-echelon form of A is at most _____. Why?

   (d) If A is a 5×3 matrix, then the number of parameters in the general solution of $A\mathbf{x} = \mathbf{0}$ is at most _____. Why?

19. (a) If A is a 3×5 matrix, then the rank of A is at most _____. Why?

   (b) If A is a 3×5 matrix, then the nullity of A is at most _____. Why?

   (c) If A is a 3×5 matrix, then the rank of A^T is at most _____. Why?

   (d) If A is a 3×5 matrix, then the nullity of A^T is at most _____. Why?

CHAPTER 5
Supplementary Exercises

1. In each part, the solution space is a subspace of R^3 and so must be a line through the origin, a plane through the origin, all of R^3, or the origin only. For each system, determine which is the case. If the subspace is a plane, find an equation for it, and if it is a line, find parametric equations.

   (a) $0x + 0y + 0z = 0$

   (b) $\begin{aligned} 2x - 3y + z &= 0 \\ 6x - 9y + 3z &= 0 \\ -4x + 6y - 2z &= 0 \end{aligned}$

   (c) $\begin{aligned} x - 2y + 7z &= 0 \\ -4x + 8y + 5z &= 0 \\ 2x - 4y + 3z &= 0 \end{aligned}$

   (d) $\begin{aligned} x + 4y + 8z &= 0 \\ 2x + 5y + 6z &= 0 \\ 3x + y - 4z &= 0 \end{aligned}$

2. For what values of s is the solution space of

$$\begin{aligned} x_1 + x_2 + sx_3 &= 0 \\ x_1 + sx_2 + x_3 &= 0 \\ sx_1 + x_2 + x_3 &= 0 \end{aligned}$$

   the origin only, a line through the origin, a plane through the origin, or all of R^3?

3. (a) Express $(4a, a - b, a + 2b)$ as a linear combination of $(4, 1, 1)$ and $(0, -1, 2)$.

   (b) Express $(3a + b + 3c, -a + 4b - c, 2a + b + 2c)$ as a linear combination of $(3, -1, 2)$ and $(1, 4, 1)$.

   (c) Express $(2a - b + 4c, 3a - c, 4b + c)$ as a linear combination of three nonzero vectors.

4. Let W be the space spanned by $\mathbf{f} = \sin x$ and $\mathbf{g} = \cos x$.

   (a) Show that for any value of θ, $\mathbf{f}_1 = \sin(x + \theta)$ and $\mathbf{g}_1 = \cos(x + \theta)$ are vectors in W.

   (b) Show that $\mathbf{f}_1$ and $\mathbf{g}_1$ form a basis for W.

5. (a) Express $\mathbf{v} = (1, 1)$ as a linear combination of $\mathbf{v}_1 = (1, -1)$, $\mathbf{v}_2 = (3, 0)$, and $\mathbf{v}_3 = (2, 1)$ in two different ways.

   (b) Explain why this does not violate Theorem 5.4.1.

6. Let A be an $n \times n$ matrix, and let $\mathbf{v}_1, \mathbf{v}_2, \ldots, \mathbf{v}_n$ be linearly independent vectors in R^n expressed as $n \times 1$ matrices. What must be true about A for $A\mathbf{v}_1, A\mathbf{v}_2, \ldots, A\mathbf{v}_n$ to be linearly independent?

7. Must a basis for P_n contain a polynomial of degree k for each $k = 0, 1, 2, \ldots, n$? Justify your answer.

8. For purposes of this problem, let us define a "checkerboard matrix" to be a square matrix $A = [a_{ij}]$ such that

$$a_{ij} = \begin{cases} 1 & \text{if } i + j \text{ is even} \\ 0 & \text{if } i + j \text{ is odd} \end{cases}$$

   Find the rank and nullity of the following checkerboard matrices:

   (a) the 3×3 checkerboard matrix

   (b) the 4×4 checkerboard matrix

   (c) the $n \times n$ checkerboard matrix

9. For purposes of this exercise, let us define an "X-matrix" to be a square matrix with an odd number of rows and columns that has 0's everywhere except on the two diagonals, where it has 1's. Find the rank and nullity of the following X-matrices:

   (a) $\begin{bmatrix} 1 & 0 & 1 \\ 0 & 1 & 0 \\ 1 & 0 & 1 \end{bmatrix}$ (b) $\begin{bmatrix} 1 & 0 & 0 & 0 & 1 \\ 0 & 1 & 0 & 1 & 0 \\ 0 & 0 & 1 & 0 & 0 \\ 0 & 1 & 0 & 1 & 0 \\ 1 & 0 & 0 & 0 & 1 \end{bmatrix}$

   (c) the X-matrix of size $(2n + 1) \times (2n + 1)$

10. In each part, show that the set of polynomials is a subspace of P_n and find a basis for it.

    (a) all polynomials in P_n such that $p(-x) = p(x)$

    (b) all polynomials in P_n such that $p(0) = 0$

11. **(For Readers Who Have Studied Calculus)** Show that the set of all polynomials in P_n that have a horizontal tangent at $x = 0$ is a subspace of P_n. Find a basis for this subspace.

12. (a) Find a basis for the vector space of all 3×3 symmetric matrices.

    (b) Find a basis for the vector space of all 3×3 skew-symmetric matrices.

13. In advanced linear algebra, one proves the following determinant criterion for rank: *The rank of a matrix A is r if and only if A has some $r \times r$ submatrix with a nonzero determinant, and all square submatrices of larger size have determinant zero.* (A submatrix of A is any matrix obtained by deleting rows or columns of A. The matrix A itself is also considered to be a submatrix of A.) In each part, use this criterion to find the rank of the matrix.

    (a) $\begin{bmatrix} 1 & 2 & 0 \\ 2 & 4 & -1 \end{bmatrix}$ (b) $\begin{bmatrix} 1 & 2 & 3 \\ 2 & 4 & 6 \end{bmatrix}$

    (c) $\begin{bmatrix} 1 & 0 & 1 \\ 2 & -1 & 3 \\ 3 & -1 & 4 \end{bmatrix}$ (d) $\begin{bmatrix} 1 & -1 & 2 & 0 \\ 3 & 1 & 0 & 0 \\ -1 & 2 & 4 & 0 \end{bmatrix}$

14. Use the result in Exercise 13 to find the possible ranks for matrices of the form

$$\begin{bmatrix} 0 & 0 & 0 & 0 & 0 & a_{16} \\ 0 & 0 & 0 & 0 & 0 & a_{26} \\ 0 & 0 & 0 & 0 & 0 & a_{36} \\ 0 & 0 & 0 & 0 & 0 & a_{46} \\ a_{51} & a_{52} & a_{53} & a_{54} & a_{55} & a_{56} \end{bmatrix}$$

15. Prove: If S is a basis for a vector space V, then for any vectors $\mathbf{u}$ and $\mathbf{v}$ in V and any scalar k, the following relationships hold:

(a) $(\mathbf{u} + \mathbf{v})_S = (\mathbf{u})_S + (\mathbf{v})_S$ (b) $(k\mathbf{u})_S = k(\mathbf{u})_S$

CHAPTER 5
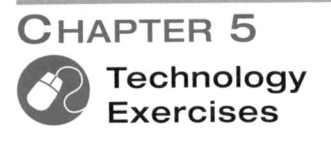
Technology Exercises

The following exercises are designed to be solved using a technology utility. Typically, this will be MATLAB, *Mathematica*, Maple, Derive, or Mathcad, but it may also be some other type of linear algebra software or a scientific calculator with some linear algebra capabilities. For each exercise you will need to read the relevant documentation for the particular utility you are using. The goal of these exercises is to provide you with a basic proficiency with your technology utility. Once you have mastered the techniques in these exercises, you will be able to use your technology utility to solve many of the problems in the regular exercise sets.

Section 5.2   **T1.** (a) Some technology utilities do not have direct commands for finding linear combinations of vectors in R^n. However, you can use matrix multiplication to calculate a linear combination by creating a matrix A with the vectors as columns and a column vector $\mathbf{x}$ with the coefficients as entries. Use this method to compute the vector

$$\mathbf{v} = 6(8, -2, 1, -4) + 17(-3, 9, 11, 6) - 9(0, -1, 2, 4)$$

Check your work by hand.

(b) Use your technology utility to determine whether the vector $(9, 1, 0)$ is a linear combination of the vectors $(1, 2, 3)$, $(1, 4, 6)$, and $(2, -3, -5)$.

Section 5.3   **T1.** Use your technology utility to perform the Wronskian test of linear independence on the sets in Exercise 20.

Section 5.4   **T1.** **(Linear Independence)** Devise three different procedures for using your technology utility to determine whether a set of n vectors in R^n is linearly independent, and use all of your procedures to determine whether the vectors

$$\mathbf{v}_1 = (4, -5, 2, 6), \quad \mathbf{v}_2 = (2, -2, 1, 3), \quad \mathbf{v}_3 = (6, -3, 3, 9), \quad \mathbf{v}_4 = (4, -1, 5, 6)$$

are linearly independent.

T2. **(Dimension)** Devise three different procedures for using your technology utility to determine the dimension of the subspace spanned by a set of vectors in R^n, and use all of your procedures to determine the dimension of the subspace of R^5 spanned by the vectors

$$\mathbf{v}_1 = (2, 2, -1, 0, 1), \quad\quad \mathbf{v}_2 = (-1, -1, 2, -3, 1),$$
$$\mathbf{v}_3 = (1, 1, -2, 0, -1), \quad\quad \mathbf{v}_4 = (0, 0, 1, 1, 1)$$

Section 5.5   **T1.** **(Basis for Row Space)** Some technology utilities provide a command for finding a basis for the row space of a matrix. If your utility has this capability, read the documentation and then use your utility to find a basis for the row space of the matrix in Example 6.

T2. **(Basis for Column Space)** Some technology utilities provide a command for finding a basis for the column space of a matrix. If your utility has this capability, read the documentation and then use your utility to find a basis for the column space of the matrix in Example 6.

T3. **(Nullspace)** Some technology utilities provide a command for finding a basis for the nullspace of a matrix. If your utility has this capability, read the documentation and then check your understanding of the procedure by finding a basis for the nullspace of the matrix A in Example 4. Use this result to find the general solution of the homogeneous system $A\mathbf{x} = \mathbf{0}$.

Section 5.6   **T1.** **(Rank and Nullity)** Read your documentation on finding the rank of a matrix, and then use your utility to find the rank of the matrix A in Example 1. Find the nullity of the matrix using Theorem 5.6.3 and the rank.

T2. There is a result, called **Sylvester's inequality**, which states that if A and B are $n \times n$ matrices with rank r_A and r_B, respectively, then the rank r_{AB} of AB satisfies the inequality $r_A + r_B - n \le r_{AB} \le \min(r_A, r_B)$, where $\min(r_A, r_B)$ denotes the smaller of r_A and r_B or their common value if the two ranks are the same. Use your technology utility to confirm this result for some matrices of your choice.

CHAPTER 6

Applications of Linear Algebra

CHAPTER CONTENTS

INTRODUCTION: This chapter consists of nine applications of linear algebra. Each application is in its own independent section, so sections can be deleted or permuted as desired. Each topic begins with a list of linear algebra prerequisites.

Because our primary objective in this chapter is to present applications of linear algebra, proofs are often omitted. Whenever results from other fields are needed, they are stated precisely, with motivation where possible, but usually without proof.

6.1

CONSTRUCTING CURVES AND SURFACES THROUGH SPECIFIED POINTS

In this section we describe a technique that uses determinants to construct lines, circles, and general conic sections through specified points in the plane. The procedure is also used to pass planes and spheres in 3-space through fixed points.

PREREQUISITES: Linear Systems
Determinants
Analytic Geometry

The following theorem follows from Theorem 2.3.6.

THEOREM 6.1.1

A homogeneous linear system with as many equations as unknowns has a nontrivial solution if and only if the determinant of the coefficient matrix is zero.

We shall now show how this result can be used to determine equations of various curves and surfaces through specified points.

A Line through Two Points

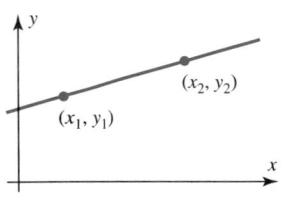

Figure 6.1.1

Suppose that (x_1, y_1) and (x_2, y_2) are two distinct points in the plane. There exists a unique line

$$c_1 x + c_2 y + c_3 = 0 \tag{1}$$

that passes through these two points (Figure 11.1.1). Note that c_1, c_2, and c_3 are not all zero and that these coefficients are unique only up to a multiplicative constant. Because (x_1, y_1) and (x_2, y_2) lie on the line, substituting them in (1) gives the two equations

$$c_1 x_1 + c_2 y_1 + c_3 = 0 \tag{2}$$
$$c_1 x_2 + c_2 y_2 + c_3 = 0 \tag{3}$$

The three equations, (1), (2), and (3), can be grouped together and rewritten as

$$x c_1 + y c_2 + c_3 = 0$$
$$x_1 c_1 + y_1 c_2 + c_3 = 0$$
$$x_2 c_1 + y_2 c_2 + c_3 = 0$$

which is a homogeneous linear system of three equations for c_1, c_2, and c_3. Because c_1, c_2, and c_3 are not all zero, this system has a nontrivial solution, so the determinant of the system must be zero. That is,

$$\begin{vmatrix} x & y & 1 \\ x_1 & y_1 & 1 \\ x_2 & y_2 & 1 \end{vmatrix} = 0 \tag{4}$$

Consequently, every point (x, y) on the line satisfies (4); conversely, it can be shown that every point (x, y) that satisfies (4) lies on the line.

EXAMPLE 1   Equation of a Line

Find the equation of the line that passes through the two points (2, 1) and (3, 7).

Solution

Substituting the coordinates of the two points into Equation (4) gives

$$\begin{vmatrix} x & y & 1 \\ 2 & 1 & 1 \\ 3 & 7 & 1 \end{vmatrix} = 0$$

The cofactor expansion of this determinant along the first row then gives

$$-6x + y + 11 = 0 \quad \blacklozenge$$

A Circle through Three Points

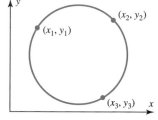

Figure 6.1.2

Suppose that there are three distinct points in the plane, (x_1, y_1), (x_2, y_2), and (x_3, y_3), not all lying on a straight line. From analytic geometry we know that there is a unique circle, say,

$$c_1(x^2 + y^2) + c_2 x + c_3 y + c_4 = 0 \tag{5}$$

that passes through them (Figure 6.1.2). Substituting the coordinates of the three points into this equation gives

$$c_1(x_1^2 + y_1^2) + c_2 x_1 + c_3 y_1 + c_4 = 0 \tag{6}$$
$$c_1(x_2^2 + y_2^2) + c_2 x_2 + c_3 y_2 + c_4 = 0 \tag{7}$$
$$c_1(x_3^2 + y_3^2) + c_2 x_3 + c_3 y_3 + c_4 = 0 \tag{8}$$

As before, Equations (5) through (8) form a homogeneous linear system with a nontrivial solution for c_1, c_2, c_3, and c_4. Thus the determinant of the coefficient matrix is zero:

$$\begin{vmatrix} x^2 + y^2 & x & y & 1 \\ x_1^2 + y_1^2 & x_1 & y_1 & 1 \\ x_2^2 + y_2^2 & x_2 & y_2 & 1 \\ x_3^2 + y_3^2 & x_3 & y_3 & 1 \end{vmatrix} = 0 \tag{9}$$

This is a determinant form for the equation of the circle.

EXAMPLE 2 Equation of a Circle

Find the equation of the circle that passes through the three points $(1, 7)$, $(6, 2)$, and $(4, 6)$.

Solution

Substituting the coordinates of the three points into Equation (9) gives

$$\begin{vmatrix} x^2 + y^2 & x & y & 1 \\ 50 & 1 & 7 & 1 \\ 40 & 6 & 2 & 1 \\ 52 & 4 & 6 & 1 \end{vmatrix} = 0$$

which reduces to

$$10(x^2 + y^2) - 20x - 40y - 200 = 0$$

In standard form this is

$$(x - 1)^2 + (y - 2)^2 = 5^2$$

Thus the circle has center $(1, 2)$ and radius 5. $\blacklozenge$

A General Conic Section through Five Points

The general equation of a conic section in the plane (a parabola, hyperbola, or ellipse, or degenerate forms of these curves) is given by

$$c_1 x^2 + c_2 xy + c_3 y^2 + c_4 x + c_5 y + c_6 = 0$$

This equation contains six coefficients, but we can reduce the number to five if we divide through by any one of them that is not zero. Thus only five coefficients must be determined, so five distinct points in the plane are sufficient to determine the equation of the conic section (Figure 6.1.3). As before, the equation can be put in determinant form (see Exercise 7):

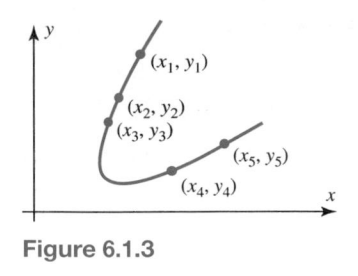

Figure 6.1.3

$$\begin{vmatrix} x^2 & xy & y^2 & x & y & 1 \\ x_1^2 & x_1 y_1 & y_1^2 & x_1 & y_1 & 1 \\ x_2^2 & x_2 y_2 & y_2^2 & x_2 & y_2 & 1 \\ x_3^2 & x_3 y_3 & y_3^2 & x_3 & y_3 & 1 \\ x_4^2 & x_4 y_4 & y_4^2 & x_4 & y_4 & 1 \\ x_5^2 & x_5 y_5 & y_5^2 & x_5 & y_5 & 1 \end{vmatrix} = 0 \qquad (10)$$

EXAMPLE 3 Equation of an Orbit

An astronomer who wants to determine the orbit of an asteroid about the sun sets up a Cartesian coordinate system in the plane of the orbit with the sun at the origin. Astronomical units of measurement are used along the axes (1 astronomical unit = mean distance of earth to sun = 93 million miles). By Kepler's first law, the orbit must be an ellipse, so the astronomer makes five observations of the asteroid at five different times and finds five points along the orbit to be

$$(8.025, 8.310), \ (10.170, 6.355), \ (11.202, 3.212), \ (10.736, 0.375), \ (9.092, -2.267)$$

Find the equation of the orbit.

Solution

Substituting the coordinates of the five given points into (10) gives

$$\begin{vmatrix} x^2 & xy & y^2 & x & y & 1 \\ 64.401 & 66.688 & 69.056 & 8.025 & 8.310 & 1 \\ 103.429 & 64.630 & 40.386 & 10.170 & 6.355 & 1 \\ 125.485 & 35.981 & 10.317 & 11.202 & 3.212 & 1 \\ 115.262 & 4.026 & 0.141 & 10.736 & 0.375 & 1 \\ 82.664 & -20.612 & 5.139 & 9.092 & -2.267 & 1 \end{vmatrix} = 0$$

The cofactor expansion of this determinant along the first row is

$$386.799 x^2 - 102.896 xy + 446.026 y^2 - 2476.409 x - 1427.971 y - 17109.378 = 0$$

Figure 6.1.4 is an accurate diagram of the orbit, together with the five given points. ◆

A Plane through Three Points

In Exercise 8 we ask the reader to show the following: The plane in 3-space with equation

$$c_1 x + c_2 y + c_3 z + c_4 = 0$$

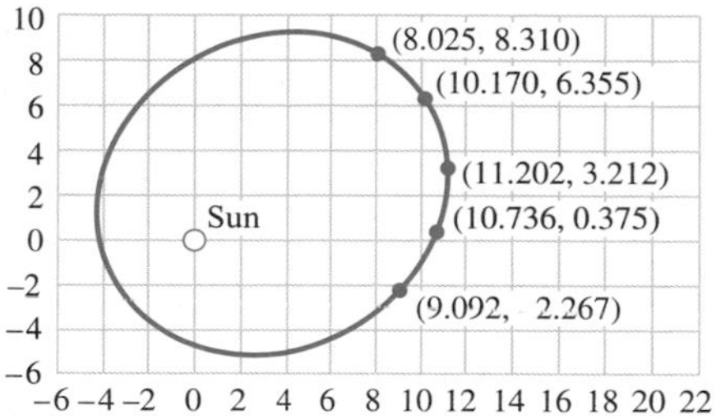

Figure 6.1.4

that passes through three noncollinear points (x_1, y_1, z_1), (x_2, y_2, z_2), and (x_3, y_3, z_3) is given by the determinant equation

$$\begin{vmatrix} x & y & z & 1 \\ x_1 & y_1 & z_1 & 1 \\ x_2 & y_2 & z_2 & 1 \\ x_3 & y_3 & z_3 & 1 \end{vmatrix} = 0 \qquad (11)$$

EXAMPLE 4 Equation of a Plane

The equation of the plane that passes through the three noncollinear points $(1, 1, 0)$, $(2, 0, -1)$, and $(2, 9, 2)$ is

$$\begin{vmatrix} x & y & z & 1 \\ 1 & 1 & 0 & 1 \\ 2 & 0 & -1 & 1 \\ 2 & 9 & 2 & 1 \end{vmatrix} = 0$$

which reduces to

$$2x - y + 3z - 1 = 0 \quad \blacklozenge$$

A Sphere through Four Points

In Exercise 9 we ask the reader to show the following: The sphere in 3-space with equation

$$c_1(x^2 + y^2 + z^2) + c_2 x + c_3 y + c_4 z + c_5 = 0$$

that passes through four noncoplanar points (x_1, y_1, z_1), (x_2, y_2, z_2), (x_3, y_3, z_3), and (x_4, y_4, z_4) is given by the following determinant equation:

$$\begin{vmatrix} x^2 + y^2 + z^2 & x & y & z & 1 \\ x_1^2 + y_1^2 + z_1^2 & x_1 & y_1 & z_1 & 1 \\ x_2^2 + y_2^2 + z_2^2 & x_2 & y_2 & z_2 & 1 \\ x_3^2 + y_3^2 + z_3^2 & x_3 & y_3 & z_3 & 1 \\ x_4^2 + y_4^2 + z_4^2 & x_4 & y_4 & z_4 & 1 \end{vmatrix} = 0 \qquad (12)$$

EXAMPLE 5 Equation of a Sphere

The equation of the sphere that passes through the four points $(0, 3, 2)$, $(1, -1, 1)$, $(2, 1, 0)$, and $(5, 1, 3)$ is

$$\begin{vmatrix} x^2 + y^2 + z^2 & x & y & z & 1 \\ 13 & 0 & 3 & 2 & 1 \\ 3 & 1 & -1 & 1 & 1 \\ 5 & 2 & 1 & 0 & 1 \\ 35 & 5 & 1 & 3 & 1 \end{vmatrix} = 0$$

This reduces to

$$x^2 + y^2 + z^2 - 4x - 2y - 6z + 5 = 0$$

which in standard form is

$$(x - 2)^2 + (y - 1)^2 + (z - 3)^2 = 9 \quad \blacklozenge$$

EXERCISE SET 6.1

1. Find the equations of the lines that pass through the following points:
   (a) $(1, -1), (2, 2)$ (b) $(0, 1), (1, -1)$

2. Find the equations of the circles that pass through the following points:
   (a) $(2, 6), (2, 0), (5, 3)$ (b) $(2, -2), (3, 5), (-4, 6)$

3. Find the equation of the conic section that passes through the points $(0, 0), (0, -1), (2, 0)$, $(2, -5)$, and $(4, -1)$.

4. Find the equations of the planes in 3-space that pass through the following points:
   (a) $(1, 1, -3), (1, -1, 1), (0, -1, 2)$ (b) $(2, 3, 1), (2, -1, -1), (1, 2, 1)$

5. (a) Alter Equation (11) so that it determines the plane that passes through the origin and is parallel to the plane that passes through three specified noncollinear points.
   (b) Find the two planes described in part (a) corresponding to the triplets of points in Exercises 4(a) and 4(b).

6. Find the equations of the spheres in 3-space that pass through the following points:
   (a) $(1, 2, 3), (-1, 2, 1), (1, 0, 1), (1, 2, -1)$
   (b) $(0, 1, -2), (1, 3, 1), (2, -1, 0), (3, 1, -1)$

7. Show that Equation (10) is the equation of the conic section that passes through five given distinct points in the plane.

8. Show that Equation (11) is the equation of the plane in 3-space that passes through three given noncollinear points.

9. Show that Equation (12) is the equation of the sphere in 3-space that passes through four given noncoplanar points.

10. Find a determinant equation for the parabola of the form

$$c_1 y + c_2 x^2 + c_3 x + c_4 = 0$$

   that passes through three given noncollinear points in the plane.

11. What does Equation (9) become if the three distinct points are collinear?

12. What does Equation (11) become if the three distinct points are collinear?

13. What does Equation (12) become if the four points are coplanar?

SECTION 6.1
Technology Exercises

The following exercises are designed to be solved using a technology utility. Typically, this will be MATLAB, *Mathematica*, Maple, Derive, or Mathcad, but it may also be some other type of linear algebra software or a scientific calculator with some linear algebra capabilities. For each exercise you will need to read the relevant documentation for the particular utility you are using. The goal of these exercises is to provide you with a basic proficiency with your technology utility. Once you have mastered the techniques in these exercises, you will be able to use your technology utility to solve many of the problems in the regular exercise sets.

T1. The general equation of a quadric surface is given by

$$a_1x^2 + a_2y^2 + a_3z^2 + a_4xy + a_5xz + a_6yz + a_7x + a_8y + a_9z + a_{10} = 0$$

Given nine points on this surface, it may be possible to determine its equation.

(a) Show that if the nine points (x_i, y_i) for $i = 1, 2, 3, \ldots, 9$ lie on this surface, and if they determine uniquely the equation of this surface, then its equation can be written in determinant form as

$$\begin{vmatrix} x^2 & y^2 & z^2 & xy & xz & yz & x & y & z & 1 \\ x_1^2 & y_1^2 & z_1^2 & x_1y_1 & x_1z_1 & y_1z_1 & x_1 & y_1 & z_1 & 1 \\ x_2^2 & y_2^2 & z_2^2 & x_2y_2 & x_2z_2 & y_2z_2 & x_2 & y_2 & z_2 & 1 \\ x_3^2 & y_3^2 & z_3^2 & x_3y_3 & x_3z_3 & y_3z_3 & x_3 & y_3 & z_3 & 1 \\ x_4^2 & y_4^2 & z_4^2 & x_4y_4 & x_4z_4 & y_4z_4 & x_4 & y_4 & z_4 & 1 \\ x_5^2 & y_5^2 & z_5^2 & x_5y_5 & x_5z_5 & y_5z_5 & x_5 & y_5 & z_5 & 1 \\ x_6^2 & y_6^2 & z_6^2 & x_6y_6 & x_6z_6 & y_6z_6 & x_6 & y_6 & z_6 & 1 \\ x_7^2 & y_7^2 & z_7^2 & x_7y_7 & x_7z_7 & y_7z_7 & x_7 & y_7 & z_7 & 1 \\ x_8^2 & y_8^2 & z_8^2 & x_8y_8 & x_8z_8 & y_8z_8 & x_8 & y_8 & z_8 & 1 \\ x_9^2 & y_9^2 & z_9^2 & x_9y_9 & x_9z_9 & y_9z_9 & x_9 & y_9 & z_9 & 1 \end{vmatrix} = 0$$

(b) Use the result in part (a) to determine the equation of the quadric surface that passes through the points $(1, 2, 3)$, $(2, 1, 7)$, $(0, 4, 6)$, $(3, -1, 4)$, $(3, 0, 11)$, $(-1, 5, 8)$, $(9, -8, 3)$, $(4, 5, 3)$, and $(-2, 6, 10)$.

T2. (a) A hyperplane in the n-dimensional Euclidean space R^n has an equation of the form

$$a_1x_1 + a_2x_2 + a_3x_3 + \cdots + a_nx_n + a_{n+1} = 0$$

where $a_i, i = 1, 2, 3, \ldots, n + 1$, are constants, not all zero, and $x_i, i = 1, 2, 3, \ldots, n$, are variables for which

$$(x_1, x_2, x_3, \ldots, x_n) \in R^n$$

A point

$$(x_{10}, x_{20}, x_{30}, \ldots, x_{n0}) \in R^n$$

lies on this hyperplane if

$$a_1x_{10} + a_2x_{20} + a_3x_{30} + \cdots + a_nx_{n0} + a_{n+1} = 0$$

Given that the n points $(x_{1i}, x_{2i}, x_{3i}, \ldots, x_{ni}), i = 1, 2, 3, \ldots, n$, lie on this hyperplane and that they uniquely determine the equation of the hyperplane, show that the equation of the hyperplane can be written in determinant form as

$$\begin{vmatrix} x_1 & x_2 & x_3 & \cdots & x_n & 1 \\ x_{11} & x_{21} & x_{31} & \cdots & x_{n1} & 1 \\ x_{12} & x_{22} & x_{32} & \cdots & x_{n2} & 1 \\ x_{13} & x_{23} & x_{33} & \cdots & x_{n3} & 1 \\ \vdots & \vdots & \vdots & \ddots & \vdots & \vdots \\ x_{1n} & x_{2n} & x_{3n} & \cdots & x_{nn} & 1 \end{vmatrix} = 0$$

(b) Determine the equation of the hyperplane in R^9 that goes through the following nine points:

$$(1, 2, 3, 4, 5, 6, 7, 8, 9) \quad (2, 3, 4, 5, 6, 7, 8, 9, 1) \quad (3, 4, 5, 6, 7, 8, 9, 1, 2)$$
$$(4, 5, 6, 7, 8, 9, 1, 2, 3) \quad (5, 6, 7, 8, 9, 1, 2, 3, 4) \quad (6, 7, 8, 9, 1, 2, 3, 4, 5)$$
$$(7, 8, 9, 1, 2, 3, 4, 5, 6) \quad (8, 9, 1, 2, 3, 4, 5, 6, 7) \quad (9, 1, 2, 3, 4, 5, 6, 7, 8)$$

6.2
ELECTRICAL NETWORKS

In this section basic laws of electrical circuits are discussed, and it is shown how these laws can be used to obtain systems of linear equations whose solutions yield the currents flowing in an electrical circuit.

PREREQUISITES: Linear Systems

The simplest electrical circuits consist of two basic components:

| electrical sources | denoted by |
| resistors | denoted by |

Electrical sources, such as batteries, create currents in an electrical circuit. Resistors, such as lightbulbs, limit the magnitudes of the currents.

There are three basic quantities associated with electrical circuits: *electrical potential* (E), *resistance* (R), and *current* (I). These are commonly measured in the following units:

$$E \quad \text{in volts} \quad \text{(V)}$$
$$R \quad \text{in ohms} \quad (\Omega)$$
$$I \quad \text{in amperes} \quad \text{(A)}$$

Electrical potential is associated with two points in an electrical circuit and is measured in practice by connecting those points to a device called a *voltmeter*. For example, a common AA battery is rated at 1.5 volts, which means that this is the electrical potential across its positive and negative terminals (Figure 6.2.1).

In an electrical circuit the electrical potential between two points is called the *voltage drop* between these points. As we shall see, currents and voltage drops can be either positive or negative.

The flow of current in an electrical circuit is governed by three basic principles:

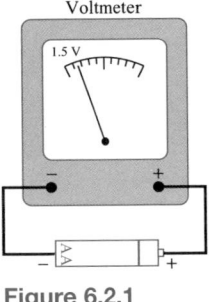

Voltmeter

Figure 6.2.1

1. *Ohm's Law* The voltage drop across a resistor is the product of the current passing through it and its resistance; that is, $E = IR$.

2. *Kirchhoff's Current Law* The sum of the currents flowing into any point equals the sum of the currents flowing out from the point.

3. *Kirchhoff's Voltage Law* Around any closed loop, the algebraic sum of the voltage drops is zero.

EXAMPLE 1 Finding Currents in a Circuit

Find the unknown currents I_1, I_2, and I_3 in the circuit shown in Figure 6.2.2.

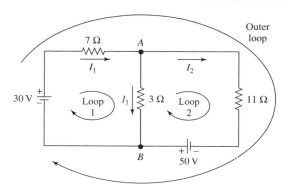

Figure 6.2.2

Solution

The flow directions for the currents I_1, I_2, and I_3 (marked by the arrowheads) were picked arbitrarily. Any of these currents that turn out to be negative actually flow opposite to the direction selected.

Applying Kirchhoff's current law to points A and B yields

$$I_1 = I_2 + I_3 \quad \text{(Point } A\text{)}$$
$$I_3 + I_2 = I_1 \quad \text{(Point } B\text{)}$$

Since these equations both simplify to the same linear equation

$$I_1 - I_2 - I_3 = 0 \tag{1}$$

we still need two more equations to determine I_1, I_2, and I_3 uniquely. We will obtain them using Kirchhoff's voltage law.

To apply Kirchhoff's voltage law to a loop, select a positive direction around the loop (say clockwise) and make the following sign conventions:

- A current passing through a resistor produces a positive voltage drop if it flows in the positive direction of the loop and a negative voltage drop if it flows in the negative direction of the loop.

- A current passing through an electrical source produces a positive voltage drop if the positive direction of the loop is from + to − and a negative voltage drop if the positive direction of the loop is from − to +.

Applying Kirchhoff's voltage law and Ohm's law to loop 1 in Figure 6.2.2 yields

$$7I_1 + 3I_3 - 30 = 0 \tag{2}$$

and applying them to loop 2 yields

$$11I_2 - 3I_3 - 50 = 0 \tag{3}$$

Combining (1), (2), and (3) yields the linear system

$$
\begin{aligned}
I_1 - \quad I_2 - \quad I_3 &= \ 0 \\
7I_1 \qquad\quad + 3I_3 &= 30 \\
11I_2 - 3I_3 &= 50
\end{aligned}
$$

Solving this linear system yields the following values for the currents:

$$I_1 = \tfrac{570}{131} \text{ (A)}, \qquad I_2 = \tfrac{590}{131} \text{ (A)}, \qquad I_3 = -\tfrac{20}{131} \text{ (A)}$$

Note that I_3 is negative, which means that this current flows opposite to the direction indicated in Figure 6.2.2. Also note that we could have applied Kirchhoff's voltage law to the outer loop of the circuit. However, this produces a redundant equation (try it). ◆

EXERCISE SET 6.2

In Exercises 1–4 find the currents in the circuits.

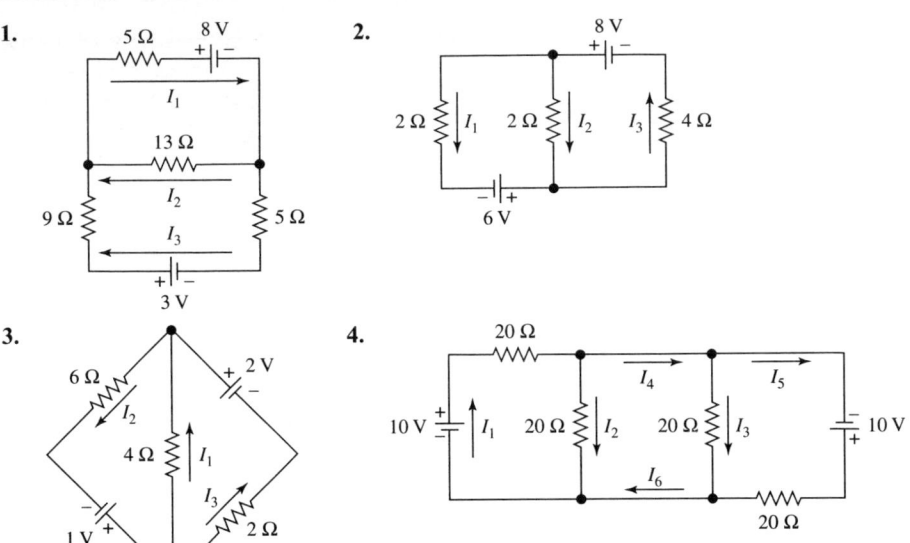

1.

2.

3.

4.

5. Show that if the current I_5 in the circuit of the accompanying figure is zero, then $R_4 = R_3 R_2 / R_1$.

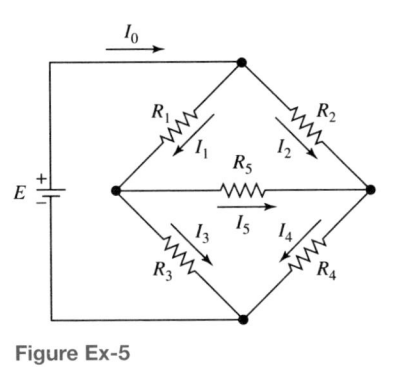

Figure Ex-5

REMARK This circuit, called a Wheatstone bridge circuit, is used for the precise measurement of resistance. Here, R_4 is an unknown resistance and R_1, R_2, and R_3 are adjustable calibrated resistors. R_5 represents a galvanometer—a device for measuring current. After the operator varies the resistances R_1, R_2, and R_3 until the galvanometer reading is zero, the formula $R_4 = R_3 R_2 / R_1$ determines the unknown resistance R_4.

6. Show that if the two currents labeled I in the circuits of the accompanying figure are equal, then $R = \dfrac{1}{\dfrac{1}{R_1} + \dfrac{1}{R_2}}$.

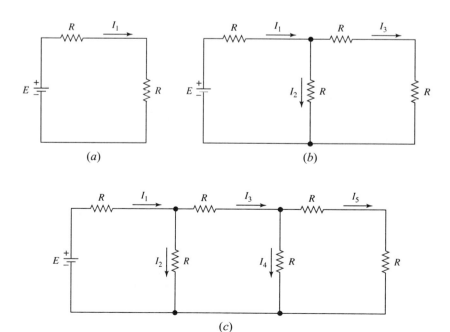

Figure Ex-6

Technology Exercises

The following exercises are designed to be solved using a technology utility. Typically, this will be MATLAB, *Mathematica*, Maple, Derive, or Mathcad, but it may also be some other type of linear algebra software or a scientific calculator with some linear algebra capabilities. For each exercise you will need to read the relevant documentation for the particular utility you are using. The goal of these exercises is to provide you with a basic proficiency with your technology utility. Once you have mastered the techniques in these exercises, you will be able to use your technology utility to solve many of the problems in the regular exercise sets.

T1. The accompanying figure shows a sequence of different circuits.

(a) Solve for the current I_1, for the circuit in part (a) of the figure.

(b) Solve for the currents I_1 through I_3, for the circuit in part (b) of the figure.

(c) Solve for the currents I_1 through I_5, for the circuit in part (c) of the figure.

(d) Continue this process until you discover a pattern in the values of $I_1, I_2, I_3, \ldots$.

(e) Investigate the sequence of values for I_1 in each of the circuits in parts (a), (b), (c), and so on, and numerically show that the limit of this sequence approaches the value

$$\left(\frac{\sqrt{5}-1}{2}\right)\frac{E}{R} \approx (0.6180)\frac{E}{R}$$

Figure Ex-T1

T2. The accompanying figure shows a sequence of different circuits.

(a) Solve for the current I_1, for the circuit in part (*a*) of the figure.

(b) Solve for the current I_1, for the circuit in part (*b*) of the figure.

(c) Solve for the current I_1, for the circuit in part (*c*) of the figure.

(d) Continue this process until you discover a pattern in the values of I_1.

(e) Investigate the sequence of values for I_1 in each of the circuits in parts (a), (b), (c), and so on, and numerically show that the limit of this sequence approaches the value

$$\left(\frac{\sqrt{5}-1}{2}\right)\frac{E}{R} \approx (0.6180)\frac{E}{R}$$

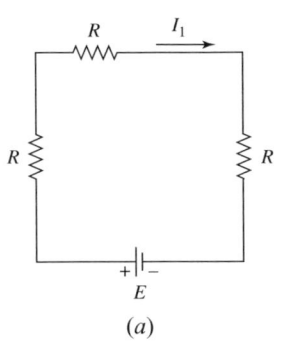

(a)

(b)

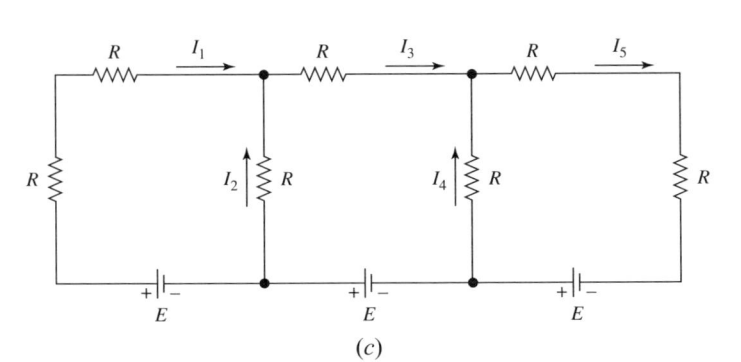

(c)

Figure Ex-T2

6.3

GEOMETRIC LINEAR PROGRAMMING

In this section we describe a geometric technique for maximizing or minimizing a linear expression in two variables subject to a set of linear constraints.

> **PREREQUISITES:** Linear Systems
> Linear Inequalities

Linear Programming

The study of linear programming theory has expanded greatly since the pioneering work of George Dantzig in the late 1940s. Today, linear programming is applied to a wide variety of problems in industry and science. In this section we present a geometric approach to the solution of simple linear programming problems. Let us begin with some examples.

EXAMPLE 1 Maximizing Sales Revenue

A candy manufacturer has 130 pounds of chocolate-covered cherries and 170 pounds of chocolate-covered mints in stock. He decides to sell them in the form of two different mixtures. One mixture will contain half cherries and half mints by weight and will sell for $2.00 per pound. The other mixture will contain one-third cherries and two-thirds mints by weight and will sell for $1.25 per pound. How many pounds of each mixture should the candy manufacturer prepare in order to maximize his sales revenue?

Solution

Let us first formulate this problem mathematically. Let the mixture of half cherries and half mints be called mix A, and let x_1 be the number of pounds of this mixture to be prepared. Let the mixture of one-third cherries and two-thirds mints be called mix B, and let x_2 be the number of pounds of this mixture to be prepared. Since mix A sells for $2.00 per pound and mix B sells for $1.25 per pound, the total sales z (in dollars) will be

$$z = 2.00x_1 + 1.25x_2$$

Since each pound of mix A contains $\frac{1}{2}$ pound of cherries and each pound of mix B contains $\frac{1}{3}$ pound of cherries, the total number of pounds of cherries used in both mixtures is

$$\tfrac{1}{2}x_1 + \tfrac{1}{3}x_2$$

Similarly, since each pound of mix A contains $\frac{1}{2}$ pound of mints and each pound of mix B contains $\frac{2}{3}$ pound of mints, the total number of pounds of mints used in both mixtures is

$$\tfrac{1}{2}x_1 + \tfrac{2}{3}x_2$$

Because the manufacturer can use at most 130 pounds of cherries and 170 pounds of mints, we must have

$$\tfrac{1}{2}x_1 + \tfrac{1}{3}x_2 \leq 130$$
$$\tfrac{1}{2}x_1 + \tfrac{2}{3}x_2 \leq 170$$

Furthermore, since x_1 and x_2 cannot be negative numbers, we must have

$$x_1 \geq 0 \quad \text{and} \quad x_2 \geq 0$$

The problem can therefore be formulated mathematically as follows: Find values of x_1 and x_2 that maximize

$$z = 2.00x_1 + 1.25x_2$$

subject to

$$\tfrac{1}{2}x_1 + \tfrac{1}{3}x_2 \leq 130$$
$$\tfrac{1}{2}x_1 + \tfrac{2}{3}x_2 \leq 170$$
$$x_1 \geq 0$$
$$x_2 \geq 0$$

Later in this section we shall show how to solve this type of mathematical problem geometrically. ◆

EXAMPLE 2 Maximizing Annual Yield

A woman has up to $10,000 to invest. Her broker suggests investing in two bonds, A and B. Bond A is a rather risky bond with an annual yield of 10%, and bond B is a rather safe bond with an annual yield of 7%. After some consideration, she decides to invest at most $6000 in bond A, to invest at least $2000 in bond B, and to invest at least as much in bond A as in bond B. How should she invest her $10,000 in order to maximize her annual yield?

Solution

To formulate this problem mathematically, let x_1 be the number of dollars to be invested in bond A, and let x_2 be the number of dollars to be invested in bond B. Since each dollar invested in bond A earns $.10 per year and each dollar invested in bond B earns $.07 per year, the total dollar amount z earned each year by both bonds is

$$z = .10x_1 + .07x_2$$

The constraints imposed can be formulated mathematically as follows:

Invest no more than $10,000:	$x_1 + x_2 \leq 10{,}000$
Invest at most $6000 in bond A:	$x_1 \leq 6000$
Invest at least $2000 in bond B:	$x_2 \geq 2000$
Invest at least as much in bond A as in bond B:	$x_1 \geq x_2$

We also have the implicit assumption that x_1 and x_2 are nonnegative:

$$x_1 \geq 0 \quad \text{and} \quad x_2 \geq 0$$

Thus the complete mathematical formulation of the problem is as follows: Find values of x_1 and x_2 that maximize

$$z = .10x_1 + .07x_2$$

subject to

$$x_1 + x_2 \leq 10{,}000$$
$$x_1 \leq 6000$$
$$x_2 \geq 2000$$
$$x_1 - x_2 \geq 0$$
$$x_1 \geq 0$$
$$x_2 \geq 0 \quad \blacklozenge$$

EXAMPLE 3 Minimizing Cost

A student desires to design a breakfast of corn flakes and milk that is as economical as possible. On the basis of what he eats during his other meals, he decides that his breakfast should supply him with at least 9 grams of protein, at least $\frac{1}{3}$ the recommended daily allowance (RDA) of vitamin D, and at least $\frac{1}{4}$ the RDA of calcium. He finds the following nutrition information on the milk and corn flakes containers:

	Milk ($\frac{1}{2}$ cup)	Corn Flakes (1 ounce)
Cost	7.5 cents	5.0 cents
Protein	4 grams	2 grams
Vitamin D	$\frac{1}{8}$ of RDA	$\frac{1}{10}$ of RDA
Calcium	$\frac{1}{6}$ of RDA	None

In order not to have his mixture too soggy or too dry, the student decides to limit himself to mixtures that contain 1 to 3 ounces of corn flakes per cup of milk, inclusive. What quantities of milk and corn flakes should he use to minimize the cost of his breakfast?

Solution

For the mathematical formulation of this problem, let x_1 be the quantity of milk used (measured in $\frac{1}{2}$-cup units), and let x_2 be the quantity of corn flakes used (measured in 1-ounce units). Then if z is the cost of the breakfast in cents, we may write the following.

Cost of breakfast:	$z = 7.5x_1 + 5.0x_2$
At least 9 grams protein:	$4x_1 + 2x_2 \geq 9$
At least $\frac{1}{3}$ RDA vitamin D:	$\frac{1}{8}x_1 + \frac{1}{10}x_2 \geq \frac{1}{3}$
At least $\frac{1}{4}$ RDA calcium:	$\frac{1}{6}x_1 \geq \frac{1}{4}$
At least 1 ounce corn flakes per cup (two $\frac{1}{2}$-cups) of milk:	$\dfrac{x_2}{x_1} \geq \dfrac{1}{2}$ (or $x_1 - 2x_2 \leq 0$)
At most 3 ounces corn flakes per cup (two $\frac{1}{2}$-cups) of milk:	$\dfrac{x_2}{x_1} \leq \dfrac{3}{2}$ (or $3x_1 - 2x_2 \geq 0$)

As before, we also have the implicit assumption that $x_1 \geq 0$ and $x_2 \geq 0$. Thus the complete mathematical formulation of the problem is as follows: Find values of x_1 and x_2 that minimize

$$z = 7.5x_1 + 5.0x_2$$

subject to

$$4x_1 + 2x_2 \geq 9$$
$$\tfrac{1}{8}x_1 + \tfrac{1}{10}x_2 \geq \tfrac{1}{3}$$
$$\tfrac{1}{6}x_1 \geq \tfrac{1}{4}$$
$$x_1 - 2x_2 \leq 0$$
$$3x_1 - 2x_2 \geq 0$$
$$x_1 \geq 0$$
$$x_2 \geq 0 \quad \blacklozenge$$

Geometric Solution of Linear Programming Problems

Each of the preceding three examples is a special case of the following problem.

Problem: Find values of x_1 and x_2 that either maximize or minimize

$$z = c_1x_1 + c_2x_2 \tag{1}$$

subject to

$$a_{11}x_1 + a_{12}x_2 \ (\leq)(\geq)(=) \ b_1$$
$$a_{21}x_1 + a_{22}x_2 \ (\leq)(\geq)(=) \ b_2$$
$$\vdots \qquad \vdots \qquad \qquad \vdots \tag{2}$$
$$a_{m1}x_1 + a_{m2}x_2 \ (\leq)(\geq)(=) \ b_m$$

and

$$x_1 \geq 0, \qquad x_2 \geq 0 \tag{3}$$

In each of the m conditions of (2), any one of the symbols $\leq$, $\geq$, and $=$ may be used.

The problem above is called the **general linear programming problem** in two variables. The linear function z in (1) is called the **objective function**. Equations (2) and (3) are called the **constraints**; in particular, the equations in (3) are called the **nonnegativity constraints** on the variables x_1 and x_2.

We shall now show how to solve a linear programming problem in two variables graphically. A pair of values (x_1, x_2) that satisfy all of the constraints is called a **feasible solution**. The set of all feasible solutions determines a subset of the x_1x_2-plane called the **feasible region**. Our desire is to find a feasible solution that maximizes the objective function. Such a solution is called an **optimal solution**.

To examine the feasible region of a linear programming problem, let us note that each constraint of the form

$$a_{i1}x_1 + a_{i2}x_2 = b_i$$

defines a line in the x_1x_2-plane, whereas each constraint of the form

$$a_{i1}x_1 + a_{i2}x_2 \leq b_i \quad \text{or} \quad a_{i1}x_1 + a_{i2}x_2 \geq b_i$$

defines a half-plane that includes its boundary line

$$a_{i1}x_1 + a_{i2}x_2 = b_i$$

Thus the feasible region is always an intersection of finitely many lines and half-planes. For example, the four constraints

$$\tfrac{1}{2}x_1 + \tfrac{1}{3}x_2 \leq 130$$
$$\tfrac{1}{2}x_1 + \tfrac{2}{3}x_2 \leq 170$$
$$x_1 \geq 0$$
$$x_2 \geq 0$$

of Example 1 define the half-planes illustrated in parts (*a*), (*b*), (*c*), and (*d*) of Figure 6.3.1. The feasible region of this problem is thus the intersection of these four half-planes, which is illustrated in Figure 6.3.1*e*.

It can be shown that the feasible region of a linear programming problem has a boundary consisting of a finite number of straight line segments. If the feasible region can be enclosed in a sufficiently large circle, it is called **bounded** (Figure 6.3.1*e*); otherwise, it is called **unbounded** (see Figure 6.3.5). If the feasible region is *empty* (contains no points), then the constraints are inconsistent and the linear programming problem has no solution (see Figure 6.3.6).

Those boundary points of a feasible region that are intersections of two of the straight line boundary segments are called **extreme points**. (They are also called *corner points* and *vertex points*.) For example, in Figure 6.3.1*e*, we see that the feasible region of Example 1 has four extreme points:

$$(0, 0), \quad (0, 255), \quad (180, 120), \quad (260, 0) \tag{4}$$

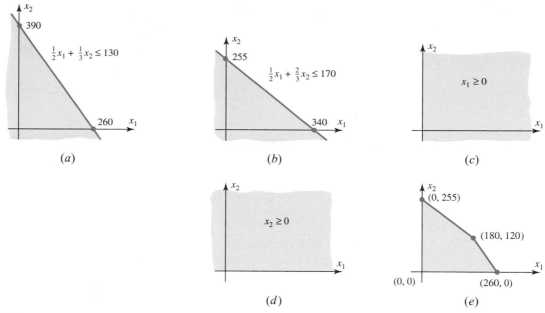

Figure 6.3.1

The importance of the extreme points of a feasible region is shown by the following theorem.

THEOREM 6.3.1

Maximum and Minimum Values

If the feasible region of a linear programming problem is nonempty and bounded, then the objective function attains both a maximum and a minimum value, and these occur at extreme points of the feasible region. If the feasible region is unbounded, then the objective function may or may not attain a maximum or minimum value; however, if it attains a maximum or minimum value, it does so at an extreme point.

Figure 6.3.2 suggests the idea behind the proof of this theorem. Since the objective function

$$z = c_1 x_1 + c_2 x_2$$

of a linear programming problem is a linear function of x_1 and x_2, its level curves (the curves along which z has constant values) are straight lines. As we move in a direction perpendicular to these level curves, the objective function either increases or decreases monotonically. Within a bounded feasible region, the maximum and minimum values of z must therefore occur at extreme points, as Figure 6.3.2 indicates.

In the next few examples we use Theorem 6.3.1 to solve several linear programming problems and illustrate the variations in the nature of the solutions that may occur.

EXAMPLE 4   Example 1 Revisited

Figure 6.3.1e shows that the feasible region of Example 1 is bounded. Consequently, from Theorem 6.3.1 the objective function

$$z = 2.00x_1 + 1.25x_2$$

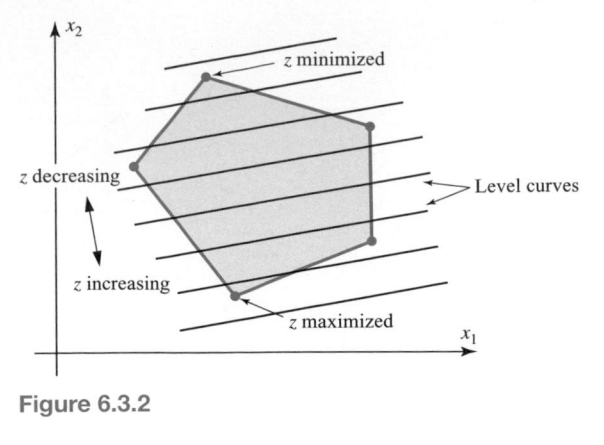

Figure 6.3.2

attains both its minimum and maximum values at extreme points. The four extreme points and the corresponding values of z are given in the following table.

Extreme Point (x_1, x_2)	Value of $z = 2.00x_1 + 1.25x_2$
(0, 0)	0
(0, 255)	318.75
(180, 120)	510.00
(260, 0)	520.00

We see that the largest value of z is 520.00 and the corresponding optimal solution is (260, 0). Thus the candy manufacturer attains maximum sales of \$520 when he produces 260 pounds of mixture A and none of mixture B. ◆

EXAMPLE 5 Using Theorem 6.3.1

Find values of x_1 and x_2 that maximize

$$z = x_1 + 3x_2$$

subject to

$$2x_1 + 3x_2 \leq 24$$
$$x_1 - x_2 \leq 7$$
$$x_2 \leq 6$$
$$x_1 \geq 0$$
$$x_2 \geq 0$$

Solution

In Figure 6.3.3 we have drawn the feasible region of this problem. Since it is bounded, the maximum value of z is attained at one of the five extreme points. The values of the objective function at the five extreme points are given in the following table.
From this table, the maximum value of z is 21, which is attained at $x_1 = 3$ and $x_2 = 6$. ◆

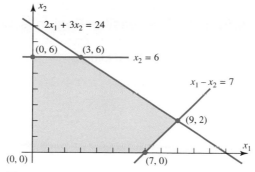

Figure 6.3.3

Extreme Point (x_1, x_2)	Value of $z = x_1 + 3x_2$
(0, 6)	18
(3, 6)	21
(9, 2)	15
(7, 0)	7
(0, 0)	0

EXAMPLE 6 Using Theorem 6.3.1

Find values of x_1 and x_2 that maximize

$$z = 4x_1 + 6x_2$$

subject to

$$2x_1 + 3x_2 \leq 24$$
$$x_1 - x_2 \leq 7$$
$$x_2 \leq 6$$
$$x_1 \geq 0$$
$$x_2 \geq 0$$

Solution

The constraints in this problem are identical to the constraints in Example 5, so the feasible region of this problem is also given by Figure 6.3.3. The values of the objective function at the extreme points are given in the following table.

Extreme Point (x_1, x_2)	Value of $z = 4x_1 + 6x_2$
(0, 6)	36
(3, 6)	48
(9, 2)	48
(7, 0)	28
(0, 0)	0

We see that the objective function attains a maximum value of 48 at two adjacent extreme points, (3, 6) and (9, 2). This shows that an optimal solution to a linear programming problem need not be unique. As we ask the reader to show in Exercise 10, if the objective function has the same value at two adjacent extreme points, it has the same value at all points on the straight line boundary segment connecting the two extreme points. Thus, in this example the maximum value of z is attained at all points on the straight line segment connecting the extreme points (3, 6) and (9, 2). ◆

EXAMPLE 7   The Feasible Region Is a Line Segment

Find values of x_1 and x_2 that minimize

$$z = 2x_1 - x_2$$

subject to

$$2x_1 + 3x_2 = 12$$
$$2x_1 - 3x_2 \geq 0$$
$$x_1 \geq 0$$
$$x_2 \geq 0$$

Solution

In Figure 6.3.4 we have drawn the feasible region of this problem. Because one of the constraints is an equality constraint, the feasible region is a straight line segment with two extreme points. The values of z at the two extreme points are given in the following table.

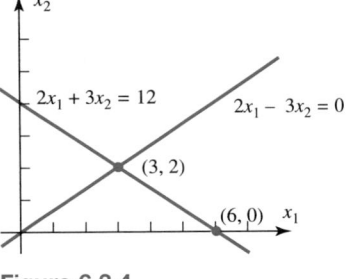

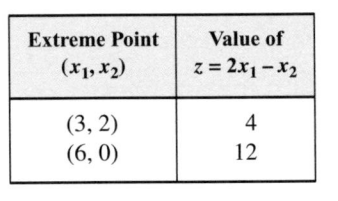

Extreme Point (x_1, x_2)	Value of $z = 2x_1 - x_2$
(3, 2)	4
(6, 0)	12

Figure 6.3.4

The minimum value of z is thus 4 and is attained at $x_1 = 3$ and $x_2 = 2$. ◆

EXAMPLE 8   Using Theorem 6.3.1

Find values of x_1 and x_2 that maximize

$$z = 2x_1 + 5x_2$$

subject to

$$2x_1 + x_2 \geq 8$$
$$-4x_1 + x_2 \leq 2$$
$$2x_1 - 3x_2 \leq 0$$
$$x_1 \geq 0$$
$$x_2 \geq 0$$

Solution

The feasible region of this linear programming problem is illustrated in Figure 6.3.5. Since it is unbounded, we are not assured by Theorem 6.3.1 that the objective function attains a maximum value. In fact, it is easily seen that since the feasible region contains points for which both x_1 and x_2 are arbitrarily large and positive, the objective function

$$z = 2x_1 + 5x_2$$

can be made arbitrarily large and positive. This problem has no optimal solution. Instead, we say the problem has an ***unbounded solution***. ◆

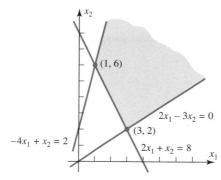

Figure 6.3.5

EXAMPLE 9 Using Theorem 6.3.1

Find values of x_1 and x_2 that maximize

$$z = -5x_1 + x_2$$

subject to

$$2x_1 + x_2 \geq 8$$
$$-4x_1 + x_2 \leq 2$$
$$2x_1 - 3x_2 \leq 0$$
$$x_1 \geq 0$$
$$x_2 \geq 0$$

Solution

The above constraints are the same as those in Example 8, so the feasible region of this problem is also given by Figure 6.3.5. In Exercise 11 we ask the reader to show that the objective function of this problem attains a maximum within the feasible region. By Theorem 6.3.1, this maximum must be attained at an extreme point. The values of z at the two extreme points of the feasible region are given in the following table.

Extreme Point (x_1, x_2)	Value of $z = -5x_1 + x_2$
(1, 6)	1
(3, 2)	−13

The maximum value of z is thus 1 and is attained at the extreme point $x_1 = 1$, $x_2 = 6$. ◆

EXAMPLE 10   Inconsistent Constraints

Find values of x_1 and x_2 that minimize

$$z = 3x_1 - 8x_2$$

subject to

$$2x_1 - x_2 \leq 4$$
$$3x_1 + 11x_2 \leq 33$$
$$3x_1 + 4x_2 \geq 24$$
$$x_1 \geq 0$$
$$x_2 \geq 0$$

Solution

As can be seen from Figure 6.3.6, the intersection of the five half-planes defined by the five constraints is empty. This linear programming problem has no feasible solutions since the constraints are inconsistent. ◆

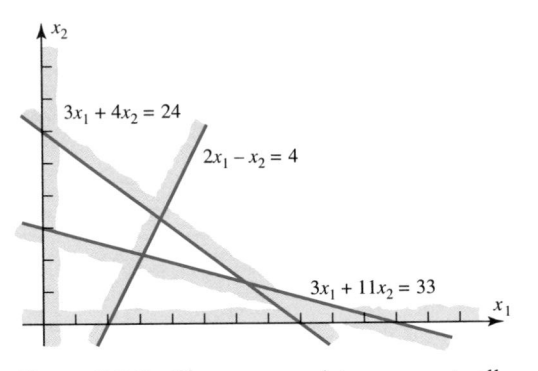

Figure 6.3.6   There are no points common to all five shaded half-planes.

EXERCISE SET 6.3

1. Find values of x_1 and x_2 that maximize

$$z = 3x_1 + 2x_2$$

subject to

$$2x_1 + 3x_2 \leq 6$$
$$2x_1 - x_2 \geq 0$$
$$x_1 \leq 2$$
$$x_2 \leq 1$$
$$x_1 \geq 0$$
$$x_2 \geq 0$$

2. Find values of x_1 and x_2 that minimize

$$z = 3x_1 - 5x_2$$

subject to

$$2x_1 - x_2 \leq -2$$
$$4x_1 - x_2 \geq 0$$

$$x_2 \leq 3$$
$$x_1 \geq 0$$
$$x_2 \geq 0$$

3. Find values of x_1 and x_2 that minimize

$$z = -3x_1 + 2x_2$$

subject to

$$3x_1 - x_2 \geq -5$$
$$-x_1 + x_2 \geq 1$$
$$2x_1 + 4x_2 \geq 12$$
$$x_1 \geq 0$$
$$x_2 \geq 0$$

4. Solve the linear programming problem posed in Example 2.

5. Solve the linear programming problem posed in Example 3.

6. In Example 5 the constraint $x_1 - x_2 \leq 7$ is said to be *nonbinding* because it can be removed from the problem without affecting the solution. Likewise, the constraint $x_2 \leq 6$ is said to be *binding* because removing it will change the solution.

   (a) Which of the remaining constraints are nonbinding and which are binding?

   (b) For what values of the righthand side of the nonbinding constraint $x_1 - x_2 \leq 7$ will this constraint become binding? For what values will the resulting feasible set be empty?

   (c) For what values of the righthand side of the binding constraints $x_2 \leq 6$ will this constraint become nonbinding? For what values will the resulting feasible set be empty?

7. A trucking firm ships the containers of two companies, A and B. Each container from company A weighs 40 pounds and is 2 cubic feet in volume. Each container from company B weighs 50 pounds and is 3 cubic feet in volume. The trucking firm charges company A $2.20 for each container shipped and charges company B $3.00 for each container shipped. If one of the firm's trucks cannot carry more than 37,000 pounds and cannot hold more than 2000 cubic feet, how many containers from companies A and B should a truck carry to maximize the shipping charges?

8. Repeat Exercise 7 if the trucking firm raises its price for shipping a container from company A to $2.50.

9. A manufacturer produces sacks of chicken feed from two ingredients, A and B. Each sack is to contain at least 10 ounces of nutrient N_1, at least 8 ounces of nutrient N_2, and at least 12 ounces of nutrient N_3. Each pound of ingredient A contains 2 ounces of nutrient N_1, 2 ounces of nutrient N_2, and 6 ounces of nutrient N_3. Each pound of ingredient B contains 5 ounces of nutrient N_1, 3 ounces of nutrient N_2, and 4 ounces of nutrient N_3. If ingredient A costs 8 cents per pound and ingredient B costs 9 cents per pound, how much of each ingredient should the manufacturer use in each sack of feed to minimize his costs?

10. If the objective function of a linear programming problem has the same value at two adjacent extreme points, show that it has the same value at all points on the straight line segment connecting the two extreme points.

   Hint If (x_1', x_2') and (x_1'', x_2'') are any two points in the plane, a point (x_1, x_2) lies on the straight line segment connecting them if

$$x_1 = tx_1' + (1 - t)x_1''$$

and

$$x_2 = tx_2' + (1 - t)x_2''$$

where t is a number in the interval $[0, 1]$.

11. Show that the objective function in Example 10 attains a maximum value in the feasible set.

   Hint Examine the level curves of the objective function.

SECTION 6.3
Technology Exercises

The following exercises are designed to be solved using a technology utility. Typically, this will be MATLAB, *Mathematica*, Maple, Derive, or Mathcad, but it may also be some other type of linear algebra software or a scientific calculator with some linear algebra capabilities. For each exercise you will need to read the relevant documentation for the particular utility you are using. The goal of these exercises is to provide you with a basic proficiency with your technology utility. Once you have mastered the techniques in these exercises, you will be able to use your technology utility to solve many of the problems in the regular exercise sets.

T1. Consider the feasible region consisting of $0 \leq x$, $0 \leq y$ along with the set of inequalities

$$x \cos \left(\frac{(2k+1)\pi}{4n} \right) + y \sin \left(\frac{(2k+1)\pi}{4n} \right) \leq \cos \left(\frac{\pi}{4n} \right)$$

for $k = 0, 1, 2, \ldots, n-1$. Maximize the objective function

$$z = 3x + 4y$$

assuming that (a) $n = 1$, (b) $n = 2$, (c) $n = 3$, (d) $n = 4$, (e) $n = 5$, (f) $n = 6$, (g) $n = 7$, (h) $n = 8$, (i) $n = 9$, (j) $n = 10$, and (k) $n = 11$. (l) Next, maximize this objective function using the nonlinear feasible region, $0 \leq x$, $0 \leq y$, and

$$x^2 + y^2 \leq 1$$

(m) Let the results of parts (a) through (k) begin a sequence of values for $z_{\max}$. Do these values approach the value determined in part (l)? Explain.

T2. Repeat Exercise T1 using the objective function $z = x + y$.

6.4
THE SIMPLEX METHOD

In this section, we describe a procedure involving matrix algebra for maximizing a linear expression of two or more variables subject to a set of linear constraints.

PREREQUISITES: Linear Systems

All linear optimizing (also known as linear programming) problems can be solved using an algorithm called the simplex method. This procedure essentially uses matrices and elementary row operations. Its applications in everyday life—transportation logistics, production planning to name just two—are as frequent as they are varied.

Standard Maximizing Problem

In this text, we shall focus our attention on the standard linear programming problem which consists of **maximizing** a linear function that is accompanied by constraints whose inequalities are of the "≤" type. In addition, all variables must be assumed nonnegative.

Let us begin with an example, which we shall use to introduce the vocabulary and steps associated to the simplex algorithm.

EXAMPLE 1   Maximizing a Two-Variable Problem

Find the values of variables x and y that maximize the objective function

$$P = 4x + 5y$$

subject to

$$x + 3y \le 24$$
$$x - y \le 4$$
$$x \le 6$$
$$x \le 0$$
$$y \ge 0$$

The following graph illustrates this problem's feasible region. We shall refer to it frequently to demonstrate the effect of the simplex method's steps.

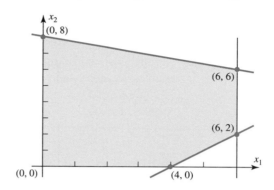

Solution

Step 1. **Converting the Problem into Standard Dorm**

Convert all constraints into equations by adding a "slack" variable to their left-hand side.

Rewrite the objective function with all variables on the left-hand side.

constraint 1:	$x + 3y \le 24$	$\Rightarrow$	$x + 3y + s_1 = 24$
constraint 2:	$x - y \le 4$	$\Rightarrow$	$x - y + s_2 = 4$
constraint 3:	$x \le 6$	$\Rightarrow$	$x + s_3 = 6$
objective:	$P = 4x + 5y$	$\Rightarrow$	$-4x - 5y + P = 0$

The new variables s_1, s_2, and s_3 are called **slack variables**. They are assumed to be nonnegative and are added to the left-hand side of each constraint to create equations rather than inequalities.

For instance, the coordinate point $(6, 2)$ of the feasible region satisfies the constraint $x + 3y \le 24$ since $6 + 3(2) = 12 \le 24$. There is, however, a difference of 12 between the two sides of the inequality. Point $(6, 2)$ would satisfy the equation $x + 3y + x_1 = 24$ were we to let the slack variable $s_1 = 12$.

Coordinate point $(6, 6)$ also satisfies the inequality $x + 3y \le 24$ since $6 + 3(6) = 24 \le 24$. This time, notice that there is no difference between the two sides of the inequality. Point $(6, 6)$ would therefore satisfy the equation $x + 3y + s_1 = 24$ were we to let the slack variable $s_1 = 0$.

Including slack variables seems artificial at first but keep in mind that we have never performed matrix operations involving inequalities. Matrices are designed to represent equations, which explains the necessity to "create" them ourselves in this case.

Step 2. **Set Up the Simplex Matrix**

Rewrite all equations (constraints and objective function) in their *standard* form into a matrix, placing the objective function in the last row. As always, the matrix will keep track only of each variable's coefficients.

$$
\begin{array}{cccccc}
x & y & s_1 & s_2 & s_3 & P \\
\end{array}
$$

$$
\left[
\begin{array}{cccccc|c}
1 & 3 & 1 & 0 & 0 & 0 & 24 \\
1 & -1 & 0 & 1 & 0 & 0 & 4 \\
1 & 0 & 0 & 0 & 1 & 0 & 6 \\
\hline
-4 & -5 & 0 & 0 & 0 & 1 & 0 \\
\end{array}
\right]
$$

Let us now introduce some vocabulary.

We shall refer to as **basic** all variables whose columns are filled with a single 1 and with zeros everywhere else. One will notice these variables are somewhat comparable to "leading variables." All other variables are considered **nonbasic**.

From our previous simplex matrix, we can deduce that s_1, s_2, s_3, and P are all basic variables, leaving x and y as nonbasic variables.

$$
\begin{array}{cccccc}
x & y & s_1 & s_2 & s_3 & P \\
\end{array}
$$

$$
\left[
\begin{array}{cccccc|c}
1 & 3 & ① & 0 & 0 & 0 & 24 \\
1 & -1 & 0 & ① & 0 & 0 & 4 \\
1 & 0 & 0 & 0 & ① & 0 & 6 \\
\hline
-4 & -5 & 0 & 0 & 0 & ① & 0 \\
\end{array}
\right]
$$

By convention, we must assign to nonbasic variables the value 0. For instance, since the above matrix depicts x and y as being nonbasic variables, we must consider that $x = 0$, and that $y = 0$. Once this has been established, the other variables' values can be read directly from the matrix:

Row 1: $s_1 = 24$ Row 3: $s_3 = 6$

Row 2: $s_2 = 4$ Row 4: $P = 0$

Graphically, we are currently located at corner ($x = 0$, $y = 0$) of the feasible region. It makes sense that $P = 0$ since the objective function is defined as $4x + 5y$. One can guess that this is certainly not the maximum value function P can attain.

So, how does one know if and when a maximum has been reached?

When the objective function has been maximized, no negative numbers should appear at the bottom row of the simplex matrix. One can observe that "-4" and "-5" appear in the initial matrix of our example. Clearly, the maximum solution has not yet been reached. To accomplish this, operations must be done to the matrix.

Step 3. **Simplex Operations (Pivoting)**

Identify the column with the largest negative number in the bottom row.

Considering only strictly positive numbers, identify the entry in the chosen column whose ratio with the constants in the rightmost column is smallest. The ratio is the result of the constant in the rightmost column divided by the corresponding entry in the chosen column.

Transform this entry into a "pivoting 1"; use it to make all numbers above and below it zeros.

In our example, "-5" is the largest negative number in the bottom row and is found in the y-column. Only one number, that is the circled 3, in this column is neither zero nor negative. Its corresponding ratio with the right-hand column is 8 (no comparison need be done with other ratios since 3 is the only positive number in the y-column).

$$
\begin{array}{cccccc}
x & y & s_1 & s_2 & s_3 & P \\
\end{array}
$$

$$
\left[\begin{array}{cccccc|c}
1 & ③ & 1 & 0 & 0 & 0 & 24 \\
1 & -1 & 0 & 1 & 0 & 0 & 4 \\
1 & 0 & 0 & 0 & 1 & 0 & 6 \\
\hline
-4 & -5 & 0 & 0 & 0 & 1 & 0
\end{array}\right]
\quad \longleftarrow \quad \textit{ratio} \;\; \frac{24}{3} = 8
$$

We must transform the 3 into a pivoting 1 by multiplying row 1 by $\frac{1}{3}$.

$$
\begin{array}{cccccc}
x & y & s_1 & s_2 & s_3 & P \\
\end{array}
$$

$$
\left[\begin{array}{cccccc|c}
\frac{1}{3} & 1 & \frac{1}{3} & 0 & 0 & 0 & 8 \\
1 & -1 & 0 & 1 & 0 & 0 & 4 \\
1 & 0 & 0 & 0 & 1 & 0 & 6 \\
\hline
-4 & -5 & 0 & 0 & 0 & 1 & 0
\end{array}\right]
$$

Row operations are then necessary to replace the remaining entries of column y by zeros. Note that these operations must involve using the new "pivoting 1." After having added row 1 to row 2, and 5 times row 1 to row 4, we obtain the following matrix:

$$
\begin{array}{cccccc}
x & y & s_1 & s_2 & s_3 & P \\
\end{array}
$$

$$
\left[\begin{array}{cccccc|c}
\frac{1}{3} & ① & \frac{1}{3} & 0 & 0 & 0 & 8 \\
\frac{4}{3} & 0 & \frac{1}{3} & ① & 0 & 0 & 12 \\
1 & 0 & 0 & 0 & ① & 0 & 6 \\
\hline
-\frac{7}{3} & 0 & \frac{5}{3} & 0 & 0 & ① & 40
\end{array}\right]
$$

Before we move on, one can observe that the basic variables are now y, s_2, s_3 and P. Variables x and s_1 are nonbasic. Recall that nonbasic variables are systematically attributed the value 0. As a result, the current value of each basic variable is given by

Row 1: $\quad y = 8$ $\qquad$ Row 3: $\quad s_3 = 6$

Row 2: $\quad s_2 = 12$ $\qquad$ Row 4: $\quad P = 40$

We are now located at corner $(x = 0, y = 8)$ of the feasible region.

Our objective, which is to maximize function P, has not yet been reached since a negative still appears at the bottom row. More operations will need to be done. We shall repeat step 3 until no more negatives appear in the last row.

Where will we now insert a new leading 1?

The column containing the largest (and only) negative number in its bottom row is the x-column. From this column, we then must identify the positive entry whose ratio with the rightmost constant is smallest.

$$
\begin{array}{c}
\begin{array}{cccccc} x & y & s_1 & s_2 & s_3 & P \end{array} \\
\left[
\begin{array}{cccccc|c}
\frac{1}{3} & 1 & \frac{1}{3} & 0 & 0 & 0 & 8 \\
\frac{4}{3} & 0 & \frac{1}{3} & 1 & 0 & 0 & 12 \\
\boxed{①} & 0 & 0 & 0 & 1 & 0 & 6 \\
\hline
-\frac{7}{3} & 0 & \frac{5}{3} & 0 & 0 & 1 & 40
\end{array}
\right]
\end{array}
\qquad
\begin{array}{l}
\textit{ratio} \\[4pt]
8 \div \frac{1}{3} = 24 \\[6pt]
12 \div \frac{4}{3} = 9 \\[6pt]
\longleftarrow \quad 6 \div 1 = 6
\end{array}
$$

Since the boxed entry is already a 1, we can immediately begin creating zeros above and below it. Adding appropriate multiples of row 3 to the others, we obtain

$$
\begin{array}{c}
\begin{array}{cccccc} x & y & s_1 & s_2 & s_3 & P \end{array} \\
\left[
\begin{array}{cccccc|c}
0 & ① & \frac{1}{3} & 0 & -\frac{1}{3} & 0 & 6 \\
0 & 0 & \frac{1}{3} & ① & -\frac{4}{3} & 0 & 4 \\
① & 0 & 0 & 0 & 1 & 0 & 6 \\
\hline
0 & 0 & \frac{5}{3} & 0 & \frac{7}{3} & ① & 54
\end{array}
\right]
\end{array}
$$

This time, no negatives appear in the last row. We have therefore reached the optimal solution.

What solution does this final matrix table represent?

Observe that x, y, s_2 and P are basic variables. The others, s_1 and s_3, are nonbasic and hence equal 0. By inspecting each row, one can identify the values of the basic variables:

Row 1: $y = 6$ Row 3: $x = 6$

Row 2: $s_2 = 4$ Row 4: $P = 54$

Hence, we are now located at corner ($x = 6$, $y = 6$) of the feasible region. One will observe that, essentially, the effect of our simplex operations was to move from one corner of the feasible region to the next, until a maximum was finally reached.

The solution we have obtained can be confirmed through a quick verification. If $x = 6$ and $y = 6$, we should obtain

$P = 4(6) + 5(6) = 24 + 30 = 54$

Constraint 1: $(6) + 3(6) = 24 \leq 24$ (no slack, so $s_1 = 0$)

Constraint 2: $(6) - (6) = 0 \leq 4$ (difference of 4, so $s_2 = 4$)

Constraint 3: $(6) \leq 6$ (no slack, so $s_3 = 0$)   ◆

The following example will underline one of the advantages of the simplex method; it can easily be generalized to problems containing more than two variables. This cannot be said about the graphical method for obvious reasons.

EXAMPLE 2   Maximizing a Three-Variable Problem

Find the values of variables x, y and z that maximize the objective function

$$P = 9x + 6y + 2z$$

subject to

$$x + y + z \leq 100$$
$$3x + 2y + 5z \leq 120$$
$$2x + z \leq 60$$
$$(x, y, z \geq 0)$$

Solution

First, we must convert the constraints and objective function into *standard* form

constraint 1:	$x + y + z \leq 100$	$\Rightarrow$ $x + y + z + s_1 = 100$
constraint 2:	$3x + 2y + 5z \leq 120$	$\Rightarrow$ $3x + 2y + 5z + s_2 = 120$
constraint 3:	$2x + z \leq 60$	$\Rightarrow$ $2x + z + s_3 = 60$
objective:	$P = 9x + 6y + 2z$	$\Rightarrow$ $-9x - 6y - 2z + P = 0$

The initial simplex matrix is obtained from the above equations.

$$
\begin{array}{ccccccc}
x & y & z & s_1 & s_2 & s_3 & P \\
\end{array}
$$
$$
\left[
\begin{array}{ccccccc|c}
1 & 1 & 1 & 1 & 0 & 0 & 0 & 100 \\
3 & 2 & 5 & 0 & 1 & 0 & 0 & 120 \\
2 & 0 & 1 & 0 & 0 & 1 & 0 & 60 \\
\hline
-9 & -6 & -2 & 0 & 0 & 0 & 1 & 0 \\
\end{array}
\right]
$$

Because negative numbers appear in the bottom row, we know a maximum value has not yet been attained. We must therefore perform simplex operations in order to improve our current situation. The x-column has the largest negative number in its bottom row.

$$
\begin{array}{ccccccc}
x & y & z & s_1 & s_2 & s_3 & P \\
\end{array}
$$
$$
\left[
\begin{array}{ccccccc|c}
1 & 1 & 1 & 1 & 0 & 0 & 0 & 100 \\
3 & 2 & 5 & 0 & 1 & 0 & 0 & 120 \\
② & 0 & 1 & 0 & 0 & 1 & 0 & 60 \\
\hline
-9 & -6 & -2 & 0 & 0 & 0 & 1 & 0 \\
\end{array}
\right]
$$

ratio
100
40
30

We have found that the circled 2 is to be turned into a pivoting 1, which can be accomplished by multiplying row 3 by $\frac{1}{2}$. Multiples of row 3 can be then be used to replace the remaining entries of the x-column by zeros. The result of these operations are shown in the following matrix:

$$
\begin{array}{ccccccc}
x & y & z & s_1 & s_2 & s_3 & P \\
\end{array}
$$
$$
\left[
\begin{array}{ccccccc|c}
0 & 1 & \frac{1}{2} & 1 & 0 & -\frac{1}{2} & 0 & 70 \\
0 & ② & \frac{7}{2} & 0 & 1 & -\frac{3}{2} & 0 & 30 \\
1 & 0 & \frac{1}{2} & 0 & 0 & \frac{1}{2} & 0 & 30 \\
\hline
0 & -6 & \frac{5}{2} & 0 & 0 & \frac{9}{2} & 1 & 270 \\
\end{array}
\right]
$$

ratio
70
15

A negative number still appears at the bottom of the y-column, which means the objective function can still be increased. By comparing ratios, it is obvious we must now make the

2 appearing in the y-column our new pivoting 1. Elementary row operations involving this pivoting 1 are used to create zeros in the remaining entries of column y.

$$
\begin{array}{ccccccc}
x & y & z & s_1 & s_2 & s_3 & P \\
\end{array}
$$

$$
\left[
\begin{array}{ccccccc|c}
0 & 0 & -\frac{5}{4} & 1 & -\frac{1}{2} & \frac{1}{4} & 0 & 55 \\
0 & 1 & \frac{7}{4} & 0 & \frac{1}{2} & -\frac{3}{4} & 0 & 15 \\
1 & 0 & \frac{1}{2} & 0 & 0 & \frac{1}{2} & 0 & 30 \\
\hline
0 & 0 & 13 & 0 & 3 & 0 & 1 & 360
\end{array}
\right]
$$

The fact no more negative values appear in the bottom row indicates function P has reached its maximum value. According to this final matrix, $z = 0$, $s_2 = 0$, and $s_3 = 0$ since they are nonbasic variables. The remaining variables' values are

$$x = 30, \quad y = 15, \quad s_1 = 55, \quad P = 360 \quad \blacklozenge$$

Post-Optimal Analysis

Once a problem has been solved by using the simplex algorithm, it is of utmost importance and utility to be able to interpret the meaning of the final matrix. This table displays the optimal solution, but it can also be used to

- determine how this solution would vary if constraints were to change;
- identify which constraints are binding or nonbinding;
- detect the existence of other solutions.

Interpreting the Slack Variables' Columns

Let us take a another look at the final matrix obtained in Example 2.

$$
\begin{array}{ccccccc}
x & y & z & s_1 & s_2 & s_3 & P \\
\end{array}
$$

$$
\left[
\begin{array}{ccccccc|c}
0 & 0 & -\frac{5}{4} & 1 & -\frac{1}{2} & \frac{1}{4} & 0 & 55 \\
0 & 1 & \frac{7}{4} & 0 & \frac{1}{2} & -\frac{3}{4} & 0 & 15 \\
1 & 0 & \frac{1}{2} & 0 & 0 & \frac{1}{2} & 0 & 30 \\
\hline
0 & 0 & 13 & 0 & 3 & 0 & 1 & 360
\end{array}
\right]
$$

Useless in appearance, the entries in the slack variables' columns actually provide invaluable information. Every number indicates how the basic variable in the corresponding row will be affected when the right-hand side of that constraint is increased by one unit.

This means, for instance, that if the right-hand side of contraint 2 were to be increased by one unit:

- the basic variable in row 1 (s_1) would decrease by $\frac{1}{2}$;
- the basic variable in row 2 (y) would increase by $\frac{1}{2}$;
- the basic variable in row 3 (x) would be unaffected;
- the basic variable in row 4 (P) would increase by 3.

Unless nonnegativity restrictions are not observed, all nonbasic variables remain nonbasic.

EXAMPLE 3   Effect of a Changing Constraint

Let us assume that the values of x, y, and z that maximize the objective function

$$P = 9x + 6y + 2z$$

subject to the constraints

$$x + y + z \leq 100$$
$$3x + 2y + 5z \leq 120$$
$$2x + z \leq 60$$
$$(x, y, z \geq 0)$$

are $x = 30$, $y = 15$, $z = 0$, and that the following is the final matrix obtained after using the simplex method:

$$
\begin{array}{ccccccc}
x & y & z & s_1 & s_2 & s_3 & P \\
\end{array}
$$

$$
\left[
\begin{array}{ccccccc|c}
0 & 0 & -\frac{5}{4} & 1 & -\frac{1}{2} & \frac{1}{4} & 0 & 55 \\
0 & 1 & \frac{7}{4} & 0 & \frac{1}{2} & -\frac{3}{4} & 0 & 15 \\
1 & 0 & \frac{1}{2} & 0 & 0 & \frac{1}{2} & 0 & 30 \\
\hline
0 & 0 & 13 & 0 & 3 & 0 & 1 & 360 \\
\end{array}
\right]
$$

Find the new optimal solution if constraint 2 were changed from $3x + 2y + 5z \leq 120$ to $3x + 2y + 5z \leq 124$.

Solution

We shall denote the change to the right-hand side of constraint 2 by $\Delta_2 = 4$. Recall that the entries in column s_2 indicate how each basic variable in the corresponding row will be affected when constraint 2 is increased by one unit. Of course, the effect will be 4 times greater in our current situation since the right hand of the latter has undergone an increase of 4 units.

Rather than looking at each basic variable's change individually, it is convenient to use vector forms, as shown below, to determine the new optimal solution.

$$
\underbrace{
\begin{bmatrix}
s_1 \\
y \\
x \\
P
\end{bmatrix}
=
\begin{bmatrix}
55 \\
15 \\
30 \\
360
\end{bmatrix}
}_{\substack{\text{original} \\ \text{solution}}}
+ \Delta_2
\underbrace{
\begin{bmatrix}
-\frac{1}{2} \\
\frac{1}{2} \\
0 \\
3
\end{bmatrix}
}_{\substack{\text{column} \\ s_2}}
=
\begin{bmatrix}
55 \\
15 \\
30 \\
360
\end{bmatrix}
+ 4
\begin{bmatrix}
-\frac{1}{2} \\
\frac{1}{2} \\
0 \\
3
\end{bmatrix}
=
\underbrace{
\begin{bmatrix}
53 \\
17 \\
30 \\
372
\end{bmatrix}
}_{\substack{\text{new} \\ \text{optimal} \\ \text{solution}}}
$$

All original basic variables have remained nonnegative. All variables that were previously nonbasic remain as such (thus $z = 0$). ◆

If more than one constraint is modified, the effects of each change are added. Changing many constraints therefore requires practically the same amount of work as changing only one.

EXAMPLE 4   Effect of Changing Multiple Constraints

Consider once again the optimizing problem discussed in Example 2. What would be the new optimal solution if all three constraints were changed as follows:

$$x + y + z \leq 100 \xrightarrow{\Delta_1 = -10} x + y + z \leq 90$$
$$3x + 2y + 5z \leq 120 \xrightarrow{\Delta_2 = 8} 3x + 2y + 5z \leq 128$$
$$2x + z \leq 60 \xrightarrow{\Delta_3 = -4} 2x + z \leq 56$$

Solution

The s_1, s_2, and s_3 columns of the final simplex table will be needed in this case since all their corresponding constraints are undergoing changes. The total variation to the original solution will be obtained by adding the impact caused by each constraint's change, that is

$$
\begin{bmatrix} s_1 \\ y \\ x \\ P \end{bmatrix} = \begin{bmatrix} 55 \\ 15 \\ 30 \\ 360 \end{bmatrix} + \Delta_1 \begin{bmatrix} 1 \\ 0 \\ 0 \\ 0 \end{bmatrix} + \Delta_2 \begin{bmatrix} -\frac{1}{2} \\ \frac{1}{2} \\ 0 \\ 3 \end{bmatrix} + \Delta_3 \begin{bmatrix} \frac{1}{4} \\ -\frac{3}{4} \\ \frac{1}{2} \\ 0 \end{bmatrix}
$$

Replacing the Δ's by their respective values, we obtain

$$
\begin{bmatrix} s_1 \\ y \\ x \\ P \end{bmatrix} = \begin{bmatrix} 55 \\ 15 \\ 30 \\ 360 \end{bmatrix} + (-10) \begin{bmatrix} 1 \\ 0 \\ 0 \\ 0 \end{bmatrix} + (8) \begin{bmatrix} -\frac{1}{2} \\ \frac{1}{2} \\ 0 \\ 3 \end{bmatrix} + (-4) \begin{bmatrix} \frac{1}{4} \\ -\frac{3}{4} \\ \frac{1}{2} \\ 0 \end{bmatrix} = \begin{bmatrix} 40 \\ 22 \\ 28 \\ 384 \end{bmatrix}
$$

Note that all nonbasic variables (z, s_2, and s_3) remain nonbasic. ◆

Identifying Binding Constraints

A constraint is binding when its limiting value is attained at the optimal solution. In other words, no slack remains between the left- and right-hand side of this particular constraint.

EXAMPLE 5   Finding Binding Constraints

Let us consider the final matrix obtained from Example 2.

$$
\begin{array}{ccccccc}
x & y & z & s_1 & s_2 & s_3 & P \\
\end{array}
$$
$$
\left[\begin{array}{ccccccc|c}
0 & 0 & -\frac{5}{4} & 1 & -\frac{1}{2} & \frac{1}{4} & 0 & 55 \\
0 & 1 & \frac{7}{4} & 0 & \frac{1}{2} & -\frac{3}{4} & 0 & 15 \\
1 & 0 & \frac{1}{2} & 0 & 0 & \frac{1}{2} & 0 & 30 \\
\hline
0 & 0 & 13 & 0 & 3 & 0 & 1 & 360
\end{array} \right]
$$

The optimal solution depicts s_2 and s_3 as nonbasic variables. This means

$$s_2 = 0 \quad \rightarrow \quad \text{constraint 2 is binding}$$
$$s_3 = 0 \quad \rightarrow \quad \text{constraint 3 is binding}$$

However, one will notice that s_1 is basic. More specifically, $s_1 = 55$. This means the optimal solution does not occur at the boundary value of constraint 1. It is therefore a nonbinding constraint. In an economic context, this would mean it is not advantageous to use all the resources represented by constraint 1 in order to maximize the objective function (profit, for example). ◆

Existence of Multiple Solutions

One can detect the existence of multiple solutions of a linear optimizing problem if the objective function reaches its maximum value at different points of the feasible region. In terms of the simplex matrix, this means that basic variables can be changed without altering the objective's maximum value. The following example will illustrate how one can detect the existence of multiple solutions, and how the latter can be found.

EXAMPLE 6

According to the final table of Example 2, shown below, the objective function (P) is maximized when $x = 30$, $y = 15$, $z = 0$.

$$\begin{array}{ccccccc|c}
x & y & z & s_1 & s_2 & s_3 & P & \\
\hline
0 & 0 & -\frac{5}{4} & 1 & -\frac{1}{2} & \frac{1}{4} & 0 & 55 \\
0 & 1 & \frac{7}{4} & 0 & \frac{1}{2} & -\frac{3}{4} & 0 & 15 \\
1 & 0 & \frac{1}{2} & 0 & 0 & \frac{1}{2} & 0 & 30 \\
\hline
0 & 0 & 13 & 0 & 3 & 0 & 1 & 360
\end{array}$$

Explain why there are multiple optimal solutions to this linear programming problem, and find one alternative solution.

Solution

According to the final table's last row, $0x + 0y + 13z + 0s_1 + 3s_2 + 0s_3 + 1P = 360$.
Isolating P from the previous equation, we obtain

$$P = 360 - 0x - 0y - 13z - 0s_1 - 3s_2 - 0s_3$$

Variable z is nonbasic, and for good reason. If z were increased by 1 unit, the value of P would be reduced by 13. Since our objective is to maximize function P, having z as a basic variable would not be wise. Variable s_2 is also nonbasic. Function P would be reduced by 3 for every unit increase of s_2. It is therefore not profitable to make s_2 a basic variable. One will quickly conclude that nothing can be done to make the value of P any greater than it already is (360). However, although s_3 is nonbasic, the 0 coefficient at the bottom of its column indicates there will be no effect to the maximum if s_3's value were to be changed. In other words, s_3 could be made basic without diminishing the current value of P.

In short, the existence of multiple optimal solutions can be detected when one of the nonbasic variables has a "0" in its bottom row. An alternative optimal solution is obtained by making this variable basic through a regular pivot operation. We shall treat this 0 the same way we treated negative numbers in the bottom row earlier. By identifying the lowest ratio, one can identify the location of a new pivoting 1:

$$\begin{array}{ccccccc|c}
x & y & z & s_1 & s_2 & s_3 & P & \\
\hline
0 & 0 & -\frac{5}{4} & 1 & -\frac{1}{2} & \frac{1}{4} & 0 & 55 \\
0 & 1 & \frac{7}{4} & 0 & \frac{1}{2} & -\frac{3}{4} & 0 & 15 \\
1 & 0 & \frac{1}{2} & 0 & 0 & \boxed{\tfrac{1}{2}} & 0 & 30 \\
\hline
0 & 0 & 13 & 0 & 3 & 0 & 1 & 360
\end{array}$$

ratio
220

60

Multiplying row 3 by 2, the $\frac{1}{2}$ is transformed into a pivoting 1.

$$\begin{array}{ccccccc|c}
x & y & z & s_1 & s_2 & s_3 & P & \\
\hline
0 & 0 & -\frac{5}{4} & 1 & -\frac{1}{2} & \frac{1}{4} & 0 & 55 \\
0 & 1 & \frac{7}{4} & 0 & \frac{1}{2} & -\frac{3}{4} & 0 & 15 \\
2 & 0 & 1 & 0 & 0 & \boxed{1} & 0 & 60 \\
\hline
0 & 0 & 13 & 0 & 3 & 0 & 1 & 360
\end{array}$$

The reader can verify that appropriate row operations involving the new pivoting 1 reveals the following alternative final table:

$$
\begin{array}{ccccccc}
x & y & z & s_1 & s_2 & s_3 & P \\
\end{array}
$$

$$
\left[
\begin{array}{ccccccc|c}
-\frac{1}{2} & 0 & -\frac{3}{2} & 1 & -\frac{1}{2} & 0 & 0 & 40 \\
\frac{3}{2} & 1 & \frac{5}{2} & 0 & \frac{1}{2} & 0 & 0 & 60 \\
2 & 0 & 1 & 0 & 0 & 1 & 0 & 60 \\
\hline
0 & 0 & 13 & 0 & 3 & 0 & 1 & 360 \\
\end{array}
\right]
$$

According to the above table, the objective's maximum value of 360 is obtained when $x = 0$, $y = 60$, $z = 0$.

One can show that the objective function remains equal to 360 at any point of the segment joining the original solution ($x = 30$, $y = 15$, $z = 0$) and the one we have just obtained. In addition, all constraints are respected along this segment. ◆

EXERCISE SET 6.4

In Exercises 1–5, solve the given standard linear programming problem using the simplex method.

1. Maximize

$$P = 5x + 3y$$

Subject to the constraints

$$x + 2y \le 20$$
$$x - y \le 5$$
$$x \le 8$$
$$(x, y \ge 0)$$

2. Maximize

$$f = 3x + 2y$$

Subject to the constraints

$$x + 2y \le 16$$
$$x + y \le 10$$
$$2x + y \le 16$$
$$(x, y \ge 0)$$

3. Maximize

$$f = 5x + 2y$$

Subject to the constraints

$$x + 4y \le 36$$
$$x - y \le 6$$
$$x \le 8$$
$$(x, y \ge 0)$$

4. Maximize

$$P = 7x + 5y + 6z$$

Subject to the constraints

$$x + y - z \le 3$$
$$x + 2y + z \le 8$$
$$x + y \le 5$$
$$(x, y, z \ge 0)$$

5. Maximize

$$P = x + 2y + 5z + t$$

Subject to the constraints

$$x + 2y + z - t \leq 20$$
$$-x + y + z + t \leq 12$$
$$2x + y + z - t \leq 30$$
$$(x, y, z, t \geq 0)$$

6. Consider the linear programming problem described in Exercise 2.

    (a) Identify all constraints that are binding?

    (b) How many solutions does this problem have?

    (c) What are the values of x, y, and P that produce the optimal solution if the second constraint were changed to $x + y \leq 8$?

7. RiverFlo produces two types of River Kayaks. The *Coolit* model is designed for the "drifters," that is for leisurely promenades on the water. The *MoveOver* model is designed for the more adventurous kayakers. There are constraints to be observed during the production process. The company orders 500 m² of carbon fibre and 200 m² of fibreglass every week. The company disposes of a maximum of 240 man-hours (for example, 6 people who work 40 hours each) per week to complete its production. The *MoveOver* model requires more production time because of handcrafted elements.

    The following table describes the material composition as well as the required production time for each model.

Model	Carbon Fibre (m²)	Fibreglass (m²)	Production Time (hours)
Coolit	15	10	4
MoveOver	20	5	6

The cost of every square metre of carbon fibre is $6, whereas $4 is the cost for every square metre of fibreglass. Suppose all employees are paid at an hourly wage of $20. A *Coolit* kayak is sold at a price of $255. The *Moveover* kayak is sold for $300.

The objective is to determine how many kayaks of each model should be produced in order to maximize profits.

    (a) Define the variables for this problem.

    (b) Set up the objective function and the constraints in terms of these variables.

    (c) Solve the problem using the simplex method. What is the maximum profit the company can obtain?

    (d) Identify all binding constraints.

8. The following is the optimal table obtained after solving a linear programming problem through the simplex method.

$$
\begin{array}{ccccccc}
x & y & z & s_1 & s_2 & s_3 & P \\
\end{array}
$$

$$
\left[
\begin{array}{ccccccc|c}
1 & 0 & 2 & 0 & 1 & 1.5 & 0 & 24 \\
0 & 1 & 1 & 0 & -1 & 1.5 & 0 & 16 \\
0 & 0 & 1 & 1 & 0.5 & 2 & 0 & 6 \\
\hline
0 & 0 & 0 & 0 & 2 & 5 & 1 & 100 \\
\end{array}
\right]
$$

    (a) If constraint 3 were increased by 2 units, how will the optimal solution be affected?

    (b) Explain why a new solution cannot be obtained by post-optimal analysis if constraint 3 were reduced by 4.

(c) Explain how one knows there are many optimal solutions to the linear programming problem.

(d) If constraint 2 were increased by 6 and constraint 3 were decreased by 2 units, what values would x, y, z, and P take at the new optimal solution?

(e) The z variable is currently nonbasic, which means we have chosen to produce no units of product z. Find two other solutions with the same optimal profit but where z is basic.

9. Consider the following linear programming problem:

$$
\begin{aligned}
\text{Maximize} \qquad & P = 2x + 5y \\
\text{Subject to} \qquad & x + 2y \le 22 \\
& 2x + y \le 20 \\
& x \ge 2 \\
& y \ge 3
\end{aligned}
$$

The above is not a standard maximizing problem due to the form of the last two constraints. Standardize and solve the problem by replacing x by $u + 2$ and y by $v + 3$.

10. A small Montreal-based airline, Queeject, is providing flights to Canada's other major cities: Toronto, Calgary, and Vancouver. Daily profits are generated by computing the difference between the company's revenues (from ticket sales) and its costs (fuel, employees on board, meals provided, rent paid to airport, etc.). The company's daily profit is described as follows.

$$
P = 400t + 360c + 450v
$$

where the variables t, c, and v are

t: number of planes used for Montreal-Toronto round-trips

c: number of planes used for Montreal-Calgary round-trips

v: number of planes used for Montreal-Vancouver round-trips

The following table shows the staffing requirements for each destination, as well as the demand to be fulfilled.

	Pilots	On-board staff	Minimum number of flights per day
Toronto	2	4	3
Calgary	3	6	2
Vancouver	3	8	1

Queeject employs 56 pilots, has a total personnel of 120 attendants to provide service during the flights, and has a fleet of 18 airplanes at its disposal.

(a) How many planes should Queeject send to each city if it wishes to maximize its profits?

(b) Are there many solutions to this optimizing problem?

(c) The company wants to expand and considers buying 2 more planes and hiring 12 extra on-board attendants. How will the optimal solution be affected?

(d) According to the Queeject Pilots Union, the pilots are already overworked. In their opinion, the changes proposed in part (c) will worsen their working conditions. Is the union exaggerating?

SECTION 6.4

Technology Exercises

The following exercises are designed to be solved using a technology utility. Typically, this will be *Mathematica*, Maple, or Microsoft Excel (using its Solver add-in). For each exercise, you will need to read the relevant documentation for the particular utility you are using. The goal of these exercises is to provide you with a basic proficiency with your technology utility. Once you have mastered the techniques in these exercises, you will be able to use your technology utility to solve many of the problems in the regular exercise sets.

T1. A financial advisor must invest all of the $500,000 submitted to him by a client. There are 4 major investment sectors from which he may choose, all of which are presented in the table below along with their average yearly return.

Investment Sector	Average Return (μ)
Government bonds	4%
New technologies	15%
Natural resources	6%
International investing	8%

Based on return alone, one would invest everything in new technologies. However, it is an internal policy of the advisor's firm to place at least 30% of the total investment in the safer sectors, bonds, and natural resources. International investing is a tempting sector, but legislation imposes that it represent no more than 10% of one's portfolio.

Let $x_{GB}, x_{NT}, x_{NR}, x_{II}$ represent the amounts that will be invested in each sector. How should the financial advisor distribute the sum to maximize the profit function

$$P = 0.04x_{GB} + 0.15x_{NT} + 0.06x_{NR} + 0.08x_{II}$$

while respecting all constraints?

T2. In Exercise T1, the profit function depended only on the average return of each investment sector, without consideration for their possible variations. Recent experience has shown that certain sectors are much riskier than others, but can also be very rewarding. The table below shows each sector's standard deviation.

Investment Sector	Standard Deviation (σ)
Government bonds	0%
New technologies	18%
Natural resources	3%
International investing	6%

One will notice that the return on government bonds does not vary. In contrast, it would not be unusual for investments in new technologies to vary by 18% from their average return. An aggressive investor may be interested in this sector because of its high-reward possibility. A conservative investor may perceive this sector as very risky. In taking into account the different perceptions of clients, financial planners have modeled the following *utility* function

$$U = 0.04x_{GB} + 0.15x_{NT} + 0.06x_{NR} + 0.08x_{II}$$
$$+ k\left[(0x_{GB})^2 + (0.18x_{NT})^2 + (0.03x_{NR})^2 + (0.06x_{II})^2\right]$$

The value of k depends on the investor's tolerance to risk. The aggressive investor is attributed the value $k = 1$, whereas this parameter's value is -1 for the conservative investor. We use $k = 0$ for investors who are ambivalent to risk (this is the case you have treated in T1).

For aggressive and conservative investors, how should the financial advisor distribute the sum to maximize the utility function, while respecting all constraints?

6.5
MARKOV CHAINS

In this section we describe a general model of a system that changes from state to state. We then apply the model to several concrete problems.

> PREREQUISITES: Linear Systems
> Matrices
> Intuitive Understanding of Limits

A Markov Process

Suppose a physical or mathematical system undergoes a process of change such that at any moment it can occupy one of a finite number of states. For example, the weather in a certain city could be in one of three possible states: sunny, cloudy, or rainy. Or an individual could be in one of four possible emotional states: happy, sad, angry, or apprehensive. Suppose that such a system changes with time from one state to another and at scheduled times the state of the system is observed. If the state of the system at any observation cannot be predicted with certainty, but the probability that a given state occurs can be predicted by just knowing the state of the system at the preceding observation, then the process of change is called a *Markov chain* or *Markov process*.

> ### DEFINITION
>
> If a Markov chain has k possible states, which we label as $1, 2, \ldots, k$, then the probability that the system is in state i at any observation after it was in state j at the preceding observation is denoted by p_{ij} and is called the *transition probability* from state j to state i. The matrix $P = [p_{ij}]$ is called the *transition matrix of the Markov chain*.

For example, in a three-state Markov chain, the transition matrix has the form

Preceding State

$$
\begin{array}{ccc}
1 & 2 & 3
\end{array}
$$

$$
\begin{bmatrix}
p_{11} & p_{12} & p_{13} \\
p_{21} & p_{22} & p_{23} \\
p_{31} & p_{32} & p_{33}
\end{bmatrix}
\begin{array}{l}
1 \\
2 \quad \text{New State} \\
3
\end{array}
$$

In this matrix, p_{32} is the probability that the system will change from state 2 to state 3, p_{11} is the probability that the system will still be in state 1 if it was previously in state 1, and so forth.

EXAMPLE 1 Transition Matrix of the Markov Chain

A car rental agency has three rental locations, denoted by 1, 2, and 3. A customer may rent a car from any of the three locations and return the car to any of the three locations. The manager finds that customers return the cars to the various locations according to the following probabilities:

Rented from Location

$$
\begin{array}{ccc}
1 & 2 & 3
\end{array}
$$

$$
\begin{bmatrix}
.8 & .3 & .2 \\
.1 & .2 & .6 \\
.1 & .5 & .2
\end{bmatrix}
\begin{array}{l}
1 \\
2 \\
3
\end{array}
\begin{array}{l}
\textbf{Returned} \\
\textbf{to} \\
\textbf{Location}
\end{array}
$$

This matrix is the transition matrix of the system considered as a Markov chain. From this matrix, the probability is .6 that a car rented from location 3 will be returned to location 2, the probability is .8 that a car rented from location 1 will be returned to location 1, and so forth. ◆

EXAMPLE 2 Transition Matrix of the Markov Chain

By reviewing its donation records, the alumni office of a college finds that 80% of its alumni who contribute to the annual fund one year will also contribute the next year, and 30% of those who do not contribute one year will contribute the next. This can be viewed as a Markov chain with two states: state 1 corresponds to an alumnus giving a donation in any one year, and state 2 corresponds to the alumnus not giving a donation in that year. The transition matrix is

$$
P = \begin{bmatrix} .8 & .3 \\ .2 & .7 \end{bmatrix} \blacklozenge
$$

In the examples above, the transition matrices of the Markov chains have the property that the entries in any column sum to 1. This is not accidental. If $P = [p_{ij}]$ is the transition matrix of any Markov chain with k states, then for each j we must have

$$
p_{1j} + p_{2j} + \cdots + p_{kj} = 1 \tag{1}
$$

because if the system is in state j at one observation, it is certain to be in one of the k possible states at the next observation.

A matrix with property (1) is called a ***stochastic matrix***, a ***probability matrix***, or a ***Markov matrix***. From the preceding discussion, it follows that the transition matrix for a Markov chain must be a stochastic matrix.

In a Markov chain, the state of the system at any observation time cannot generally be determined with certainty. The best one can usually do is specify probabilities for each of the possible states. For example, in a Markov chain with three states, we might describe the possible state of the system at some observation time by a column vector

$$
\mathbf{x} = \begin{bmatrix} x_1 \\ x_2 \\ x_3 \end{bmatrix}
$$

in which x_1 is the probability that the system is in state 1, x_2 the probability that it is in state 2, and x_3 the probability that it is in state 3. In general we make the following definition.

DEFINITION

The ***state vector*** for an observation of a Markov chain with k states is a column vector $\mathbf{x}$ whose ith component x_i is the probability that the system is in the ith state at that time.

Observe that the entries in any state vector for a Markov chain are nonnegative and have a sum of 1. (Why?) A column vector that has this property is called a ***probability vector***.

Let us suppose now that we know the state vector $\mathbf{x}^{(0)}$ for a Markov chain at some initial observation. The following theorem will enable us to determine the state vectors

$$\mathbf{x}^{(1)}, \mathbf{x}^{(2)}, \ldots, \mathbf{x}^{(n)}, \ldots$$

at the subsequent observation times.

THEOREM 6.5.1

> *If P is the transition matrix of a Markov chain and $\mathbf{x}^{(n)}$ is the state vector at the nth observation, then $\mathbf{x}^{(n+1)} = P\mathbf{x}^{(n)}$.*

The proof of this theorem involves ideas from probability theory and will not be given here. From this theorem, it follows that

$$\mathbf{x}^{(1)} = P\mathbf{x}^{(0)}$$
$$\mathbf{x}^{(2)} = P\mathbf{x}^{(1)} = P^2\mathbf{x}^{(0)}$$
$$\mathbf{x}^{(3)} = P\mathbf{x}^{(2)} = P^3\mathbf{x}^{(0)}$$
$$\vdots$$
$$\mathbf{x}^{(n)} = P\mathbf{x}^{(n-1)} = P^n\mathbf{x}^{(0)}$$

In this way, the initial state vector $\mathbf{x}^{(0)}$ and the transition matrix P determine $\mathbf{x}^{(n)}$ for $n = 1, 2, \ldots$.

EXAMPLE 3   Example 2 Revisited

The transition matrix in Example 2 was

$$P = \begin{bmatrix} .8 & .3 \\ .2 & .7 \end{bmatrix}$$

We now construct the probable future donation record of a new graduate who did not give a donation in the initial year after graduation. For such a graduate the system is initially in state 2 with certainty, so the initial state vector is

$$\mathbf{x}^{(0)} = \begin{bmatrix} 0 \\ 1 \end{bmatrix}$$

From Theorem 6.5.1 we then have

$$\mathbf{x}^{(1)} = P\mathbf{x}^{(0)} = \begin{bmatrix} .8 & .3 \\ .2 & .7 \end{bmatrix} \begin{bmatrix} 0 \\ 1 \end{bmatrix} = \begin{bmatrix} .3 \\ .7 \end{bmatrix}$$

$$\mathbf{x}^{(2)} = P\mathbf{x}^{(1)} = \begin{bmatrix} .8 & .3 \\ .2 & .7 \end{bmatrix} \begin{bmatrix} .3 \\ .7 \end{bmatrix} = \begin{bmatrix} .45 \\ .55 \end{bmatrix}$$

$$\mathbf{x}^{(3)} = P\mathbf{x}^{(2)} = \begin{bmatrix} .8 & .3 \\ .2 & .7 \end{bmatrix} \begin{bmatrix} .45 \\ .55 \end{bmatrix} = \begin{bmatrix} .525 \\ .475 \end{bmatrix}$$

Thus, after three years the alumnus can be expected to make a donation with probability .525. Beyond three years, we find the following state vectors (to three decimal places):

$$\mathbf{x}^{(4)} = \begin{bmatrix} .563 \\ .438 \end{bmatrix}, \qquad \mathbf{x}^{(5)} = \begin{bmatrix} .581 \\ .419 \end{bmatrix}, \qquad \mathbf{x}^{(6)} = \begin{bmatrix} .591 \\ .409 \end{bmatrix}, \qquad \mathbf{x}^{(7)} = \begin{bmatrix} .595 \\ .405 \end{bmatrix}$$

$$\mathbf{x}^{(8)} = \begin{bmatrix} .598 \\ .402 \end{bmatrix}, \qquad \mathbf{x}^{(9)} = \begin{bmatrix} .599 \\ .401 \end{bmatrix}, \qquad \mathbf{x}^{(10)} = \begin{bmatrix} .599 \\ .401 \end{bmatrix}, \qquad \mathbf{x}^{(11)} = \begin{bmatrix} .600 \\ .400 \end{bmatrix}$$

For all n beyond 11, we have

$$\mathbf{x}^{(n)} = \begin{bmatrix} .600 \\ .400 \end{bmatrix}$$

to three decimal places. In other words, the state vectors converge to a fixed vector as the number of observations increases. (We shall discuss this further below.) ◆

EXAMPLE 4 Example 1 Revisited

The transition matrix in Example 1 was

$$\begin{bmatrix} .8 & .3 & .2 \\ .1 & .2 & .6 \\ .1 & .5 & .2 \end{bmatrix}$$

If a car is rented initially from location 2, then the initial state vector is

$$\mathbf{x}^{(0)} = \begin{bmatrix} 0 \\ 1 \\ 0 \end{bmatrix}$$

Using this vector and Theorem 6.5.1, one obtains the later state vectors listed in Table 1.

Table 1

n \ $x^{(n)}$	0	1	2	3	4	5	6	7	8	9	10	11
$x_1^{(n)}$	0	.300	.400	.477	.511	.533	.544	.550	.553	.555	.556	.557
$x_2^{(n)}$	1	.200	.370	.252	.261	.240	.238	.233	.232	.231	.230	.230
$x_3^{(n)}$	0	.500	.230	.271	.228	.227	.219	.217	.215	.214	.214	.213

For all values of n greater than 11, all state vectors are equal to $\mathbf{x}^{(11)}$ to three decimal places.

Two things should be observed in this example. First, it was not necessary to know how long a customer kept the car. That is, in a Markov process the time period between observations need not be regular. Second, the state vectors approach a fixed vector as n increases, just as in the first example. ◆

Figure 6.5.1

EXAMPLE 5 Using Theorem 6.5.1

A traffic officer is assigned to control the traffic at the eight intersections indicated in Figure 6.5.1. She is instructed to remain at each intersection for an hour and then to either remain at the same intersection or move to a neighboring intersection. To avoid establishing a pattern, she is told to choose her new intersection on a random basis, with each possible choice equally likely. For example, if she is at intersection 5, her next intersection can be 2, 4, 5, or 8, each with probability $\frac{1}{4}$. Every day she starts at the location where she stopped the day before. The transition matrix for this Markov chain is

Old Intersection

$$
\begin{array}{c}
\;1\;\;\;\;2\;\;\;\;3\;\;\;\;4\;\;\;\;5\;\;\;\;6\;\;\;\;7\;\;\;\;8
\end{array}
$$

$$
\begin{bmatrix}
\frac{1}{3} & \frac{1}{3} & 0 & \frac{1}{5} & 0 & 0 & 0 & 0 \\
\frac{1}{3} & \frac{1}{3} & 0 & 0 & \frac{1}{4} & 0 & 0 & 0 \\
0 & 0 & \frac{1}{3} & \frac{1}{5} & 0 & \frac{1}{3} & 0 & 0 \\
\frac{1}{3} & 0 & \frac{1}{3} & \frac{1}{5} & \frac{1}{4} & 0 & \frac{1}{4} & 0 \\
0 & \frac{1}{3} & 0 & \frac{1}{5} & \frac{1}{4} & 0 & 0 & \frac{1}{3} \\
0 & 0 & \frac{1}{3} & 0 & 0 & \frac{1}{3} & \frac{1}{4} & 0 \\
0 & 0 & 0 & \frac{1}{5} & 0 & \frac{1}{3} & \frac{1}{4} & \frac{1}{3} \\
0 & 0 & 0 & 0 & \frac{1}{4} & 0 & \frac{1}{4} & \frac{1}{3}
\end{bmatrix}
\begin{array}{l}
1 \\ 2 \\ 3 \\ 4 \\ 5 \\ 6 \\ 7 \\ 8
\end{array}
\begin{array}{l}
\\ \\ \\ \text{New} \\ \text{Intersection} \\ \\ \\
\end{array}
$$

If the traffic officer begins at intersection 5, her probable locations, hour by hour, are given by the state vectors given in Table 2. For all values of n greater than 22, all state vectors are equal to $\mathbf{x}^{(22)}$ to three decimal places. Thus, as with the first two examples, the state vectors approach a fixed vector as n increases. ◆

Table 2

$x^{(n)}$ ＼ n	0	1	2	3	4	5	10	15	20	22
$x_1^{(n)}$	0	.000	.133	.116	.130	.123	.113	.109	.108	.107
$x_2^{(n)}$	0	.250	.146	.163	.140	.138	.115	.109	.108	.107
$x_3^{(n)}$	0	.000	.050	.039	.067	.073	.100	.106	.107	.107
$x_4^{(n)}$	0	.250	.113	.187	.162	.178	.178	.179	.179	.179
$x_5^{(n)}$	1	.250	.279	.190	.190	.168	.149	.144	.143	.143
$x_6^{(n)}$	0	.000	.000	.050	.056	.074	.099	.105	.107	.107
$x_7^{(n)}$	0	.000	.133	.104	.131	.125	.138	.142	.143	.143
$x_8^{(n)}$	0	.250	.146	.152	.124	.121	.108	.107	.107	.107

Limiting Behavior of the State Vectors

In our examples we saw that the state vectors approached some fixed vector as the number of observations increased. We now ask whether the state vectors always approach a fixed vector in a Markov chain. A simple example shows that this is not the case.

EXAMPLE 6 System Oscillates between Two State Vectors

Let

$$
P = \begin{bmatrix} 0 & 1 \\ 1 & 0 \end{bmatrix} \quad \text{and} \quad \mathbf{x}^{(0)} = \begin{bmatrix} 1 \\ 0 \end{bmatrix}
$$

Then, because $P^2 = I$ and $P^3 = P$, we have that

$$
\mathbf{x}^{(0)} = \mathbf{x}^{(2)} = \mathbf{x}^{(4)} = \cdots = \begin{bmatrix} 1 \\ 0 \end{bmatrix}
$$

and

$$\mathbf{x}^{(1)} = \mathbf{x}^{(3)} = \mathbf{x}^{(5)} = \cdots = \begin{bmatrix} 0 \\ 1 \end{bmatrix}$$

This system oscillates indefinitely between the two state vectors $\begin{bmatrix} 1 \\ 0 \end{bmatrix}$ and $\begin{bmatrix} 0 \\ 1 \end{bmatrix}$, so it does not approach any fixed vector. ◆

However, if we impose a mild condition on the transition matrix, we can show that a fixed limiting state vector is approached. This condition is described by the following definition.

DEFINITION

A transition matrix is ***regular*** if some integer power of it has all positive entries.

Thus, for a regular transition matrix P, there is some positive integer m such that all entries of P^m are positive. This is the case with the transition matrices of Examples 1 and 2 for $m = 1$. In Example 5 it turns out that P^4 has all positive entries. Consequently, in all three examples the transition matrices are regular.

A Markov chain that is governed by a regular transition matrix is called a ***regular Markov chain***. We shall see that every regular Markov chain has a fixed state vector $\mathbf{q}$ such that $P^n \mathbf{x}^{(0)}$ approaches $\mathbf{q}$ as n increases for any choice of $\mathbf{x}^{(0)}$. This result is of major importance in the theory of Markov chains. It is based on the following theorem.

THEOREM 6.5.2

> **Behavior of P^n as $n \to \infty$**
>
> *If P is a regular transition matrix, then as $n \to $,*
>
> $$P^n \to \begin{bmatrix} q_1 & q_1 & \cdots & q_1 \\ q_2 & q_2 & \cdots & q_2 \\ \vdots & \vdots & & \vdots \\ q_k & q_k & \cdots & q_k \end{bmatrix}$$
>
> *where the q_i are positive numbers such that $q_1 + q_2 + \cdots + q_k = 1$.*

We will not prove this theorem here. The interested reader is referred to a more specialized text, such as J. Kemeny and J. Snell, *Finite Markov Chains* (New York: Springer-Verlag, 1976).

Let us set

$$Q = \begin{bmatrix} q_1 & q_1 & \cdots & q_1 \\ q_2 & q_2 & \cdots & q_2 \\ \vdots & \vdots & & \vdots \\ q_k & q_k & \cdots & q_k \end{bmatrix} \quad \text{and} \quad \mathbf{q} = \begin{bmatrix} q_1 \\ q_2 \\ \vdots \\ q_k \end{bmatrix}$$

Thus, Q is a transition matrix, all of whose columns are equal to the probability vector $\mathbf{q}$. Q has the property that if $\mathbf{x}$ is any probability vector, then

$$Q\mathbf{x} = \begin{bmatrix} q_1 & q_1 & \cdots & q_1 \\ q_2 & q_2 & \cdots & q_2 \\ \vdots & \vdots & & \vdots \\ q_k & q_k & \cdots & q_k \end{bmatrix} \begin{bmatrix} x_1 \\ x_2 \\ \vdots \\ x_k \end{bmatrix} = \begin{bmatrix} q_1 x_1 + q_1 x_2 + \cdots + q_1 x_k \\ q_2 x_1 + q_2 x_2 + \cdots + q_2 x_k \\ \vdots & \vdots & & \vdots \\ q_k x_1 + q_k x_2 + \cdots + q_k x_k \end{bmatrix}$$

$$= (x_1 + x_2 + \cdots + x_k) \begin{bmatrix} q_1 \\ q_2 \\ \vdots \\ q_k \end{bmatrix} = (1)\mathbf{q} = \mathbf{q}$$

That is, Q transforms any probability vector $\mathbf{x}$ into the fixed probability vector $\mathbf{q}$. This result leads to the following theorem.

THEOREM 6.5.3

> **Behavior of $P^n\mathbf{x}$ as $n \to \infty$**
>
> *If P is a regular transition matrix and $\mathbf{x}$ is any probability vector, then as $n \to$,*
>
> $$P^n\mathbf{x} \to \begin{bmatrix} q_1 \\ q_2 \\ \vdots \\ q_k \end{bmatrix} = \mathbf{q}$$
>
> *where $\mathbf{q}$ is a fixed probability vector, independent of n, all of whose entries are positive.*

This result holds since Theorem 6.5.2 implies that $P^n \to Q$ as $n \to$. This in turn implies that $P^n\mathbf{x} \to Q\mathbf{x} = \mathbf{q}$ as $n \to$. Thus, for a regular Markov chain, the system eventually approaches a fixed state vector $\mathbf{q}$. The vector $\mathbf{q}$ is called the ***steady-state vector*** of the regular Markov chain.

For systems with many states, usually the most efficient technique of computing the steady-state vector $\mathbf{q}$ is simply to calculate $P^n\mathbf{x}$ for some large n. Our examples illustrate this procedure. Each is a regular Markov process, so that convergence to a steady-state vector is ensured. Another way of computing the steady-state vector is to make use of the following theorem.

THEOREM 6.5.4

> **Steady-State Vector**
>
> *The steady-state vector $\mathbf{q}$ of a regular transition matrix P is the unique probability vector that satisfies the equation $P\mathbf{q} = \mathbf{q}$.*

To see this, consider the matrix identity $PP^n = P^{n+1}$. By Theorem 6.5.2, both P^n and P^{n+1} approach Q as $n \to$. Thus, we have $PQ = Q$. Any one column of this matrix equation gives $P\mathbf{q} = \mathbf{q}$. To show that $\mathbf{q}$ is the only probability vector that satisfies this equation, suppose $\mathbf{r}$ is another probability vector such that $P\mathbf{r} = \mathbf{r}$. Then also $P^n\mathbf{r} = \mathbf{r}$ for $n = 1, 2, \ldots$. When we let $n \to$, Theorem 6.5.3 leads to $\mathbf{q} = \mathbf{r}$.

Theorem 6.5.4 can also be expressed by the statement that the homogeneous linear system

$$(I - P)\mathbf{q} = \mathbf{0}$$

has a unique solution vector $\mathbf{q}$ with nonnegative entries that satisfy the condition $q_1 + q_2 + \cdots + q_k = 1$. We can apply this technique to the computation of the steady-state vectors for our examples.

EXAMPLE 7 Example 2 Revisited

In Example 2 the transition matrix was

$$P = \begin{bmatrix} .8 & .3 \\ .2 & .7 \end{bmatrix}$$

so the linear system $(I - P)\mathbf{q} = \mathbf{0}$ is

$$\begin{bmatrix} .2 & -.3 \\ -.2 & .3 \end{bmatrix} \begin{bmatrix} q_1 \\ q_2 \end{bmatrix} = \begin{bmatrix} 0 \\ 0 \end{bmatrix} \tag{2}$$

This leads to the single independent equation

$$.2q_1 - .3q_2 = 0$$

or

$$q_1 = 1.5q_2$$

Thus, when we set $q_2 = s$, any solution of (2) is of the form

$$\mathbf{q} = s \begin{bmatrix} 1.5 \\ 1 \end{bmatrix}$$

where s is an arbitrary constant. To make the vector $\mathbf{q}$ a probability vector, we set $s = 1/(1.5 + 1) = .4$. Consequently,

$$\mathbf{q} = \begin{bmatrix} .6 \\ .4 \end{bmatrix}$$

is the steady-state vector of this regular Markov chain. This means that over the long run, 60% of the alumni will give a donation in any one year, and 40% will not. Observe that this agrees with the result obtained numerically in Example 3. ◆

EXAMPLE 8 Example 1 Revisited

In Example 1 the transition matrix was

$$P = \begin{bmatrix} .8 & .3 & .2 \\ .1 & .2 & .6 \\ .1 & .5 & .2 \end{bmatrix}$$

so the linear system $(I - P)\mathbf{q} = \mathbf{0}$ is

$$\begin{bmatrix} .2 & -.3 & -.2 \\ -.1 & .8 & -.6 \\ -.1 & -.5 & .8 \end{bmatrix} \begin{bmatrix} q_1 \\ q_2 \\ q_3 \end{bmatrix} = \begin{bmatrix} 0 \\ 0 \\ 0 \end{bmatrix}$$

The reduced row-echelon form of the coefficient matrix is (verify)

$$\begin{bmatrix} 1 & 0 & -\frac{34}{13} \\ 0 & 1 & -\frac{14}{13} \\ 0 & 0 & 0 \end{bmatrix}$$

so the original linear system is equivalent to the system

$$q_1 = \left(\tfrac{34}{13}\right)q_3$$
$$q_2 = \left(\tfrac{14}{13}\right)q_3$$

When we set $q_3 = s$, any solution of the linear system is of the form

$$\mathbf{q} = s \begin{bmatrix} \tfrac{34}{13} \\ \tfrac{14}{13} \\ 1 \end{bmatrix}$$

To make this a probability vector, we set

$$s = \frac{1}{\tfrac{34}{13} + \tfrac{14}{13} + 1} = \frac{13}{61}$$

Thus, the steady-state vector of the system is

$$\mathbf{q} = \begin{bmatrix} \tfrac{34}{61} \\ \tfrac{14}{61} \\ \tfrac{13}{61} \end{bmatrix} = \begin{bmatrix} .5573\ldots \\ .2295\ldots \\ .2131\ldots \end{bmatrix}$$

This agrees with the result obtained numerically in Table 1. The entries of $\mathbf{q}$ give the long-run probabilities that any one car will be returned to location 1, 2, or 3, respectively. If the car rental agency has a fleet of 1000 cars, it should design its facilities so that there are at least 558 spaces at location 1, at least 230 spaces at location 2, and at least 214 spaces at location 3. ◆

EXAMPLE 9 Example 5 Revisited

We will not give the details of the calculations but simply state that the unique probability vector solution of the linear system $(I - P)\mathbf{q} = \mathbf{0}$ is

$$\mathbf{q} = \begin{bmatrix} \tfrac{3}{28} \\ \tfrac{3}{28} \\ \tfrac{3}{28} \\ \tfrac{5}{28} \\ \tfrac{4}{28} \\ \tfrac{3}{28} \\ \tfrac{4}{28} \\ \tfrac{3}{28} \end{bmatrix} = \begin{bmatrix} .1071\ldots \\ .1071\ldots \\ .1071\ldots \\ .1785\ldots \\ .1428\ldots \\ .1071\ldots \\ .1428\ldots \\ .1071\ldots \end{bmatrix}$$

The entries in this vector indicate the proportion of time the traffic officer spends at each intersection over the long term. Thus, if the objective is for her to spend the same proportion of time at each intersection, then the strategy of random movement with equal probabilities from one intersection to another is not a good one. (See Exercise 5.) ◆

EXERCISE SET 6.5

1. Consider the transition matrix

$$P = \begin{bmatrix} .4 & .5 \\ .6 & .5 \end{bmatrix}$$

(a) Calculate $\mathbf{x}^{(n)}$ for $n = 1, 2, 3, 4, 5$, if $\mathbf{x}^{(0)} = \begin{bmatrix} 1 \\ 0 \end{bmatrix}$.

(b) State why P is regular and find its steady-state vector.

2. Consider the transition matrix

$$P = \begin{bmatrix} .2 & .1 & .7 \\ .6 & .4 & .2 \\ .2 & .5 & .1 \end{bmatrix}$$

(a) Calculate $\mathbf{x}^{(1)}$, $\mathbf{x}^{(2)}$, and $\mathbf{x}^{(3)}$ to three decimal places if

$$\mathbf{x}^{(0)} = \begin{bmatrix} 0 \\ 0 \\ 1 \end{bmatrix}$$

(b) State why P is regular and find its steady-state vector.

3. Find the steady-state vectors of the following regular transition matrices:

(a) $\begin{bmatrix} \frac{1}{3} & \frac{3}{4} \\ \frac{2}{3} & \frac{1}{4} \end{bmatrix}$ (b) $\begin{bmatrix} .81 & .26 \\ .19 & .74 \end{bmatrix}$ (c) $\begin{bmatrix} \frac{1}{3} & \frac{1}{2} & 0 \\ \frac{1}{3} & 0 & \frac{1}{4} \\ \frac{1}{3} & \frac{1}{2} & \frac{3}{4} \end{bmatrix}$

4. Let P be the transition matrix

$$\begin{bmatrix} \frac{1}{2} & 0 \\ \frac{1}{2} & 1 \end{bmatrix}$$

(a) Show that P is not regular.

(b) Show that as n increases, $P^n \mathbf{x}^{(0)}$ approaches $\begin{bmatrix} 0 \\ 1 \end{bmatrix}$ for any initial state vector $\mathbf{x}^{(0)}$.

(c) What conclusion of Theorem 6.5.3 is not valid for the steady state of this transition matrix?

5. Verify that if P is a $k \times k$ regular transition matrix all of whose row sums are equal to 1, then the entries of its steady-state vector are all equal to $1/k$.

6. Show that the transition matrix

$$P = \begin{bmatrix} 0 & \frac{1}{2} & \frac{1}{2} \\ \frac{1}{2} & \frac{1}{2} & 0 \\ \frac{1}{2} & 0 & \frac{1}{2} \end{bmatrix}$$

is regular, and use Exercise 5 to find its steady-state vector.

7. John is either happy or sad. If he is happy one day, then he is happy the next day four times out of five. If he is sad one day, then he is sad the next day one time out of three. Over the long term, what are the chances that John is happy on any given day?

8. A country is divided into three demographic regions. It is found that each year 5% of the residents of region 1 move to region 2, and 5% move to region 3. Of the residents of region 2, 15% move to region 1 and 10% move to region 3. And of the residents of region 3, 10% move to region 1 and 5% move to region 2. What percentage of the population resides in each of the three regions after a long period of time?

SECTION 6.5

Technology Exercises

The following exercises are designed to be solved using a technology utility. Typically, this will be MATLAB, *Mathematica*, Maple, Derive, or Mathcad, but it may also be some other type of linear algebra software or a scientific calculator with some linear algebra capabilities. For each exercise you will need to read the relevant documentation for the particular utility you are using. The goal of these exercises is to provide you with a basic proficiency with your technology utility. Once you have mastered the techniques in these exercises, you will be able to use your technology utility to solve many of the problems in the regular exercise sets.

T1. Consider the sequence of transition matrices

$$\{P_2, P_3, P_4, \ldots\}$$

with

$$P_2 = \begin{bmatrix} 0 & \frac{1}{2} \\ 1 & \frac{1}{2} \end{bmatrix}, \qquad P_3 = \begin{bmatrix} 0 & 0 & \frac{1}{3} \\ 0 & \frac{1}{2} & \frac{1}{3} \\ 1 & \frac{1}{2} & \frac{1}{3} \end{bmatrix},$$

$$P_4 = \begin{bmatrix} 0 & 0 & 0 & \frac{1}{4} \\ 0 & 0 & \frac{1}{3} & \frac{1}{4} \\ 0 & \frac{1}{2} & \frac{1}{3} & \frac{1}{4} \\ 1 & \frac{1}{2} & \frac{1}{3} & \frac{1}{4} \end{bmatrix}, \qquad P_5 = \begin{bmatrix} 0 & 0 & 0 & 0 & \frac{1}{5} \\ 0 & 0 & 0 & \frac{1}{4} & \frac{1}{5} \\ 0 & 0 & \frac{1}{3} & \frac{1}{4} & \frac{1}{5} \\ 0 & \frac{1}{2} & \frac{1}{3} & \frac{1}{4} & \frac{1}{5} \\ 1 & \frac{1}{2} & \frac{1}{3} & \frac{1}{4} & \frac{1}{5} \end{bmatrix},$$

and so on.

(a) Use a computer to show that each of these four matrices is regular by computing their squares.

(b) Verify Theorem 6.5.2 by computing the 100th power of P_k for $k = 2, 3, 4, 5$. Then make a conjecture as to the limiting value of P_k^n as $n \to$ for all $k = 2, 3, 4, \ldots$.

(c) Verify that the common column $\mathbf{q}_k$ of the limiting matrix you found in part (b) satisfies the equation $P_k \mathbf{q}_k = \mathbf{q}_k$, as required by Theorem 6.5.4.

T2. A mouse is placed in a box with nine rooms as shown in the accompanying figure. Assume that it is equally likely that the mouse goes through any door in the room or stays in the room.

(a) Construct the 9×9 transition matrix for this problem and show that it is regular.

(b) Determine the steady-state vector for the matrix.

(c) Use a symmetry argument to show that this problem may be solved using only a 3×3 matrix.

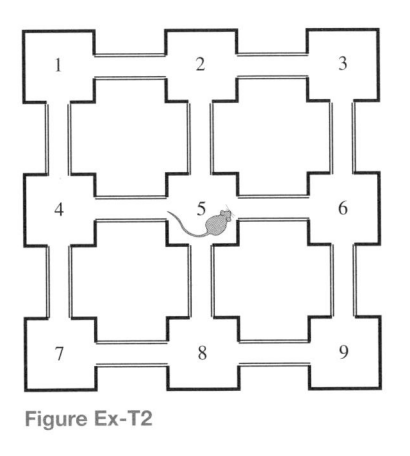

Figure Ex-T2

6.6
GRAPH THEORY

In this section we introduce matrix representations of relations among members of a set. We use matrix arithmetic to analyze these relationships.

PREREQUISITES: Matrix Addition and Multiplication

Relations among Members of a Set

There are countless examples of sets with finitely many members in which some relation exists among members of the set. For example, the set could consist of a collection of people, animals, countries, companies, sports teams, or cities; and the relation between two members, A and B, of such a set could be that person A dominates person B, animal A feeds on animal B, country A militarily supports country B, company A sells its product to company B, sports team A consistently beats sports team B, or city A has a direct airline flight to city B.

We shall now show how the theory of *directed graphs* can be used to mathematically model relations such as those in the preceding examples.

Directed Graphs

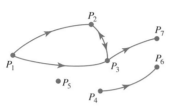

Figure 6.6.1

A **directed graph** is a finite set of elements, $\{P_1, P_2, \ldots, P_n\}$, together with a finite collection of ordered pairs (P_i, P_j) of distinct elements of this set, with no ordered pair being repeated. The elements of the set are called **vertices**, and the ordered pairs are called **directed edges**, of the directed graph. We use the notation $P_i \rightarrow P_j$ (which is read "P_i is connected to P_j") to indicate that the directed edge (P_i, P_j) belongs to the directed graph. Geometrically, we can visualize a directed graph (Figure 6.6.1) by representing the vertices as points in the plane and representing the directed edge $P_i \rightarrow P_j$ by drawing a line or arc from vertex P_i to vertex P_j, with an arrow pointing from P_i to P_j. If both $P_i \rightarrow P_j$ and $P_j \rightarrow P_i$ hold (denoted $P_i \leftrightarrow P_j$), we draw a single line between P_i and P_j with two oppositely pointing arrows (as with P_2 and P_3 in the figure).

As in Figure 6.6.1, for example, a directed graph may have separate "components" of vertices that are connected only among themselves; and some vertices, such as P_5, may not be connected with any other vertex. Also, because $P_i \rightarrow P_i$ is not permitted in a directed graph, a vertex cannot be connected with itself by a single arc that does not pass through any other vertex.

Figure 6.6.2 shows diagrams representing three more examples of directed graphs. With a directed graph having n vertices, we may associate an $n \times n$ matrix $M = [m_{ij}]$, called the **vertex matrix** of the directed graph. Its elements are defined by

$$m_{ij} = \begin{cases} 1, & \text{if } P_i \rightarrow P_j \\ 0, & \text{otherwise} \end{cases}$$

for $i, j = 1, 2, \ldots, n$. For the three directed graphs in Figure 6.6.2, the corresponding vertex matrices are

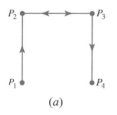

(a)

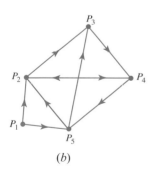

(b)

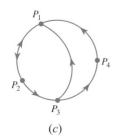

(c)

Figure 6.6.2

Figure 6.6.2a:
$$M = \begin{bmatrix} 0 & 1 & 0 & 0 \\ 0 & 0 & 1 & 0 \\ 0 & 1 & 0 & 1 \\ 0 & 0 & 0 & 0 \end{bmatrix}$$

Figure 6.6.2b:
$$M = \begin{bmatrix} 0 & 1 & 0 & 0 & 1 \\ 0 & 0 & 1 & 1 & 0 \\ 0 & 0 & 0 & 1 & 0 \\ 0 & 1 & 0 & 0 & 1 \\ 0 & 1 & 1 & 0 & 0 \end{bmatrix}$$

Figure 6.6.2c:
$$M = \begin{bmatrix} 0 & 1 & 0 & 0 \\ 1 & 0 & 1 & 0 \\ 1 & 0 & 0 & 1 \\ 1 & 0 & 0 & 0 \end{bmatrix}$$

By their definition, vertex matrices have the following two properties:

(i)   All entries are either 0 or 1.

(ii) All diagonal entries are 0.

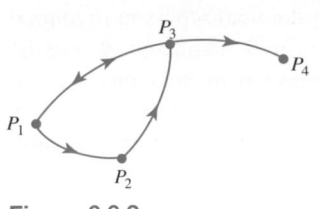

Figure 6.6.3

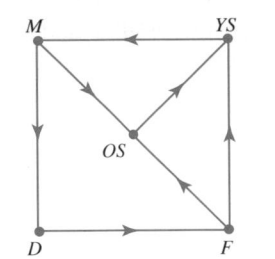

Figure 6.6.4

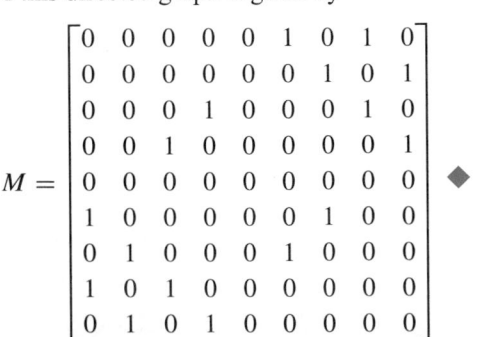

Figure 6.6.5

Conversely, any matrix with these two properties determines a unique directed graph having the given matrix as its vertex matrix. For example, the matrix

$$M = \begin{bmatrix} 0 & 1 & 1 & 0 \\ 0 & 0 & 1 & 0 \\ 1 & 0 & 0 & 1 \\ 0 & 0 & 0 & 0 \end{bmatrix}$$

determines the directed graph in Figure 6.6.3.

EXAMPLE 1   Influences within a Family

A certain family consists of a mother, father, daughter, and two sons. The family members have influence, or power, over each other in the following ways: the mother can influence the daughter and the oldest son; the father can influence the two sons; the daughter can influence the father; the oldest son can influence the youngest son; and the youngest son can influence the mother. We may model this family influence pattern with a directed graph whose vertices are the five family members. If family member A influences family member B, we write $A \rightarrow B$. Figure 6.6.4 is the resulting directed graph, where we have used obvious letter designations for the five family members. The vertex matrix of this directed graph is

$$
\begin{array}{c}
 \\
M \\
F \\
D \\
OS \\
YS
\end{array}
\begin{array}{c}
\begin{array}{ccccc} M & F & D & OS & YS \end{array} \\
\begin{bmatrix} 0 & 0 & 1 & 1 & 0 \\ 0 & 0 & 0 & 1 & 1 \\ 0 & 1 & 0 & 0 & 0 \\ 0 & 0 & 0 & 0 & 1 \\ 1 & 0 & 0 & 0 & 0 \end{bmatrix}
\end{array}
\blacklozenge
$$

EXAMPLE 2   Vertex Matrix: Moves on a Chessboard

In chess the knight moves in an "L"-shaped pattern about the chessboard. For the board in Figure 6.6.5 it may move horizontally two squares and then vertically one square, or it may move vertically two squares and then horizontally one square. Thus, from the center square in the figure, the knight may move to any of the eight marked shaded squares. Suppose that the knight is restricted to the nine numbered squares in Figure 6.6.6. If by $i \rightarrow j$ we mean that the knight may move from square i to square j, the directed graph in Figure 6.6.7 illustrates all possible moves that the knight may make among these nine squares. In Figure 6.6.8 we have "unraveled" Figure 6.6.7 to make the pattern of possible moves clearer.

The vertex matrix of this directed graph is given by

$$M = \begin{bmatrix} 0 & 0 & 0 & 0 & 0 & 1 & 0 & 1 & 0 \\ 0 & 0 & 0 & 0 & 0 & 0 & 1 & 0 & 1 \\ 0 & 0 & 0 & 1 & 0 & 0 & 0 & 1 & 0 \\ 0 & 0 & 1 & 0 & 0 & 0 & 0 & 0 & 1 \\ 0 & 0 & 0 & 0 & 0 & 0 & 0 & 0 & 0 \\ 1 & 0 & 0 & 0 & 0 & 0 & 1 & 0 & 0 \\ 0 & 1 & 0 & 0 & 0 & 1 & 0 & 0 & 0 \\ 1 & 0 & 1 & 0 & 0 & 0 & 0 & 0 & 0 \\ 0 & 1 & 0 & 1 & 0 & 0 & 0 & 0 & 0 \end{bmatrix} \blacklozenge$$

1	2	3
4	5	6
7	8	9

Figure 6.6.6

Figure 6.6.7

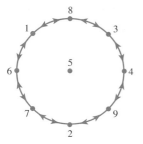

Figure 6.6.8

In Example 1 the father cannot directly influence the mother; that is, $F \rightarrow M$ is not true. But he can influence the youngest son, who can then influence the mother. We write this as $F \rightarrow YS \rightarrow M$ and call it a *2-step connection* from F to M. Analogously, we call $M \rightarrow D$ a *1-step connection*, $F \rightarrow OS \rightarrow YS \rightarrow M$ a *3-step connection*, and so forth. Let us now consider a technique for finding the number of all possible r-step connections ($r = 1, 2, \ldots$) from one vertex P_i to another vertex P_j of an arbitrary directed graph. (This will include the case when P_i and P_j are the same vertex.) The number of 1-step connections from P_i to P_j is simply m_{ij}. That is, there is either zero or one 1-step connection from P_i to P_j, depending on whether m_{ij} is zero or one. For the number of 2-step connections, we consider the square of the vertex matrix. If we let $m_{ij}^{(2)}$ be the (i, j)-th element of M^2, we have

$$m_{ij}^{(2)} = m_{i1}m_{1j} + m_{i2}m_{2j} + \cdots + m_{in}m_{nj} \tag{1}$$

Now, if $m_{i1} = m_{1j} = 1$, there is a 2-step connection $P_i \rightarrow P_1 \rightarrow P_j$ from P_i to P_j. But if either m_{i1} or m_{1j} is zero, such a 2-step connection is not possible. Thus $P_i \rightarrow P_1 \rightarrow P_j$ is a 2-step connection if and only if $m_{i1}m_{1j} = 1$. Similarly, for any $k = 1, 2, \ldots, n$, $P_i \rightarrow P_k \rightarrow P_j$ is a 2-step connection from P_i to P_j if and only if the term $m_{ik}m_{kj}$ on the right side of (1) is one; otherwise, the term is zero. Thus, the right side of (1) is the total number of two 2-step connections from P_i to P_j.

A similar argument will work for finding the number of 3-, 4-, $\ldots$, n-step connections from P_i to P_j. In general, we have the following result.

THEOREM 6.6.1

> *Let M be the vertex matrix of a directed graph and let $m_{ij}^{(r)}$ be the (i, j)-th element of M^r. Then $m_{ij}^{(r)}$ is equal to the number of r-step connections from P_i to P_j.*

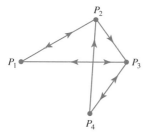

Figure 6.6.9

EXAMPLE 3   Using Theorem 6.6.1

Figure 6.6.9 is the route map of a small airline that services the four cities P_1, P_2, P_3, P_4. As a directed graph, its vertex matrix is

$$M = \begin{bmatrix} 0 & 1 & 1 & 0 \\ 1 & 0 & 1 & 0 \\ 1 & 0 & 0 & 1 \\ 0 & 1 & 1 & 0 \end{bmatrix}$$

We have that

$$M^2 = \begin{bmatrix} 2 & 0 & 1 & 1 \\ 1 & 1 & 1 & 1 \\ 0 & 2 & 2 & 0 \\ 2 & 0 & 1 & 1 \end{bmatrix} \quad \text{and} \quad M^3 = \begin{bmatrix} 1 & 3 & 3 & 1 \\ 2 & 2 & 3 & 1 \\ 4 & 0 & 2 & 2 \\ 1 & 3 & 3 & 1 \end{bmatrix}$$

If we are interested in connections from city P_4 to city P_3, we may use Theorem 6.6.1 to find their number. Because $m_{43} = 1$, there is one 1-step connection; because $m_{43}^{(2)} = 1$, there is one 2-step connection; and because $m_{43}^{(3)} = 3$, there are three 3-step connections. To verify this, from Figure 6.6.9 we find

$$\text{1-step connections from } P_4 \text{ to } P_3: \quad P_4 \rightarrow P_3$$

$$\text{2-step connections from } P_4 \text{ to } P_3: \quad P_4 \rightarrow P_2 \rightarrow P_3$$

3-step connections from P_4 to P_3: $P_4 \rightarrow P_3 \rightarrow P_4 \rightarrow P_3$

$$P_4 \rightarrow P_2 \rightarrow P_1 \rightarrow P_3$$

$$P_4 \rightarrow P_3 \rightarrow P_1 \rightarrow P_3 \quad \blacklozenge$$

Cliques

In everyday language a "clique" is a closely knit group of people (usually three or more) that tends to communicate within itself and has no place for outsiders. In graph theory this concept is given a more precise meaning.

DEFINITION

A subset of a directed graph is called a **_clique_** if it satisfies the following three conditions:

  (i)   The subset contains at least three vertices.

  (ii)   For each pair of vertices P_i and P_j in the subset, both $P_i \rightarrow P_j$ and $P_j \rightarrow P_i$ are true.

  (iii) The subset is as large as possible; that is, it is not possible to add another vertex to the subset and still satisfy condition (ii).

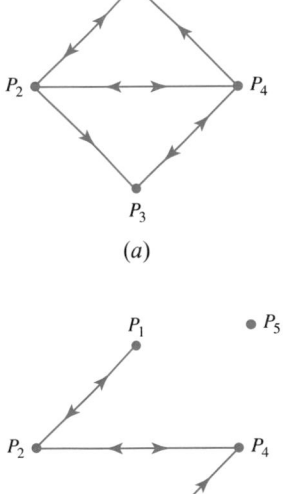

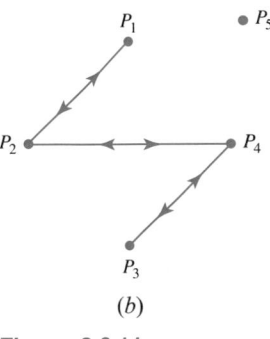

Figure 6.6.10

This definition suggests that cliques are maximal subsets that are in perfect "communication" with each other. For example, if the vertices represent cities, and $P_i \rightarrow P_j$ means that there is a direct airline flight from city P_i to city P_j, then there is a direct flight between any two cities within a clique in either direction.

EXAMPLE 4 A Directed Graph with Two Cliques

The directed graph illustrated in Figure 6.6.10 (which might represent the route map of an airline) has two cliques:

$$\{P_1, P_2, P_3, P_4\} \quad \text{and} \quad \{P_3, P_4, P_6\}$$

This example shows that a directed graph may contain several cliques and that a vertex may simultaneously belong to more than one clique. $\blacklozenge$

For simple directed graphs, cliques can be found by inspection. But for large directed graphs, it would be desirable to have a systematic procedure for detecting cliques. For this purpose, it will be helpful to define a matrix $S = [s_{ij}]$ related to a given directed graph as follows:

$$s_{ij} = \begin{cases} 1, & \text{if } P_i \leftrightarrow P_j \\ 0, & \text{otherwise} \end{cases}$$

The matrix S determines a directed graph that is the same as the given directed graph, with the exception that the directed edges with only one arrow are deleted. For example, if the original directed graph is given by Figure 6.6.11a, the directed graph that has S as its vertex matrix is given in Figure 6.6.11b. The matrix S may be obtained from the vertex matrix M of the original directed graph by setting $s_{ij} = 1$ if $m_{ij} = m_{ji} = 1$ and setting $s_{ij} = 0$ otherwise.

The following theorem, which uses the matrix S, is helpful for identifying cliques.

(a)

(b)

Figure 6.6.11

THEOREM 6.6.2

Identifying Cliques

Let $s_{ij}^{(3)}$ be the (i, j)-th element of S^3. Then a vertex P_i belongs to some clique if and only if $s_{ii}^{(3)} \neq 0$.

Proof If $s_{ii}^{(3)} \neq 0$, then there is at least one 3-step connection from P_i to itself in the modified directed graph determined by S. Suppose it is $P_i \rightarrow P_j \rightarrow P_k \rightarrow P_i$. In the modified directed graph, all directed relations are two-way, so we also have the connections $P_i \leftrightarrow P_j \leftrightarrow P_k \leftrightarrow P_i$. But this means that $\{P_i, P_j, P_k\}$ is either a clique or a subset of a clique. In either case, P_i must belong to some clique. The converse statement, "if P_i belongs to a clique, then $s_{ij}^{(3)} \neq 0$," follows in a similar manner. ∎

EXAMPLE 5 Using Theorem 6.6.2

Suppose that a directed graph has as its vertex matrix

$$M = \begin{bmatrix} 0 & 1 & 1 & 1 \\ 1 & 0 & 1 & 0 \\ 0 & 1 & 0 & 1 \\ 1 & 0 & 0 & 0 \end{bmatrix}$$

Then

$$S = \begin{bmatrix} 0 & 1 & 0 & 1 \\ 1 & 0 & 1 & 0 \\ 0 & 1 & 0 & 0 \\ 1 & 0 & 0 & 0 \end{bmatrix} \quad \text{and} \quad S^3 = \begin{bmatrix} 0 & 3 & 0 & 2 \\ 3 & 0 & 2 & 0 \\ 0 & 2 & 0 & 1 \\ 2 & 0 & 1 & 0 \end{bmatrix}$$

Because all diagonal entries of S^3 are zero, it follows from Theorem 6.6.2 that the directed graph has no cliques. ◆

EXAMPLE 6 Using Theorem 6.6.2

Suppose that a directed graph has as its vertex matrix

$$M = \begin{bmatrix} 0 & 1 & 0 & 1 & 1 \\ 1 & 0 & 0 & 1 & 0 \\ 1 & 1 & 0 & 1 & 0 \\ 1 & 1 & 0 & 0 & 0 \\ 1 & 0 & 0 & 1 & 0 \end{bmatrix}$$

Then

$$S = \begin{bmatrix} 0 & 1 & 0 & 1 & 1 \\ 1 & 0 & 0 & 1 & 0 \\ 0 & 0 & 0 & 0 & 0 \\ 1 & 1 & 0 & 0 & 0 \\ 1 & 0 & 0 & 0 & 0 \end{bmatrix} \quad \text{and} \quad S^3 = \begin{bmatrix} 2 & 4 & 0 & 4 & 3 \\ 4 & 2 & 0 & 3 & 1 \\ 0 & 0 & 0 & 0 & 0 \\ 4 & 3 & 0 & 2 & 1 \\ 3 & 1 & 0 & 1 & 0 \end{bmatrix}$$

The nonzero diagonal entries of S^3 are $s_{11}^{(3)}$, $s_{22}^{(3)}$, and $s_{44}^{(3)}$. Consequently, in the given directed graph, P_1, P_2, and P_4 belong to cliques. Because a clique must contain at least three vertices, the directed graph has only one clique, $\{P_1, P_2, P_4\}$. ◆

Dominance-Directed Graphs

In many groups of individuals or animals, there is a definite "pecking order" or dominance relation between any two members of the group. That is, given any two individuals A and B, either A dominates B or B dominates A, but not both. In terms of a directed graph in which $P_i \to P_j$ means P_i dominates P_j, this means that for all distinct pairs, either $P_i \to P_j$ or $P_j \to P_i$, but not both. In general, we have the following definition.

> **DEFINITION**
>
> A **dominance-directed graph** is a directed graph such that for any distinct pair of vertices P_i and P_j, either $P_i \to P_j$ or $P_j \to P_i$, but not both.

An example of a directed graph satisfying this definition is a league of n sports teams that play each other exactly one time, as in one round of a round-robin tournament in which no ties are allowed. If $P_i \to P_j$ means that team P_i beat team P_j in their single match, it is easy to see that the definition of a dominance-directed group is satisfied. For this reason, dominance-directed graphs are sometimes called **tournaments**.

Figure 6.6.12 illustrates some dominance-directed graphs with three, four, and five vertices, respectively. In these three graphs, the circled vertices have the following interesting property: from each one there is either a 1-step or a 2-step connection to any other vertex in its graph. In a sports tournament, these vertices would correspond to the most "powerful" teams in the sense that these teams either beat any given team or beat some other team that beat the given team. We can now state and prove a theorem that guarantees that any dominance-directed graph has at least one vertex with this property.

THEOREM 6.6.3

> **Connections in Dominance-Directed Graphs**
>
> *In any dominance-directed graph, there is at least one vertex from which there is a 1-step or 2-step connection to any other vertex.*

Proof Consider a vertex (there may be several) with the largest total number of 1-step and 2-step connections to other vertices in the graph. By renumbering the vertices, we may assume that P_1 is such a vertex. Suppose there is some vertex P_i such that there is no 1-step or 2-step connection from P_1 to P_i. Then, in particular, $P_1 \to P_i$ is not true, so that by definition of a dominance-directed graph, it must be that $P_i \to P_1$. Next, let P_k be any vertex such that $P_1 \to P_k$ is true. Then we cannot have $P_k \to P_i$, as then $P_1 \to P_k \to P_i$ would be a 2-step connection from P_1 to P_i. Thus, it must be that $P_i \to P_k$. That is, P_i has 1-step connections to all the vertices to which P_1 has 1-step connections. The vertex P_i must then also have 2-step connections to all the vertices to which P_1 has 2-step connections. But because, in addition, we have that $P_i \to P_1$, this means that P_i has more 1-step and 2-step connections to other vertices than does P_1. However, this contradicts the way in which P_1 was chosen. Hence, there can be no vertex P_i to which P_1 has no 1-step or 2-step connection. ∎

This proof shows that a vertex with the largest total number of 1-step and 2-step connections to other vertices has the property stated in the theorem. There is a simple way of finding such vertices using the vertex matrix M and its square M^2. The sum of the entries in the ith row of M is the total number of 1-step connections from P_i to other vertices, and the sum of the entries of the ith row of M^2 is the total number of 2-step

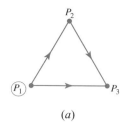

(a)

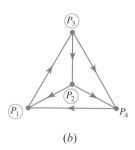

(b)

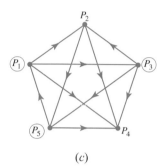

(c)

Figure 6.6.12

connections from P_i to other vertices. Consequently, the sum of the entries of the ith row of the matrix $A = M + M^2$ is the total number of 1-step and 2-step connections from P_i to other vertices. In other words, a row of $A = M + M^2$ with the largest row sum identifies a vertex having the property stated in Theorem 6.6.3.

EXAMPLE 7   Using Theorem 6.6.3

Suppose that five baseball teams play each other exactly once, and the results are as indicated in the dominance-directed graph of Figure 6.6.13. The vertex matrix of the graph is

$$M = \begin{bmatrix} 0 & 0 & 1 & 1 & 0 \\ 1 & 0 & 1 & 0 & 1 \\ 0 & 0 & 0 & 1 & 0 \\ 0 & 1 & 0 & 0 & 0 \\ 1 & 0 & 1 & 1 & 0 \end{bmatrix}$$

so

$$A = M + M^2 = \begin{bmatrix} 0 & 0 & 1 & 1 & 0 \\ 1 & 0 & 1 & 0 & 1 \\ 0 & 0 & 0 & 1 & 0 \\ 0 & 1 & 0 & 0 & 0 \\ 1 & 0 & 1 & 1 & 0 \end{bmatrix} + \begin{bmatrix} 0 & 1 & 0 & 1 & 0 \\ 1 & 0 & 2 & 3 & 0 \\ 0 & 1 & 0 & 0 & 0 \\ 1 & 0 & 1 & 0 & 1 \\ 0 & 1 & 1 & 2 & 0 \end{bmatrix} = \begin{bmatrix} 0 & 1 & 1 & 2 & 0 \\ 2 & 0 & 3 & 3 & 1 \\ 0 & 1 & 0 & 1 & 0 \\ 1 & 1 & 1 & 0 & 1 \\ 1 & 1 & 2 & 3 & 0 \end{bmatrix}$$

The row sums of A are

1st row sum $= 4$

2nd row sum $= 9$

3rd row sum $= 2$

4th row sum $= 4$

5th row sum $= 7$

Because the second row has the largest row sum, the vertex P_2 must have a 1-step or 2-step connection to any other vertex. This is easily verified from Figure 6.6.13. ◆

We have informally suggested that a vertex with the largest number of 1-step and 2-step connections to other vertices is a "powerful" vertex. We can formalize this concept with the following definition.

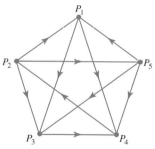

Figure 6.6.13

> **DEFINITION**
>
> The ***power*** of a vertex of a dominance-directed graph is the total number of 1-step and 2-step connections from it to other vertices. Alternatively, the power of a vertex P_i is the sum of the entries of the ith row of the matrix $A = M + M^2$, where M is the vertex matrix of the directed graph.

EXAMPLE 8   Example 7 Revisited

Let us rank the five baseball teams in Example 7 according to their powers. From the calculations for the row sums in that example, we have

Power of team $P_1 = 4$

Power of team $P_2 = 9$

Power of team $P_3 = 2$

Power of team $P_4 = 4$

Power of team $P_5 = 7$

Hence, the ranking of the teams according to their powers would be

$$P_2 \text{ (first)}, \quad P_5 \text{ (second)}, \quad P_1 \text{ and } P_4 \text{ (tied for third)}, \quad P_3 \text{ (last)} \quad \blacklozenge$$

EXERCISE SET 6.6

1. Construct the vertex matrix for each of the directed graphs illustrated in the accompanying figure.

2. Draw a diagram of the directed graph corresponding to each of the following vertex matrices.

(a) $\begin{bmatrix} 0 & 1 & 1 & 0 \\ 1 & 0 & 0 & 0 \\ 0 & 0 & 0 & 1 \\ 1 & 0 & 1 & 0 \end{bmatrix}$

(b) $\begin{bmatrix} 0 & 0 & 1 & 0 & 0 \\ 1 & 0 & 0 & 0 & 1 \\ 0 & 1 & 0 & 1 & 1 \\ 0 & 0 & 0 & 0 & 0 \\ 1 & 1 & 1 & 0 & 0 \end{bmatrix}$

(c) $\begin{bmatrix} 0 & 1 & 0 & 1 & 0 & 1 \\ 1 & 0 & 0 & 0 & 1 & 0 \\ 0 & 0 & 0 & 0 & 0 & 0 \\ 1 & 1 & 0 & 0 & 1 & 0 \\ 0 & 0 & 0 & 1 & 0 & 1 \\ 0 & 1 & 0 & 0 & 1 & 0 \end{bmatrix}$

3. Let M be the following vertex matrix of a directed graph:

$$\begin{bmatrix} 0 & 1 & 1 & 1 \\ 1 & 0 & 0 & 0 \\ 0 & 1 & 0 & 1 \\ 0 & 1 & 1 & 0 \end{bmatrix}$$

(a) Draw a diagram of the directed graph.

(b) Use Theorem 6.6.1 to find the number of 1-, 2-, and 3-step connections from the vertex P_1 to the vertex P_2. Verify your answer by listing the various connections as in Example 3.

(c) Repeat part (b) for the 1-, 2-, and 3-step connections from P_1 to P_4.

4. (a) Compute the matrix product $M^T M$ for the vertex matrix M in Example 1.

(b) Verify that the kth diagonal entry of $M^T M$ is the number of family members who influence the kth family member. Why is this true?

(c) Find a similar interpretation for the values of the nondiagonal entries of $M^T M$.

5. By inspection, locate all cliques in each of the directed graphs illustrated in the accompanying figure.

Figure Ex-1

(a)

(b)

(c)

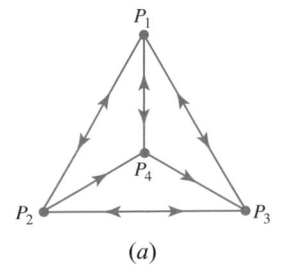

(a)

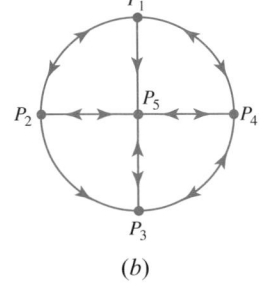

(b)

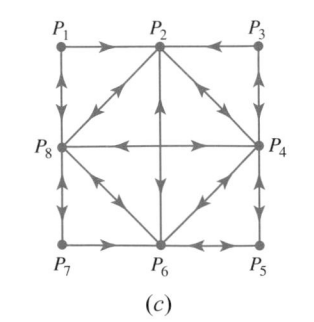

(c)

Figure Ex-5

6. For each of the following vertex matrices, use Theorem 6.6.2 to find all cliques in the corresponding directed graphs.

$$
\text{(a)} \quad
\begin{bmatrix}
0 & 1 & 0 & 1 & 0 \\
1 & 0 & 1 & 0 & 1 \\
0 & 1 & 0 & 1 & 1 \\
1 & 0 & 0 & 0 & 1 \\
1 & 0 & 1 & 1 & 0
\end{bmatrix}
\qquad
\text{(b)} \quad
\begin{bmatrix}
0 & 1 & 0 & 1 & 1 & 0 \\
1 & 0 & 1 & 0 & 1 & 1 \\
0 & 1 & 0 & 1 & 0 & 1 \\
1 & 0 & 1 & 0 & 1 & 1 \\
0 & 1 & 0 & 1 & 0 & 0 \\
0 & 0 & 1 & 1 & 1 & 0
\end{bmatrix}
$$

7. For the dominance-directed graph illustrated in the accompanying figure, construct the vertex matrix and find the power of each vertex.

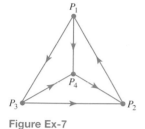

P_1

P_4

P_3 P_2

Figure Ex-7

8. Five baseball teams play each other one time with the following results:

A beats B, C, D

B beats C, E

C beats D, E

D beats B

E beats A, D

Rank the five baseball teams in accordance with the powers of the vertices they correspond to in the dominance-directed graph representing the outcomes of the games.

SECTION 6.6

Technology Exercises

The following exercises are designed to be solved using a technology utility. Typically, this will be MATLAB, *Mathematica*, Maple, Derive, or Mathcad, but it may also be some other type of linear algebra software or a scientific calculator with some linear algebra capabilities. For each exercise you will need to read the relevant documentation for the particular utility you are using. The goal of these exercises is to provide you with a basic proficiency with your technology utility. Once you have mastered the techniques in these exercises, you will be able to use your technology utility to solve many of the problems in the regular exercise sets.

T1. A graph having n vertices such that every vertex is connected to every other vertex has vertex matrix given by

$$
M_n =
\begin{bmatrix}
0 & 1 & 1 & 1 & 1 & \cdots & 1 \\
1 & 0 & 1 & 1 & 1 & \cdots & 1 \\
1 & 1 & 0 & 1 & 1 & \cdots & 1 \\
1 & 1 & 1 & 0 & 1 & \cdots & 1 \\
1 & 1 & 1 & 1 & 0 & \cdots & 1 \\
\vdots & \vdots & \vdots & \vdots & \vdots & \ddots & \vdots \\
1 & 1 & 1 & 1 & 1 & \cdots & 0
\end{bmatrix}
$$

In this problem we develop a formula for M_n^k whose (i, j)-th entry equals the number of k-step connections from P_i to P_j.

(a) Use a computer to compute the eight matrices M_n^k for $n = 2, 3$ and for $k = 2, 3, 4, 5$.

(b) Use the results in part (a) and symmetry arguments to show that M_n^k can be written as

$$
M_n^k =
\begin{bmatrix}
0 & 1 & 1 & 1 & 1 & \cdots & 1 \\
1 & 0 & 1 & 1 & 1 & \cdots & 1 \\
1 & 1 & 0 & 1 & 1 & \cdots & 1 \\
1 & 1 & 1 & 0 & 1 & \cdots & 1 \\
1 & 1 & 1 & 1 & 0 & \cdots & 1 \\
\vdots & \vdots & \vdots & \vdots & \vdots & \ddots & \vdots \\
1 & 1 & 1 & 1 & 1 & \cdots & 0
\end{bmatrix}^k
=
\begin{bmatrix}
\alpha_k & \beta_k & \beta_k & \beta_k & \beta_k & \cdots & \beta_k \\
\beta_k & \alpha_k & \beta_k & \beta_k & \beta_k & \cdots & \beta_k \\
\beta_k & \beta_k & \alpha_k & \beta_k & \beta_k & \cdots & \beta_k \\
\beta_k & \beta_k & \beta_k & \alpha_k & \beta_k & \cdots & \beta_k \\
\beta_k & \beta_k & \beta_k & \beta_k & \alpha_k & \cdots & \beta_k \\
\vdots & \vdots & \vdots & \vdots & \vdots & \ddots & \vdots \\
\beta_k & \beta_k & \beta_k & \beta_k & \beta_k & \cdots & \alpha_k
\end{bmatrix}
$$

(c) Using the fact that $M_n^k = M_n M_n^{k-1}$, show that

$$
\begin{bmatrix} \alpha_k \\ \beta_k \end{bmatrix}
=
\begin{bmatrix} 0 & n-1 \\ 1 & n-2 \end{bmatrix}
\begin{bmatrix} \alpha_{k-1} \\ \beta_{k-1} \end{bmatrix}
$$

with

$$\begin{bmatrix} \alpha_1 \\ \beta_1 \end{bmatrix} = \begin{bmatrix} 0 \\ 1 \end{bmatrix}$$

(d) Using part (c), show that

$$\begin{bmatrix} \alpha_k \\ \beta_k \end{bmatrix} = \begin{bmatrix} 0 & n-1 \\ 1 & n-2 \end{bmatrix}^{k-1} \begin{bmatrix} 0 \\ 1 \end{bmatrix}$$

T2. Consider a round-robin tournament among n players (labeled $a_1, a_2, a_3, \ldots, a_n$) where a_1 beats a_2, a_2 beats a_3, a_3 beats a_4, $\ldots$, a_{n-1} beats a_n, and a_n beats a_1. Compute the "power" of each player, showing that they all have the same power; then determine that common power.

Hint Use a computer to study the cases $n = 3, 4, 5, 6$; then make a conjecture and prove your conjecture to be true.

6.7
LEONTIEF ECONOMIC MODELS

In this section we discuss two linear models for economic systems. Some results about nonnegative matrices are applied to determine equilibrium price structures and outputs necessary to satisfy demand.

> **PREREQUISITES:** Linear Systems
> Matrices

Economic Systems

Matrix theory has been very successful in describing the interrelations among prices, outputs, and demands in economic systems. In this section we discuss some simple models based on the ideas of Nobel laureate Wassily Leontief. We examine two different but related models: the closed or input–output model, and the open or production model. In each, we are given certain economic parameters that describe the interrelations between the "industries" in the economy under consideration. Using matrix theory, we then evaluate certain other parameters, such as prices or output levels, in order to satisfy a desired economic objective. We begin with the closed model.

Leontief Closed (Input–Output) Model

First we present a simple example; then we proceed to the general theory of the model.

EXAMPLE 1 An Input–Output Model

Three homeowners—a carpenter, an electrician, and a plumber—agree to make repairs in their three homes. They agree to work a total of 10 days each according to the following schedule:

	Work Performed by		
	Carpenter	Electrician	Plumber
Days of Work in Home of Carpenter	2	1	6
Days of Work in Home of Electrician	4	5	1
Days of Work in Home of Plumber	4	4	3

For tax purposes, they must report and pay each other a reasonable daily wage, even for the work each does on his or her own home. Their normal daily wages are about $100, but they agree to adjust their respective daily wages so that each homeowner will come out even—that is, so that the total amount paid out by each is the same as the total amount each receives. We can set

$$p_1 = \text{daily wage of carpenter}$$

$$p_2 = \text{daily wage of electrician}$$

$$p_3 = \text{daily wage of plumber}$$

To satisfy the "equilibrium" condition that each homeowner comes out even, we require that

$$\text{total expenditures} = \text{total income}$$

for each of the homeowners for the 10-day period. For example, the carpenter pays a total of $2p_1 + p_2 + 6p_3$ for the repairs in his own home and receives a total income of $10p_1$ for the repairs that he performs on all three homes. Equating these two expressions then gives the first of the following three equations:

$$2p_1 + p_2 + 6p_3 = 10p_1$$
$$4p_1 + 5p_2 + p_3 = 10p_2$$
$$4p_1 + 4p_2 + 3p_3 = 10p_3$$

The remaining two equations are the equilibrium equations for the electrician and the plumber. Dividing these equations by 10 and rewriting them in matrix form yields

$$\begin{bmatrix} .2 & .1 & .6 \\ .4 & .5 & .1 \\ .4 & .4 & .3 \end{bmatrix} \begin{bmatrix} p_1 \\ p_2 \\ p_3 \end{bmatrix} = \begin{bmatrix} p_1 \\ p_2 \\ p_3 \end{bmatrix} \tag{1}$$

Equation (1) can be rewritten as a homogeneous system by subtracting the left side from the right side to obtain

$$\begin{bmatrix} .8 & -.1 & -.6 \\ -.4 & .5 & -.1 \\ -.4 & -.4 & .7 \end{bmatrix} \begin{bmatrix} p_1 \\ p_2 \\ p_3 \end{bmatrix} = \begin{bmatrix} 0 \\ 0 \\ 0 \end{bmatrix}$$

The solution of this homogeneous system is found to be (verify)

$$\begin{bmatrix} p_1 \\ p_2 \\ p_3 \end{bmatrix} = s \begin{bmatrix} 31 \\ 32 \\ 36 \end{bmatrix}$$

where s is an arbitrary constant. This constant is a scale factor, which the homeowners may choose for their convenience. For example, they may set $s = 3$ so that the corresponding daily wages—$93, $96, and $108—are about $100. ◆

This example illustrates the salient features of the Leontief input–output model of a closed economy. In the basic Equation (1), each column sum of the coefficient matrix is 1, corresponding to the fact that each of the homeowners' "output" of labor is completely distributed among these same homeowners in the proportions given by the entries in the column. Our problem is to determine suitable "prices" for these outputs so as to put the system in equilibrium—that is, so that each homeowner's total expenditures equal his or her total income.

In the general model we have an economic system consisting of a finite number of "industries," which we number as industries $1, 2, \ldots, k$. Over some fixed period of time, each industry produces an "output" of some good or service that is completely utilized in a predetermined manner by the k industries. An important problem is to find suitable "prices" to be charged for these k outputs so that for each industry, total expenditures equal total income. Such a price structure represents an equilibrium position for the economy.

For the fixed time period in question, let us set

p_i = price charged by the ith industry for its total output

e_{ij} = fraction of the total output of the jth industry purchased by the ith industry

for $i, j = 1, 2, \ldots, k$. By definition, we have

(i) $p_i \geq 0, \qquad i = 1, 2, \ldots, k$

(ii) $e_{ij} \geq 0, \qquad i, j = 1, 2, \ldots, k$

(iii) $e_{1j} + e_{2j} + \cdots + e_{kj} = 1, \qquad j = 1, 2, \ldots, k$

With these quantities, we form the **price vector**

$$\mathbf{p} = \begin{bmatrix} p_1 \\ p_2 \\ \vdots \\ p_k \end{bmatrix}$$

and the **exchange matrix** or **input–output matrix**

$$E = \begin{bmatrix} e_{11} & e_{12} & \cdots & e_{1k} \\ e_{21} & e_{22} & \cdots & e_{2k} \\ \vdots & \vdots & & \vdots \\ e_{k1} & e_{k2} & \cdots & e_{kk} \end{bmatrix}$$

Condition (iii) expresses the fact that all the column sums of the exchange matrix are 1.

As in the example, in order that the expenditures of each industry be equal to its income, the following matrix equation must be satisfied [see (1)]:

$$E\mathbf{p} = \mathbf{p} \tag{2}$$

or

$$(I - E)\mathbf{p} = \mathbf{0} \tag{3}$$

Equation (3) is a homogeneous linear system for the price vector $\mathbf{p}$. It will have a nontrivial solution if and only if the determinant of its coefficient matrix $I - E$ is zero. In Exercise 7 we ask the reader to show that this is the case for any exchange matrix E. Thus, (3) always has nontrivial solutions for the price vector $\mathbf{p}$.

Actually, for our economic model to make sense, we need more than just the fact that (3) has nontrivial solutions for $\mathbf{p}$. We also need the prices p_i of the k outputs to be nonnegative numbers. We express this condition as $\mathbf{p} \geq 0$. (In general, if A is any vector or matrix, the notation $A \geq 0$ means that every entry of A is nonnegative, and the notation $A > 0$ means that every entry of A is positive. Similarly, $A \geq B$ means $A - B \geq 0$, and $A > B$ means $A - B > 0$.) To show that (3) has a nontrivial solution for which $\mathbf{p} \geq 0$ is a bit more difficult than showing merely that some nontrivial solution exists. But it is true, and we state this fact without proof in the following theorem.

THEOREM 6.7.1

If E is an exchange matrix, then $E\mathbf{p} = \mathbf{p}$ always has a nontrivial solution $\mathbf{p}$ whose entries are nonnegative.

Let us consider a few simple examples of this theorem.

EXAMPLE 2   Using Theorem 6.7.1

Let

$$E = \begin{bmatrix} \frac{1}{2} & 0 \\ \frac{1}{2} & 1 \end{bmatrix}$$

Then $(I - E)\mathbf{p} = \mathbf{0}$ is

$$\begin{bmatrix} \frac{1}{2} & 0 \\ -\frac{1}{2} & 0 \end{bmatrix} \begin{bmatrix} p_1 \\ p_2 \end{bmatrix} = \begin{bmatrix} 0 \\ 0 \end{bmatrix}$$

which has the general solution

$$\mathbf{p} = s \begin{bmatrix} 0 \\ 1 \end{bmatrix}$$

where s is an arbitrary constant. We then have nontrivial solutions $\mathbf{p} \geq 0$ for any $s > 0$. ◆

EXAMPLE 3   Using Theorem 6.7.1

Let

$$E = \begin{bmatrix} 1 & 0 \\ 0 & 1 \end{bmatrix}$$

Then $(I - E)\mathbf{p} = \mathbf{0}$ has the general solution

$$\mathbf{p} = s \begin{bmatrix} 1 \\ 0 \end{bmatrix} + t \begin{bmatrix} 0 \\ 1 \end{bmatrix}$$

where s and t are independent arbitrary constants. Nontrivial solutions $\mathbf{p} \geq 0$ then result from any $s \geq 0$ and $t \geq 0$, not both zero. ◆

Example 2 indicates that in some situations one of the prices must be zero in order to satisfy the equilibrium condition. Example 3 indicates that there may be several linearly independent price structures available. Neither of these situations describes a truly interdependent economic structure. The following theorem gives sufficient conditions for both cases to be excluded.

THEOREM 6.7.2

Let E be an exchange matrix such that for some positive integer m all the entries of E^m are positive. Then there is exactly one linearly independent solution of $(I - E)\mathbf{p} = \mathbf{0}$, and it may be chosen so that all its entries are positive.

We will not give a proof of this theorem. The reader who has read Section 6.5 on Markov chains may observe that this theorem is essentially the same as Theorem 6.5.4. What we are calling exchange matrices in this section were called stochastic or Markov matrices in Section 6.5.

EXAMPLE 4 Using Theorem 6.7.2

The exchange matrix in Example 1 was

$$E = \begin{bmatrix} .2 & .1 & .6 \\ .4 & .5 & .1 \\ .4 & .4 & .3 \end{bmatrix}$$

Because $E > 0$, the condition $E^m > 0$ in Theorem 6.7.2 is satisfied for $m = 1$. Consequently, we are guaranteed that there is exactly one linearly independent solution of $(I - E)\mathbf{p} = \mathbf{0}$, and it can be chosen so that $\mathbf{p} > 0$. In that example, we found that

$$\mathbf{p} = \begin{bmatrix} 31 \\ 32 \\ 36 \end{bmatrix}$$

is such a solution. ◆

Leontief Open (Production) Model

In contrast with the closed model, in which the outputs of k industries are distributed only among themselves, the open model attempts to satisfy an outside demand for the outputs. Portions of these outputs may still be distributed among the industries themselves, to keep them operating, but there is to be some excess, some net production, with which to satisfy the outside demand. In the closed model the outputs of the industries are fixed, and our objective is to determine prices for these outputs so that the equilibrium condition, that expenditures equal incomes, is satisfied. In the open model it is the prices that are fixed, and our objective is to determine levels of the outputs of the industries needed to satisfy the outside demand. We will measure the levels of the outputs in terms of their economic values using the fixed prices. To be precise, over some fixed period of time, let

x_i = monetary value of the total output of the ith industry

d_i = monetary value of the output of the ith industry needed to satisfy the outside demand

c_{ij} = monetary value of the output of the ith industry needed by the jth industry to produce one unit of monetary value of its own output

With these quantities, we define the ***production vector***

$$\mathbf{x} = \begin{bmatrix} x_1 \\ x_2 \\ \vdots \\ x_k \end{bmatrix}$$

the ***demand vector***

$$\mathbf{d} = \begin{bmatrix} d_1 \\ d_2 \\ \vdots \\ d_k \end{bmatrix}$$

and the ***consumption matrix***

$$C = \begin{bmatrix} c_{11} & c_{12} & \cdots & c_{1k} \\ c_{21} & c_{22} & \cdots & c_{2k} \\ \vdots & \vdots & & \vdots \\ c_{k1} & c_{k2} & \cdots & c_{kk} \end{bmatrix}$$

By their nature, we have that

$$\mathbf{x} \geq 0, \quad \mathbf{d} \geq 0, \quad \text{and} \quad C \geq 0$$

From the definition of c_{ij} and x_j, it can be seen that the quantity

$$c_{i1}x_1 + c_{i2}x_2 + \cdots + c_{ik}x_k$$

is the value of the output of the ith industry needed by all k industries to produce a total output specified by the production vector $\mathbf{x}$. Because this quantity is simply the ith entry of the column vector $C\mathbf{x}$, we can say further that the ith entry of the column vector

$$\mathbf{x} - C\mathbf{x}$$

is the value of the excess output of the ith industry available to satisfy the outside demand. The value of the outside demand for the output of the ith industry is the ith entry of the demand vector $\mathbf{d}$. Consequently, we are led to the following equation

$$\mathbf{x} - C\mathbf{x} = \mathbf{d}$$

or

$$(I - C)\mathbf{x} = \mathbf{d} \tag{4}$$

for the demand to be exactly met, without any surpluses or shortages. Thus, given C and $\mathbf{d}$, our objective is to find a production vector $\mathbf{x} \geq 0$ that satisfies Equation (4).

EXAMPLE 5　Production Vector for a Town

A town has three main industries: a coal-mining operation, an electric power-generating plant, and a local railroad. To mine $1 of coal, the mining operation must purchase $.25 of electricity to run its equipment and $.25 of transportation for its shipping needs. To produce $1 of electricity, the generating plant requires $.65 of coal for fuel, $.05 of its own electricity to run auxiliary equipment, and $.05 of transportation. To provide $1 of transportation, the railroad requires $.55 of coal for fuel and $.10 of electricity for its auxiliary equipment. In a certain week the coal-mining operation receives orders for $50,000 of coal from outside the town, and the generating plant receives orders for $25,000 of electricity from outside. There is no outside demand for the local railroad. How much must each of the three industries produce in that week to exactly satisfy their own demand and the outside demand?

Solution

For the one-week period let

$$x_1 = \text{value of total output of coal-mining operation}$$
$$x_2 = \text{value of total output of power-generating plant}$$
$$x_3 = \text{value of total output of local railroad}$$

From the information supplied, the consumption matrix of the system is

$$C = \begin{bmatrix} 0 & .65 & .55 \\ .25 & .05 & .10 \\ .25 & .05 & 0 \end{bmatrix}$$

The linear system $(I - C)\mathbf{x} = \mathbf{d}$ is then

$$\begin{bmatrix} 1.00 & -.65 & -.55 \\ -.25 & .95 & -.10 \\ -.25 & -.05 & 1.00 \end{bmatrix} \begin{bmatrix} x_1 \\ x_2 \\ x_3 \end{bmatrix} = \begin{bmatrix} 50,000 \\ 25,000 \\ 0 \end{bmatrix}$$

The coefficient matrix on the left is invertible, and the solution is given by

$$\mathbf{x} = (I - C)^{-1}\mathbf{d} = \frac{1}{503} \begin{bmatrix} 756 & 542 & 470 \\ 220 & 690 & 190 \\ 200 & 170 & 630 \end{bmatrix} \begin{bmatrix} 50{,}000 \\ 25{,}000 \\ 0 \end{bmatrix} = \begin{bmatrix} 102{,}087 \\ 56{,}163 \\ 28{,}330 \end{bmatrix}$$

Thus, the total output of the coal-mining operation should be \$102,087, the total output of the power-generating plant should be \$56,163, and the total output of the railroad should be \$28,330. ◆

Let us reconsider Equation (4):

$$(I - C)\mathbf{x} = \mathbf{d}$$

If the square matrix $I - C$ is invertible, we can write

$$\mathbf{x} = (I - C)^{-1}\mathbf{d} \tag{5}$$

In addition, if the matrix $(I - C)^{-1}$ has only nonnegative entries, then we are guaranteed that for any $\mathbf{d} \geq 0$, Equation (5) has a unique nonnegative solution for $\mathbf{x}$. This is a particularly desirable situation, as it means that any outside demand can be met. The terminology used to describe this case is given in the following definition.

DEFINITION

A consumption matrix C is said to be *productive* if $(I - C)^{-1}$ exists and

$$(I - C)^{-1} \geq 0$$

We will now consider some simple criteria that guarantee that a consumption matrix is productive. The first is given in the following theorem.

THEOREM 6.7.3

Productive Consumption Matrix

A consumption matrix C is productive if and only if there is some production vector $\mathbf{x} \geq 0$ such that $\mathbf{x} > C\mathbf{x}$.

(The proof is outlined in Exercise 9.) The condition $\mathbf{x} > C\mathbf{x}$ means that there is some production schedule possible such that each industry produces more than it consumes.

Theorem 6.7.3 has two interesting corollaries. Suppose that all the row sums of C are less than 1. If

$$\mathbf{x} = \begin{bmatrix} 1 \\ 1 \\ \vdots \\ 1 \end{bmatrix}$$

then $C\mathbf{x}$ is a column vector whose entries are these row sums. Therefore, $\mathbf{x} > C\mathbf{x}$, and the condition of Theorem 6.7.3 is satisfied. Thus, we arrive at the following corollary:

COROLLARY 6.7.4

A consumption matrix is productive if each of its row sums is less than 1.

As we ask the reader to show in Exercise 8, this corollary leads to the following:

COROLLARY 6.7.5

> *A consumption matrix is productive if each of its column sums is less than* 1.

Recalling the definition of the entries of the consumption matrix C, we see that the jth column sum of C is the total value of the outputs of all k industries needed to produce one unit of value of output of the jth industry. The jth industry is thus said to be **profitable** if that jth column sum is less than 1. In other words, Corollary 6.7.5 says that a consumption matrix is productive if all k industries in the economic system are profitable.

EXAMPLE 6 Using Corollary 6.7.5

The consumption matrix in Example 5 was

$$C = \begin{bmatrix} 0 & .65 & .55 \\ .25 & .05 & .10 \\ .25 & .05 & 0 \end{bmatrix}$$

All three column sums in this matrix are less than 1 and so all three industries are profitable. Consequently, by Corollary 6.7.5, the consumption matrix C is productive. This can also be seen in the calculations in Example 5, as $(I - C)^{-1}$ is nonnegative.

◆

EXERCISE SET 6.7

1. For the following exchange matrices, find nonnegative price vectors that satisfy the equilibrium condition (3).

    (a) $\begin{bmatrix} \frac{1}{2} & \frac{1}{3} \\ \frac{1}{2} & \frac{2}{3} \end{bmatrix}$
    (b) $\begin{bmatrix} \frac{1}{2} & 0 & \frac{1}{2} \\ \frac{1}{3} & 0 & \frac{1}{2} \\ \frac{1}{6} & 1 & 0 \end{bmatrix}$
    (c) $\begin{bmatrix} .35 & .50 & .30 \\ .25 & .20 & .30 \\ .40 & .30 & .40 \end{bmatrix}$

2. Using Theorem 6.7.3 and its corollaries, show that each of the following consumption matrices is productive.

    (a) $\begin{bmatrix} .8 & .1 \\ .3 & .6 \end{bmatrix}$
    (b) $\begin{bmatrix} .70 & .30 & .25 \\ .20 & .40 & .25 \\ .05 & .15 & .25 \end{bmatrix}$
    (c) $\begin{bmatrix} .7 & .3 & .2 \\ .1 & .4 & .3 \\ .2 & .4 & .1 \end{bmatrix}$

3. Using Theorem 6.7.2, show that there is only one linearly independent price vector for the closed economic system with exchange matrix

    $$E = \begin{bmatrix} 0 & .2 & .5 \\ 1 & .2 & .5 \\ 0 & .6 & 0 \end{bmatrix}$$

4. Three neighbors have backyard vegetable gardens. Neighbor A grows tomatoes, neighbor B grows corn, and neighbor C grows lettuce. They agree to divide their crops among themselves as follows: A gets $\frac{1}{2}$ of the tomatoes, $\frac{1}{3}$ of the corn, and $\frac{1}{4}$ of the lettuce. B gets $\frac{1}{3}$ of the tomatoes, $\frac{1}{3}$ of the corn, and $\frac{1}{4}$ of the lettuce. C gets $\frac{1}{6}$ of the tomatoes, $\frac{1}{3}$ of the corn, and $\frac{1}{2}$ of the lettuce. What prices should the neighbors assign to their respective crops if the equilibrium condition of a closed economy is to be satisfied, and if the lowest-priced crop is to have a price of $100?

5. Three engineers—a civil engineer (CE), an electrical engineer (EE), and a mechanical engineer (ME)—each have a consulting firm. The consulting they do is of a multidisciplinary nature, so they buy a portion of each others' services. For each $1 of consulting the CE does, she buys $.10 of the EE's services and $.30 of the ME's services. For each $1 of consulting the EE does, she buys $.20 of the CE's services and $.40 of the ME's services. And for each $1 of consulting the ME does, she buys $.30 of the CE's services and $.40 of the EE's services. In a certain week the CE receives outside consulting orders of $500, the EE receives outside consulting orders of $700, and the ME receives outside consulting orders of $600. What dollar amount of consulting does each engineer perform in that week?

6. (a) Suppose that the demand d_i for the output of the ith industry increases by one unit. Explain why the ith column of the matrix $(I - C)^{-1}$ is the increase that must be made to the production vector $\mathbf{x}$ to satisfy this additional demand.

   (b) Referring to Example 5, use the result in part (a) to determine the increase in the value of the output of the coal-mining operation needed to satisfy a demand of one additional unit in the value of the output of the power-generating plant.

7. Using the fact that the column sums of an exchange matrix E are all 1, show that the column sums of $I - E$ are zero. From this, show that $I - E$ has zero determinant, and so $(I - E)\mathbf{p} = \mathbf{0}$ has nontrivial solutions for $\mathbf{p}$.

8. Show that Corollary 6.7.5 follows from Corollary 6.7.4.

   Hint   Use the fact that $(A^T)^{-1} = (A^{-1})^T$ for any invertible matrix A.

9. **(For Readers Who Have Studied Calculus)** Prove Theorem 6.7.3 as follows:

   (a) Prove the "only if " part of the theorem; that is, show that if C is a productive consumption matrix, then there is a vector $\mathbf{x} \geq 0$ such that $\mathbf{x} > C\mathbf{x}$.

   (b) Prove the "if " part of the theorem as follows:

   Step 1.   Show that if there is a vector $\mathbf{x}^* \geq 0$ such that $C\mathbf{x}^* < \mathbf{x}^*$, then $\mathbf{x}^* > 0$.

   Step 2.   Show that there is a number λ such that $0 < \lambda < 1$ and $C\mathbf{x}^* < \lambda\mathbf{x}^*$.

   Step 3.   Show that $C^n\mathbf{x}^* < \lambda^n\mathbf{x}^*$ for $n = 1, 2, \ldots$.

   Step 4.   Show that $C^n \to 0$ as $n \to$.

   Step 5.   By multiplying out, show that

   $$(I - C)(I + C + C^2 + \cdots + C^{n-1}) = I - C^n$$

   for $n = 1, 2, \ldots$.

   Step 6.   By letting $n \to$ in Step 5, show that the matrix infinite sum

   $$S = I + C + C^2 + \cdots$$

   exists and that $(I - C)S = I$.

   Step 7.   Show that $S \geq 0$ and that $S = (I - C)^{-1}$.

   Step 8.   Show that C is a productive consumption matrix.

SECTION 6.7
Technology
Exercises

The following exercises are designed to be solved using a technology utility. Typically, this will be MATLAB, *Mathematica*, Maple, Derive, or Mathcad, but it may also be some other type of linear algebra software or a scientific calculator with some linear algebra capabilities. For each exercise you will need to read the relevant documentation for the particular utility you are using. The goal of these exercises is to provide you with a basic proficiency with your technology utility. Once you have mastered the techniques in these exercises, you will be able to use your technology utility to solve many of the problems in the regular exercise sets.

T1. Consider a sequence of exchange matrices $\{E_2, E_3, E_4, E_5, \ldots, E_n\}$, where

$$E_2 = \begin{bmatrix} 0 & \frac{1}{2} \\ 1 & \frac{1}{2} \end{bmatrix}, \qquad E_3 = \begin{bmatrix} 0 & \frac{1}{2} & \frac{1}{3} \\ 1 & 0 & \frac{1}{3} \\ 0 & \frac{1}{2} & \frac{1}{3} \end{bmatrix},$$

$$E_4 = \begin{bmatrix} 0 & \frac{1}{2} & \frac{1}{3} & \frac{1}{4} \\ 1 & 0 & \frac{1}{3} & \frac{1}{4} \\ 0 & \frac{1}{2} & 0 & \frac{1}{4} \\ 0 & 0 & \frac{1}{3} & \frac{1}{4} \end{bmatrix}, \qquad E_5 = \begin{bmatrix} 0 & \frac{1}{2} & \frac{1}{3} & \frac{1}{4} & \frac{1}{5} \\ 1 & 0 & \frac{1}{3} & \frac{1}{4} & \frac{1}{5} \\ 0 & \frac{1}{2} & 0 & \frac{1}{4} & \frac{1}{5} \\ 0 & 0 & \frac{1}{3} & 0 & \frac{1}{5} \\ 0 & 0 & 0 & \frac{1}{4} & \frac{1}{5} \end{bmatrix}$$

and so on. Use a computer to show that $E_2^2 > 0_2$, $E_3^3 > 0_3$, $E_4^4 > 0_4$, $E_5^5 > 0_5$, and make the conjecture that although $E_n^n > 0_n$ is true, $E_n^k > 0_n$ is not true for $k = 1, 2, 3, \ldots, n - 1$. Next, use a computer to determine the vectors $\mathbf{p}_n$ such that $E_n \mathbf{p}_n = \mathbf{p}_n$ (for $n = 2, 3, 4, 5, 6$), and then see if you can discover a pattern that would allow you to compute $\mathbf{p}_{n+1}$ easily from $\mathbf{p}_n$. Test your discovery by first constructing $\mathbf{p}_8$ from

$$\mathbf{p}_7 = \begin{bmatrix} 2520 \\ 3360 \\ 1890 \\ 672 \\ 175 \\ 36 \\ 7 \end{bmatrix}$$

and then checking to see whether $E_8 \mathbf{p}_8 = \mathbf{p}_8$.

T2. Consider an open production model having n industries with $n > 1$. In order to produce $\$1$ of its own output, the jth industry must spend $\$(1/n)$ for the output of the ith industry (for all $i \neq j$), but the jth industry (for all $j = 1, 2, 3, \ldots, n$) spends nothing for its own output. Construct the consumption matrix C_n, show that it is productive, and determine an expression for $(I_n - C_n)^{-1}$. In determining an expression for $(I_n - C_n)^{-1}$, use a computer to study the cases when $n = 2, 3, 4,$ and 5; then make a conjecture and prove your conjecture to be true.

Hint If $F_n = [1]_{n \times n}$ (that is, the $n \times n$ matrix with every entry equal to 1), first show that

$$F_n^2 = nF_n$$

and then express your value of $(I_n - C_n)^{-1}$ in terms of n, I_n, and F_n.

6.8
COMPUTER GRAPHICS

In this section we assume that a view of a three-dimensional object is displayed on a video screen and show how matrix algebra can be used to obtain new views of the object by rotation, translation, and scaling.

PREREQUISITES: Matrix Algebra
Analytic Geometry

Visualization of a Three-Dimensional Object

Suppose that we want to visualize a three-dimensional object by displaying various views of it on a video screen. The object we have in mind to display is to be determined by a finite number of straight line segments. As an example, consider the truncated right pyramid with hexagonal base illustrated in Figure 6.8.1. We first introduce an xyz-coordinate system in which to embed the object. As in Figure 6.8.1, we orient the coordinate system so that its origin is at the center of the video screen and the xy-plane coincides with the plane of the screen. Consequently, an observer will see only the projection of the view of the three-dimensional object onto the two-dimensional xy-plane.

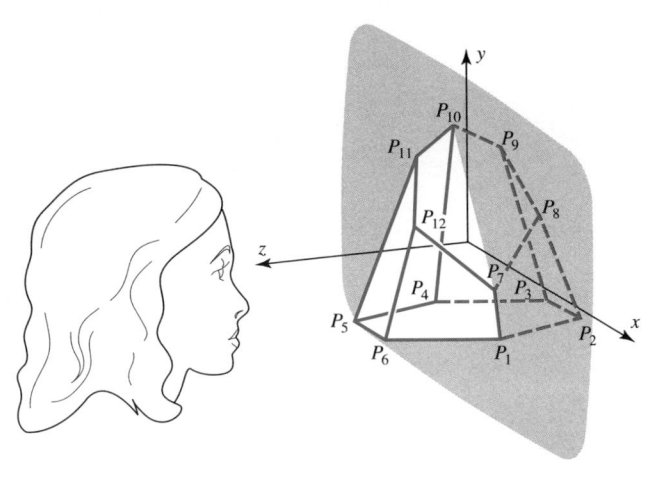

Figure 6.8.1

In the xyz-coordinate system, the endpoints $P_1, P_2, \ldots, P_n$ of the straight line segments that determine the view of the object will have certain coordinates—say,

$$(x_1, y_1, z_1), \quad (x_2, y_2, z_2), \ldots, \quad (x_n, y_n, z_n)$$

These coordinates, together with a specification of which pairs are to be connected by straight line segments, are to be stored in the memory of the video display system. For example, assume that the 12 vertices of the truncated pyramid in Figure 6.8.1 have the following coordinates (the screen is 4 units wide by 3 units high):

$$P_1: (1.000, -.800, .000), \qquad P_2: (.500, -.800, -.866),$$
$$P_3: (-.500, -.800, -.866), \qquad P_4: (-1.000, -.800, .000),$$
$$P_5: (-.500, -.800, .866), \qquad P_6: (.500, -.800, .866),$$
$$P_7: (.840, -.400, .000), \qquad P_8: (.315, .125, -.546),$$
$$P_9: (-.210, .650, -.364), \qquad P_{10}: (-.360, .800, .000),$$
$$P_{11}: (-.210, .650, .364), \qquad P_{12}: (.315, .125, .546)$$

These 12 vertices are connected pairwise by 18 straight line segments as follows, where $P_i \leftrightarrow P_j$ denotes that point P_i is connected to point P_j:

$$P_1 \leftrightarrow P_2, \quad P_2 \leftrightarrow P_3, \quad P_3 \leftrightarrow P_4, \quad P_4 \leftrightarrow P_5, \quad P_5 \leftrightarrow P_6, \quad P_6 \leftrightarrow P_1,$$
$$P_7 \leftrightarrow P_8, \quad P_8 \leftrightarrow P_9, \quad P_9 \leftrightarrow P_{10}, \quad P_{10} \leftrightarrow P_{11}, \quad P_{11} \leftrightarrow P_{12}, \quad P_{12} \leftrightarrow P_7,$$
$$P_1 \leftrightarrow P_7, \quad P_2 \leftrightarrow P_8, \quad P_3 \leftrightarrow P_9, \quad P_4 \leftrightarrow P_{10}, \quad P_5 \leftrightarrow P_{11}, \quad P_6 \leftrightarrow P_{12}$$

In View 1 these 18 straight line segments are shown as they would appear on the video screen. It should be noticed that only the x- and y-coordinates of the vertices are needed by the video display system to draw the view, because only the projection of the object onto the xy-plane is displayed. However, we must keep track of the z-coordinates to carry out certain transformations discussed later.

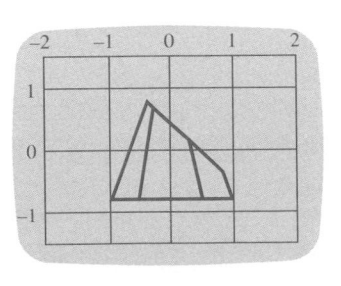

View 1

We now show how to form new views of the object by scaling, translating, or rotating the initial view. We first construct a $3 \times n$ matrix P, referred to as the *coordinate matrix of the view*, whose columns are the coordinates of the n points of a view:

$$P = \begin{bmatrix} x_1 & x_2 & \cdots & x_n \\ y_1 & y_2 & \cdots & y_n \\ z_1 & z_2 & \cdots & z_n \end{bmatrix}$$

For example, the coordinate matrix P corresponding to View 1 is the 3×12 matrix

$$\begin{bmatrix} 1.000 & .500 & -.500 & -1.000 & -.500 & .500 & .840 & .315 & -.210 & -.360 & -.210 & .315 \\ -.800 & -.800 & -.800 & -.800 & -.800 & -.800 & -.400 & .125 & .650 & .800 & .650 & .125 \\ .000 & -.866 & -.866 & .000 & .866 & .866 & .000 & -.546 & -.364 & .000 & .364 & .546 \end{bmatrix}$$

We will show below how to transform the coordinate matrix P of a view to a new coordinate matrix P' corresponding to a new view of the object. The straight line segments connecting the various points move with the points as they are transformed. In this way, each view is uniquely determined by its coordinate matrix once we have specified which pairs of points in the original view are to be connected by straight lines.

Scaling

The first type of transformation we consider consists of scaling a view along the x, y, and z directions by factors of α, β, and γ, respectively. By this we mean that if a point P_i has coordinates (x_i, y_i, z_i) in the original view, it is to move to a new point P_i' with coordinates $(\alpha x_i, \beta y_i, \gamma z_i)$ in the new view. This has the effect of transforming a unit cube in the original view to a rectangular parallelepiped of dimensions $\alpha \times \beta \times \gamma$ (Figure 6.8.2). Mathematically, this may be accomplished with matrix multiplication as follows. Define a 3×3 diagonal matrix

$$S = \begin{bmatrix} \alpha & 0 & 0 \\ 0 & \beta & 0 \\ 0 & 0 & \gamma \end{bmatrix}$$

Then, if a point P_i in the original view is represented by the column vector

$$\begin{bmatrix} x_i \\ y_i \\ z_i \end{bmatrix}$$

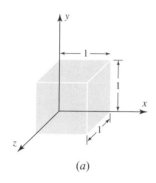

(a)

then the transformed point P_i' is represented by the column vector

$$\begin{bmatrix} x_i' \\ y_i' \\ z_i' \end{bmatrix} = \begin{bmatrix} \alpha & 0 & 0 \\ 0 & \beta & 0 \\ 0 & 0 & \gamma \end{bmatrix} \begin{bmatrix} x_i \\ y_i \\ z_i \end{bmatrix}$$

Using the coordinate matrix P, which contains the coordinates of all n points of the original view as its columns, we can transform these n points simultaneously to produce the coordinate matrix P' of the scaled view, as follows:

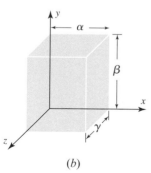

(b)

Figure 6.8.2

$$SP = \begin{bmatrix} \alpha & 0 & 0 \\ 0 & \beta & 0 \\ 0 & 0 & \gamma \end{bmatrix} \begin{bmatrix} x_1 & x_2 & \cdots & x_n \\ y_1 & y_2 & \cdots & y_n \\ z_1 & z_2 & \cdots & z_n \end{bmatrix}$$

$$= \begin{bmatrix} \alpha x_1 & \alpha x_2 & \cdots & \alpha x_n \\ \beta y_1 & \beta y_2 & \cdots & \beta y_n \\ \gamma z_1 & \gamma z_2 & \cdots & \gamma z_n \end{bmatrix} = P'$$

The new coordinate matrix can then be entered into the video display system to produce the new view of the object. As an example, View 2 is View 1 scaled by setting $\alpha = 1.8$, $\beta = 0.5$, and $\gamma = 3.0$. Note that the scaling $\gamma = 3.0$ along the z-axis is not visible in View 2, since we see only the projection of the object onto the xy-plane.

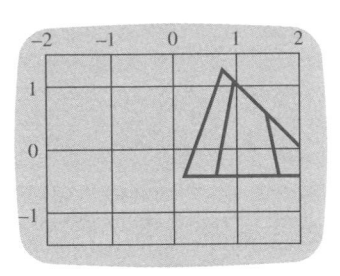

View 2   View 1 scaled by $\alpha = 1.8$, $\beta = 0.5$, $\gamma = 3.0$.

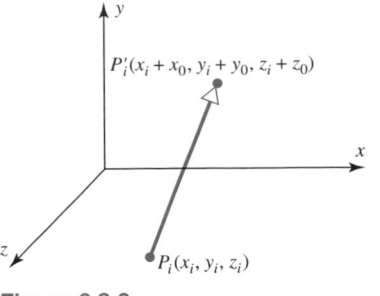

Figure 6.8.3

Translation

We next consider the transformation of translating or displacing an object to a new position on the screen. Referring to Figure 6.8.3, suppose we desire to change an existing view so that each point P_i with coordinates (x_i, y_i, z_i) moves to a new point P_i' with coordinates $(x_i + x_0, y_i + y_0, z_i + z_0)$. The vector

$$\begin{bmatrix} x_0 \\ y_0 \\ z_0 \end{bmatrix}$$

is called the **translation vector** of the transformation. By defining a $3 \times n$ matrix T as

$$T = \begin{bmatrix} x_0 & x_0 & \cdots & x_0 \\ y_0 & y_0 & \cdots & y_0 \\ z_0 & z_0 & \cdots & z_0 \end{bmatrix}$$

we can translate all n points of the view determined by the coordinate matrix P by matrix addition via the equation

$$P' = P + T$$

The coordinate matrix P' then specifies the new coordinates of the n points. For example, if we wish to translate View 1 according to the translation vector

$$\begin{bmatrix} 1.2 \\ 0.4 \\ 1.7 \end{bmatrix}$$

the result is View 3. Note, again, that the translation $z_0 = 1.7$ along the z-axis does not show up explicitly in View 3.

In Exercise 7, a technique of performing translations by matrix multiplication rather than by matrix addition is explained.

View 3   View 1 translated by $x_0 = 1.2$, $y_0 = 0.4$, $z_0 = 1.7$.

Rotation

A more complicated type of transformation is a rotation of a view about one of the three coordinate axes. We begin with a rotation about the z-axis (the axis perpendicular to the screen) through an angle θ. Given a point P_i in the original view with coordinates (x_i, y_i, z_i), we wish to compute the new coordinates (x_i', y_i', z_i') of the rotated point P_i'.

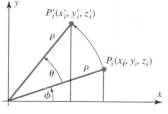

Figure 6.8.4

Referring to Figure 6.8.4 and using a little trigonometry, the reader should be able to derive the following:

$$x_i' = \rho \cos(\phi + \theta) = \rho \cos \phi \cos \theta - \rho \sin \phi \sin \theta = x_i \cos \theta - y_i \sin \theta$$
$$y_i' = \rho \sin(\phi + \theta) = \rho \cos \phi \sin \theta + \rho \sin \phi \cos \theta = x_i \sin \theta + y_i \cos \theta$$
$$z_i' = z_i$$

These equations can be written in matrix form as

$$\begin{bmatrix} x_i' \\ y_i' \\ z_i' \end{bmatrix} = \begin{bmatrix} \cos \theta & -\sin \theta & 0 \\ \sin \theta & \cos \theta & 0 \\ 0 & 0 & 1 \end{bmatrix} \begin{bmatrix} x_i \\ y_i \\ z_i \end{bmatrix}$$

If we let R denote the 3×3 matrix in this equation, all n points can be rotated by the matrix product

$$P' = RP$$

to yield the coordinate matrix P' of the rotated view.

Rotations about the x- and y-axes can be accomplished analogously, and the resulting rotation matrices are given with Views 4, 5, and 6. These three new views of the truncated pyramid correspond to rotations of View 1 about the x-, y-, and z-axes, respectively, each through an angle of $90°$.

Rotation about the x-axis

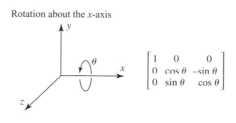

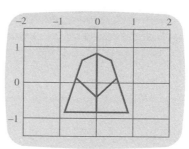

View 4   View 1 rotated $90°$ about the x-axis.

Rotation about the y-axis

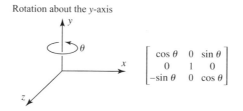

View 5   View 1 rotated $90°$ about the y-axis.

Rotations about three coordinate axes may be combined to give oblique views of an object. For example, View 7 is View 1 rotated first about the x-axis through $30°$, then about the y-axis through $-70°$, and finally about the z-axis through $-27°$. Mathematically, these three successive rotations can be embodied in the single transformation

Rotation about the z-axis

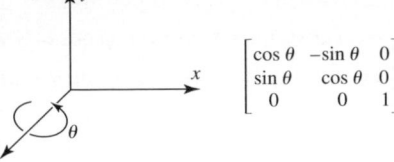

$$\begin{bmatrix} \cos\theta & -\sin\theta & 0 \\ \sin\theta & \cos\theta & 0 \\ 0 & 0 & 1 \end{bmatrix}$$

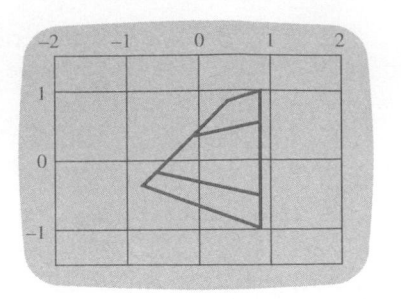

View 6 View 1 rotated 90° about the z-axis.

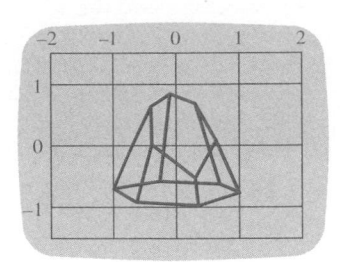

View 7 Oblique view of truncated pyramid.

equation $P' = RP$, where R is the product of three individual rotation matrices:

$$R_1 = \begin{bmatrix} 1 & 0 & 0 \\ 0 & \cos(30°) & -\sin(30°) \\ 0 & \sin(30°) & \cos(30°) \end{bmatrix}$$

$$R_2 = \begin{bmatrix} \cos(-70°) & 0 & \sin(-70°) \\ 0 & 1 & 0 \\ -\sin(-70°) & 0 & \cos(-70°) \end{bmatrix}$$

$$R_3 = \begin{bmatrix} \cos(-27°) & -\sin(-27°) & 0 \\ \sin(-27°) & \cos(-27°) & 0 \\ 0 & 0 & 1 \end{bmatrix}$$

in the order

$$R = R_3 R_2 R_1 = \begin{bmatrix} .305 & -.025 & -.952 \\ -.155 & .985 & -.076 \\ .940 & .171 & .296 \end{bmatrix}$$

As a final illustration, in View 8 we have two separate views of the truncated pyramid, which constitute a stereoscopic pair. They were produced by first rotating View 7 about the y-axis through an angle of $-3°$ and translating it to the right, then rotating the same View 7 about the y-axis through an angle of $+3°$ and translating it to the left. The translation distances were chosen so that the stereoscopic views are about $2\frac{1}{2}$ inches apart—the approximate distance between a pair of eyes.

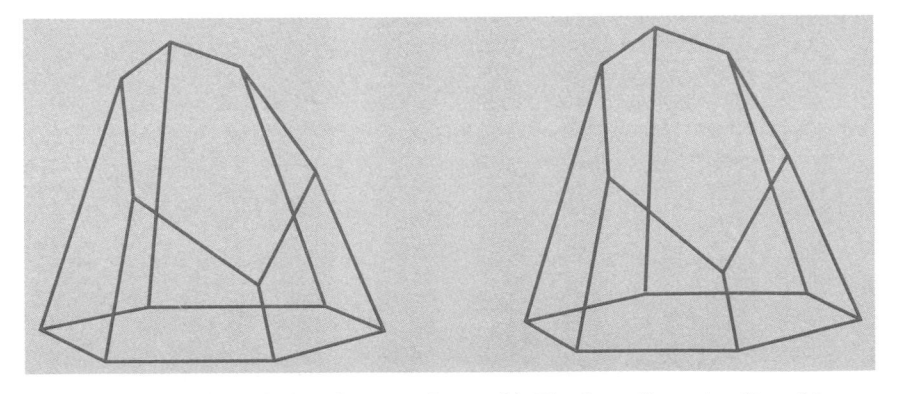

View 8 Stereoscopic figure of truncated pyramid. The three-dimensionality of the diagram can be seen by holding the book about one foot away and focusing on a distant object. Then by shifting gaze to View 8 without refocusing, you can make the two views of the stereoscopic pair merge together and produce the desired effect.

EXERCISE SET
6.8

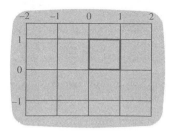

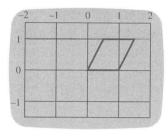

View 9   Square with vertices $(0, 0, 0)$, $(1, 0, 0)$, $(1, 1, 0)$, and $(0, 1, 0)$ (Exercises 1 and 2).

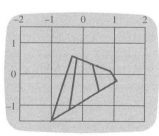

View 10   View 9 sheared along the x-axis by $\frac{1}{2}$ with respect to the y-coordinate (Exercise 2).

1. View 9 is a view of a square with vertices $(0, 0, 0)$, $(1, 0, 0)$, $(1, 1, 0)$, and $(0, 1, 0)$.

   (a)   What is the coordinate matrix of View 9?

   (b)   What is the coordinate matrix of View 9 after it is scaled by a factor $1\frac{1}{2}$ in the x-direction and $\frac{1}{2}$ in the y-direction? Draw a sketch of the scaled view.

   (c)   What is the coordinate matrix of View 9 after it is translated by the following vector?

$$\begin{bmatrix} -2 \\ -1 \\ 3 \end{bmatrix}$$

   Draw a sketch of the translated view.

   (d)   What is the coordinate matrix of View 9 after it is rotated through an angle of $-30°$ about the z-axis? Draw a sketch of the rotated view.

2. (a)   If the coordinate matrix of View 9 is multiplied by the matrix

$$\begin{bmatrix} 1 & \frac{1}{2} & 0 \\ 0 & 1 & 0 \\ 0 & 0 & 1 \end{bmatrix}$$

   the result is the coordinate matrix of View 10. Such a transformation is called a *shear in the x-direction with factor $\frac{1}{2}$ with respect to the y-coordinate*. Show that under such a transformation, a point with coordinates (x_i, y_i, z_i) has new coordinates $(x_i + \frac{1}{2}y_i, y_i, z_i)$.

   (b)   What are the coordinates of the four vertices of the shear square in View 10?

   (c)   The matrix

$$\begin{bmatrix} 1 & 0 & 0 \\ .6 & 1 & 0 \\ 0 & 0 & 1 \end{bmatrix}$$

   determines a *shear in the y-direction with factor .6 with respect to the x-coordinate* (an example appears in View 11). Sketch a view of the square in View 9 after such a shearing transformation, and find the new coordinates of its four vertices.

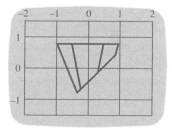

View 11   View 1 sheared along the y-axis by .6 with respect to the x-coordinate (Exercise 2).

View 12   View 1 reflected about the xz-plane (Exercise 3).

3. (a)   The *reflection about the xz-plane* is defined as the transformation that takes a point (x_i, y_i, z_i) to the point $(x_i, -y_i, z_i)$ (e.g., View 12). If P and P' are the coordinate matrices of a view and its reflection about the xz-plane, respectively, find a matrix M such that $P' = MP$.

   (b)   Analogous to part (a), define the *reflection about the yz-plane* and construct the corresponding transformation matrix. Draw a sketch of View 1 reflected about the yz-plane.

   (c)   Analogous to part (a), define the *reflection about the xy-plane* and construct the corresponding transformation matrix. Draw a sketch of View 1 reflected about the xy-plane.

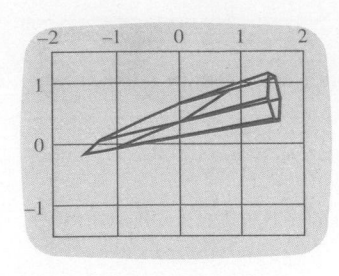

View 13 View 1 scaled, translated, and rotated (Exercise 4).

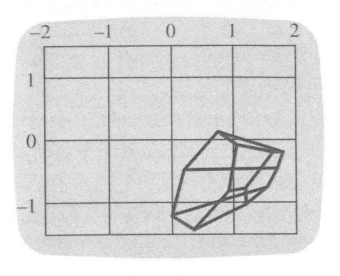

View 14 View 1 scaled, translated, and rotated (Exercise 5).

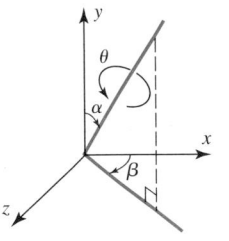

Figure Ex-6

4. (a) View 13 is View 1 subject to the following five transformations:

   1. Scale by a factor of $\frac{1}{2}$ in the x-direction, 2 in the y-direction, and $\frac{1}{3}$ in the z-direction.

   2. Translate $\frac{1}{2}$ unit in the x-direction.

   3. Rotate $20°$ about the x-axis.

   4. Rotate $-45°$ about the y-axis.

   5. Rotate $90°$ about the z-axis.

   Construct the five matrices M_1, M_2, M_3, M_4, and M_5 associated with these five transformations.

(b) If P is the coordinate matrix of View 1 and P' is the coordinate matrix of View 13, express P' in terms of M_1, M_2, M_3, M_4, M_5, and P.

5. (a) View 14 is View 1 subject to the following seven transformations:

   1. Scale by a factor of .3 in the x-direction and by a factor of .5 in the y-direction.

   2. Rotate $45°$ about the x-axis.

   3. Translate 1 unit in the x-direction.

   4. Rotate $35°$ about the y-axis.

   5. Rotate $-45°$ about the z-axis.

   6. Translate 1 unit in the z-direction.

   7. Scale by a factor of 2 in the x-direction.

   Construct the matrices M_1, M_2, $\ldots$, M_7 associated with these seven transformations.

(b) If P is the coordinate matrix of View 1 and P' is the coordinate matrix of View 14, express P' in terms of M_1, M_2, $\ldots$, M_7, and P.

6. Suppose that a view with coordinate matrix P is to be rotated through an angle θ about an axis through the origin and specified by two angles α and β (see Figure Ex-6). If P' is the coordinate matrix of the rotated view, find rotation matrices R_1, R_2, R_3, R_4, and R_5 such that

$$P' = R_5 R_4 R_3 R_2 R_1 P$$

Hint The desired rotation can be accomplished in the following five steps:

   1. Rotate through an angle of β about the y-axis.

   2. Rotate through an angle of α about the z-axis.

   3. Rotate through an angle of θ about the y-axis.

   4. Rotate through an angle of $-\alpha$ about the z-axis.

   5. Rotate through an angle of $-\beta$ about the y-axis.

7. This exercise illustrates a technique for translating a point with coordinates (x_i, y_i, z_i) to a point with coordinates $(x_i + x_0, y_i + y_0, z_i + z_0)$ by matrix multiplication rather than matrix addition.

(a) Let the point (x_i, y_i, z_i) be associated with the column vector

$$\mathbf{v}_i = \begin{bmatrix} x_i \\ y_i \\ z_i \\ 1 \end{bmatrix}$$

and let the point $(x_i + x_0, y_i + y_0, z_i + z_0)$ be associated with the column vector

$$\mathbf{v}'_i = \begin{bmatrix} x_i + x_0 \\ y_i + y_0 \\ z_i + z_0 \\ 1 \end{bmatrix}$$

Find a 4×4 matrix M such that $\mathbf{v}'_i = M\mathbf{v}_i$.

(b)   Find the specific 4×4 matrix of the above form that will effect the translation of the point $(4, -2, 3)$ to the point $(-1, 7, 0)$.

8.   For the three rotation matrices given with Views 4, 5, and 6, show that

$$R^{-1} = R^T$$

(A matrix with this property is called an ***orthogonal matrix***.)

SECTION 6.8

Technology Exercises

The following exercises are designed to be solved using a technology utility. Typically, this will be MATLAB, *Mathematica*, Maple, Derive, or Mathcad, but it may also be some other type of linear algebra software or a scientific calculator with some linear algebra capabilities. For each exercise you will need to read the relevant documentation for the particular utility you are using. The goal of these exercises is to provide you with a basic proficiency with your technology utility. Once you have mastered the techniques in these exercises, you will be able to use your technology utility to solve many of the problems in the regular exercise sets.

T1.   Let (a, b, c) be a unit vector normal to the plane $ax + by + cz = 0$, and let $\mathbf{r} = (x, y, z)$ be a vector. It can be shown that the mirror image of the vector $\mathbf{r}$ through the above plane has coordinates $\mathbf{r}_m = (x_m, y_m, z_m)$, where

$$\begin{bmatrix} x_m \\ y_m \\ z_m \end{bmatrix} = M \begin{bmatrix} x \\ y \\ z \end{bmatrix}$$

with

$$M = I - 2\mathbf{n}\mathbf{n}^T = \begin{bmatrix} 1 & 0 & 0 \\ 0 & 1 & 0 \\ 0 & 0 & 1 \end{bmatrix} - 2 \begin{bmatrix} a \\ b \\ c \end{bmatrix} \begin{bmatrix} a & b & c \end{bmatrix}$$

(a)   Show that $M^2 = I$ and give a physical reason why this must be so.

   Hint   Use the fact that (a, b, c) is a unit vector to show that $\mathbf{n}^T\mathbf{n} = 1$.

(b)   Use a computer to show that $\det(M) = -1$.

(c)   The eigenvectors of M satisfy the equation

$$\begin{bmatrix} x_m \\ y_m \\ z_m \end{bmatrix} = M \begin{bmatrix} x \\ y \\ z \end{bmatrix} = \lambda \begin{bmatrix} x \\ y \\ z \end{bmatrix}$$

and therefore correspond to those vectors whose direction is not affected by a reflection through the plane. Use a computer to determine the eigenvectors and eigenvalues of M, and then give a physical argument to support your answer.

T2.   A vector $\mathbf{v} = (x, y, z)$ is rotated by an angle θ about an axis having unit vector (a, b, c), thereby forming the rotated vector $\mathbf{v}_R = (x_R, y_R, z_R)$. It can be shown that

$$\begin{bmatrix} x_R \\ y_R \\ z_R \end{bmatrix} = R(\theta) \begin{bmatrix} x \\ y \\ z \end{bmatrix}$$

with

$$R(\theta) = \cos(\theta) \begin{bmatrix} 1 & 0 & 0 \\ 0 & 1 & 0 \\ 0 & 0 & 1 \end{bmatrix} + (1 - \cos(\theta)) \begin{bmatrix} a \\ b \\ c \end{bmatrix} \begin{bmatrix} a & b & c \end{bmatrix}$$

$$+ \sin(\theta) \begin{bmatrix} 0 & -c & b \\ c & 0 & -a \\ -b & a & 0 \end{bmatrix}$$

(a) Use a computer to show that $R(\theta)R(\varphi) = R(\theta + \varphi)$, and then give a physical reason why this must be so. Depending on the sophistication of the computer you are using, you may have to experiment using different values of a, b, and

$$c = \sqrt{1 - a^2 - b^2}$$

(b) Show also that $R^{-1}(\theta) = R(-\theta)$ and give a physical reason why this must be so.

(c) Use a computer to show that $\det(R(\theta)) = +1$.

6.9 CRYPTOGRAPHY

In this section we present a method of encoding and decoding messages. We also examine modular arithmetic and show how Gaussian elimination can sometimes be used to break an opponent's code.

> PREREQUISITES: Matrices
> Gaussian Elimination
> Matrix Operations
> Linear Independence
> Linear Transformations

Ciphers

The study of encoding and decoding secret messages is called ***cryptography***. Although secret codes date to the earliest days of written communication, there has been a recent surge of interest in the subject because of the need to maintain the privacy of information transmitted over public lines of communication. In the language of cryptography, codes are called ***ciphers***, uncoded messages are called ***plaintext***, and coded messages are called ***ciphertext***. The process of converting from plaintext to ciphertext is called ***enciphering***, and the reverse process of converting from ciphertext to plaintext is called ***deciphering***.

The simplest ciphers, called ***substitution ciphers***, are those that replace each letter of the alphabet by a different letter. For example, in the substitution cipher

Plain   *A B C D E F G H I J K L M N O P Q R S T U V W X Y Z*
Cipher *D E F G H I J K L M N O P Q R S T U V W X Y Z A B C*

the plaintext letter *A* is replaced by *D*, the plaintext letter *B* by *E*, and so forth. With this cipher the plaintext message

ROME WAS NOT BUILT IN A DAY

becomes

URPH ZDV QRW EXLOW LQ D GDB

Hill Ciphers

A disadvantage of substitution ciphers is that they preserve the frequencies of individual letters, making it relatively easy to break the code by statistical methods. One way to overcome this problem is to divide the plaintext into groups of letters and encipher the plaintext group by group, rather than one letter at a time. A system of cryptography in which the plaintext is divided into sets of n letters, each of which is replaced by a set of n cipher letters, is called a ***polygraphic system***. In this section we will study a class of polygraphic systems based on matrix transformations. (The ciphers that we will discuss are called ***Hill ciphers*** after Lester S. Hill, who introduced them in two papers: "Cryptography in an Algebraic Alphabet," *American Mathematical Monthly, 36* (June–July 1929), pp. 306–312; and "Concerning Certain Linear Transformation Apparatus of Cryptography," *American Mathematical Monthly, 38* (March 1931), pp. 135–154.)

In the discussion to follow, we assume that each plaintext and ciphertext letter except *Z* is assigned the numerical value that specifies its position in the standard alphabet (Table 1). For reasons that will become clear later, *Z* is assigned a value of zero.

Table 1

A	B	C	D	E	F	G	H	I	J	K	L	M	N	O	P	Q	R	S	T	U	V	W	X	Y	Z
1	2	3	4	5	6	7	8	9	10	11	12	13	14	15	16	17	18	19	20	21	22	23	24	25	0

In the simplest Hill ciphers, successive *pairs* of plaintext are transformed into ciphertext by the following procedure:

Step 1. Choose a 2×2 matrix with integer entries

$$A = \begin{bmatrix} a_{11} & a_{12} \\ a_{21} & a_{22} \end{bmatrix}$$

to perform the encoding. Certain additional conditions on *A* will be imposed later.

Step 2. Group successive plaintext letters into pairs, adding an arbitrary "dummy" letter to fill out the last pair if the plaintext has an odd number of letters, and replace each plaintext letter by its numerical value.

Step 3. Successively convert each plaintext pair $p_1 p_2$ into a column vector

$$\mathbf{p} - \begin{bmatrix} p_1 \\ p_2 \end{bmatrix}$$

and form the product $A\mathbf{p}$. We will call $\mathbf{p}$ a *plaintext vector* and $A\mathbf{p}$ the corresponding *ciphertext vector*.

Step 4. Convert each ciphertext vector into its alphabetic equivalent.

EXAMPLE 1   Hill Cipher of a Message

Use the matrix

$$\begin{bmatrix} 1 & 2 \\ 0 & 3 \end{bmatrix}$$

to obtain the Hill cipher for the plaintext message

I AM HIDING

Solution

If we group the plaintext into pairs and add the dummy letter *G* to fill out the last pair, we obtain

$$IA \quad\quad MH \quad\quad ID \quad\quad IN \quad\quad GG$$

or, equivalently, from Table 1,

$$9\ 1 \quad\quad 13\ 8 \quad\quad 9\ 4 \quad\quad 9\ 14 \quad\quad 7\ 7$$

To encipher the pair *IA*, we form the matrix product

$$\begin{bmatrix} 1 & 2 \\ 0 & 3 \end{bmatrix} \begin{bmatrix} 9 \\ 1 \end{bmatrix} = \begin{bmatrix} 11 \\ 3 \end{bmatrix}$$

which, from Table 1, yields the ciphertext *KC*.

To encipher the pair *MH*, we form the product

$$\begin{bmatrix} 1 & 2 \\ 0 & 3 \end{bmatrix} \begin{bmatrix} 13 \\ 8 \end{bmatrix} = \begin{bmatrix} 29 \\ 24 \end{bmatrix} \tag{1}$$

However, there is a problem here, because the number 29 has no alphabet equivalent (Table 1). To resolve this problem, we make the following agreement:

> *Whenever an integer greater than* 25 *occurs, it will be replaced by the remainder that results when this integer is divided by* 26.

Because the remainder after division by 26 is one of the integers 0, 1, 2, ..., 25, this procedure will always yield an integer with an alphabet equivalent.

Thus, in (1) we replace 29 by 3, which is the remainder after dividing 29 by 26. It now follows from Table 1 that the ciphertext for the pair *MH* is *CX*.

The computations for the remaining ciphertext vectors are

$$\begin{bmatrix} 1 & 2 \\ 0 & 3 \end{bmatrix} \begin{bmatrix} 9 \\ 4 \end{bmatrix} = \begin{bmatrix} 17 \\ 12 \end{bmatrix}$$

$$\begin{bmatrix} 1 & 2 \\ 0 & 3 \end{bmatrix} \begin{bmatrix} 9 \\ 14 \end{bmatrix} = \begin{bmatrix} 37 \\ 42 \end{bmatrix} \quad \text{or} \quad \begin{bmatrix} 11 \\ 16 \end{bmatrix}$$

$$\begin{bmatrix} 1 & 2 \\ 0 & 3 \end{bmatrix} \begin{bmatrix} 7 \\ 7 \end{bmatrix} = \begin{bmatrix} 21 \\ 21 \end{bmatrix}$$

These correspond to the ciphertext pairs *QL*, *KP*, and *UU*, respectively. In summary, the entire ciphertext message is

$$KC \quad\quad CX \quad\quad QL \quad\quad KP \quad\quad UU$$

which would usually be transmitted as a single string without spaces:

$$KCCXQLKPUU \quad \blacklozenge$$

Because the plaintext was grouped in pairs and enciphered by a 2×2 matrix, the Hill cipher in Example 1 is referred to as a ***Hill 2-cipher***. It is obviously also possible to group the plaintext in triples and encipher by a 3×3 matrix with integer entries; this is called a ***Hill 3-cipher***. In general, for a ***Hill n-cipher***, plaintext is grouped into sets of *n* letters and enciphered by an $n \times n$ matrix with integer entries.

Modular Arithmetic

In Example 1, integers greater than 25 were replaced by their remainders after division by 26. This technique of working with remainders is at the core of a body of mathematics called *modular arithmetic*. Because of its importance in cryptography, we will digress for a moment to touch on some of the main ideas in this area.

In modular arithmetic we are given a positive integer *m*, called the ***modulus***, and any two integers whose difference is an integer multiple of the modulus are regarded as "equal" or "equivalent" with respect to the modulus. More precisely, we make the following definition.

DEFINITION

If *m* is a positive integer and *a* and *b* are any integers, then we say that *a* is ***equivalent*** to *b* modulo *m*, written

$$a = b \pmod{m}$$

if $a - b$ is an integer multiple of *m*.

EXAMPLE 2 Various Equivalences

$$7 = 2 \quad (\text{mod } 5)$$
$$19 = 3 \quad (\text{mod } 2)$$
$$-1 = 25 \quad (\text{mod } 26)$$
$$12 = 0 \quad (\text{mod } 4) \quad \blacklozenge$$

For any modulus m it can be proved that every integer a is equivalent, modulo m, to exactly one of the integers

$$0, 1, 2, \ldots, m - 1$$

We call this integer the **residue** of a modulo m, and we write

$$Z_m = \{0, 1, 2, \ldots, m - 1\}$$

to denote the set of residues modulo m.

If a is a *nonnegative* integer, then its residue modulo m is simply the remainder that results when a is divided by m. For an arbitrary integer a, the residue can be found using the following theorem.

THEOREM 6.9.1

For any integer a and modulus m, let

$$R = \text{remainder of } \frac{|a|}{m}$$

Then the residue r of a modulo m is given by

$$r = \begin{cases} R & \text{if } a \geq 0 \\ m - R & \text{if } a < 0 \quad \text{and} \quad R \neq 0 \\ 0 & \text{if } a < 0 \quad \text{and} \quad R = 0 \end{cases}$$

EXAMPLE 3 Residues mod 26

Find the residue modulo 26 of (a) 87, (b) -38, and (c) -26.

Solution (a)

Dividing $|87| = 87$ by 26 yields a remainder of $R = 9$, so $r = 9$. Thus,

$$87 = 9 \quad (\text{mod } 26)$$

Solution (b)

Dividing $|-38| = 38$ by 26 yields a remainder of $R = 12$, so $r = 26 - 12 = 14$. Thus,

$$-38 = 14 \quad (\text{mod } 26)$$

Solution (c)

Dividing $|-26| = 26$ by 26 yields a remainder of $R = 0$. Thus,

$$-26 = 0 \quad (\text{mod } 26) \quad \blacklozenge$$

In ordinary arithmetic every nonzero number a has a *reciprocal* or *multiplicative inverse*, denoted by a^{-1}, such that

$$aa^{-1} = a^{-1}a = 1$$

In modular arithmetic we have the following corresponding concept:

DEFINITION

If a is a number in Z_m, then a number a^{-1} in Z_m is called a *reciprocal* or *multiplicative inverse* of a modulo m if $aa^{-1} = a^{-1}a = 1 \pmod{m}$.

It can be proved that if a and m have no common prime factors, then a has a unique reciprocal modulo m; conversely, if a and m have a common prime factor, then a has no reciprocal modulo m.

EXAMPLE 4   Reciprocal of 3 mod 26

The number 3 has a reciprocal modulo 26 because 3 and 26 have no common prime factors. This reciprocal can be obtained by finding the number x in Z_{26} that satisfies the modular equation

$$3x = 1 \pmod{26}$$

Although there are general methods for solving such modular equations, it would take us too far afield to study them. However, because 26 is relatively small, this equation can be solved by trying the possible solutions, 0 to 25, one at a time. With this approach we find that $x = 9$ is the solution, because

$$3 \cdot 9 = 27 = 1 \pmod{26}$$

Thus,

$$3^{-1} = 9 \pmod{26} \quad \blacklozenge$$

EXAMPLE 5   A Number with No Reciprocal mod 26

The number 4 has no reciprocal modulo 26, because 4 and 26 have 2 as a common prime factor (see Exercise 8). $\blacklozenge$

For future reference, we provide the following table of reciprocals modulo 26:

Table 2   Reciprocals Modulo 26

a	1	3	5	7	9	11	15	17	19	21	23	25
a^{-1}	1	9	21	15	3	19	7	23	11	5	17	25

Deciphering

Every useful cipher must have a procedure for decipherment. In the case of a Hill cipher, decipherment uses the inverse (mod 26) of the enciphering matrix. To be precise, if m is a positive integer, then a square matrix A with entries in Z_m is said to be *invertible modulo m* if there is a matrix B with entries in Z_m such that

$$AB = BA = I \pmod{m}$$

Suppose now that

$$A = \begin{bmatrix} a_{11} & a_{12} \\ a_{21} & a_{22} \end{bmatrix}$$

is invertible modulo 26 and this matrix is used in a Hill 2-cipher. If

$$\mathbf{p} = \begin{bmatrix} p_1 \\ p_2 \end{bmatrix}$$

is a plaintext vector, then

$$\mathbf{c} = A\mathbf{p} \quad (\text{mod } 26)$$

is the corresponding ciphertext vector and

$$\mathbf{p} = A^{-1}\mathbf{c} \quad (\text{mod } 26)$$

Thus, each plaintext vector can be recovered from the corresponding ciphertext vector by multiplying it on the left by A^{-1} (mod 26).

In cryptography it is important to know which matrices are invertible modulo 26 and how to obtain their inverses. We now investigate these questions.

In ordinary arithmetic, a square matrix A is invertible if and only if $\det(A) \neq 0$, or, equivalently, if and only if $\det(A)$ has a reciprocal. The following theorem is the analog of this result in modular arithmetic.

THEOREM 6.9.2

> *A square matrix A with entries in Z_m is invertible modulo m if and only if the residue of $\det(A)$ modulo m has a reciprocal modulo m.*

Because the residue of $\det(A)$ modulo m will have a reciprocal modulo m if and only if this residue and m have no common prime factors, we have the following corollary.

COROLLARY 6.9.3

> *A square matrix A with entries in Z_m is invertible modulo m if and only if m and the residue of $\det(A)$ modulo m have no common prime factors.*

Because the only prime factors of $m = 26$ are 2 and 13, we have the following corollary, which is useful in cryptography.

COROLLARY 6.9.4

> *A square matrix A with entries in Z_{26} is invertible modulo 26 if and only if the residue of $\det(A)$ modulo 26 is not divisible by 2 or 13.*

We leave it for the reader to verify that if

$$A = \begin{bmatrix} a & b \\ c & d \end{bmatrix}$$

has entries in Z_{26} and the residue of $\det(A) = ad - bc$ modulo 26 is not divisible by 2 or 13, then the inverse of A (mod 26) is given by

$$A^{-1} = (ad - bc)^{-1} \begin{bmatrix} d & -b \\ -c & a \end{bmatrix} \quad (\text{mod } 26) \tag{2}$$

where $(ad - bc)^{-1}$ is the reciprocal of the residue of $ad - bc$ (mod 26).

EXAMPLE 6   Inverse of a Matrix mod 26

Find the inverse of

$$A = \begin{bmatrix} 5 & 6 \\ 2 & 3 \end{bmatrix}$$

modulo 26.

Solution

$$\det(A) = ad - bc = 5 \cdot 3 - 6 \cdot 2 = 3$$

so from Table 2,

$$(ad - bc)^{-1} = 3^{-1} = 9 \quad (\text{mod } 26)$$

Thus, from (2),

$$A^{-1} = 9 \begin{bmatrix} 3 & -6 \\ -2 & 5 \end{bmatrix} = \begin{bmatrix} 27 & -54 \\ -18 & 45 \end{bmatrix} = \begin{bmatrix} 1 & 24 \\ 8 & 19 \end{bmatrix} \quad (\text{mod } 26)$$

As a check,

$$AA^{-1} = \begin{bmatrix} 5 & 6 \\ 2 & 3 \end{bmatrix} \begin{bmatrix} 1 & 24 \\ 8 & 19 \end{bmatrix} = \begin{bmatrix} 53 & 234 \\ 26 & 105 \end{bmatrix} = \begin{bmatrix} 1 & 0 \\ 0 & 1 \end{bmatrix} \quad (\text{mod } 26)$$

Similarly, $A^{-1}A = I$. ◆

EXAMPLE 7   Decoding a Hill 2-Cipher

Decode the following Hill 2-cipher, which was enciphered by the matrix in Example 6:

GTNKGKDUSK

Solution

From Table 1 the numerical equivalent of this ciphertext is

$$7 \ 20 \qquad 14 \ 11 \qquad 7 \ 11 \qquad 4 \ 21 \qquad 19 \ 11$$

To obtain the plaintext pairs, we multiply each ciphertext vector by the inverse of A (obtained in Example 6):

$$\begin{bmatrix} 1 & 24 \\ 8 & 19 \end{bmatrix} \begin{bmatrix} 7 \\ 20 \end{bmatrix} = \begin{bmatrix} 487 \\ 436 \end{bmatrix} = \begin{bmatrix} 19 \\ 20 \end{bmatrix} \quad (\text{mod } 26)$$

$$\begin{bmatrix} 1 & 24 \\ 8 & 19 \end{bmatrix} \begin{bmatrix} 14 \\ 11 \end{bmatrix} = \begin{bmatrix} 278 \\ 321 \end{bmatrix} = \begin{bmatrix} 18 \\ 9 \end{bmatrix} \quad (\text{mod } 26)$$

$$\begin{bmatrix} 1 & 24 \\ 8 & 19 \end{bmatrix} \begin{bmatrix} 7 \\ 11 \end{bmatrix} = \begin{bmatrix} 271 \\ 265 \end{bmatrix} = \begin{bmatrix} 11 \\ 5 \end{bmatrix} \quad (\text{mod } 26)$$

$$\begin{bmatrix} 1 & 24 \\ 8 & 19 \end{bmatrix} \begin{bmatrix} 4 \\ 21 \end{bmatrix} = \begin{bmatrix} 508 \\ 431 \end{bmatrix} = \begin{bmatrix} 14 \\ 15 \end{bmatrix} \quad (\text{mod } 26)$$

$$\begin{bmatrix} 1 & 24 \\ 8 & 19 \end{bmatrix} \begin{bmatrix} 19 \\ 11 \end{bmatrix} = \begin{bmatrix} 283 \\ 361 \end{bmatrix} = \begin{bmatrix} 23 \\ 23 \end{bmatrix} \quad (\text{mod } 26)$$

From Table 1, the alphabet equivalents of these vectors are

ST   RI   KE   NO   WW

which yields the message

STRIKE   NOW ◆

Breaking a Hill Cipher

Because the purpose of enciphering messages and information is to prevent "opponents" from learning their contents, cryptographers are concerned with the *security* of their ciphers—that is, how readily they can be broken (deciphered by their opponents). We will conclude this section by discussing one technique for breaking Hill ciphers.

Suppose that you are able to obtain some corresponding plaintext and ciphertext from an opponent's message. For example, on examining some intercepted ciphertext, you may be able to deduce that the message is a letter that begins *DEAR SIR*. We will show that with a small amount of such data, it may be possible to determine the deciphering matrix of a Hill code and consequently obtain access to the rest of the message.

It is a basic result in linear algebra that a linear transformation is completely determined by its values at a basis. This principle suggests that if we have a Hill *n*-cipher, and if

$$\mathbf{p}_1, \mathbf{p}_2, \dots, \mathbf{p}_n$$

are linearly independent plaintext vectors whose corresponding ciphertext vectors

$$A\mathbf{p}_1, A\mathbf{p}_2, \dots, A\mathbf{p}_n$$

are known, then there is enough information available to determine the matrix A and hence A^{-1} (mod m).

The following theorem, whose proof is discussed in the exercises, provides a way to do this.

THEOREM 6.9.5

> ### Determining the Deciphering Matrix
>
> *Let $\mathbf{p}_1, \mathbf{p}_2, \dots, \mathbf{p}_n$ be linearly independent plaintext vectors, and let $\mathbf{c}_1, \mathbf{c}_2, \dots, \mathbf{c}_n$ be the corresponding ciphertext vectors in a Hill n-cipher. If*
>
> $$P = \begin{bmatrix} \mathbf{p}_1^T \\ \mathbf{p}_2^T \\ \vdots \\ \mathbf{p}_n^T \end{bmatrix}$$
>
> *is the $n \times n$ matrix with row vectors $\mathbf{p}_1^T, \mathbf{p}_2^T, \dots, \mathbf{p}_n^T$ and if*
>
> $$C = \begin{bmatrix} \mathbf{c}_1^T \\ \mathbf{c}_2^T \\ \vdots \\ \mathbf{c}_n^T \end{bmatrix}$$
>
> *is the $n \times n$ matrix with row vectors $\mathbf{c}_1^T, \mathbf{c}_2^T, \dots, \mathbf{c}_n^T$, then the sequence of elementary row operations that reduces C to I transforms P to $(A^{-1})^T$.*

This theorem tells us that to find the transpose of the deciphering matrix A^{-1}, we must find a sequence of row operations that reduces C to I and then perform this same sequence of operations on P. The following example illustrates a simple algorithm for doing this.

EXAMPLE 8   Using Theorem 6.9.5

The following Hill 2-cipher is intercepted:

$$IOSBTGXESPXHOPDE$$

Decipher the message, given that it starts with the word *DEAR*.

Solution

From Table 1, the numerical equivalent of the known plaintext is

$$\begin{array}{cc} DE & AR \\ 4\ 5 & 1\ 18 \end{array}$$

and the numerical equivalent of the corresponding ciphertext is

$$\begin{array}{cc} IO & SB \\ 9\ 15 & 19\ 2 \end{array}$$

so the corresponding plaintext and ciphertext vectors are

$$\mathbf{p}_1 = \begin{bmatrix} 4 \\ 5 \end{bmatrix} \leftrightarrow \mathbf{c}_1 = \begin{bmatrix} 9 \\ 15 \end{bmatrix}$$

$$\mathbf{p}_2 = \begin{bmatrix} 1 \\ 18 \end{bmatrix} \leftrightarrow \mathbf{c}_2 = \begin{bmatrix} 19 \\ 2 \end{bmatrix}$$

We want to reduce

$$C = \begin{bmatrix} \mathbf{c}_1^T \\ \mathbf{c}_2^T \end{bmatrix} = \begin{bmatrix} 9 & 15 \\ 19 & 2 \end{bmatrix}$$

to I by elementary row operations and simultaneously apply these operations to

$$P = \begin{bmatrix} \mathbf{p}_1^T \\ \mathbf{p}_2^T \end{bmatrix} = \begin{bmatrix} 4 & 5 \\ 1 & 18 \end{bmatrix}$$

to obtain $(A^{-1})^T$ (the transpose of the deciphering matrix). This can be accomplished by adjoining P to the right of C and applying row operations to the resulting matrix $[C \mid P]$ until the left side is reduced to I. The final matrix will then have the form $[I \mid (A^{-1})^T]$. The computations can be carried out as follows:

$$\begin{bmatrix} 9 & 15 & | & 4 & 5 \\ 19 & 2 & | & 1 & 18 \end{bmatrix}$$ ⟵ ———— We formed the matrix $[C \mid P]$.

$$\begin{bmatrix} 1 & 45 & | & 12 & 15 \\ 19 & 2 & | & 1 & 18 \end{bmatrix}$$ ⟵ ———— We multiplied the first row by $9^{-1} = 3$.

$$\begin{bmatrix} 1 & 19 & | & 12 & 15 \\ 19 & 2 & | & 1 & 18 \end{bmatrix}$$ ⟵ ———— We replaced 45 by its residue modulo 26.

$$\begin{bmatrix} 1 & 19 & | & 12 & 15 \\ 0 & -359 & | & -227 & -267 \end{bmatrix}$$ ⟵ ———— We added -19 times the first row to the second.

$$\begin{bmatrix} 1 & 19 & | & 12 & 15 \\ 0 & 5 & | & 7 & 19 \end{bmatrix}$$ ⟵ We replaced the entries in the second row by their residues modulo 26.

$$\begin{bmatrix} 1 & 19 & | & 12 & 15 \\ 0 & 1 & | & 147 & 399 \end{bmatrix}$$ ⟵ We multiplied the second row by $5^{-1} = 21$.

$$\begin{bmatrix} 1 & 19 & | & 12 & 15 \\ 0 & 1 & | & 17 & 9 \end{bmatrix}$$ ⟵ We replaced the entries in the second row by their residues modulo 26.

$$\begin{bmatrix} 1 & 0 & | & -311 & -156 \\ 0 & 1 & | & 17 & 9 \end{bmatrix}$$ ⟵ We added -19 times the second row to the first.

$$\begin{bmatrix} 1 & 0 & | & 1 & 0 \\ 0 & 1 & | & 17 & 9 \end{bmatrix}$$ ⟵ We replaced the entries in the first row by their residues modulo 26.

Thus,

$$(A^{-1})^T = \begin{bmatrix} 1 & 0 \\ 17 & 9 \end{bmatrix}$$

so the deciphering matrix is

$$A^{-1} = \begin{bmatrix} 1 & 17 \\ 0 & 9 \end{bmatrix}$$

To decipher the message, we first group the ciphertext into pairs and find the numerical equivalent of each letter:

IO	SB	TG	XE	SP	XH	OP	DE
9 15	19 2	20 7	24 5	19 16	24 8	15 16	4 5

Next, we multiply successive ciphertext vectors on the left by A^{-1} and find the alphabet equivalents of the resulting plaintext pairs:

$$\begin{bmatrix} 1 & 17 \\ 0 & 9 \end{bmatrix}\begin{bmatrix} 9 \\ 15 \end{bmatrix} = \begin{bmatrix} 4 \\ 5 \end{bmatrix} \quad \begin{matrix} D \\ E \end{matrix}$$

$$\begin{bmatrix} 1 & 17 \\ 0 & 9 \end{bmatrix}\begin{bmatrix} 19 \\ 2 \end{bmatrix} = \begin{bmatrix} 1 \\ 18 \end{bmatrix} \quad \begin{matrix} A \\ R \end{matrix}$$

$$\begin{bmatrix} 1 & 17 \\ 0 & 9 \end{bmatrix}\begin{bmatrix} 20 \\ 7 \end{bmatrix} = \begin{bmatrix} 9 \\ 11 \end{bmatrix} \quad \begin{matrix} I \\ K \end{matrix}$$

$$\begin{bmatrix} 1 & 17 \\ 0 & 9 \end{bmatrix}\begin{bmatrix} 24 \\ 5 \end{bmatrix} = \begin{bmatrix} 5 \\ 19 \end{bmatrix} \quad \begin{matrix} E \\ S \end{matrix}$$

$$\begin{bmatrix} 1 & 17 \\ 0 & 9 \end{bmatrix}\begin{bmatrix} 19 \\ 16 \end{bmatrix} = \begin{bmatrix} 5 \\ 14 \end{bmatrix} \quad \begin{matrix} E \\ N \end{matrix} \quad (\mathrm{mod}\ 26)$$

$$\begin{bmatrix} 1 & 17 \\ 0 & 9 \end{bmatrix}\begin{bmatrix} 24 \\ 8 \end{bmatrix} = \begin{bmatrix} 4 \\ 20 \end{bmatrix} \quad \begin{matrix} D \\ T \end{matrix}$$

$$\begin{bmatrix} 1 & 17 \\ 0 & 9 \end{bmatrix}\begin{bmatrix} 15 \\ 16 \end{bmatrix} = \begin{bmatrix} 1 \\ 14 \end{bmatrix} \quad \begin{matrix} A \\ N \end{matrix}$$

$$\begin{bmatrix} 1 & 17 \\ 0 & 9 \end{bmatrix}\begin{bmatrix} 4 \\ 5 \end{bmatrix} = \begin{bmatrix} 11 \\ 19 \end{bmatrix} \quad \begin{matrix} K \\ S \end{matrix}$$

Finally, we construct the message from the plaintext pairs:

DE   AR   IK   ES   EN   DT   AN   KS

DEAR IKE SEND TANKS ◆

FURTHER READINGS

Readers interested in learning more about mathematical cryptography are referred to the following books, the first of which is elementary and the second more advanced.

1. ABRAHAM SINKOV, *Elementary Cryptanalysis, a Mathematical Approach* (Mathematical Association of America, Mathematical Library, 1966).

2. ALAN G. KONHEIM, *Cryptography, a Primer* (New York: Wiley-Interscience, 1981).

EXERCISE SET 6.9

1. Obtain the Hill cipher of the message

   DARK NIGHT

   for each of the following enciphering matrices:

   (a) $\begin{bmatrix} 1 & 3 \\ 2 & 1 \end{bmatrix}$ (b) $\begin{bmatrix} 4 & 3 \\ 1 & 2 \end{bmatrix}$

2. In each part determine whether the matrix is invertible modulo 26. If so, find its inverse modulo 26 and check your work by verifying that $AA^{-1} = A^{-1}A = I$ (mod 26).

   (a) $A = \begin{bmatrix} 9 & 1 \\ 7 & 2 \end{bmatrix}$ (b) $A = \begin{bmatrix} 3 & 1 \\ 5 & 3 \end{bmatrix}$ (c) $A = \begin{bmatrix} 8 & 11 \\ 1 & 9 \end{bmatrix}$

   (d) $A = \begin{bmatrix} 2 & 1 \\ 1 & 7 \end{bmatrix}$ (e) $A = \begin{bmatrix} 3 & 1 \\ 6 & 2 \end{bmatrix}$ (f) $A = \begin{bmatrix} 1 & 8 \\ 1 & 3 \end{bmatrix}$

3. Decode the message

   SAKNOXAOJX

   given that it is a Hill cipher with enciphering matrix

   $$\begin{bmatrix} 4 & 1 \\ 3 & 2 \end{bmatrix}$$

4. A Hill 2-cipher is intercepted that starts with the pairs

   SL   HK

   Find the deciphering and enciphering matrices, given that the plaintext is known to start with the word *ARMY*.

5. Decode the following Hill 2-cipher if the last four plaintext letters are known to be *ATOM*.

   LNGIHGYBVRENJYQO

6. Decode the following Hill 3-cipher if the first nine plaintext letters are *IHAVECOME*:

   HPAFQGGDUGDDHPGODYNOR

7. All of the results of this section can be generalized to the case where the plaintext is a binary message; that is, it is a sequence of 0's and 1's. In this case we do all of our modular arithmetic using modulus 2 rather than modulus 26. Thus, for example, $1 + 1 = 0$ (mod 2). Suppose we want to encrypt the message 110101111. Let us first break it into triplets to form the

   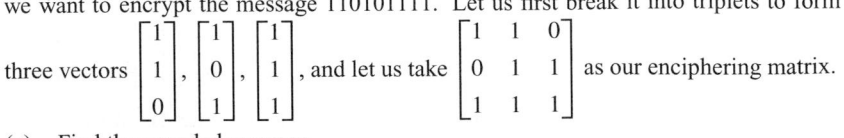

   three vectors $\begin{bmatrix} 1 \\ 1 \\ 0 \end{bmatrix}$, $\begin{bmatrix} 1 \\ 0 \\ 1 \end{bmatrix}$, $\begin{bmatrix} 1 \\ 1 \\ 1 \end{bmatrix}$, and let us take $\begin{bmatrix} 1 & 1 & 0 \\ 0 & 1 & 1 \\ 1 & 1 & 1 \end{bmatrix}$ as our enciphering matrix.

   (a) Find the encoded message.

   (b) Find the inverse modulo 2 of the enciphering matrix, and verify that it decodes your encoded message.

8. If, in addition to the standard alphabet, a period, comma, and question mark were allowed, then 29 plaintext and ciphertext symbols would be available and all matrix arithmetic would be done modulo 29. Under what conditions would a matrix with entries in Z_{29} be invertible modulo 29?

9. Show that the modular equation $4x = 1 \pmod{26}$ has no solution in Z_{26} by successively substituting the values $x = 0, 1, 2, \ldots, 25$.

10. (a) Let P and C be the matrices in Theorem 6.9.5. Show that $P = C(A^{-1})^T$.

    (b) To prove Theorem 6.9.5, let $E_1, E_2, \ldots, E_n$ be the elementary matrices that correspond to the row operations that reduce C to I, so

    $$E_n \cdots E_2 E_1 C = I$$

    Show that

    $$E_n \cdots E_2 E_1 P = (A^{-1})^T$$

    from which it follows that the same sequence of row operations that reduces C to I converts P to $(A^{-1})^T$.

11. (a) If A is the enciphering matrix of a Hill n-cipher, show that

    $$A^{-1} = (C^{-1}P)^T \quad (\bmod\ 26)$$

    where C and P are the matrices defined in Theorem 6.9.5.

    (b) Instead of using Theorem 6.9.5 as in the text, find the deciphering matrix A^{-1} of Example 8 by using the result in part (a) and Equation (2) to compute C^{-1}.

    Note Although this method is practical for Hill 2-ciphers, Theorem 6.9.5 is more efficient for Hill n-ciphers with $n > 2$.

SECTION 6.9

 **Technology Exercises**

The following exercises are designed to be solved using a technology utility. Typically, this will be MATLAB, *Mathematica*, Maple, Derive, or Mathcad, but it may also be some other type of linear algebra software or a scientific calculator with some linear algebra capabilities. For each exercise you will need to read the relevant documentation for the particular utility you are using. The goal of these exercises is to provide you with a basic proficiency with your technology utility. Once you have mastered the techniques in these exercises, you will be able to use your technology utility to solve many of the problems in the regular exercise sets.

T1. Two integers that have no common factors (except 1) are said to be relatively prime. Given a positive integer n, let $S_n = \{a_1, a_2, a_3, \ldots, a_m\}$, where $a_1 < a_2 < a_3 < \cdots < a_m$, be the set of all positive integers less than n and relatively prime to n. For example, if $n = 9$, then

$$S_9 = \{a_1, a_2, a_3, \ldots, a_6\} = \{1, 2, 4, 5, 7, 8\}$$

(a) Construct a table consisting of n and S_n for $n = 2, 3, \ldots, 15$, and then compute

$$\sum_{k=1}^{m} a_k \quad \text{and} \quad \left(\sum_{k=1}^{m} a_k\right) \quad (\bmod\ n)$$

in each case. Draw a conjecture for $n > 15$ and prove your conjecture to be true.

Hint Use the fact that if a is relatively prime to n, then $n - a$ is also relatively prime to n.

(b) Given a positive integer n and the set S_n, let P_n be the $m \times m$ matrix

$$P_n = \begin{bmatrix} a_1 & a_2 & a_3 & \cdots & a_{m-1} & a_m \\ a_2 & a_3 & a_4 & \cdots & a_m & a_1 \\ a_3 & a_4 & a_5 & \cdots & a_1 & a_2 \\ \vdots & \vdots & \vdots & \ddots & \vdots & \vdots \\ a_{m-1} & a_m & a_1 & \cdots & a_{m-3} & a_{m-2} \\ a_m & a_1 & a_2 & \cdots & a_{m-2} & a_{m-1} \end{bmatrix}$$

so that, for example,

$$P_9 = \begin{bmatrix} 1 & 2 & 4 & 5 & 7 & 8 \\ 2 & 4 & 5 & 7 & 8 & 1 \\ 4 & 5 & 7 & 8 & 1 & 2 \\ 5 & 7 & 8 & 1 & 2 & 4 \\ 7 & 8 & 1 & 2 & 4 & 5 \\ 8 & 1 & 2 & 4 & 5 & 7 \end{bmatrix}$$

Use a computer to compute $\det(P_n)$ and $\det(P_n) \pmod{n}$ for $n = 2, 3, \ldots, 15$, and then use these results to construct a conjecture.

(c) Use the results of part (a) to prove your conjecture to be true.

Hint Add the first $m - 1$ rows of P_n to its last row and then use Theorem 2.2.3.

What do these results imply about the inverse of $P_n \pmod{n}$?

T2. Given a positive integer n greater than 1, the number of positive integers less than n and relatively prime to n is called the **Euler phi function** of n and is denoted by $\varphi(n)$. For example, $\varphi(6) = 2$ since only two positive integers (1 and 5) are less than 6 and have no common factor with 6.

(a) Using a computer, for each value of $n = 2, 3, \ldots, 25$ compute and print out all positive integers that are less than n and relatively prime to n. Then use these integers to determine the values of $\varphi(n)$ for $n = 2, 3, \ldots, 25$. Can you discover a pattern in the results?

(b) It can be shown that if $\{p_1, p_2, p_3, \ldots, p_m\}$ are all the distinct prime factors of n, then

$$\varphi(n) = n \left(1 - \frac{1}{p_1} \right) \left(1 - \frac{1}{p_2} \right) \left(1 - \frac{1}{p_3} \right) \cdots \left(1 - \frac{1}{p_m} \right)$$

For example, since $\{2, 3\}$ are the distinct prime factors of 12, we have

$$\varphi(12) = 12 \left(1 - \frac{1}{2} \right) \left(1 - \frac{1}{3} \right) = 4$$

which agrees with the fact that $\{1, 5, 7, 11\}$ are the only positive integers less than 12 and relatively prime to 12. Using a computer, print out all the prime factors of n for $n = 2, 3, \ldots, 25$. Then compute $\varphi(n)$ using the formula above and compare it to your results in part (a).

ANSWERS TO EXERCISES

Exercise Set 1.1
(page 6)

1. (a), (c), (f)

3. (a) $x = \frac{3}{7} + \frac{5}{7}t$
   $y = t$

   (b) $x_1 = \frac{5}{3}s - \frac{4}{3}t + \frac{7}{3}$ $x_1 = \frac{1}{4}r - \frac{5}{8}s + \frac{3}{4}t - \frac{1}{8}$ $v = \frac{8}{3}q - \frac{2}{3}r + \frac{1}{3}s - \frac{4}{3}t$
   $x_2 = s$ $x_2 = r$ $w = q$
   $x_3 = t$ $x_3 = s$ $x = r$
                        $x_4 = t$ $y = s$
                                                     $z = t$

4. (a) $\begin{bmatrix} 3 & -2 & -1 \\ 4 & 5 & 3 \\ 7 & 3 & 2 \end{bmatrix}$ (b) $\begin{bmatrix} 2 & 0 & 2 & 1 \\ 3 & -1 & 4 & 7 \\ 6 & 1 & -1 & 0 \end{bmatrix}$

   (c) $\begin{bmatrix} 1 & 2 & 0 & -1 & 1 & 1 \\ 0 & 3 & 1 & 0 & -1 & 2 \\ 0 & 0 & 1 & 7 & 0 & 1 \end{bmatrix}$ (d) $\begin{bmatrix} 1 & 0 & 0 & 1 \\ 0 & 1 & 0 & 2 \\ 0 & 0 & 1 & 3 \end{bmatrix}$

5. (a) $2x_1 \qquad\qquad = 0$ (b) $3x_1 \qquad - 2x_3 = 5$
   $\quad 3x_1 - 4x_2 = 0$ $7x_1 + x_2 + 4x_3 = -3$
   $\qquad\qquad x_2 = 1$ $- 2x_2 + x_3 = 7$

   (c) $7x_1 + 2x_2 + x_3 - 3x_4 = 5$ (d) $x_1 \qquad\qquad = 7$
   $\quad x_1 + 2x_2 + 4x_3 \qquad = 1$ $x_2 \qquad = -2$
                                           $x_3 \qquad = 3$
                                           $x_4 = 4$

6. (a) $x - 2y = 5$ (b) Let $x = t$; then $t - 2y = 5$. Solving for y yields $y = \frac{1}{2}t - \frac{5}{2}$.

12. (a) The lines have no common point of intersection.
    (b) The lines intersect in exactly one point.   (c) The three lines coincide.

Exercise Set 1.2
(page 19)

1. (a), (b), (c), (d), (h), (i), (j)

3. (a) Both   (b) Neither   (c) Both
   (d) Row-echelon   (e) Neither   (f) Both

4. (a) $x_1 = -3, \ x_2 = 0, \ x_3 = 7$
   (b) $x_1 = 7t + 8, \ x_2 = -3t + 2, \ x_3 = -t - 5, \ x_4 = t$
   (c) $x_1 = 6s - 3t - 2, \ x_2 = s, \ x_3 = -4t + 7, \ x_4 = -5t + 8, \ x_5 = t$
   (d) Inconsistent

6. (a) $x_1 = 3, \ x_2 = 1, \ x_3 = 2$ (b) $x_1 = -\frac{1}{7} - \frac{3}{7}t, \ x_2 = \frac{1}{7} - \frac{4}{7}, \ x_3 = t$
   (c) $x = t - 1, \ y = 2s, \ z = s, \ w = t$ (d) Inconsistent

8. (a) Inconsistent   (b) $x_1 = -4, \ x_2 = 2, \ x_3 = 7$
   (c) $x_1 = 3 + 2t, \ x_2 = t$ (d) $x = \frac{8}{5} - \frac{3}{5}t - \frac{3}{5}s, \ y = \frac{1}{10} + \frac{2}{5}t - \frac{1}{10}s, \ z = t, \ w = s$

12. (a), (c), (d)

13. (a) $x_1 = 0, \ x_2 = 0, \ x_3 = 0$ (b) $x_1 = -s, \ x_2 = -t - s, \ x_3 = 4s, \ x_4 = t$
    (c) $w = t, \ x = -t, \ y = t, \ z = 0$

14. **(a)** Only the trivial solution   **(b)** $u = 7s - 5t$, $v = -6s + 4t$, $w = 2s$, $x = 2t$
    (c) Only the trivial solution

15. **(a)** $I_1 = -1$, $I_2 = 0$, $I_3 = 1$, $I_4 = 2$
    (b) $Z_1 = -s - t$, $Z_2 = s$, $Z_3 = -t$, $Z_4 = 0$, $Z_5 = t$

19. $\begin{bmatrix} 1 & 3 \\ 0 & 1 \end{bmatrix}$ and $\begin{bmatrix} 1 & 0 \\ 0 & 1 \end{bmatrix}$ are possible answers.   **20.** $\alpha = \pi/2$, $\beta = \pi$, $\gamma = 0$

23. If $\lambda = 1$, then $x_1 = x_2 = -\frac{1}{2}s$, $x_3 = s$
    If $\lambda = 2$, then $x_1 = -\frac{1}{2}s$, $x_2 = 0$, $x_3 = s$

24. $x = -13/7$, $y = 91/54$, $z = -91/8$ **25.** $a = 1$, $b = -6$, $c = 2$, $d = 10$

30. **(a)** Three lines, at least two of which are distinct   **(b)** Three identical lines

32. **(a)** False   **(b)** False   **(c)** False   **(d)** False

Exercise Set 1.3
(page 34)

1. **(a)** Undefined   **(b)** 4×2 **(c)** Undefined   **(d)** Undefined
   (e) 5×5 **(f)** 5×2 **(g)** Undefined   **(h)** 5×2

2. $a = 5$, $b = -3$, $c = 4$, $d = 1$

4. **(a)** $\begin{bmatrix} 7 & 2 & 4 \\ 3 & 5 & 7 \end{bmatrix}$ **(b)** $\begin{bmatrix} -5 & 0 & -1 \\ 4 & -1 & 1 \\ -1 & -1 & 1 \end{bmatrix}$ **(c)** $\begin{bmatrix} -5 & 0 & -1 \\ 4 & -1 & 1 \\ -1 & -1 & 1 \end{bmatrix}$

   (d) Undefined   **(e)** $\begin{bmatrix} -\frac{1}{4} & \frac{3}{2} \\ \frac{9}{4} & 0 \\ \frac{3}{4} & \frac{9}{4} \end{bmatrix}$ **(f)** $\begin{bmatrix} 0 & -1 \\ 1 & 0 \end{bmatrix}$

   (g) $\begin{bmatrix} 9 & 1 & -1 \\ -13 & 2 & -4 \\ 0 & 1 & -6 \end{bmatrix}$ **(h)** $\begin{bmatrix} 9 & -13 & 0 \\ 1 & 2 & 1 \\ -1 & -4 & -6 \end{bmatrix}$

5. **(a)** $\begin{bmatrix} 12 & -3 \\ -4 & 5 \\ 4 & 1 \end{bmatrix}$ **(b)** Undefined   **(c)** $\begin{bmatrix} 42 & 108 & 75 \\ 12 & -3 & 21 \\ 36 & 78 & 63 \end{bmatrix}$

   (d) $\begin{bmatrix} 3 & 45 & 9 \\ 11 & -11 & 17 \\ 7 & 17 & 13 \end{bmatrix}$ **(e)** $\begin{bmatrix} 3 & 45 & 9 \\ 11 & -11 & 17 \\ 7 & 17 & 13 \end{bmatrix}$ **(f)** $\begin{bmatrix} 21 & 17 \\ 17 & 35 \end{bmatrix}$

   (g) $\begin{bmatrix} 0 & -2 & 11 \\ 12 & 1 & 8 \end{bmatrix}$ **(h)** $\begin{bmatrix} 12 & 6 & 9 \\ 48 & -20 & 14 \\ 24 & 8 & 16 \end{bmatrix}$ **(i)** 61   **(j)** 35   **(k)** (28)

8. **(a)** $\begin{bmatrix} 67 \\ 64 \\ 63 \end{bmatrix} = 6\begin{bmatrix} 3 \\ 6 \\ 0 \end{bmatrix} + 0\begin{bmatrix} -2 \\ 5 \\ 4 \end{bmatrix} + 7\begin{bmatrix} 7 \\ 4 \\ 9 \end{bmatrix}$ **(b)** $\begin{bmatrix} 6 \\ 6 \\ 63 \end{bmatrix} = 3\begin{bmatrix} 6 \\ 0 \\ 7 \end{bmatrix} + 6\begin{bmatrix} -2 \\ 1 \\ 7 \end{bmatrix} + 0\begin{bmatrix} 4 \\ 3 \\ 5 \end{bmatrix}$

   $\begin{bmatrix} 41 \\ 21 \\ 67 \end{bmatrix} = -2\begin{bmatrix} 3 \\ 6 \\ 0 \end{bmatrix} + 1\begin{bmatrix} -2 \\ 5 \\ 4 \end{bmatrix} + 7\begin{bmatrix} 7 \\ 4 \\ 9 \end{bmatrix}$ $\begin{bmatrix} -6 \\ 17 \\ 41 \end{bmatrix} = -2\begin{bmatrix} 6 \\ 0 \\ 7 \end{bmatrix} + 5\begin{bmatrix} -2 \\ 1 \\ 7 \end{bmatrix} + 4\begin{bmatrix} 4 \\ 3 \\ 5 \end{bmatrix}$

   $\begin{bmatrix} 41 \\ 59 \\ 57 \end{bmatrix} = 4\begin{bmatrix} 3 \\ 6 \\ 0 \end{bmatrix} + 3\begin{bmatrix} -2 \\ 5 \\ 4 \end{bmatrix} + 5\begin{bmatrix} 7 \\ 4 \\ 9 \end{bmatrix}$ $\begin{bmatrix} 70 \\ 31 \\ 122 \end{bmatrix} = 7\begin{bmatrix} 6 \\ 0 \\ 7 \end{bmatrix} + 4\begin{bmatrix} -2 \\ 1 \\ 7 \end{bmatrix} + 9\begin{bmatrix} 4 \\ 3 \\ 5 \end{bmatrix}$

13. **(a)** $A = \begin{bmatrix} 2 & -3 & 5 \\ 9 & -1 & 1 \\ 1 & 5 & 4 \end{bmatrix}$, $\mathbf{x} = \begin{bmatrix} x_1 \\ x_2 \\ x_3 \end{bmatrix}$, $\mathbf{b} = \begin{bmatrix} 7 \\ -1 \\ 0 \end{bmatrix}$

 (b) $A = \begin{bmatrix} 4 & 0 & -3 & 1 \\ 5 & 1 & 0 & -8 \\ 2 & -5 & 9 & -1 \\ 0 & 3 & -1 & 7 \end{bmatrix}$, $\mathbf{x} = \begin{bmatrix} x_1 \\ x_2 \\ x_3 \\ x_4 \end{bmatrix}$, $\mathbf{b} = \begin{bmatrix} 1 \\ 3 \\ 0 \\ 2 \end{bmatrix}$

16. **(a)** $\begin{bmatrix} -3 & -15 & -11 \\ 21 & -15 & 44 \end{bmatrix}$ **(b)** $\begin{bmatrix} 4 & -7 & -19 & -43 \\ 2 & 2 & 18 & 17 \\ 0 & 5 & 25 & 35 \\ 2 & 3 & 23 & 24 \end{bmatrix}$ **(c)** $\begin{bmatrix} 3 & 3 \\ -1 & 4 \\ 1 & 5 \\ 4 & -4 \\ 0 & 14 \end{bmatrix}$

17. **(a)** A_{11} is a 2×3 matrix and B_{11} is a 2×2 matrix. $A_{11}B_{11}$ does not exist.

 (b) $\begin{bmatrix} -1 & 23 & -10 \\ 37 & -13 & 8 \\ 29 & 23 & 41 \end{bmatrix}$

21. **(a)** $\begin{bmatrix} a_{11} & 0 & 0 & 0 & 0 & 0 \\ 0 & a_{22} & 0 & 0 & 0 & 0 \\ 0 & 0 & a_{33} & 0 & 0 & 0 \\ 0 & 0 & 0 & a_{44} & 0 & 0 \\ 0 & 0 & 0 & 0 & a_{55} & 0 \\ 0 & 0 & 0 & 0 & 0 & a_{66} \end{bmatrix}$ **(b)** $\begin{bmatrix} a_{11} & a_{12} & a_{13} & a_{14} & a_{15} & a_{16} \\ 0 & a_{22} & a_{23} & a_{24} & a_{25} & a_{26} \\ 0 & 0 & a_{33} & a_{34} & a_{35} & a_{36} \\ 0 & 0 & 0 & a_{44} & a_{45} & a_{46} \\ 0 & 0 & 0 & 0 & a_{55} & a_{56} \\ 0 & 0 & 0 & 0 & 0 & a_{66} \end{bmatrix}$

 (c) $\begin{bmatrix} a_{11} & 0 & 0 & 0 & 0 & 0 \\ a_{21} & a_{22} & 0 & 0 & 0 & 0 \\ a_{31} & a_{32} & a_{33} & 0 & 0 & 0 \\ a_{41} & a_{42} & a_{43} & a_{44} & 0 & 0 \\ a_{51} & a_{52} & a_{53} & a_{54} & a_{55} & 0 \\ a_{61} & a_{62} & a_{63} & a_{64} & a_{65} & a_{66} \end{bmatrix}$ **(d)** $\begin{bmatrix} a_{11} & a_{12} & 0 & 0 & 0 & 0 \\ a_{21} & a_{22} & a_{23} & 0 & 0 & 0 \\ 0 & a_{32} & a_{33} & a_{34} & 0 & 0 \\ 0 & 0 & a_{43} & a_{44} & a_{45} & 0 \\ 0 & 0 & 0 & a_{54} & a_{55} & a_{56} \\ 0 & 0 & 0 & 0 & a_{65} & a_{66} \end{bmatrix}$

27. One; namely, $A = \begin{bmatrix} 1 & 1 & 0 \\ 1 & -1 & 0 \\ 0 & 0 & 0 \end{bmatrix}$

30. **(a)** Yes; for example, $\begin{bmatrix} 0 & 1 \\ 0 & 0 \end{bmatrix}$ **(b)** Yes; for example, $\begin{bmatrix} 1 & 0 \\ 0 & 0 \end{bmatrix}$

32. **(a)** True **(b)** False; for example, $A = \begin{bmatrix} 1 & -1 \\ 1 & -1 \end{bmatrix}$ **(c)** True **(d)** True

Exercise Set 1.4 (page 48)

4. $A^{-1} = \begin{bmatrix} 2 & -1 \\ -5 & 3 \end{bmatrix}$, $B^{-1} = \begin{bmatrix} \frac{1}{5} & \frac{3}{20} \\ -\frac{1}{5} & \frac{1}{10} \end{bmatrix}$, $C^{-1} = \begin{bmatrix} -\frac{1}{2} & -2 \\ 1 & 3 \end{bmatrix}$, $D^{-1} = \begin{bmatrix} \frac{1}{2} & 0 \\ 0 & \frac{1}{3} \end{bmatrix}$

7. **(a)** $A = \begin{bmatrix} \frac{5}{13} & \frac{1}{13} \\ -\frac{3}{13} & \frac{2}{13} \end{bmatrix}$ **(b)** $A = \begin{bmatrix} \frac{2}{7} & 1 \\ \frac{1}{7} & \frac{3}{7} \end{bmatrix}$

 (c) $A = \begin{bmatrix} -\frac{2}{5} & 1 \\ -\frac{1}{5} & \frac{3}{5} \end{bmatrix}$ **(d)** $A = \begin{bmatrix} -\frac{9}{13} & \frac{1}{13} \\ \frac{2}{13} & -\frac{6}{13} \end{bmatrix}$

9. **(a)** $p(A) = \begin{bmatrix} 1 & 1 \\ 2 & -1 \end{bmatrix}$ **(b)** $p(A) = \begin{bmatrix} 20 & 7 \\ 14 & 6 \end{bmatrix}$ **(c)** $p(A) = \begin{bmatrix} 39 & 13 \\ 26 & 13 \end{bmatrix}$

11. $\begin{bmatrix} \cos\theta & -\sin\theta \\ \sin\theta & \cos\theta \end{bmatrix}$ 13. $A^{-1} = \begin{bmatrix} \frac{1}{a_{11}} & 0 & \cdots & 0 \\ 0 & \frac{1}{a_{22}} & \cdots & 0 \\ \vdots & \vdots & & \vdots \\ 0 & 0 & \cdots & \frac{1}{a_{nn}} \end{bmatrix}$ 18. $C = -A^{-1}BA^{-1}$

19. **(a)** $\begin{bmatrix} \frac{1}{2} & -\frac{1}{2} & 0 & 0 \\ \frac{1}{2} & \frac{1}{2} & 0 & 0 \\ 0 & 0 & \frac{1}{2} & -\frac{1}{2} \\ -1 & 0 & \frac{1}{2} & \frac{1}{2} \end{bmatrix}$ **(b)** $\begin{bmatrix} 1 & -1 & 0 & 0 \\ 0 & 1 & 0 & 0 \\ 0 & 0 & 1 & -1 \\ 0 & 0 & 0 & 1 \end{bmatrix}$

20. **(a)** One example is $\begin{bmatrix} 1 & 2 & 3 \\ 2 & 1 & 4 \\ 3 & 4 & 5 \end{bmatrix}$. **(b)** One example is $\begin{bmatrix} 0 & -1 & -1 \\ 1 & 0 & -1 \\ 1 & 1 & 0 \end{bmatrix}$.

22. Yes 23. $A^{-1} = \begin{bmatrix} \frac{1}{2} & \frac{1}{2} & -\frac{1}{2} \\ -\frac{1}{2} & \frac{1}{2} & \frac{1}{2} \\ \frac{1}{2} & -\frac{1}{2} & \frac{1}{2} \end{bmatrix}$ 33. $\begin{bmatrix} \pm 1 & 0 & 0 \\ 0 & \pm 1 & 0 \\ 0 & 0 & \pm 1 \end{bmatrix}$

34. **(a)** If A is invertible, then A^T is invertible. **(b)** True

Exercise Set 1.5
(page 57)

1. (a), (c), (d), (f)

3. **(a)** $\begin{bmatrix} 0 & 0 & 1 \\ 0 & 1 & 0 \\ 1 & 0 & 0 \end{bmatrix}$ **(b)** $\begin{bmatrix} 0 & 0 & 1 \\ 0 & 1 & 0 \\ 1 & 0 & 0 \end{bmatrix}$ **(c)** $\begin{bmatrix} 1 & 0 & 0 \\ 0 & 1 & 0 \\ -2 & 0 & 1 \end{bmatrix}$ **(d)** $\begin{bmatrix} 1 & 0 & 0 \\ 0 & 1 & 0 \\ 2 & 0 & 1 \end{bmatrix}$

6. **(a)** $\begin{bmatrix} -7 & 4 \\ 2 & -1 \end{bmatrix}$ **(b)** $\begin{bmatrix} -\frac{5}{39} & \frac{2}{13} \\ \frac{4}{39} & \frac{1}{13} \end{bmatrix}$ **(c)** Not invertible

8. **(a)** $\begin{bmatrix} 1 & 3 & 1 \\ 0 & 1 & -1 \\ -2 & 2 & 0 \end{bmatrix}$ **(b)** $\begin{bmatrix} \frac{\sqrt{2}}{26} & \frac{-3\sqrt{2}}{26} & 0 \\ \frac{4\sqrt{2}}{26} & \frac{\sqrt{2}}{26} & 0 \\ 0 & 0 & 1 \end{bmatrix}$ **(c)** $\begin{bmatrix} 1 & 0 & 0 & 0 \\ -\frac{1}{3} & \frac{1}{3} & 0 & 0 \\ 0 & -\frac{1}{5} & \frac{1}{5} & 0 \\ 0 & 0 & -\frac{1}{7} & \frac{1}{7} \end{bmatrix}$

(d) Not invertible **(e)** $\begin{bmatrix} -\frac{4}{5} & \frac{3}{5} & \frac{1}{5} & \frac{1}{5} \\ \frac{3}{2} & 0 & -1 & 0 \\ \frac{1}{2} & 0 & 0 & 0 \\ \frac{4}{5} & \frac{2}{5} & -\frac{1}{5} & -\frac{1}{5} \end{bmatrix}$

10. **(a)** $E_1 = \begin{bmatrix} 1 & 0 \\ 5 & 1 \end{bmatrix}$, $E_2 = \begin{bmatrix} 1 & 0 \\ 0 & \frac{1}{2} \end{bmatrix}$ **(b)** $A^{-1} = E_2 E_1$ **(c)** $A = E_1^{-1} E_2^{-1}$

11. **(a)** $\begin{bmatrix} 1 & -4 & 7 \\ 4 & 5 & -3 \\ 2 & -1 & 0 \end{bmatrix}$ **(b)** $\begin{bmatrix} 2 & -1 & 0 \\ \frac{4}{3} & \frac{5}{3} & -1 \\ 1 & -4 & 7 \end{bmatrix}$ **(c)** $\begin{bmatrix} 10 & 9 & -6 \\ 4 & 5 & -3 \\ 1 & -4 & 7 \end{bmatrix}$

14. $\begin{bmatrix} 0 & 1 & 0 \\ 1 & 0 & 0 \\ 0 & 0 & 1 \end{bmatrix} \begin{bmatrix} 1 & 0 & 0 \\ 0 & 1 & 0 \\ -2 & 0 & 1 \end{bmatrix} \begin{bmatrix} 1 & 0 & 0 \\ 0 & 1 & 0 \\ 0 & 1 & 1 \end{bmatrix} \begin{bmatrix} 1 & 3 & 3 & 8 \\ 0 & 1 & 7 & 8 \\ 0 & 0 & 0 & 0 \end{bmatrix}$

19. **(b)** Add -1 times the first row to the second row.
   Add -1 times the first row to the third row.
   Add -1 times the second row to the first row.
   Add the second row to the third row.

24. In general, no. Try $b = 1$, $a = c = d = 0$.

1. $x_1 = 3$, $x_2 = -1$ 4. $x_1 = 1$, $x_2 = -11$, $x_3 = 16$

6. $w = -6$, $x = 1$, $y = 10$, $z = -7$

9. **(a)** $x_1 = \frac{16}{3}$, $x_2 = -\frac{4}{3}$, $x_3 = -\frac{11}{3}$ **(b)** $x_1 = -\frac{5}{3}$, $x_2 = \frac{5}{3}$, $x_3 = \frac{10}{3}$

   (c) $x_1 = 3$, $x_2 = 0$, $x_3 = -4$

11. **(a)** $x_1 = \frac{22}{17}$, $x_2 = \frac{1}{17}$ **(b)** $x_1 = \frac{21}{17}$, $x_2 = \frac{11}{17}$

13. **(a)** $x_1 = \frac{7}{15}$, $x_2 = \frac{4}{15}$ **(b)** $x_1 = \frac{34}{15}$, $x_2 = \frac{28}{15}$

   (c) $x_1 = \frac{19}{15}$, $x_2 = \frac{13}{15}$ **(d)** $x_1 = -\frac{1}{5}$, $x_2 = \frac{3}{5}$

15. **(a)** $x_1 = -12 - 3t$, $x_2 = -5 - t$, $x_3 = t$ **(b)** $x_1 = 7 - 3t$, $x_2 = 3 - t$, $x_3 = t$

19. $b_1 = b_3 + b_4$, $b_2 = 2b_3 + b_4$ 21. $X = \begin{bmatrix} 11 & 12 & -3 & 27 & 26 \\ -6 & -8 & 1 & -18 & -17 \\ -15 & -21 & 9 & -38 & -35 \end{bmatrix}$

22. **(a)** Only the trivial solution $x_1 = x_2 = x_3 = x_4 = 0$; invertible
   (b) Infinitely many solutions; not invertible

28. **(a)** $I - A$ is invertible.   **(b)** $\mathbf{x} = (I - A)^{-1}\mathbf{b}$

30. Yes, for nonsquare matrices

1. **(a)** $\begin{bmatrix} \frac{1}{2} & 0 \\ 0 & -\frac{1}{5} \end{bmatrix}$ **(b)** Not invertible   **(c)** $\begin{bmatrix} -1 & 0 & 0 \\ 0 & \frac{1}{2} & 0 \\ 0 & 0 & 3 \end{bmatrix}$

3. **(a)** $A^2 = \begin{bmatrix} 1 & 0 \\ 0 & 4 \end{bmatrix}$, $A^{-2} = \begin{bmatrix} 1 & 0 \\ 0 & \frac{1}{4} \end{bmatrix}$, $A^{-k} = \begin{bmatrix} 1 & 0 \\ 0 & 1/(-2)^k \end{bmatrix}$

   (b) $A^2 = \begin{bmatrix} \frac{1}{4} & 0 & 0 \\ 0 & \frac{1}{9} & 0 \\ 0 & 0 & \frac{1}{16} \end{bmatrix}$, $A^{-2} = \begin{bmatrix} 4 & 0 & 0 \\ 0 & 9 & 0 \\ 0 & 0 & 16 \end{bmatrix}$, $A^{-k} = \begin{bmatrix} 2^k & 0 & 0 \\ 0 & 3^k & 0 \\ 0 & 0 & 4^k \end{bmatrix}$

5. **(a)**   7. $a = 2$, $b = -1$

10. **(a)** $\begin{bmatrix} 1 & 0 & 0 \\ 0 & -1 & 0 \\ 0 & 0 & -1 \end{bmatrix}$ **(b)** $\begin{bmatrix} \pm\frac{1}{3} & 0 & 0 \\ 0 & \pm\frac{1}{2} & 0 \\ 0 & 0 & \pm1 \end{bmatrix}$

11. **(a)** $\begin{bmatrix} a_{11} & a_{12} & a_{13} \\ a_{21} & a_{22} & a_{23} \\ a_{31} & a_{32} & a_{33} \end{bmatrix} \begin{bmatrix} 3 & 0 & 0 \\ 0 & 5 & 0 \\ 0 & 0 & 7 \end{bmatrix}$ **(b)** No

16. **(b)** Yes   17. Yes

19. $\begin{bmatrix} 4 & 0 & 0 \\ 0 & 4 & 0 \\ 0 & 0 & 4 \end{bmatrix}$, $\begin{bmatrix} 4 & 0 & 0 \\ 0 & 4 & 0 \\ 0 & 0 & -1 \end{bmatrix}$, $\begin{bmatrix} 4 & 0 & 0 \\ 0 & -1 & 0 \\ 0 & 0 & 4 \end{bmatrix}$, $\begin{bmatrix} -1 & 0 & 0 \\ 0 & 4 & 0 \\ 0 & 0 & 4 \end{bmatrix}$,

   $\begin{bmatrix} -1 & 0 & 0 \\ 0 & -1 & 0 \\ 0 & 0 & 4 \end{bmatrix}$, $\begin{bmatrix} -1 & 0 & 0 \\ 0 & 4 & 0 \\ 0 & 0 & -1 \end{bmatrix}$, $\begin{bmatrix} 4 & 0 & 0 \\ 0 & -1 & 0 \\ 0 & 0 & -1 \end{bmatrix}$, $\begin{bmatrix} -1 & 0 & 0 \\ 0 & -1 & 0 \\ 0 & 0 & -1 \end{bmatrix}$

20. **(a)** Yes   **(b)** No (unless $n = 1$)   **(c)** Yes   **(d)** No (unless $n = 1$)

24. **(a)** $x_1 = \frac{7}{4}$, $x_2 = 1$, $x_3 = -\frac{1}{2}$ **(b)** $x_1 = -8$, $x_2 = -4$, $x_3 = 3$

25. $A = \begin{bmatrix} 1 & 10 \\ 0 & -2 \end{bmatrix}$ 26. $\frac{n}{2}(1+n)$

Supplementary Exercises (page 76)

1. $x' = \frac{3}{5}x + \frac{4}{5}y$, $y' = -\frac{4}{5}x + \frac{3}{5}y$

3. One possible answer is
$x_1 - 2x_2 - x_3 - x_4 = 0$
$x_1 + 5x_2 + 2x_4 \quad\quad = 0$

5. $x = 4$, $y = 2$, $z = 3$

7. **(a)** $a \neq 0$, $b \neq 2$ **(b)** $a \neq 0$, $b = 2$ **(c)** $a = 0$, $b = 2$ **(d)** $a = 0$, $b \neq 2$

9. $K = \begin{bmatrix} 0 & 2 \\ 1 & 1 \end{bmatrix}$

11. **(a)** $X = \begin{bmatrix} -1 & 3 & -1 \\ 6 & 0 & 1 \end{bmatrix}$ **(b)** $X = \begin{bmatrix} 1 & -2 \\ 3 & 1 \end{bmatrix}$ **(c)** $X = \begin{bmatrix} -\frac{113}{37} & -\frac{160}{37} \\ -\frac{20}{37} & -\frac{46}{37} \end{bmatrix}$

13. mpn multiplications and $mp(n-1)$ additions   15. $a = 1$, $b = -2$, $c = 3$

16. $a = 1$, $b = -4$, $c = -5$ 26. $A = -\frac{7}{5}$, $B = \frac{4}{5}$, $C = \frac{3}{5}$

29. **(b)** $\begin{bmatrix} a^n & 0 & 0 \\ 0 & b^n & 0 \\ d & 0 & c^n \end{bmatrix}$, where $d = \begin{cases} \dfrac{a^n - c^n}{a - c} & \text{if } a \neq c \\ na^{n-1} & \text{if } a = c \end{cases}$

Exercise Set 2.1 (page 94)

1. **(a)** $M_{11} = 29$, $M_{12} = 21$, $M_{13} = 27$, $M_{21} = -11$, $M_{22} = 13$, $M_{23} = -5$, $M_{31} = -19$, $M_{32} = -19$, $M_{33} = 19$

    (b) $C_{11} = 29$, $C_{12} = -21$, $C_{13} = 27$, $C_{21} = 11$, $C_{22} = 13$, $C_{23} = 5$, $C_{31} = -19$, $C_{32} = 19$, $C_{33} = 19$

3. 152

4. **(a)** $\text{adj}(A) = \begin{bmatrix} 29 & 11 & -19 \\ -21 & 13 & 19 \\ 27 & 5 & 19 \end{bmatrix}$ **(b)** $A^{-1} = \begin{bmatrix} \frac{29}{152} & \frac{11}{152} & -\frac{19}{152} \\ -\frac{21}{152} & \frac{13}{152} & \frac{19}{152} \\ \frac{27}{152} & \frac{5}{152} & \frac{19}{152} \end{bmatrix}$

6. -66 8. $k^3 - 8k^2 - 10k + 95$ 11. $A^{-1} = \begin{bmatrix} 3 & -5 & -5 \\ -3 & 4 & 5 \\ 2 & -2 & -3 \end{bmatrix}$

13. $A^{-1} = \begin{bmatrix} \frac{1}{2} & \frac{3}{2} & 1 \\ 0 & 1 & \frac{3}{2} \\ 0 & 0 & \frac{1}{2} \end{bmatrix}$ 15. $A^{-1} = \begin{bmatrix} -4 & 3 & 0 & -1 \\ 2 & -1 & 0 & 0 \\ -7 & 0 & -1 & 8 \\ 6 & 0 & 1 & -7 \end{bmatrix}$

16. $x_1 = 1$, $x_2 = 2$ 18. $x = -\frac{144}{55}$, $y = -\frac{61}{55}$, $z = \frac{46}{11}$

21. Cramer's rule does not apply.   22. $A^{-1} = \begin{bmatrix} \cos\theta & -\sin\theta & 0 \\ \sin\theta & \cos\theta & 0 \\ 0 & 0 & 1 \end{bmatrix}$

24. $x = 1$, $y = 0$, $z = 2$, $w = 0$ 31. $\det(A) = 10 \times (-108) = -1080$ 34. One

Exercise Set 2.2 (page 101)

2. **(a)** -30 **(b)** -2 **(c)** 0   **(d)** 0

4. 30   6. -17 8. 39   11. -2

12. **(a)** -6 **(b)** 72   **(c)** -6 **(d)** 18

16. **(a)** $\det(A) = -1$ **(b)** $\det(A) = 1$ 18. $x = 0$, -1, $\frac{1}{2}$

Exercise Set 2.3
(page 109)

1. **(a)** $\det(2A) = -40 = 2^2 \det(A)$ **(b)** $\det(-2A) = -448 = (-2)^3 \det(A)$

4. **(a)** Invertible   **(b)** Not invertible   **(c)** Not invertible   **(d)** Not invertible

6. If $x = 0$, the first and third rows are proportional.
   If $x = 2$, the first and second rows are proportional.

12. **(a)** $k = \dfrac{5 \pm \sqrt{17}}{2}$ **(b)** $k = -1$

14. **(a)** $\begin{bmatrix} \lambda - 1 & -2 \\ -2 & \lambda - 1 \end{bmatrix} \begin{bmatrix} x_1 \\ x_2 \end{bmatrix} = \begin{bmatrix} 0 \\ 0 \end{bmatrix}$ **(b)** $\begin{bmatrix} \lambda - 2 & -3 \\ -4 & \lambda - 3 \end{bmatrix} \begin{bmatrix} x_1 \\ x_2 \end{bmatrix} = \begin{bmatrix} 0 \\ 0 \end{bmatrix}$

   (c) $\begin{bmatrix} \lambda - 3 & -1 \\ 5 & \lambda + 3 \end{bmatrix} \begin{bmatrix} x_1 \\ x_2 \end{bmatrix} = \begin{bmatrix} 0 \\ 0 \end{bmatrix}$

15. (i) $\lambda^2 - 2\lambda - 3 = 0$ (ii) $\lambda = -1, \ \lambda = 3$ (iii) $\begin{bmatrix} -t \\ t \end{bmatrix}$, $\begin{bmatrix} t \\ t \end{bmatrix}$

    (i) $\lambda^2 - 5\lambda - 6 = 0$ (ii) $\lambda = -1, \ \lambda = 6$ (iii) $\begin{bmatrix} -t \\ t \end{bmatrix}$, $\begin{bmatrix} \frac{3}{4}t \\ t \end{bmatrix}$

    (i) $\lambda^2 - 4 = 0$ (ii) $\lambda = -2, \ \lambda = 2$ (iii) $\begin{bmatrix} -\frac{t}{5} \\ t \end{bmatrix}$, $\begin{bmatrix} -t \\ t \end{bmatrix}$

20. No   21. AB is singular.

22. **(a)** False   **(b)** True   **(c)** False   **(d)** True

23. **(a)** True   **(b)** True   **(c)** False   **(d)** True

Exercise Set 2.4
(page 117)

1. **(a)** 5   **(b)** 9   **(c)** 6   **(d)** 10   **(e)** 0   **(f)** 2

3. 22   5. 52   7. $a^2 - 5a + 21$ 9. -65 11. -123

13. **(a)** $\lambda = 1, \ \lambda = -3$ **(b)** $\lambda = -2, \ \lambda = 3, \ \lambda = 4$ 16. 275

17. **(a)** $= -120$ **(b)** $= -120$ 18. $x = \dfrac{3 \pm \sqrt{33}}{4}$ 22. Equals 0 if $n > 1$

Supplementary
Exercises
(page 118)

1. $x' = \frac{3}{5}x + \frac{4}{5}y, \ y' = -\frac{4}{5}x + \frac{3}{5}y$ 4. 2

5. $\cos \beta = \dfrac{c^2 + a^2 - b^2}{2ac}, \ \cos \gamma = \dfrac{a^2 + b^2 - c^2}{2ab}$ 12. $\det(B) = (-1)^{n(n-1)/2} \det(A)$

13. **(a)** The ith and jth columns will be interchanged.
    (b) The ith column will be divided by c.
    (c) $-c$ times the jth column will be added to the ith column.

15. **(a)** $\lambda^3 + (-a_{11} - a_{22} - a_{33})\lambda^2$
    $\quad + (a_{11}a_{22} + a_{11}a_{33} + a_{22}a_{33} - a_{12}a_{21} - a_{13}a_{31} - a_{23}a_{32})\lambda$
    $\quad + (a_{11}a_{23}a_{32} + a_{12}a_{21}a_{33} + a_{13}a_{22}a_{31} - a_{11}a_{22}a_{33} - a_{12}a_{23}a_{31} - a_{13}a_{21}a_{32})$

18. **(a)** $\lambda = -5, \ \lambda = 2, \ \lambda = 4;$ $\begin{bmatrix} -2t \\ t \\ t \end{bmatrix}$, $\begin{bmatrix} 5t \\ t \\ t \end{bmatrix}$, $\begin{bmatrix} 7t \\ 19t \\ t \end{bmatrix}$ **(b)** $\lambda = 1;$ $\begin{bmatrix} \frac{1}{2}t \\ -\frac{1}{2}t \\ t \end{bmatrix}$

Exercise Set 3.1
(page 130)

3. **(a)** $\overrightarrow{P_1 P_2} = (-1, -1)$ **(b)** $\overrightarrow{P_1 P_2} = (-7, -2)$ **(c)** $\overrightarrow{P_1 P_2} = (2, 1)$
   (d) $\overrightarrow{P_1 P_2} = (a, b)$ **(e)** $\overrightarrow{P_1 P_2} = (-5, 12, -6)$ **(f)** $\overrightarrow{P_1 P_2} = (1, -1, -2)$
   (g) $\overrightarrow{P_1 P_2} = (-a, -b, -c)$ **(h)** $\overrightarrow{P_1 P_2} = (a, b, c)$

5. **(a)** $P(-1, 2, -4)$ is one possible answer.   **(b)** $P(7, -2, -6)$ is one possible answer.

6. **(a)** $(-2, 1, -4)$ **(b)** $(-10, 6, 4)$ **(c)** $(-7, 1, 10)$
   (d) $(80, -20, -80)$ **(e)** $(132, -24, -72)$ **(f)** $(-77, 8, 94)$

8. $c_1 = 2$, $c_2 = -1$, $c_3 = 2$ 10. $c_1 = c_2 = c_3 = 0$

12. **(a)** $x' = 5$, $y' = 8$ **(b)** $x = -1$, $y = 3$

15. $\mathbf{u} = \left(\frac{\sqrt{3}}{2}, \frac{1}{2}\right)$, $\mathbf{v} = \left(-\frac{1}{2}, -\frac{\sqrt{3}}{2}\right)$,

$\mathbf{u} + \mathbf{v} = \left(\frac{\sqrt{3}-1}{2}, \frac{1-\sqrt{3}}{2}\right)$, $\mathbf{u} - \mathbf{v} = \left(\frac{\sqrt{3}+1}{2}, \frac{\sqrt{3}+1}{2}\right)$

Exercise Set 3.2
(page 134)

1. **(a)** 5   **(b)** $\sqrt{13}$ **(c)** 5   **(d)** $2\sqrt{3}$ **(e)** $3\sqrt{6}$ **(f)** 6

3. **(a)** $\sqrt{83}$ **(b)** $\sqrt{17} + \sqrt{26}$ **(c)** $4\sqrt{17}$ **(d)** $\sqrt{466}$
   (e) $\left(\frac{3}{\sqrt{61}}, \frac{6}{\sqrt{61}}, -\frac{4}{\sqrt{61}}\right)$ **(f)** 1

9. **(b)** $\left(\frac{3}{5}, \frac{4}{5}\right)$ **(c)** $\left(\frac{2}{7}, -\frac{3}{7}, \frac{6}{7}\right)$

10. A sphere of radius 1 centered at (x_0, y_0, z_0)

16. **(a)** $a = c = 0$ **(b)** At least one of a or c is not zero, that is, $a^2 + c^2 > 0$

17. **(a)** The distance from x to the origin is less than 1.   **(b)** $\|x - x_0\| > 1$

Exercise Set 3.3
(page 142)

1. **(a)** -11 **(b)** -24 **(c)** 0   **(d)** 0

3. **(a)** Orthogonal   **(b)** Obtuse   **(c)** Acute   **(d)** Obtuse

5. **(a)** $(6, 2)$ **(b)** $\left(-\frac{21}{13}, -\frac{14}{13}\right)$ **(c)** $\left(\frac{55}{13}, 1, -\frac{11}{13}\right)$ **(d)** $\left(\frac{73}{89}, -\frac{12}{89}, -\frac{32}{89}\right)$

8. **(b)** $(3k, 2k)$ for any scalar k **(c)** $\left(\frac{4}{5}, \frac{3}{5}\right)$, $\left(-\frac{4}{5}, -\frac{3}{5}\right)$

11. $\cos\theta_1 = \frac{\sqrt{10}}{10}$, $\cos\theta_2 = \frac{3\sqrt{10}}{10}$, $\cos\theta_3 = 0$ 13. $\pm(1/\sqrt{3}, 1/\sqrt{3}, -1/\sqrt{3})$

16. **(a)** $\frac{10}{3}$ **(b)** $-\frac{6}{5}$ **(c)** $\frac{-60+34\sqrt{3}}{33}$ **(d)** $\frac{1}{2}$

20. $\cos^{-1}\left(\frac{2}{\sqrt{6}}\right)$ 21. **(b)** $\cos\beta = \frac{b}{\|\mathbf{v}\|}$, $\cos\gamma = \frac{c}{\|\mathbf{v}\|}$

27. **(a)** The vector $\mathbf{u}$ is dotted with a scalar.   **(b)** A scalar is added to the vector $\mathbf{w}$.
    (c) Scalars do not have norms.   **(d)** The scalar k is dotted with a vector.

29. No; it merely says that $\mathbf{u}$ is orthogonal to $\mathbf{v} - \mathbf{w}$.

30. $\mathbf{r} = (\mathbf{u} \cdot \mathbf{r})\frac{\mathbf{u}}{\|\mathbf{u}\|^2} + (\mathbf{v} \cdot \mathbf{r})\frac{\mathbf{v}}{\|\mathbf{v}\|^2} + (\mathbf{w} \cdot \mathbf{r})\frac{\mathbf{w}}{\|\mathbf{w}\|^2}$ 31. Theorem of Pythagoras

Exercise Set 3.4
(page 153)

1. **(a)** $(32, -6, -4)$ **(b)** $(-14, -20, -82)$ **(c)** $(27, 40, -42)$
   (d) $(0, 176, -264)$ **(e)** $(-44, 55, -22)$ **(f)** $(-8, -3, -8)$

3. **(a)** $\sqrt{59}$ **(b)** $\sqrt{101}$ **(c)** 0

7. For example, $(1, 1, 1) \times (2, -3, 5) = (8, -3, -5)$

9. **(a)** -3 **(b)** 3   **(c)** 3   **(d)** -3 **(e)** -3 **(f)** 0

11. **(a)** No   **(b)** Yes   **(c)** No   13. $\left(\frac{6}{\sqrt{61}}, -\frac{3}{\sqrt{61}}, \frac{4}{\sqrt{61}}\right)$, $\left(-\frac{6}{\sqrt{61}}, \frac{3}{\sqrt{61}}, -\frac{4}{\sqrt{61}}\right)$

15. $2(\mathbf{v} \times \mathbf{u})$ 17. **(a)** $\frac{\sqrt{26}}{2}$ **(b)** $\frac{\sqrt{26}}{3}$ 21. **(a)** $\sqrt{122}$ **(b)** $\theta \approx 40°19''$

23. **(a)** $\mathbf{m} = (0, 1, 0)$ and $\mathbf{n} = (1, 0, 0)$ **(b)** $(-1, 0, 0)$ **(c)** $(0, 0, -1)$

28. $(-8, 0, -8)$ 31. **(a)** $\frac{2}{3}$ **(b)** $\frac{1}{2}$ 35. **(b)** $\mathbf{u} \cdot \mathbf{w} \neq 0$, $\mathbf{v} \cdot \mathbf{w} = 0$

36. No, the equation is equivalent to $\mathbf{u} \times (\mathbf{v} - \mathbf{w}) = 0$ and hence to $\mathbf{v} - \mathbf{w} = k\mathbf{u}$ for some scalar k.

38. They are collinear.

Exercise Set 3.5
(page 162)

1. **(a)** $-2(x + 1) + (y - 3) - (z + 2) = 0$ **(b)** $(x - 1) + 9(y - 1) + 8(z - 4) = 0$
   (c) $2z = 0$ **(d)** $x + 2y + 3z = 0$

3. **(a)** $(0, 0, 5)$ is a point in the plane and $\mathbf{n} = (-3, 7, 2)$ is a normal vector so that
   $-3(x - 0) + 7(y - 0) + 2(z - 5) = 0$ is a point-normal form; other points and normals
   yield other correct answers.
   (b) $(x - 0) + 0(y - 0) - 4(z - 0) = 0$ is a possibility

5. **(a)** Not parallel   **(b)** Parallel   **(c)** Parallel

9. **(a)** $x = 3 + 2t$, $y = -1 + t$, $z = 2 + 3t$ **(b)** $x = -2 + 6t$, $y = 3 - 6t$, $z = -3 - 2t$
   (c) $x = 2$, $y = 2 + t$, $z = 6$ **(d)** $x = t$, $y = -2t$, $z = 3t$

11. **(a)** $x = -12 - 7t$, $y = -41 - 23t$, $z = t$ **(b)** $x = \frac{5}{2}t$, $y = 0$, $z = t$

13. **(a)** Parallel   **(b)** Not parallel   17. $2x + 3y - 5z + 36 = 0$

19. **(a)** $z - z_0 = 0$ **(b)** $x - x_0 = 0$ **(c)** $y - y_0 = 0$ 21. $5x - 2y + z - 34 = 0$

23. $y + 2z - 9 = 0$ 27. $x + 5y + 3z - 18 = 0$

29. $4x + 13y - z - 17 = 0$ 31. $3x - y - z - 2 = 0$

37. **(a)** $x = \frac{11}{23} + \frac{7}{23}t$, $y = -\frac{41}{23} - \frac{1}{23}t$, $z = t$ **(b)** $x = -\frac{2}{5}t$, $y = 0$, $z = t$

39. **(a)** $\frac{5}{3}$ **(b)** $\frac{1}{\sqrt{29}}$ **(c)** $\frac{4}{\sqrt{3}}$

43. **(a)** $\dfrac{x-3}{2} = y+1 = \dfrac{z-2}{3}$ **(b)** $\dfrac{x+2}{6} = -\dfrac{y-3}{6} = -\dfrac{z+3}{2}$

44. **(a)** $x - 2y - 17 = 0$ and $x + 4z - 27 = 0$ is one possible answer.
    (b) $x - 2y = 0$ and $-7y + 2z = 0$ is one possible answer.

45. **(a)** $\theta \approx 35°$ **(b)** $\theta \approx 79°$ 47. They are identical.

Exercise Set 4.1
(page 178)

1. **(a)** $(-1, 9, -11, 1)$ **(b)** $(22, 53, -19, 14)$ **(c)** $(-13, 13, -36, -2)$
   (d) $(-90, -114, 60, -36)$ **(e)** $(-9, -5, -5, -3)$ **(f)** $(27, 29, -27, 9)$

3. $c_1 = 1$, $c_2 = 1$, $c_3 = -1$, $c_4 = 1$ 5. **(a)** $\sqrt{29}$ **(b)** 3   **(c)** 13   **(d)** $\sqrt{31}$

8. $k = \pm\frac{5}{7}$ 10. **(a)** $\left(\frac{1}{\sqrt{10}}, \frac{3}{\sqrt{10}}\right), \left(-\frac{1}{\sqrt{10}}, -\frac{3}{\sqrt{10}}\right)$

14. **(a)** Yes   **(b)** No   **(c)** Yes   **(d)** No   **(e)** No   **(f)** Yes

15. **(a)** $k = -3$ **(b)** $k = -2$, $k = -3$ 19. $x_1 = 1$, $x_2 = -1$, $x_3 = 2$

22. The component in the **a** direction is $\text{proj}_{\mathbf{a}}\mathbf{u} = \frac{4}{15}(-1, 1, 2, 3)$; the orthogonal component
    is $\frac{1}{15}(34, 11, 52, -27)$.

23. They do not intersect.

33. **(a)** Euclidean measure of "box" in R^n: $a_1 a_2 \cdots a_n$
    (b) Length of diagonal: $\sqrt{a_1^2 + a_2^2 + \cdots + a_n^2}$

35. **(a)** $d(\mathbf{u}, \mathbf{v}) = \sqrt{2}$

37. **(a)** True   **(b)** True   **(c)** False   **(d)** True   **(e)** True, unless $\mathbf{u} = \mathbf{0}$

Exercise Set 4.2
(page 193)

1. **(a)** Linear; $R^3 \to R^2$ **(b)** Nonlinear; $R^2 \to R^3$ **(c)** Linear; $R^3 \to R^3$
   (d) Nonlinear; $R^4 \to R^2$

3. $\begin{bmatrix} 3 & 5 & -1 \\ 4 & -1 & 1 \\ 3 & 2 & -1 \end{bmatrix}$; $T(-1, 2, 4) = (3, -2, -3)$

5. **(a)** $\begin{bmatrix} 0 & 1 \\ -1 & 0 \\ 1 & 3 \\ 1 & -1 \end{bmatrix}$ **(b)** $\begin{bmatrix} 7 & 2 & -1 & 1 \\ 0 & 1 & 1 & 0 \\ -1 & 0 & 0 & 0 \end{bmatrix}$ **(c)** $\begin{bmatrix} 0 & 0 & 0 \\ 0 & 0 & 0 \\ 0 & 0 & 0 \\ 0 & 0 & 0 \\ 0 & 0 & 0 \end{bmatrix}$

   (d) $\begin{bmatrix} 0 & 0 & 0 & 1 \\ 1 & 0 & 0 & 0 \\ 0 & 0 & 1 & 0 \\ 0 & 1 & 0 & 0 \\ 1 & 0 & -1 & 0 \end{bmatrix}$

7. **(a)** $T(-1, 4) = (5, 4)$ **(b)** $T(2, 1, -3) = (0, -2, 0)$

9. **(a)** $(2, -5, -3)$ **(b)** $(2, 5, 3)$ **(c)** $(-2, -5, 3)$

13. **(a)** $\left(-2, \frac{\sqrt{3}-2}{2}, \frac{1+2\sqrt{3}}{2}\right)$ **(b)** $(0, 1, 2\sqrt{2})$ **(c)** $(-1, -2, 2)$

15. **(a)** $\left(-2, \frac{\sqrt{3}+2}{2}, \frac{-1+2\sqrt{3}}{2}\right)$ **(b)** $(-2\sqrt{2}, 1, 0)$ **(c)** $(1, 2, 2)$

17. **(a)** $\begin{bmatrix} 0 & 0 \\ 1/2 & -\sqrt{3}/2 \end{bmatrix}$ **(b)** $\begin{bmatrix} -\sqrt{2} & \sqrt{2} \\ \sqrt{2} & \sqrt{2} \end{bmatrix}$ **(c)** $\begin{bmatrix} -1 & 0 \\ 0 & -1 \end{bmatrix}$

19. **(a)** $\begin{bmatrix} \sqrt{3}/8 & -\sqrt{3}/16 & 1/16 \\ 1/8 & 3/16 & -\sqrt{3}/16 \\ 0 & 1/8 & \sqrt{3}/8 \end{bmatrix}$ **(b)** $\begin{bmatrix} 0 & 0 & 0 \\ 0 & -1 & 0 \\ 0 & 0 & -1 \end{bmatrix}$

   (c) $\begin{bmatrix} 0 & 1 & 0 \\ 0 & 0 & -1 \\ -1 & 0 & 0 \end{bmatrix}$

21. **(a)** Yes   **(b)** No

24. $\begin{bmatrix} \frac{1}{3}(1-\cos\theta)+\cos\theta & \frac{1}{3}(1-\cos\theta)-\frac{1}{\sqrt{3}}\sin\theta & \frac{1}{3}(1-\cos\theta)-\frac{1}{\sqrt{3}}\sin\theta \\ \frac{1}{3}(1-\cos\theta)-\frac{1}{\sqrt{3}}\sin\theta & \frac{1}{3}(1-\cos\theta)+\cos\theta & \frac{1}{3}(1-\cos\theta)-\frac{1}{\sqrt{3}}\sin\theta \\ \frac{1}{3}(1-\cos\theta)-\frac{1}{\sqrt{3}}\sin\theta & \frac{1}{3}(1-\cos\theta)-\frac{1}{\sqrt{3}}\sin\theta & \frac{1}{3}(1-\cos\theta)+\cos\theta \end{bmatrix}$

28. **(c)** $90°$

29. **(a)** Twice the orthogonal projection on the x-axis
    (b) Twice the reflection about the x-axis

30. **(a)** The x-coordinate is stretched by a factor of 2 and the y-coordinate is stretched by a factor of 3.
    (b) Rotation through $30°$

31. Rotation through the angle 2θ 34. Only if $b = 0$.

Exercise Set 4.3 (page 206)

1. **(a)** Not one-to-one   **(b)** One-to-one   **(c)** One-to-one   **(d)** One-to-one
   (e) One-to-one   **(f)** One-to-one   **(g)** One-to-one

3. For example, the vector $(1, 3)$ is not in the range.

5. **(a)** One-to-one; $\begin{bmatrix} \frac{1}{3} & -\frac{2}{3} \\ \frac{1}{3} & \frac{1}{3} \end{bmatrix}$; $T^{-1}(w_1, w_2) = \left(\frac{1}{3}w_1 - \frac{2}{3}w_2, \frac{1}{3}w_1 + \frac{1}{3}w_2\right)$
   (b) Not one-to-one
   (c) One-to-one; $\begin{bmatrix} 0 & -1 \\ -1 & 0 \end{bmatrix}$; $T^{-1}(w_1, w_2) = (-w_2, -w_1)$ **(d)** Not one-to-one

7. **(a)** Reflection about the x-axis   **(b)** Rotation through the angle $-\pi/4$
   (c) Contraction by a factor of $\frac{1}{3}$ **(d)** Reflection about the yz-plane
   (e) Dilation by a factor of 5

9. **(a)** Linear   **(b)** Nonlinear   **(c)** Linear   **(d)** Nonlinear

12. **(a)** For a reflection about the y-axis, $T(\mathbf{e}_1) = \begin{bmatrix} -1 \\ 0 \end{bmatrix}$ and $T(\mathbf{e}_2) = \begin{bmatrix} 0 \\ 1 \end{bmatrix}$.

    Thus, $T = \begin{bmatrix} -1 & 0 \\ 0 & 1 \end{bmatrix}$.

    (b) For a reflection about the xz-plane, $T(\mathbf{e}_1) = \begin{bmatrix} 1 \\ 0 \\ 0 \end{bmatrix}$, $T(\mathbf{e}_2) = \begin{bmatrix} 0 \\ -1 \\ 0 \end{bmatrix}$, and

    $T(\mathbf{e}_3) = \begin{bmatrix} 0 \\ 0 \\ 1 \end{bmatrix}$. Thus, $T = \begin{bmatrix} 1 & 0 & 0 \\ 0 & -1 & 0 \\ 0 & 0 & 1 \end{bmatrix}$.

(c) For an orthogonal projection on the x-axis, $T(\mathbf{e}_1) = \begin{bmatrix} 1 \\ 0 \end{bmatrix}$ and $T(\mathbf{e}_2) = \begin{bmatrix} 0 \\ 0 \end{bmatrix}$.

Thus, $T = \begin{bmatrix} 1 & 0 \\ 0 & 0 \end{bmatrix}$.

(d) For an orthogonal projection on the yz-plane, $T(\mathbf{e}_1) = \begin{bmatrix} 0 \\ 0 \\ 0 \end{bmatrix}$, $T(\mathbf{e}_2) = \begin{bmatrix} 0 \\ 1 \\ 0 \end{bmatrix}$, and

$T(\mathbf{e}_3) = \begin{bmatrix} 0 \\ 0 \\ 1 \end{bmatrix}$. Thus, $T = \begin{bmatrix} 0 & 0 & 0 \\ 0 & 1 & 0 \\ 0 & 0 & 1 \end{bmatrix}$.

(e) For a rotation through a positive angle θ, $T(\mathbf{e}_1) = \begin{bmatrix} \cos\theta \\ \sin\theta \end{bmatrix}$ and $T(\mathbf{e}_2) = \begin{bmatrix} -\sin\theta \\ \cos\theta \end{bmatrix}$.

Thus, $T = \begin{bmatrix} \cos\theta & -\sin\theta \\ \sin\theta & \cos\theta \end{bmatrix}$.

(f) For a dilation by a factor $k \geq 1$, $T(\mathbf{e}_1) = \begin{bmatrix} k \\ 0 \\ 0 \end{bmatrix}$, $T(\mathbf{e}_2) = \begin{bmatrix} 0 \\ k \\ 0 \end{bmatrix}$, and $T(\mathbf{e}_3) = \begin{bmatrix} 0 \\ 0 \\ k \end{bmatrix}$.

Thus, $T = \begin{bmatrix} k & 0 & 0 \\ 0 & k & 0 \\ 0 & 0 & k \end{bmatrix}$.

13. (a) $T(\mathbf{e}_1) = \begin{bmatrix} -1 \\ 0 \end{bmatrix}$ and $T(\mathbf{e}_2) = \begin{bmatrix} 0 \\ 0 \end{bmatrix}$. Thus, $T = \begin{bmatrix} -1 & 0 \\ 0 & 0 \end{bmatrix}$.

(b) $T(\mathbf{e}_1) = \begin{bmatrix} 0 \\ -1 \end{bmatrix}$ and $T(\mathbf{e}_2) = \begin{bmatrix} 1 \\ 0 \end{bmatrix}$. Thus, $T = \begin{bmatrix} 0 & 1 \\ -1 & 0 \end{bmatrix}$.

(c) $T(\mathbf{e}_1) = \begin{bmatrix} 0 \\ 3 \end{bmatrix}$ and $T(\mathbf{e}_2) = \begin{bmatrix} 0 \\ 0 \end{bmatrix}$. Thus, $T = \begin{bmatrix} 0 & 0 \\ 3 & 0 \end{bmatrix}$.

16. (a) Linear transformation from $R^2 \to R^3$; one-to-one
(b) Linear transformation from $R^3 \to R^2$; not one-to-one

17. (a) $\left(\frac{1}{2}, \frac{1}{2}\right)$ (b) $\left(\frac{3}{4}, \frac{\sqrt{3}}{4}\right)$ (c) $\left(\frac{1-5\sqrt{3}}{4}, \frac{15-\sqrt{3}}{4}\right)$

19. (a) $\lambda = 1$; $\begin{bmatrix} 0 \\ s \\ t \end{bmatrix}$ $\lambda = -1$; $\begin{bmatrix} t \\ 0 \\ 0 \end{bmatrix}$ (b) $\lambda = 1$; $\begin{bmatrix} s \\ 0 \\ t \end{bmatrix}$ $\lambda = 0$; $\begin{bmatrix} 0 \\ t \\ 0 \end{bmatrix}$

(c) $\lambda = 2$; all vectors in R^3 are eigenvectors   (d) $\lambda = 1$; $\begin{bmatrix} 0 \\ 0 \\ t \end{bmatrix}$

23. (a) $\begin{bmatrix} \cos 2\theta & \sin 2\theta \\ \sin 2\theta & -\cos 2\theta \end{bmatrix}$ (b) $\left(\frac{1+5\sqrt{3}}{2}, \frac{\sqrt{3}-5}{2}\right)$

27. (a) The range of T is a proper subset of R^n.
(b) T must map infinitely many vectors to 0.

Exercise Set 4.4
(page 217)

1. (a) $x^2 + 2x - 1 - 2(3x^2 + 2) = -5x^2 + 2x - 5$ 4. (a) Yes; $A = \begin{bmatrix} 1 & 0 & 0 \\ 0 & 1 & 0 \\ 0 & 0 & 1 \\ 0 & 0 & 0 \end{bmatrix}$

7. (a) $L: P_1 \to P_1$ where L maps $ax + b$ to $(a + b)x + a - b$

9. (a) $3e^t + 3e^{-t} = 6\cosh(t)$ (b) Yes

12. $y = 2x^2$ 14. (a) $y = x^3 - x$ 15. (a) $y = 2x^3 - 2x + 2$

18. (a) No, because of the arbitrary constant of integration
    (b) No (except for P_0)

21. (a) Each $L_i(x)$ is a polynomial of degree at most n and hence so is the sum
    $y_0 L(x) + \cdots + y_n L(x)$; also, $p(x_i) = 0 + 0 + \cdots + 0 + y_i \cdot L_i(x_i) + 0 + \cdots + 0$
    $+ 0 = y_i$, showing that this function is an interpolant of degree at most n.
    (b) It is $I_{n+1}\mathbf{c} = \mathbf{y}$ where $\mathbf{c}$ is the vector of c_i values and $\mathbf{y}$ is the vector of y-values.

Exercise Set 5.1 (page 226)

1. Not a vector space. Axiom 8 fails.

3. Not a vector space. Axioms 9 and 10 fail.

5. The set is a vector space under the given operations.

7. The set is a vector space under the given operations.

9. Not a vector space. Axioms 1, 4, 5, and 6 fail.

11. The set is a vector space under the given operations.

13. The set is a vector space under the given operations.

25. No. A vector space must have a zero element.

26. No. Axioms 1, 4, and 6 will fail.

29. (1) Axiom 7   (2) Axiom 4   (3) Axiom 5   (4) Follows from statement 2
    (5) Axiom 3   (6) Axiom 5   (7) Axiom 4

32. No; $\mathbf{0}_1 = \mathbf{0}_1 + \mathbf{0}_2 = \mathbf{0}_2$

Exercise Set 5.2 (page 238)

1. (a), (c)   3. (a), (b), (d)   5. (a), (b), (d)

6. (a) Line; $x = -\frac{1}{2}t$, $y = -\frac{3}{2}t$, $z = t$ (b) Line; $x = 2t$, $y = t$, $z = 0$ (c) Origin
   (d) Origin   (e) Line; $x = -3t$, $y = -2t$, $z = t$ (f) Plane; $x - 3y + z = 0$

9. (a) $-9 - 7x - 15x^2 = -2\mathbf{p}_1 + \mathbf{p}_2 - 2\mathbf{p}_3$ (b) $6 + 11x + 6x^2 = 4\mathbf{p}_1 - 5\mathbf{p}_2 + \mathbf{p}_3$
   (c) $0 = 0\mathbf{p}_1 + 0\mathbf{p}_2 + 0\mathbf{p}_3$ (d) $7 + 8x + 9x^2 = 0\mathbf{p}_1 - 2\mathbf{p}_2 + 3\mathbf{p}_3$

11. (a) The vectors span.   (b) The vectors do not span.
    (c) The vectors do not span.   (d) The vectors span.

12. (a), (c), (e)   15. $y = z$

24. (a) They span a line if they are collinear and not both 0. They span a plane if they are
        not collinear.
    (b) If $\mathbf{u} = a\mathbf{v}$ and $\mathbf{v} = b\mathbf{u}$ for some real numbers a, b
    (c) We must have $\mathbf{b} = \mathbf{0}$ since a subspace must contain $\mathbf{x} = \mathbf{0}$ and then $\mathbf{b} = A\mathbf{0} = \mathbf{0}$.

26. (a) For example, $\begin{bmatrix} 1 & 0 \\ 0 & 0 \end{bmatrix}$, $\begin{bmatrix} 0 & 1 \\ 0 & 0 \end{bmatrix}$, $\begin{bmatrix} 0 & 0 \\ 1 & 0 \end{bmatrix}$, $\begin{bmatrix} 0 & 0 \\ 0 & 1 \end{bmatrix}$
    (b) The set of matrices having one entry equal to 1 and all other entries equal to 0

Exercise Set 5.3 (page 248)

1. (a) $\mathbf{u}_2$ is a scalar multiple of $\mathbf{u}_1$.
   (b) The vectors are linearly dependent by Theorem 5.3.3.
   (c) $\mathbf{p}_2$ is a scalar multiple of $\mathbf{p}_1$.   (d) B is a scalar multiple of A.

3. None   5. (a) They do not lie in a plane.   (b) They do lie in a plane.

7. (b) $\mathbf{v}_1 = \frac{2}{7}\mathbf{v}_2 - \frac{3}{7}\mathbf{v}_3$, $\mathbf{v}_2 = \frac{7}{2}\mathbf{v}_1 + \frac{3}{2}\mathbf{v}_3$, $\mathbf{v}_3 = -\frac{7}{3}\mathbf{v}_1 + \frac{2}{3}\mathbf{v}_2$ 9. $\lambda = -\frac{1}{2}$, $\lambda = 1$

18. If and only if the vector is not zero

19. (a) They are linearly independent since $\mathbf{v}_1$, $\mathbf{v}_2$, and $\mathbf{v}_3$ do not lie in the same plane when they
        are placed with their initial points at the origin.
    (b) They are not linearly independent since $\mathbf{v}_1$, $\mathbf{v}_2$, and $\mathbf{v}_3$ lie in the same plane when they
        are placed with their initial points at the origin.

20. (a), (d), (e), (f)

24. (a) False   (b) False   (c) True   (d) False   27. (a) Yes

Exercise Set 5.4
(page 263)

1. **(a)** A basis for R^2 has two linearly independent vectors.
   (b) A basis for R^3 has three linearly independent vectors.
   (c) A basis for P_2 has three linearly independent vectors.
   (d) A basis for M_{22} has four linearly independent vectors.

3. (a), (b)   7. **(a)** $(\mathbf{w})_S = (3, -7)$ **(b)** $(\mathbf{w})_S = \left(\frac{5}{28}, \frac{3}{14}\right)$ **(c)** $(\mathbf{w})_S = \left(a, \dfrac{b-a}{2}\right)$

9. **(a)** $(\mathbf{v})_S = (3, -2, 1)$ **(b)** $(\mathbf{v})_S = (-2, 0, 1)$ 11. $(A)_S = (-1, 1, -1, 3)$

13. Basis: $\left(-\frac{1}{4}, -\frac{1}{4}, 1, 0\right), (0, -1, 0, 1)$; dimension $= 2$

15. Basis: $(3, 1, 0), (-1, 0, 1)$; dimension $= 2$

19. **(a)** 3-dimensional   **(b)** 2-dimensional   **(c)** 1-dimensional

20. 3-dimensional

21. **(a)** $\{\mathbf{v}_1, \mathbf{v}_2, \mathbf{e}_1\}$ or $\{\mathbf{v}_1, \mathbf{v}_2, \mathbf{e}_2\}$ **(b)** $\{\mathbf{v}_1, \mathbf{v}_2, \mathbf{e}_1\}$ or $\{\mathbf{v}_1, \mathbf{v}_2, \mathbf{e}_2\}$ or $\{\mathbf{v}_1, \mathbf{v}_2, \mathbf{e}_3\}$

27. **(a)** One possible answer is $\{-1 + x - 2x^2, 3 + 3x + 6x^2, 9\}$.
    (b) One possible answer is $\{1 + x, x^2, -2 + 2x^2\}$.
    (c) One possible answer is $\{1 + x - 3x^2\}$.

29. **(a)** $(2, 0)$ **(b)** $\left(\dfrac{2}{\sqrt{3}}, -\dfrac{1}{\sqrt{3}}\right)$ **(c)** $(0, 1)$ **(d)** $\left(\dfrac{2}{\sqrt{3}}a, b - \dfrac{a}{\sqrt{3}}\right)$

31. Yes; for example, $\begin{bmatrix} 1 & 0 \\ 0 & \pm 1 \end{bmatrix}$, $\begin{bmatrix} 0 & 1 \\ \pm 1 & 0 \end{bmatrix}$

32. **(a)** n **(b)** $n(n+1)/2$ **(c)** $n(n+1)/2$

35. **(a)** The dimension is $n - 1$.
    (b) $(1, 0, 0, \ldots, 0, -1), (0, 1, 0, \ldots, 0, -1), (0, 0, 1, \ldots, 0, -1), \ldots, (0, 0, 0, \ldots, 1, -1)$
    is a basis of size $n - 1$.

Exercise Set 5.5
(page 276)

1. $\mathbf{r}_1 = (2, -1, 0, 1)$, $\mathbf{r}_2 = (3, 5, 7, -1)$, $\mathbf{r}_3 = (1, 4, 2, 7)$;

$$\mathbf{c}_1 = \begin{bmatrix} 2 \\ 3 \\ 1 \end{bmatrix}, \quad \mathbf{c}_2 = \begin{bmatrix} -1 \\ 5 \\ 4 \end{bmatrix}, \quad \mathbf{c}_3 = \begin{bmatrix} 0 \\ 7 \\ 2 \end{bmatrix}, \quad \mathbf{c}_4 = \begin{bmatrix} 1 \\ -1 \\ 7 \end{bmatrix}$$

3. **(a)** $\begin{bmatrix} -2 \\ 10 \end{bmatrix} = \begin{bmatrix} 1 \\ 4 \end{bmatrix} - \begin{bmatrix} 3 \\ -6 \end{bmatrix}$ **(b)** $\mathbf{b}$ is not in the column space of A.

   (c) $\begin{bmatrix} 1 \\ 9 \\ 1 \end{bmatrix} - 3 \begin{bmatrix} -1 \\ 3 \\ 1 \end{bmatrix} + \begin{bmatrix} 1 \\ 1 \\ 1 \end{bmatrix} = \begin{bmatrix} 5 \\ 1 \\ -1 \end{bmatrix}$

   (d) $\begin{bmatrix} 2 \\ 0 \\ 0 \end{bmatrix} = \begin{bmatrix} 1 \\ 1 \\ -1 \end{bmatrix} + (t-1) \begin{bmatrix} -1 \\ 1 \\ -1 \end{bmatrix} + t \begin{bmatrix} 1 \\ -1 \\ 1 \end{bmatrix}$

   (e) $\begin{bmatrix} 4 \\ 3 \\ 5 \\ 7 \end{bmatrix} = -26 \begin{bmatrix} 1 \\ 0 \\ 1 \\ 0 \end{bmatrix} + 13 \begin{bmatrix} 2 \\ 1 \\ 2 \\ 1 \end{bmatrix} - 7 \begin{bmatrix} 0 \\ 2 \\ 1 \\ 2 \end{bmatrix} + 4 \begin{bmatrix} 1 \\ 1 \\ 3 \\ 2 \end{bmatrix}$

5. **(a)** $\begin{bmatrix} 1 \\ 0 \end{bmatrix} + t \begin{bmatrix} 3 \\ 1 \end{bmatrix}$; $t \begin{bmatrix} 3 \\ 1 \end{bmatrix}$ **(b)** $\begin{bmatrix} -2 \\ 7 \\ 0 \end{bmatrix} + t \begin{bmatrix} -1 \\ -1 \\ 1 \end{bmatrix}$; $t \begin{bmatrix} -1 \\ -1 \\ 1 \end{bmatrix}$

(c) $\begin{bmatrix} -1 \\ 0 \\ 0 \\ 0 \end{bmatrix} + r\begin{bmatrix} 2 \\ 1 \\ 0 \\ 0 \end{bmatrix} + s\begin{bmatrix} -1 \\ 0 \\ 1 \\ 0 \end{bmatrix} + t\begin{bmatrix} -2 \\ 0 \\ 0 \\ 1 \end{bmatrix}; \quad r\begin{bmatrix} 2 \\ 1 \\ 0 \\ 0 \end{bmatrix} + s\begin{bmatrix} -1 \\ 0 \\ 1 \\ 0 \end{bmatrix} + t\begin{bmatrix} -2 \\ 0 \\ 0 \\ 1 \end{bmatrix}$

(d) $\begin{bmatrix} \frac{6}{5} \\ \frac{7}{5} \\ 0 \\ 0 \end{bmatrix} + s\begin{bmatrix} \frac{7}{5} \\ \frac{4}{5} \\ 1 \\ 0 \end{bmatrix} + t\begin{bmatrix} \frac{1}{5} \\ -\frac{3}{5} \\ 0 \\ 1 \end{bmatrix}; \quad s\begin{bmatrix} \frac{7}{5} \\ \frac{4}{5} \\ 1 \\ 0 \end{bmatrix} + t\begin{bmatrix} \frac{1}{5} \\ -\frac{3}{5} \\ 0 \\ 1 \end{bmatrix}$

7. **(a)** $\mathbf{r}_1 = [1 \quad 0 \quad 2]$, $\mathbf{r}_2 = [0 \quad 0 \quad 1]$, $\mathbf{c} = \begin{bmatrix} 1 \\ 0 \\ 0 \end{bmatrix}$, $\mathbf{c}_2 = \begin{bmatrix} 2 \\ 1 \\ 0 \end{bmatrix}$

(b) $\mathbf{r}_1 = [1 \quad -3 \quad 0 \quad 0]$, $\mathbf{r}_2 = [0 \quad 1 \quad 0 \quad 0]$, $\mathbf{c}_1 = \begin{bmatrix} 1 \\ 0 \\ 0 \\ 0 \end{bmatrix}$, $\mathbf{c}_2 = \begin{bmatrix} -3 \\ 1 \\ 0 \\ 0 \end{bmatrix}$

(c) $\mathbf{r}_1 = [1 \quad 2 \quad 4 \quad 5]$, $\mathbf{r}_2 = [0 \quad 1 \quad -3 \quad 0]$,
$\mathbf{r}_3 = [0 \quad 0 \quad 1 \quad -3]$, $\mathbf{r}_4 = [0 \quad 0 \quad 0 \quad 1]$,

$\mathbf{c}_1 = \begin{bmatrix} 1 \\ 0 \\ 0 \\ 0 \\ 0 \end{bmatrix}$, $\mathbf{c}_2 = \begin{bmatrix} 2 \\ 1 \\ 0 \\ 0 \\ 0 \end{bmatrix}$, $\mathbf{c}_3 = \begin{bmatrix} 4 \\ -3 \\ 1 \\ 0 \\ 0 \end{bmatrix}$, $\mathbf{c}_4 = \begin{bmatrix} 5 \\ 0 \\ -3 \\ 1 \\ 0 \end{bmatrix}$

(d) $\mathbf{r}_1 = [1 \quad 2 \quad -1 \quad 5]$, $\mathbf{r}_2 = [0 \quad 1 \quad 4 \quad 3]$,
$\mathbf{r}_3 = [0 \quad 0 \quad 1 \quad -7]$, $\mathbf{r}_4 = [0 \quad 0 \quad 0 \quad 1]$,

$\mathbf{c}_1 = \begin{bmatrix} 1 \\ 0 \\ 0 \\ 0 \end{bmatrix}$, $\mathbf{c}_2 = \begin{bmatrix} 2 \\ 1 \\ 0 \\ 0 \end{bmatrix}$, $\mathbf{c}_3 = \begin{bmatrix} -1 \\ 4 \\ 1 \\ 0 \end{bmatrix}$, $\mathbf{c}_4 = \begin{bmatrix} 5 \\ 3 \\ -7 \\ 1 \end{bmatrix}$

9. **(a)** $\begin{bmatrix} 1 \\ 5 \\ 7 \end{bmatrix}$, $\begin{bmatrix} -1 \\ -4 \\ -6 \end{bmatrix}$ **(b)** $\begin{bmatrix} 2 \\ 4 \\ 0 \end{bmatrix}$ **(c)** $\begin{bmatrix} 1 \\ 2 \\ -1 \end{bmatrix}$, $\begin{bmatrix} 4 \\ 1 \\ 3 \end{bmatrix}$

(d) $\begin{bmatrix} 1 \\ 3 \\ -1 \\ 2 \end{bmatrix}$, $\begin{bmatrix} 4 \\ -2 \\ 0 \\ 3 \end{bmatrix}$ **(e)** $\begin{bmatrix} 1 \\ 0 \\ 2 \\ 3 \\ -2 \end{bmatrix}$, $\begin{bmatrix} -3 \\ 3 \\ -3 \\ -6 \\ 9 \end{bmatrix}$, $\begin{bmatrix} 2 \\ 6 \\ -2 \\ 0 \\ 2 \end{bmatrix}$

11. **(a)** $(1, 1, -4, -3)$, $(0, 1, -5, -2)$, $\left(0, 0, 1, -\frac{1}{2}\right)$
(b) $\left(1, -1, 2, 0\right)$, $(0, 1, 0, 0)$, $\left(0, 0, 1, -\frac{1}{6}\right)$
(c) $(1, 1, 0, 0)$, $(0, 1, 1, 1)$, $(0, 0, 1, 1)$, $(0, 0, 0, 1)$

14. **(b)** $\begin{bmatrix} 0 & 0 & 0 \\ 0 & 1 & 0 \\ 0 & 0 & 1 \end{bmatrix}$ 17. $\begin{bmatrix} 3a & -5a \\ 3b & -5b \end{bmatrix}$ for all real numbers a, b not both 0.

Exercise Set 5.6
(page 288)

1. Rank $(A) = $ rank$(A^T) = 2$

3. **(a)** 2; 1   **(b)** 1; 2   **(c)** 2; 2   **(d)** 2; 3   **(e)** 3; 2

5. **(a)** Rank $= 4$, nullity $= 0$ **(b)** Rank $= 3$, nullity $= 2$ **(c)** Rank $= 3$, nullity $= 0$

7. (a) Yes, 0   (b) No   (c) Yes, 2   (d) Yes, 7   (e) No
    (f) Yes, 4   (g) Yes, 0

9. $b_1 = r$, $b_2 = s$, $b_3 = 4s - 3r$, $b_4 = 2r - s$, $b_5 = 8s - 7r$ 11.   No

13. Rank is 2 if $r = 2$ and $s = 1$; the rank is never 1.

16. (a) $\begin{bmatrix} 1 & 0 & 0 \\ 0 & 1 & 0 \\ 0 & 0 & 0 \end{bmatrix}$ (b) A line through the origin   (c) A plane through the origin

    (d) The nullspace is a line through the origin and the row space is a plane through the origin.

19. (a) 3   (b) 5   (c) 3   (d) 3

Supplementary Exercises (page 290)

1. (a) All of R^3 (b) Plane: $2x - 3y + z = 0$
    (c) Line: $x = 2t$, $y = t$, $z = 0$ (d) The origin: $(0, 0, 0)$

3. (a) $a(4, 1, 1) + b(0, -1, 2)$ (b) $(a + c)(3, -1, 2) + b(1, 4, 1)$
    (c) $a(2, 3, 0) + b(-1, 0, 4) + c(4, -1, 1)$

5. (a) $\mathbf{v} = (-1 + r)\mathbf{v}_1 + \left(\frac{2}{3} - r\right)\mathbf{v}_2 + r\mathbf{v}_3$; r arbitrary   7.   No

9. (a) Rank $= 2$, nullity $= 1$ (b) Rank $= 3$, nullity $= 2$
    (c) Rank $= n + 1$, nullity $= n$

11. $\{1, x^2, x^3, x^4, x^5, x^6, \ldots, x^n\}$ 13. (a) 2   (b) 1   (c) 2   (d) 3

Exercise Set 6.1 (page 300)

1. (a) $y = 3x - 4$ (b) $y = -2x + 1$

2. (a) $x^2 + y^2 - 4x - 6y + 4 = 0$ or $(x - 2)^2 + (y - 3)^2 = 9$
    (b) $x^2 + y^2 + 2x - 4y - 20 = 0$ or $(x + 1)^2 + (y - 2)^2 = 25$

3. $x^2 + 2xy + y^2 - 2x + y = 0$ (a parabola)

4. (a) $x + 2y + z = 0$ (b) $-x + y - 2z + 1 = 0$

5. (a) $\begin{vmatrix} x & y & z & 0 \\ x_1 & y_1 & z_1 & 1 \\ x_2 & y_2 & z_2 & 1 \\ x_3 & y_3 & z_3 & 1 \end{vmatrix} = 0$ (b) $x + 2y + z = 0$; $-x + y - 2z = 0$

6. (a) $x^2 + y^2 + z^2 - 2x - 4y - 2z = -2$ or $(x - 1)^2 + (y - 2)^2 + (z - 1)^2 = 4$
    (b) $x^2 + y^2 + z^2 - 2x - 2y = 3$ or $(x - 1)^2 + (y - 1)^2 + z^2 = 5$

10. $\begin{vmatrix} y & x^2 & x & 1 \\ y_1 & x_1^2 & x_1 & 1 \\ y_2 & x_2^2 & x_2 & 1 \\ y_3 & x_3^2 & x_3 & 1 \end{vmatrix} = 0$

11. The equation of the line through the three collinear points   12. $0 = 0$

13. The equation of the plane through the four coplanar points

Exercise Set 6.2 (page 304)

1. $I_1 = \frac{255}{317}$, $I_2 = \frac{97}{317}$, $I_3 = \frac{158}{317}$ 2. $I_1 = \frac{13}{5}$, $I_2 = -\frac{2}{5}$, $I_3 = \frac{11}{5}$

3. $I_1 = -\frac{5}{22}$, $I_2 = \frac{7}{22}$, $I_3 = \frac{6}{11}$ 4. $I_1 = \frac{1}{2}$, $I_2 = 0$, $I_3 = 0$, $I_4 = \frac{1}{2}$, $I_5 = \frac{1}{2}$, $I_6 = \frac{1}{2}$

Exercise Set 6.3 (page 316)

1. $x_1 = 2$, $x_2 = \frac{2}{3}$; maximum value of $z = \frac{22}{3}$

2. No feasible solutions   3.   Unbounded solution

4. Invest \$6000 in bond A and \$4000 in bond B; the annual yield is \$880.

5. $\frac{7}{9}$ cup of milk, $\frac{25}{18}$ ounces of corn flakes; minimum cost $= \frac{335}{18} = 18.6$¢

6. (a) $x_1 \geq 0$ and $x_2 \geq 0$ are nonbinding; $2x_1 + 3x_2 \leq 24$ is binding
    (b) $x_1 - x_2 \leq v$ for $v < -3$ is binding and for $v < -6$ yields the empty set.
    (c) $x_2 \leq v$ for $v < 8$ is binding and for $v < 0$ yields the empty set.

7. 550 containers from company A and 300 containers from company B; maximum shipping charges $= \$2110$

8. 925 containers from company A and no containers from company B; maximum shipping charges $= \$2312.50$

9. 0.4 pound of ingredient A and 2.4 pounds of ingredient B; minimum cost $= 24.8¢$

Exercise Set 6.4 (page 328)

1. $x = 8, y = 6, P = 58$ 2. $x = 6, y = 4, P = 26$
3. $x = 8, y = 7, f = 54$ 4. $x = 5, y = 0, z = 3, P = 53$
5. $x = 10, y = 0, z = 16, t = 6, P = 58$ (one of many solutions)
7. **(c)** Maximum profit $= \$1,180$ (12 Coolit and 16 MoveOver models)
8. **(a)** $x = 27, y = 19, P = 110$
   (b) The non negativity assumption is no longer satisfied for s_1.
   (c) z is a non basic variable with a 0 at the bottom of its column.
   (d) $x = 27, y = 7, P = 102$ **(e)** $x = 18, y = 13$
9. $x = 2, y = 10, P = 54$
10. **(a)** $t = 5, c = 2, v = 11, P = 7670$
    (b) The solution is unique since no 0's appear at the bottom of a non basic variable.
    (c) $t = 6, c = 2, v = 12, P = 8620$
    (d) Originally, the constraint on pilot resources was not binding (slack of 7). The proposed changes would reduce the slack to 2. In either case, pilots are not being overworked.

Exercise Set 6.5 (page 340)

1. **(a)** $\mathbf{x}^{(1)} = \begin{bmatrix} .4 \\ .6 \end{bmatrix}$, $\mathbf{x}^{(2)} = \begin{bmatrix} .46 \\ .54 \end{bmatrix}$, $\mathbf{x}^{(3)} = \begin{bmatrix} .454 \\ .546 \end{bmatrix}$, $\mathbf{x}^{(4)} = \begin{bmatrix} .4546 \\ .5454 \end{bmatrix}$, $\mathbf{x}^{(5)} = \begin{bmatrix} .45454 \\ .54546 \end{bmatrix}$

   (b) P is regular since all entries of P are positive; $\mathbf{q} = \begin{bmatrix} \frac{5}{11} \\ \frac{6}{11} \end{bmatrix}$

2. **(a)** $\mathbf{x}^{(1)} = \begin{bmatrix} .7 \\ .2 \\ .1 \end{bmatrix}$, $\mathbf{x}^{(2)} = \begin{bmatrix} .23 \\ .52 \\ .25 \end{bmatrix}$, $\mathbf{x}^{(3)} = \begin{bmatrix} .273 \\ .396 \\ .331 \end{bmatrix}$

   (b) P is regular, since all entries of P are positive: $\mathbf{q} = \begin{bmatrix} \frac{22}{72} \\ \frac{29}{72} \\ \frac{21}{72} \end{bmatrix}$

3. **(a)** $\begin{bmatrix} \frac{9}{17} \\ \frac{8}{17} \end{bmatrix}$ **(b)** $\begin{bmatrix} \frac{26}{45} \\ \frac{19}{45} \end{bmatrix}$ **(c)** $\begin{bmatrix} \frac{3}{19} \\ \frac{4}{19} \\ \frac{12}{19} \end{bmatrix}$

4. **(a)** $P^n = \begin{bmatrix} \left(\frac{1}{2}\right)^n & 0 \\ 1 - \left(\frac{1}{2}\right)^n & 1 \end{bmatrix}$, $n = 1, 2, \ldots$.

   Thus, no integer power of P has all positive entries.

   (b) $P^n \to \begin{bmatrix} 0 & 0 \\ 1 & 1 \end{bmatrix}$ as n increases, so $P^n\mathbf{x}^{(0)} \to \begin{bmatrix} 0 \\ 1 \end{bmatrix}$ for any $\mathbf{x}^{(0)}$ as n increases.

   (c) The entries of the limiting vector $\begin{bmatrix} 0 \\ 1 \end{bmatrix}$ are not all positive.

6. $P^2 = \begin{bmatrix} \frac{1}{2} & \frac{1}{4} & \frac{1}{4} \\ \frac{1}{4} & \frac{1}{2} & \frac{1}{4} \\ \frac{1}{4} & \frac{1}{4} & \frac{1}{2} \end{bmatrix}$ has all positive entries; $\mathbf{q} = \begin{bmatrix} \frac{1}{3} \\ \frac{1}{3} \\ \frac{1}{3} \end{bmatrix}$

7. $\frac{10}{13}$ 8. $54\frac{1}{6}\%$ in region 1, $16\frac{2}{3}\%$ in region 2, and $29\frac{1}{6}\%$ in region 3

**Exercise Set 6.6
(page 350)**

1. (a) $\begin{bmatrix} 0 & 0 & 0 & 1 \\ 1 & 0 & 1 & 1 \\ 1 & 1 & 0 & 1 \\ 0 & 0 & 0 & 0 \end{bmatrix}$ (b) $\begin{bmatrix} 0 & 1 & 1 & 0 & 0 \\ 0 & 0 & 0 & 0 & 1 \\ 1 & 0 & 0 & 1 & 0 \\ 0 & 0 & 1 & 0 & 0 \\ 0 & 0 & 1 & 0 & 0 \end{bmatrix}$ (c) $\begin{bmatrix} 0 & 1 & 0 & 1 & 0 & 0 \\ 1 & 0 & 0 & 0 & 0 & 0 \\ 0 & 1 & 0 & 1 & 1 & 1 \\ 0 & 0 & 0 & 0 & 0 & 1 \\ 0 & 0 & 0 & 0 & 0 & 1 \\ 0 & 0 & 1 & 0 & 1 & 0 \end{bmatrix}$

2. (a) 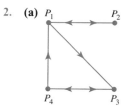   (b)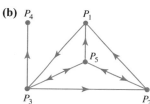

(c)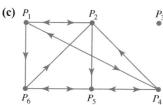

3. (a) 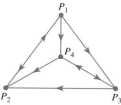   (b)   1-step: $P_1 \rightarrow P_2$

2-step: $P_1 \rightarrow P_4 \rightarrow P_2$
$P_1 \rightarrow P_3 \rightarrow P_2$

3-step: $P_1 \rightarrow P_2 \rightarrow P_1 \rightarrow P_2$
$P_1 \rightarrow P_3 \rightarrow P_4 \rightarrow P_2$
$P_1 \rightarrow P_4 \rightarrow P_3 \rightarrow P_2$

(c)   1-step: $P_1 \rightarrow P_4$

2-step: $P_1 \rightarrow P_3 \rightarrow P_4$

3-step: $P_1 \rightarrow P_2 \rightarrow P_1 \rightarrow P_4$
$P_1 \rightarrow P_4 \rightarrow P_3 \rightarrow P_4$

4. (a) $\begin{bmatrix} 1 & 0 & 0 & 0 & 0 \\ 0 & 1 & 0 & 0 & 0 \\ 0 & 0 & 1 & 1 & 0 \\ 0 & 0 & 1 & 2 & 1 \\ 0 & 0 & 0 & 1 & 2 \end{bmatrix}$

(c) The ijth entry is the number of family members who influence both the ith and jth family members.

5. (a) $\{P_1, P_2, P_3\}$ (b) $\{P_3, P_4, P_5\}$ (c) $\{P_2, P_4, P_6, P_8\}$ and $\{P_4, P_5, P_6\}$

6. (a) None   (b) $\{P_3, P_4, P_6\}$

7. $\begin{bmatrix} 0 & 0 & 1 & 1 \\ 1 & 0 & 0 & 0 \\ 0 & 1 & 0 & 1 \\ 0 & 1 & 0 & 0 \end{bmatrix}$ Power of $P_1 = 5$
Power of $P_2 = 3$
Power of $P_3 = 4$
Power of $P_4 = 2$

8. First, A; second, B and E (tie); fourth, C; fifth, D

Exercise Set 6.7
(page 359)

1. **(a)** $\begin{bmatrix} 2 \\ 3 \end{bmatrix}$ **(b)** $\begin{bmatrix} 6 \\ 5 \\ 6 \end{bmatrix}$ **(c)** $\begin{bmatrix} 78 \\ 54 \\ 79 \end{bmatrix}$

2. **(a)** Use Corollary 6.7.4; all row sums are less than one.
   (b) Use Corollary 6.7.5; all column sums are less than one.
   (c) Use Theorem 6.7.3, with $\mathbf{x} = \begin{bmatrix} 2 \\ 1 \\ 1 \end{bmatrix} > C\mathbf{x} = \begin{bmatrix} 1.9 \\ .9 \\ .9 \end{bmatrix}$.

3. E^2 has all positive entries.

4. Price of tomatoes, \$120.00; price of corn, \$100.00; price of lettuce, \$106.67

5. \$1256 for the CE, \$1448 for the EE, \$1556 for the ME 6. **(b)** $\frac{542}{503}$

Exercise Set 6.8
(page 367)

1. **(a)** $\begin{bmatrix} 0 & 1 & 1 & 0 \\ 0 & 0 & 1 & 1 \\ 0 & 0 & 0 & 0 \end{bmatrix}$ **(b)** $\begin{bmatrix} 0 & \frac{3}{2} & \frac{3}{2} & 0 \\ 0 & 0 & \frac{1}{2} & \frac{1}{2} \\ 0 & 0 & 0 & 0 \end{bmatrix}$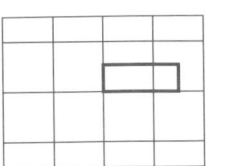

   (c) $\begin{bmatrix} -2 & -1 & -1 & -2 \\ -1 & -1 & 0 & 0 \\ 3 & 3 & 3 & 3 \end{bmatrix}$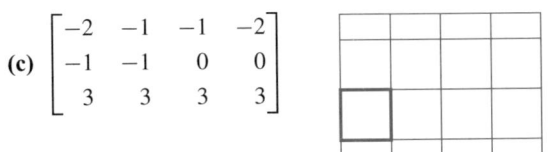

   (d) $\begin{bmatrix} 0 & .866 & 1.366 & .500 \\ 0 & -.500 & .366 & .866 \\ 0 & 0 & 0 & 0 \end{bmatrix}$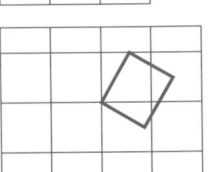

2. **(b)** $(0, 0, 0)$, $(1, 0, 0)$, $\left(1\frac{1}{2}, 1, 0\right)$, and $\left(\frac{1}{2}, 1, 0\right)$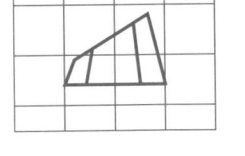
   (c) $(0, 0, 0)$, $(1, .6, 0)$, $(1, 1.6, 0)$, $(0, 1, 0)$

3. **(a)** $\begin{bmatrix} 1 & 0 & 0 \\ 0 & -1 & 0 \\ 0 & 0 & 1 \end{bmatrix}$ **(b)** $\begin{bmatrix} -1 & 0 & 0 \\ 0 & 1 & 0 \\ 0 & 0 & 1 \end{bmatrix}$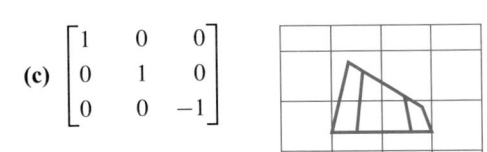

   (c) $\begin{bmatrix} 1 & 0 & 0 \\ 0 & 1 & 0 \\ 0 & 0 & -1 \end{bmatrix}$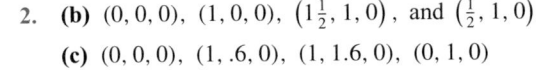

4. **(a)** $M_1 = \begin{bmatrix} \frac{1}{2} & 0 & 0 \\ 0 & 2 & 0 \\ 0 & 0 & \frac{1}{3} \end{bmatrix}$, $M_2 = \begin{bmatrix} \frac{1}{2} & \frac{1}{2} & \cdots & \frac{1}{2} \\ 0 & 0 & \cdots & 0 \\ 0 & 0 & \cdots & 0 \end{bmatrix}$, $M_3 = \begin{bmatrix} 1 & 0 & 0 \\ 0 & \cos 20° & -\sin 20° \\ 0 & \sin 20° & \cos 20° \end{bmatrix}$,

$M_4 = \begin{bmatrix} \cos(-45°) & 0 & \sin(-45°) \\ 0 & 1 & 0 \\ -\sin(-45°) & 0 & \cos(-45°) \end{bmatrix}$, $M_5 = \begin{bmatrix} 0 & -1 & 0 \\ 1 & 0 & 0 \\ 0 & 0 & 1 \end{bmatrix}$

(b) $P' = M_5 M_4 M_3 (M_1 P + M_2)$

5. **(a)** $M_1 = \begin{bmatrix} .3 & 0 & 0 \\ 0 & .5 & 0 \\ 0 & 0 & 1 \end{bmatrix}$, $M_2 = \begin{bmatrix} 1 & 0 & 0 \\ 0 & \cos 45° & -\sin 45° \\ 0 & \sin 45° & \cos 45° \end{bmatrix}$, $M_3 = \begin{bmatrix} 1 & 1 & \cdots & 1 \\ 0 & 0 & \cdots & 0 \\ 0 & 0 & \cdots & 0 \end{bmatrix}$,

$M_4 = \begin{bmatrix} \cos 35° & 0 & \sin 35° \\ 0 & 1 & 0 \\ -\sin 35° & 0 & \cos 35° \end{bmatrix}$, $M_5 = \begin{bmatrix} \cos(-45°) & -\sin(-45°) & 0 \\ \sin(-45°) & \cos(-45°) & 0 \\ 0 & 0 & 1 \end{bmatrix}$,

$M_6 = \begin{bmatrix} 0 & 0 & \cdots & 0 \\ 0 & 0 & \cdots & 0 \\ 1 & 1 & \cdots & 1 \end{bmatrix}$, $M_7 = \begin{bmatrix} 2 & 0 & 0 \\ 0 & 1 & 0 \\ 0 & 0 & 1 \end{bmatrix}$

(b) $P' = M_7(M_5 M_4(M_2 M_1 P + M_3) + M_6)$

6. $R_1 = \begin{bmatrix} \cos \beta & 0 & \sin \beta \\ 0 & 1 & 0 \\ -\sin \beta & 0 & \cos \beta \end{bmatrix}$, $R_2 = \begin{bmatrix} \cos \alpha & -\sin \alpha & 0 \\ \sin \alpha & \cos \alpha & 0 \\ 0 & 0 & 1 \end{bmatrix}$,

$R_3 = \begin{bmatrix} \cos \theta & 0 & \sin \theta \\ 0 & 1 & 0 \\ -\sin \theta & 0 & \cos \theta \end{bmatrix}$, $R_4 = \begin{bmatrix} \cos \alpha & \sin \alpha & 0 \\ -\sin \alpha & \cos \alpha & 0 \\ 0 & 0 & 1 \end{bmatrix}$,

$R_5 = \begin{bmatrix} \cos \beta & 0 & -\sin \beta \\ 0 & 1 & 0 \\ \sin \beta & 0 & \cos \beta \end{bmatrix}$

7. **(a)** $M = \begin{bmatrix} 1 & 0 & 0 & x_0 \\ 0 & 1 & 0 & y_0 \\ 0 & 0 & 1 & z_0 \\ 0 & 0 & 0 & 1 \end{bmatrix}$ **(b)** $\begin{bmatrix} 1 & 0 & 0 & -5 \\ 0 & 1 & 0 & 9 \\ 0 & 0 & 1 & -3 \\ 0 & 0 & 0 & 1 \end{bmatrix}$

**Exercise Set 6.9
(page 380)**

1. **(a)** *GIYUOKEVBH* **(b)** *SFANEFZWJH*

2. **(a)** $A^{-1} = \begin{bmatrix} 12 & 7 \\ 23 & 15 \end{bmatrix}$ **(b)** Not invertible **(c)** $A^1 = \begin{bmatrix} 1 & 19 \\ 23 & 24 \end{bmatrix}$

(d) Not invertible **(e)** Not invertible **(f)** $A^{-1} = \begin{bmatrix} 15 & 12 \\ 21 & 5 \end{bmatrix}$

3. *WE LOVE MATH*

4. Deciphering matrix $= \begin{bmatrix} 7 & 15 \\ 6 & 5 \end{bmatrix}$; enciphering matrix $= \begin{bmatrix} 7 & 5 \\ 2 & 15 \end{bmatrix}$

5. *THEY SPLIT THE ATOM* **6.** *I HAVE COME TO BURY CAESAR*

7. **(a)** 010100001 **(b)** $\begin{bmatrix} 0 & 1 & 1 \\ 1 & 1 & 1 \\ 1 & 0 & 1 \end{bmatrix}$

8. A is invertible modulo 29 if and only if $\det(A) \neq 0 \pmod{29}$.